دکتر جمشیدی عزیز

امیدوارم که از این کتاب خوشتان بیاید

از کمک‌های شما متشکرم

[illegible]

Circuit Analysis Fundamentals

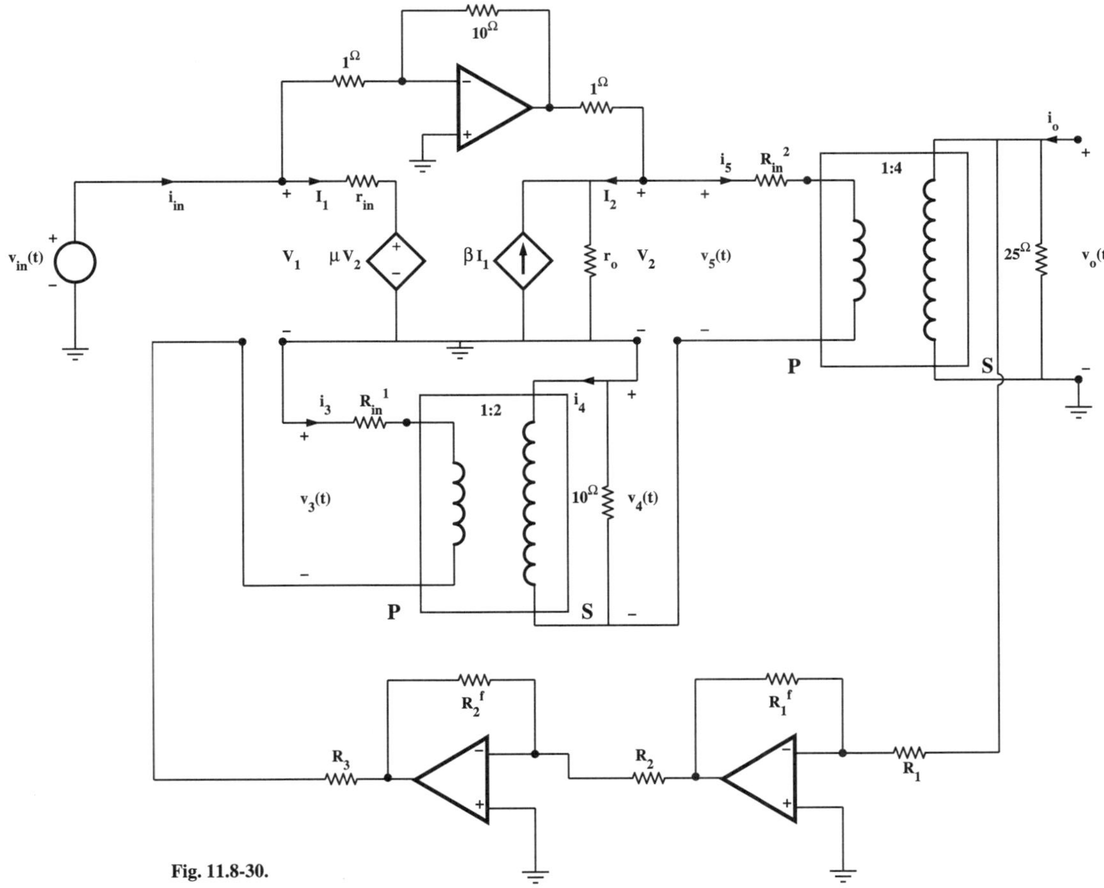

Fig. 11.8-30.

Mansour Eslami

Circuit Analysis Fundamentals

Agile Press
A Division of *C* FAR
Center For Agile Research
Chicago, Illinois

Center For Agile Research, Inc.

Corporate Headquarters

P.O. Box 1939, Chicago, IL 60690-1939

agilepress@agileresearch.com

Professor Mansour Eslami
Dept. of Electrical and Computer Engineering
The University of Illinois at Chicago
Chicago, Illinois 60607
USA
eslami@ieee.org
May 24, 2005

ISBN 0-9718239-5-2 Agile Press, Chicago, Illinois, U.S.A.
Library of Congress Control Number: 2005902650

Cataloging-in-Publication Data:
Eslami, Mansour
Circuit Analysis Fundamentals
ISBN 0-9718239-5-2
1. Circuits. 2. Electrical Circuits.
3. Active Networks. 4. Laplace Transformations.
I. Title. II. References. III. Index.

Director of Operations: Mark Ossanlo
Director of Marketing: Kayvan Mozaffari
Art Coordinator: The author
Cover Design: The author with Felicia S. Barley of Comet Press & Media Services, Chicago, IL, USA

Printed by BookMasters, Inc., Mansfield, OH, USA
Binding by Dekker Bookbinding, Grand Rapids, MI, USA

10 9 8 7 6 5 4 3 2 1

Typesetting: Camera-ready by the author
This book is set in Times Roman using Dell Inspiron-8100/Linux

0 TEXTBOOK STRUCTURE

TABLE OF CONTENTS

CHAPTER THREE

NETWORK ANALYSIS CONTINUED

CHAPTER FOUR

NUMERICAL EXAMPLES AND PROBLEMS

CHAPTER FIVE

ESSENTIAL FORMULAE AND THEOREMS

CHAPTER SIX

CIRCUITS WITH COMPLEX ELEMENTS

CHAPTER SEVEN

DIFFERENTIAL EQUATIONS AND CIRCUITS WITH SEQUENTIAL SWITCHES

CHAPTER EIGHT

MUTUAL INDUCTANCE AND IDEAL TRANSFORMERS

CHAPTER NINE

LAPLACE TRANSFORMATION

CHAPTER TEN

FREQUENCY RESPONSE

CHAPTER ELEVEN

TWO-PORT NETWORKS

CHAPTER TWELVE

POLYPHASE CIRCUITS

PREFACE

In the spirit of Omar Khayyam, the great Persian mathematician, astronomer, poet and philosopher, who said "I wonder what *could possibly* wine sellers purchase with the money which they have earned selling wines." (Or in the words of Edward Fitzgerald: "I wonder what the wine sellers [could possibly] buy half so precious as what they sell.") We may wonder what *could possibly* be missing in the many classical textbooks on circuits written by so many outstanding authors? Or, why do we need another entry-level textbook on fundamentals of circuit analysis? The answer is simple – nothing is missing in the many of these rather *encyclopedic* textbooks. However, the basic materials in circuit theory can still be restructured in a focused manner, and taught differently, in order that the students benefit the most and remember the underlying topics longer. To accomplish this goal is our mission.

On the overall objectives of this book, through many years of teaching basic circuit analysis, the author has learned that students are either familiar with the physics and mathematics behind circuits (the prerequisites), or they are not. For those few who do not have the basic understanding of the prerequisites, though they have taken them all, there is not enough time to teach both the prerequisites topics and circuit theory. Even one such lecture wastes class' time and is not beneficial. Because there are enough new materials in circuits to be covered, we cannot justify covering background materials. Further, it is very hard on those students with possibly weak background to learn both physics and circuits concurrently. Not mentioning that in most electrical engineering programs there is a new course called signals and systems, where some issues that used to be covered in circuits are now taught there. Neither do I want to repeat those topics. Therefore, the primary objective has been to cut through a body of materials, which these materials are not simple for the first timers, and present them in an orderly and algorithmically-oriented approach, in order that the overall body of materials covered in the course will be grasped by students even though their backgrounds are possibly weak. It is not the author's intention to write a textbook that crosses over several circuit courses, nor is a small portion to be used in one course. Rather, the intention has been to write a textbook that is used cover to cover for one basic course and is kept for life.

On the technical aspects of the book, based on the above simple premises, a new, informative and yet concise way to teach basics of circuits is presented, which is conceptually a hybrid approach among most classical textbooks, and has been tested in the classroom settings repeatedly. Only materials that are absolutely essential topics for a first circuit course are covered. The main approach in writing this book has been to motivate and indeed encourage students to think, and allow them to rely on the knowledge gained, and build upon that as they proceed – a real-life experience. Here, I want them to think and dream about the results and constantly strive toward reaching that illusive end-game with minimum distraction and by focusing on the materials at hand. For instance, I purposely have not included computer simulations, nor have I included extensive computational schemes, because nowadays students are literate in computer applications and they do not need such reinforcement. When they master the fundamentals of circuits as they should, computer simulations and/or other related and important materials can be augmented quickly.

A word or two on the use of the book, I hope this book becomes of use to many future students, and on that possibility I cordially ask instructors to alert students to various comments presented in the problem set sections throughout the book. I also ask students, specially those who wish to take upper-level electronic or other circuit-related courses to learn the materials in this book in its entirety. If so, many of those courses become individually a very smooth ride. Otherwise, portions of this book must be repeated.

On the homework and related issues, I believe that excessive repetition is fruitless. Again referring to Omar Khayyam who said "one letter is enough to excite a prepared mind," I adhere to this principle. I just cannot accept the repetition of one question more than twice. In fact the major contribution of this textbook, in the author's opinion, is to stay within the boundary of the covered topics. I am aware of "proctor laboratory" for this course manned by teaching assistants or computers, but I do not endorse that concept. I want the students to demand inquisitively questions for each new topic. They must be involved in the learning process and building new knowledge. They must look at problems for their reasons and relations to the covered theory and not for the sake of their solutions. There is no limit in the number of problems which can be constructed, but there is a limit on the number of possible ways to find solutions. To solve problems for the sake of solving problems without an opportunity to link them with materials covered in the course is unproductive.

A word or two with potential instructors, in many instances, I have tried to be as brief as possible, only for the sake of keeping students' attention on the subject at hand. For instance, when I explain inductor and capacitor in, say, (6.4-1) and (6.4-2), I briefly mention duality among circuit's elements in (6.4-3) which complements Comment 5.8-2. In contrast, I do not get into what is discussed in many textbooks rather elementarily as a duality discussion. Because in my experience anything more will distract students. Nor do I take this discussion into a new level by claiming that using standard symbols for mass, spring, and friction coefficient, and experience gained in circuit analysis, one can solve all mechanical engineering problems by studying their electrical duals. Such an argument has not been found fruitful and is not in the main realm of circuit analysis. Not one single electrical engineer has been able to solve one mechanical problem without taking a solid and humbling dynamics course. Thus, I leave it up to the discretion of potential instructors who may use this textbook to elaborate on collateral issues. There are many similar instances which I have labored to balance in the covered materials versus the allotted time and space for presentations. I take my chances with the audience and see how they will receive these arrangements and my lecture layouts (which will accompany this textbook for those who adopt it in their classes). All constructive suggestions to improve these presentations will be greatly appreciated and gratefully incorporated with all due credits in future editions.

Mansour Eslami
May 24, 2005

ACKNOWLEDGMENT

Writing a textbook with a sense of purpose, and preparing its camera-ready manuscript with all its drawings is a challenging task, aside from its technical and intellectual merits. Many individuals help in various manners who often are not properly compensated nor acknowledged. They do so, because they enjoy helping to create something new. Let me also say that this textbook is prepared to help students to learn materials that can be very intimidating. For most non-electrical engineering students the first circuit course is often the point of reference as to why they choose not to become electrical engineers. So, the stake has been high.

Now, please allow me to say a few words about my education in circuits. I took my first circuit course in 1968 with Professor B. Shahzadi, the second one with Professor P. Jabehdar-Maralani in 1970, at Teheran University. My first textbook was written by the late Professor W.H. Hayt and the late Dr. J.E. Kemmerly. I am among the last group of students at the University of Wisconsin-Madison (UW) who took all the available graduate-level circuit courses (graph theory, synthesis, active networks – no longer listed in the catalog) with Professor D. P. Brown and Professor Y. Sun prior to 1975. Starting in 1980, the first textbook that I then used five times to teach circuits at State University of New York at Stony Brook was written by Professor A. Budak. I also used, three times, textbooks by the late Professor M.E. Van Valkenberg to teach circuits prior to 1989. Finally regarding my education, I have immensely benefited from the generous advice of the late Professor T.J. Higgins at UW. Perhaps a touch of his independent thinking style and quite a few of his favorite words are present herein. For instance, once Professor Higgins gave the following "account" on how technical textbooks should be prepared. He said quite rightfully that when someone writes a book, or even a short essay, that person does not start from the first line on the first page and end on the last line of the document. Therefore, when an author is in Chapter Five, *per se*, and is referring to topics in "subsequent" chapters, by saying for example "we *will* explain a particular issue in Chapter Six", this conduct is not necessarily accurate, because Chapter Six may have been written prior to Chapter Five. There are so much back and forth that no one really knows the correct time line of writing. Thus, when we talk about the *location* of a topic in a technical writing, we must use the present tense and refer to the actual location by its section and/or equation number. Because from the reader's point of view the task of writing is completed, and now the final product is available for studying. All the reader wants is to locate the referenced subject and nothing else. Similar reference is also expected for the preceding chapters. According to Professor Higgins we only use non-present tense when describing a time-sensitive event, such as, we *will* study advanced topics in circuit theory in future courses. Or, Bode diagrams *were* introduced more than 60 years ago. To adhere to this style has been difficult in the contemporary technical writings, but I consider this as [an intelligent] consistency.

In Spring of 2000, while visiting Turabo University in Puerto Rico, I taught the first circuit course from heart, and repeated that at University of California at Riverside (UCR) in the Fall of the same year. At UCR, I started to write about 170 pages of hand-written notes which were put on the Web by students for their use, and became the first draft of this textbook. In the light of the generous comments that I had received from many of my students, one particular follows, I decided to write this textbook as soon as possible.

Many groups have helped me to finish this project. Starting with my former students at Stony Brook who had labored through many iterations of my notes and education to teach. My students at the University of Illinois at Chicago (UIC) were also helpful in polishing up my presentations. My students at UCR whom I refer to as "thirsty students," thirsty to learn, were also overwhelmingly helpful. The anonymous attached letter is a sample of their generosity and kindness. I hope this textbook and the one before on "Senior Design Experience: Lessons for Life," which are based on my lectures at UCR have completed my debts to them. There are also many other students, out of 4750 students worldwide which I have been blessed with throughout my career, who have made constructive suggestions and comments helpful to the completion of this project. So, I humbly thank all my former students.

Special thanks are also due to the many of my colleagues for their support and encouragement, especially Professor Armen Zemanian, Professor Chi-Tsong Chen, Professor Kenneth L. Short (all at Stony Brook), Professor Richard S. Marleau at the UW-Madison, Professor Andrew A. Frank at University of California Davis, Professor Kumpati Narendra at Yale University and Professor William R. Perkins at the University of Illinois Urbana-Champaign.

I also want to thank the faculty in the School of Electrical Engineering at Amir Kabir Technological University (formerly Teheran Polytechnic), for their support of my educational activities in Iran, especially Professor S.K.Y. Nikravesh for his additionally reading portions of the manuscript and his constructive suggestions. Special thanks are also due to many outstanding graduate students in this school for their continuous support. They are the reason for my cooperation with this premier school of engineering.

Specific individuals who have helped directly with the actual production of this book are: the late Mr. Bob Amendola of UIC, who left me with the first draft of PostScript™ programs (PostScript is a trademark of Adobe System Inc.) that he had developed in mid-1980's to draw circuits and other figures. (In 1987, the current logo for the IEEE Transactions on Circuits and Systems was designed by Bob who was a consummate professional.) Mr. Karl Doering, my former student at UCR, who during his internship at the Argonne National Laboratory helped me to process most efficiently the figures which I had drawn. Karl has liberated me from pasting figures for the rest of my life. Mr. Hasan Arslan, a former graduate student at UIC, helped me with some aspects of processing files among my various computers. Mr. Kaveh Moezzi Madani, my graduate student at the Amir Kabir Technological University in Teheran, Iran, helped me with some simulations and drawings. Also initially at Stony Brook, the text-wrapping program of Ms. Gina Yu Liu, a former staff, and the computer related work of Mr. Haidar Chamas, my former student, are appreciated. Finally, the continuous support of my colleagues at *C* FAR, especially Mr. Richard Bleier, is greatly appreciated.

These days nothing can be done without extensive support from the computer staff. I thank both current and former computer staff and my good friends at UIC, Ralph Orlick, Certa Nicholas, Artan Alickolli, Leland Luecke, Ashish Desai, Yahya Alsalqan, John Yancey, Clarence Murzyn, Richard Chang, and at UCR, Isaac Saldana, for their help in keeping my files functional and maintaining my connection trouble free. I will always remain grateful.

Finally, I am most grateful for the generous support provided to me by the charitable and educational foundation of my late brother Mr. Ali Mohammad Eslami, during my recent visit to various universities in Iran. The library that is built in his honor in Teheran is now my sanctuary. With the grace of God, I expect there will be many more scholarly projects performed by a host of visiting scholars, which are initiated from and completed in this place. As I also expect to continue shuttling between Teheran and the US in coming years.

Mansour Eslami
May 24, 2005

ACKNOWLEDGMENT Continued

Students have their own magical ways of educating us. My experience at UC Riverside during the academic year 2000-2001 was generally very rewarding, and the first circuit course that I taught there was open to all engineering students. Not everyone in that class was so happy as the one below, nor is this the unique experience in my interactions with former students. But I must share this letter, hoping that someday this student may see it and realize that I have been affected profoundly by his/her generosity and kindness. Had it not been for this letter which was dropped in my office, I would have probably not written this textbook. But I felt obliged to rewrite and polish those hand-written notes distributed at UC Riverside. Thank you for your kind words. The pleasure is all mine.

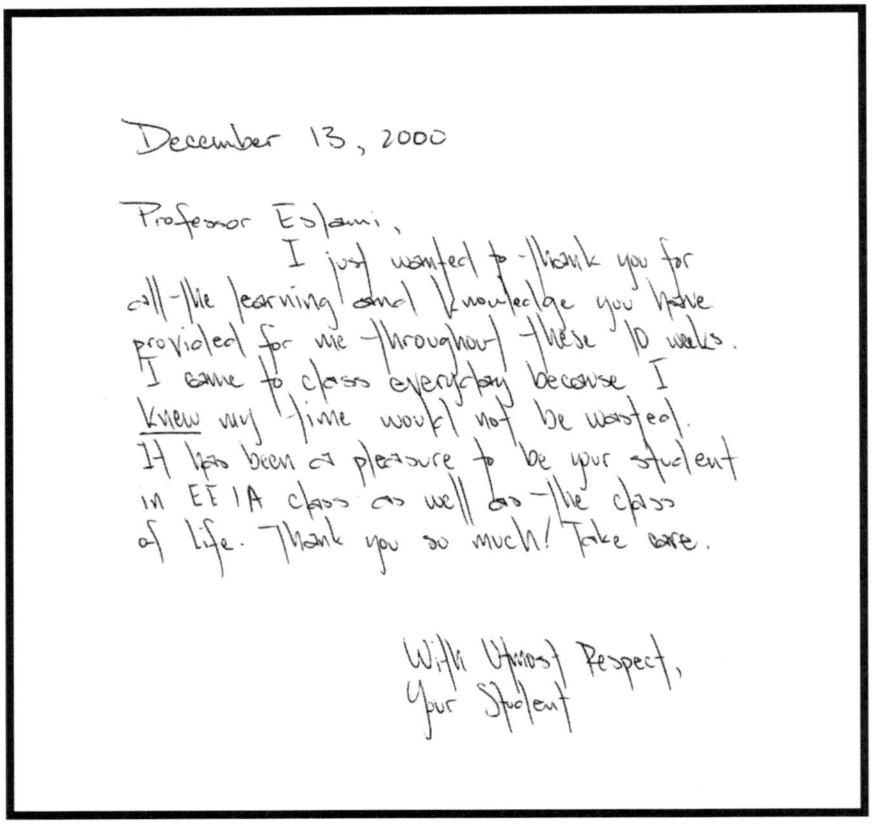

December 13, 2000

Professor Eslami,

I just wanted to thank you for all the learning and knowledge you have provided for me throughout these 10 weeks. I came to class everyday because I knew my time would not be wasted. It has been a pleasure to be your student in EE 1A class as well as the class of life. Thank you so much! Take care.

With Utmost Respect,
Your Student

1 INTRODUCTION

1.1. OVERVIEW and SCOPE of the BOOK

In this textbook we study the essentials of network analysis for an introductory electrical engineering circuit course. Here, we present:

- **a body of materials,**

which forms

- **the most essential aspects of circuit theory using circuits with only resistors and constant sources.**

Upon introducing various methodologies to analyze these circuits, we advance the analysis to cover

- **circuits with resistors, capacitors, inductors and complex sources.**

We then structure the preceding methodologies to cover the above circuits under:

- **Time-varying conditions, using**

  differential equations settings, or

  the Laplace transformation methods.

In order to analyze the entire dynamics of all linear circuits in a unified manner.

But first, we offer the chart in Table 1.1-1, which shows how to advance in the breadth and scope of the underlying topics. We show this chart with a full understanding that some of the names and topics given are to be defined subsequently.

Table 1.1-1. An elementary view of circuits complexity.

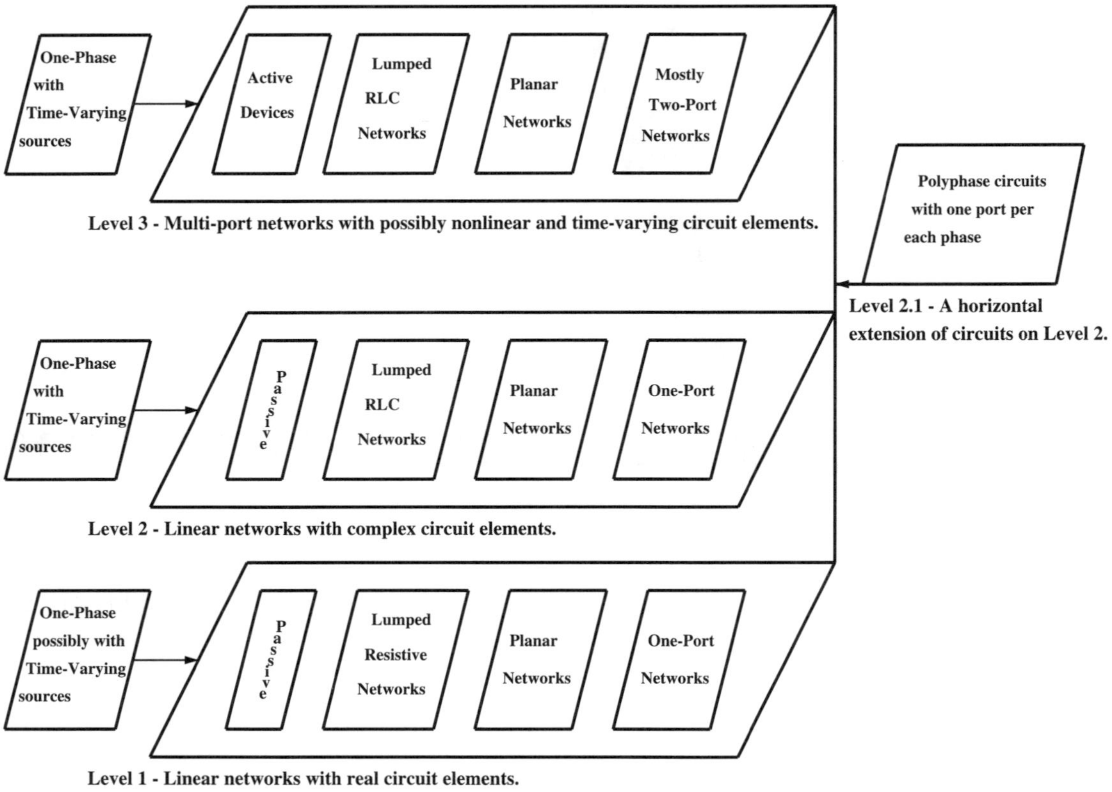

Although it may seem rather early to give this chart, the author believes strongly to have it now in order to get students involved in the learning process. It is always instructive to know where we stand as we proceed in this course. Referring to this chart, the students can monitor our progress and see what else is left to learn and how various proposed levels can be extended further horizontally and/or vertically. For instance, on Level 1, what if we had a circuit consisting of a time-varying resistor (i.e., a resistor whose value changes as a function of time), then that requires us to extend the complexity of the level for this problem vertically to those on Level 2 or possibly higher. Therefore, as we gain knowledge about these terminologies, we learn how to tackle a more complex problem than those presented herein. We elaborate on these issues in our future discussions. Perhaps it will be worthwhile if this chart is reviewed occasionally and particularly at the last lecture of this course. Finally, to learn extracting knowledge from what we have covered (the meaning of extending vertically or horizontally) is an art that we hope to inspire in our students.

In the remaining sections of this chapter, we review Kirchoff's laws, and consider this review from physics the single most important background topic needed to follow this course.

1.2. KIRCHHOFF'S LAWS

There are two fundamental laws which govern circuit theory and these were discovered by Gustav Kirchhoff (a professor of physics) when he was a 23-year old student. (For more information about this and similar historical remarks on the development of electrical engineering areas, please refer to Professor Van Valkenburg's textbook [5]). But first, let us recall a few names. For us a **circuit** is inevitably referred to an electrical circuit, which consists of a number of **elements** connected together under a specific structure also known as **topology.** The point that two or more elements are connected is called a **node.**

1.2.A. Kirchhoff's Current Law (KCL): The total currents entering a node is equal to the total currents leaving that node.

Let the incoming currents be shown by I_{in}^k, $k = 1, \ldots, n$, and let the outgoing currents be shown by I_{out}^l, $l = 1, \ldots, m$. Then KCL says that

$$\sum_{k=1}^{n} I_{in}^k \equiv \sum_{l=1}^{m} I_{out}^l. \tag{1.2-1}$$

Or,

$$\sum_{k=1}^{n} I_{in}^k - \sum_{l=1}^{m} I_{out}^l \equiv 0. \tag{1.2-2}$$

Generally, we use the sign "+" for incoming and "−" for outgoing currents. For instance, if we have only one source, then the following is the general convention to label the current direction from a source (such as a battery, or some other circuit), shown by I_s to the sink (load) or possibly another circuit.

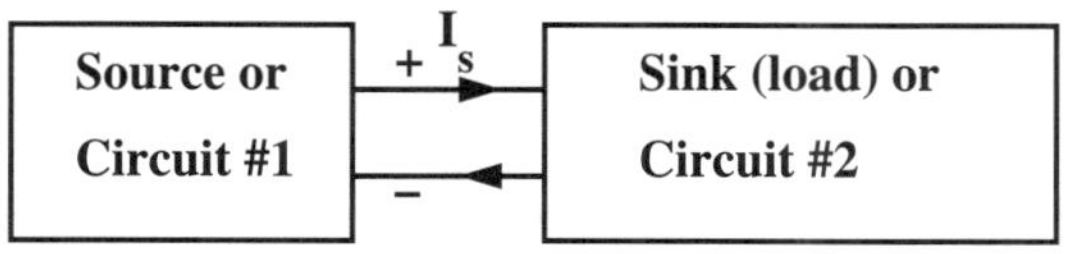

Fig. 1.2-1. Direction of current going into the load.

When we have several sinks or loads (users), we draw the corresponding configuration for labeling currents as shown in Fig. 1.2-2. The KCL becomes

$$I_s = \sum_{l=1}^{m} I_l. \tag{1.2-3}$$

In this law, the fundamental assumption is that the node n, in Fig. 1.2-2, does not drain any current, in other words, we have no hidden loss at this node. Otherwise (1.2-3) does not hold.

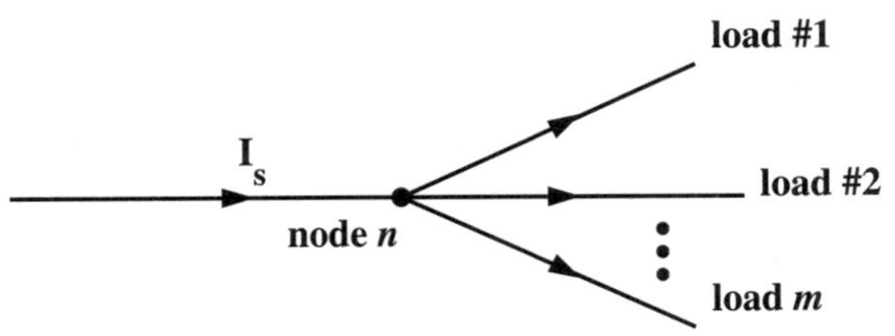

Fig. 1.2-2. Current distribution into *m* loads.

Before presenting the next Kirchhoff's law, we define the *closed path,* which in turn we need to define the *path* first. Given any circuit, such as that shown in Fig. 1.2-3, *a path refers to any deliberate selection of elements between any two nodes.* For instance, when a current arrives, say at node *m*, and we want to know how an appropriate portion of that current reaches node *n*, we look for all possible choices of elements that this current must go through in order to reach node *n*. **Each of these choices is called a path.** For example, *min* and *mkjn* are two out of many paths between node *m* and node *n*. The path between two adjacent nodes is called a *branch* or simply an element. On the other hand, a *closed path* refers to any path, which *starts and ends at the same node.* In Fig. 1.2-3, the arbitrary arrow shows the direction which we close a path that starts at node *k* and ends at the same node by going through nodes *i* and *j*.

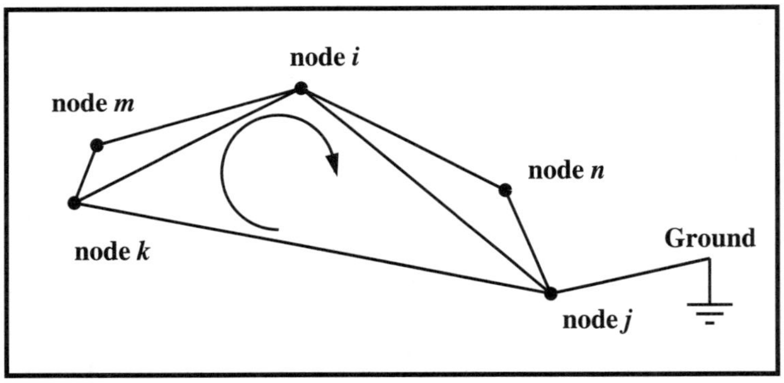

Fig. 1.2-3. A circuit with many nodes.

1.2.B. Kirchhoff's Voltage Law (KVL): The total voltages around any *closed path* is equal to zero.

This law can be expressed by the following relationship, which the summation is taken over all possible elements of the underlying closed path.

$$\sum_{k=1} V_k \equiv 0. \tag{1.2-4}$$

In Fig. 1.2-3, we need to identify each node voltage in order to express the voltage across each element. It is customary to associate each node voltage in any given circuit with respect to a reference (*datum*) node also known as *ground.* In other words, when we use $V_{\text{node } k}$, we mean this voltage value is with respect to some reference node whose voltage by definition is zero. However, if that is not the case, say the so-called ground is not at the zero voltage, but rather is at some other values with respect to the true global reference point, then what will happen to our calculations is that all our voltage values will be shifted by the same amount of the voltage for the local ground with respect to the global level. So, it is a perfectly acceptable choice to select the ground voltage in any circuit at the zero value and complete the analysis.

In Fig. 1.2-3, and based on our initial and arbitrary choice for the direction of the closed path, the KVL reads as follows.

$$(V_{\text{node } k} - V_{\text{node } i}) + (V_{\text{node } i} - V_{\text{node } j}) + (V_{\text{node } j} - V_{\text{node } k}) \equiv 0. \tag{1.2-5}$$

The first parenthesis is the voltage across element *ki*, the second parenthesis is the voltage across element *ij*, and the third parenthesis is the voltage across element *jk*. Note that this is a global definition for the voltage across any element, which depending on our starting node begins with that voltage as a positive quantity. For example, V_{ki} is positive at node k and negative at node i. If this were not true, the physics of the system will correct our initial choice. In other words, if we assume V_{ki} as a positive quantity and it turns out that we were wrong, then that value at the end of the analysis will become negative, meaning that we must reverse our initial direction. Finally, the reference or ground is shown by the symbol at the right corner of Fig. 1.2-3. In some textbooks, this symbol is used for the so-called *chassis* (a local reference), and another symbol is used for *earth* (the global reference). We only use one symbol as shown in Fig. 1.2-3 for reference node and call that ground.

For a circuit with **N** nodes, Fig. 1.2-4, to compute the voltage between any two nodes, say node i and node j, we proceed as follows. First, we compute (or using a voltmeter we measure) the voltage between node i and ground. This is shown by $V_{\text{node } i}$. Second, we compute the voltage between node j and ground, shown by $V_{\text{node } j}$. Now,

$$V_{ij} = V_{\text{node } i} - V_{\text{node } j}. \tag{1.2-6}$$

Note that,

$$V_{ji} = V_{\text{node } j} - V_{\text{node } i} = -V_{ij}. \tag{1.2-7}$$

These are simple relationships which we have reviewed here from physics for our convenience and extensive future use in this textbook.

Comment 1.2-1: Although, we have not mentioned anything about how do we compute or measure currents and voltages in **1.2.A** or **1.2.B,** these laws are valid in both the time-domain measurement (i.e., when the entire circuit is considered as a function of time and all measurements are conducted in real time); as well as the frequency-domain measurement (where, similarly all measurements are conducted in the frequency domain). In this textbook, we study these matters exhaustively, but case-by-case and systematically in each underlying problem.

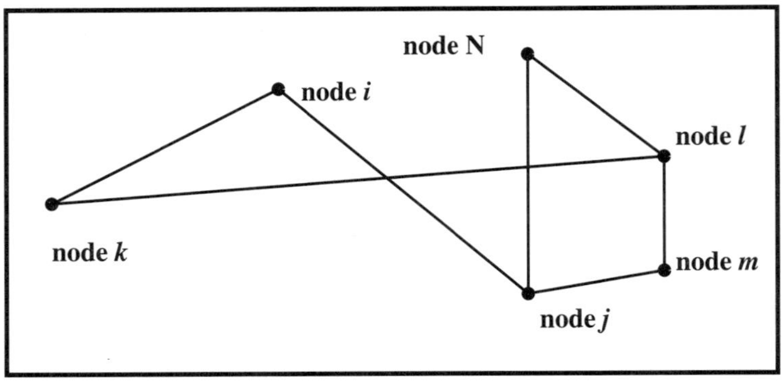

Fig. 1.2-4. A circuit with N nodes.

Remark 1.2-1: In Fig. 1.2-4, there are two crosses between nodes *k* & *l*, but there is no connection. If there were a connection, we would have shown that by a dot. For an actual circuit, we use only one dot for two or more connected elements. Non-crossing points are shown by half a circle.

Comment 1.2-2: There are two important definitions left to review before proceeding further. Suppose we have a functional circuit N in a box, Fig. 1.2-5, i.e., there are some electrical activities inside the box. We pull two wires out as shown in Fig. 1.2-5 for inspection, i.e., we connect two pieces of wires to two separate nodes inside the box and pull them out.

Fig. 1.2-5.

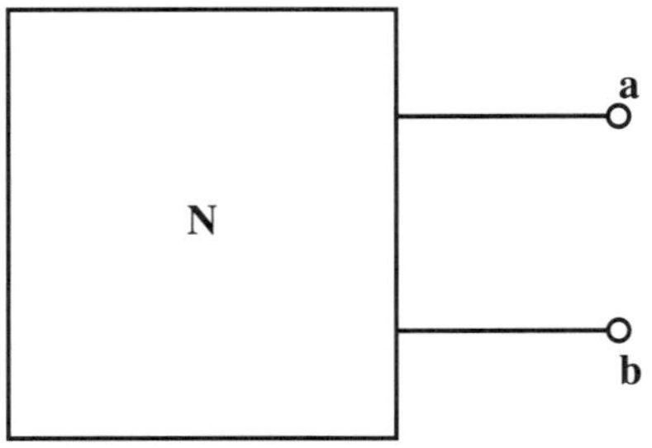

Now, we ask the following question: Is there a current going from node a to node b and outside the box? As shown no current can flow from a to b, because there is nothing connected between these two nodes outside the box. What if we connect a voltmeter V between these two nodes as shown in Fig. 1.2-6?

Fig. 1.2-6.

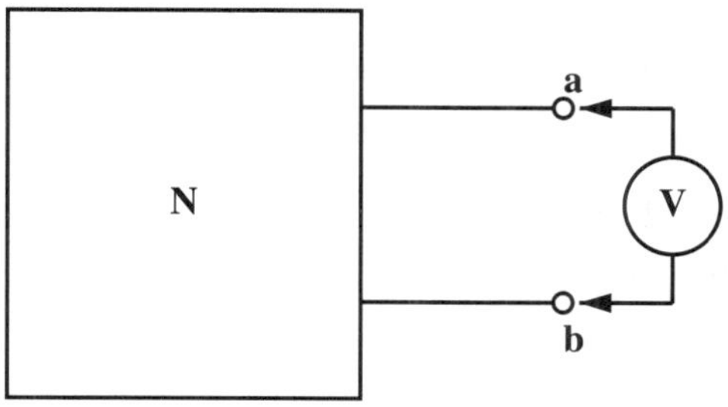

Because we start with a functional circuit, there will be some voltage between nodes a and b and from the circuitry inside of the box, which is read by the voltmeter. This voltage is called the *"open-circuit voltage"* and is shown by V_{ab}^{oc} or V_{oc} for short.

Similarly, if we put a current meter between nodes a and b as shown in Fig. 1.2-7. Then this meter creates a path outside the box N between these two nodes, which is the same as shown in Fig. 1.2-8.

Fig. 1.2-7.

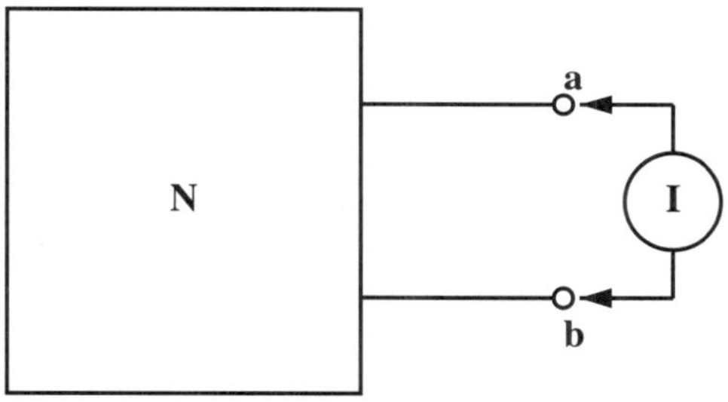

Fig. 1.2-8.

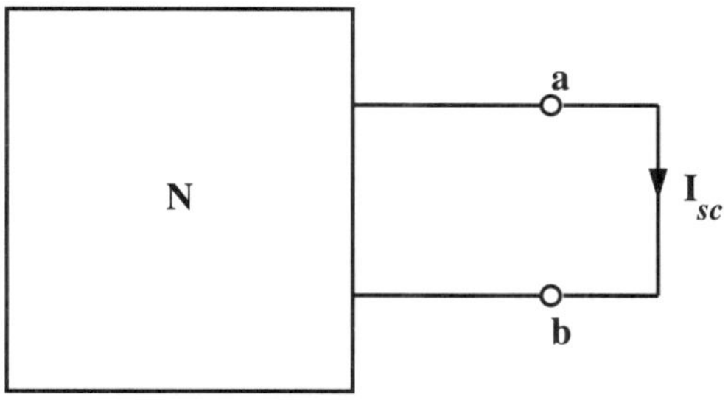

The value of current going from nodes a to b in Fig. 1.2-8, while $V_{ab} = 0$, is called the *"short-circuit current"* and is shown by I_{ab}^{sc} or I_{sc} for short.

1.3. TYPICAL CIRCUIT ELEMENTS

There are several categories of sources and sinks (loads), some possibly dependent on each other, which collectively represent most typical circuit elements.

1.3.A. [Independent] Sources: The bracket means it is understood we are considering independent sources but we do not repeat that in the following.

Constant voltage and current sources are shown by upper case letters as depicted in Fig. 1.3-1, while variable sources, which are either a function of time or frequency, are shown in Fig. 1.3-2 by mixed letters.

Normally, variable sources are shown as either v(t) or V(s) [similarly i(t) or I(s)], referring to a time- or a frequency-domain quantity. It is to be noted that quantities in the time-domain are shown by lower case letters like v(t), while those in the frequency-domain are shown by upper case letters like V(s). In the earlier circuit textbooks an independent current source was shown by two crossed circles and an arrow, Fig. 1.3-2b.

Other names to remember in order to distinguish various sources (signals) are as follows.

- *Direct current (DC),* which generally refers to a constant source (current or voltage) that is independent of time variations.
- While, *alternating current (AC)* refers to a time-varying signal (current or voltage) whose behavior is periodic in time.
- A *periodic* signal, say f(t), is a signal whose behavior in time repeats itself, i.e., $f(t) = f(t \pm kT)$ for some $T > 0$ and an integer $k \in [1, \infty)$.

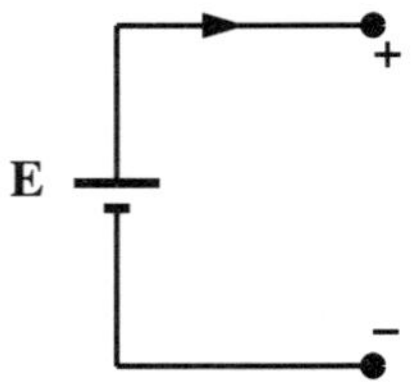

Fig. 1.3-1a. A battery.

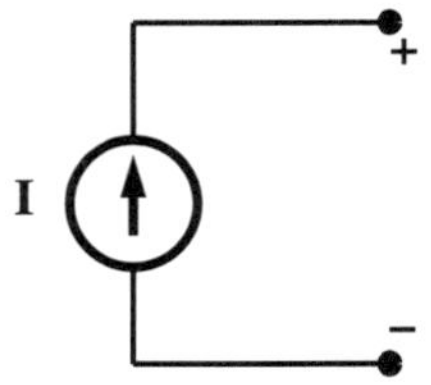

Fig. 1.3-1b. A constant current source.

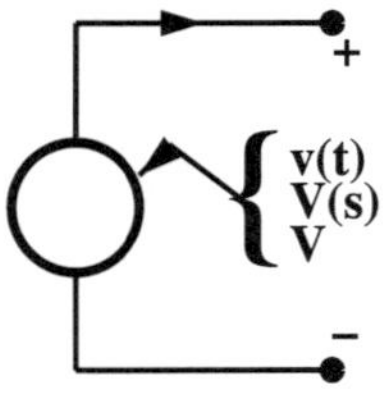

Fig. 1.3-2a. A voltage source.

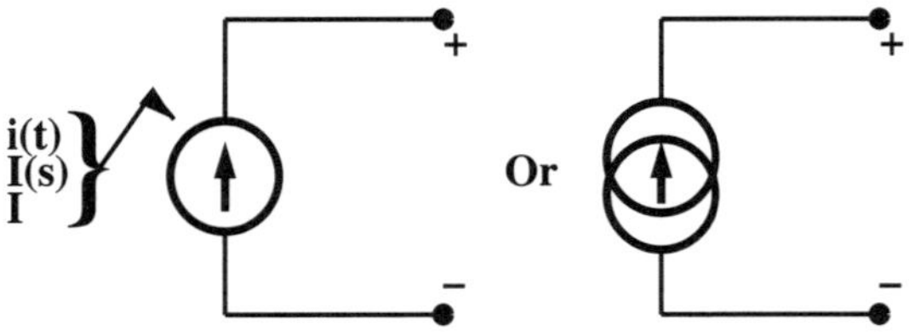

Fig. 1.3-2b. A current source.

1.3.B. Sinks or Loads: There are three [independent] elements, also known as *two-terminal, passive and lumped elements,* which are typically used in circuits and are shown in Fig. 1.3-3. The term *lumped* refers to elements whose values are concentrated and are not distributed – the only kind exclusively used herein. The inverse of resistance is called conductance and is measured in Ω^{-1}, called mho.

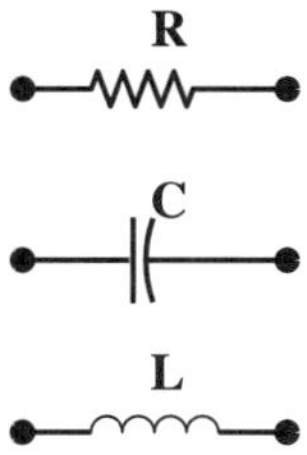

Resistor, its *resistance (R)* is measured in Ohms (Ω).

Capacitor, its *capacitance (C)* is measured in Farads (F).

Inductor, its *inductance (L)* is measured in Henrys (H).

Fig. 1.3-3. Passive elements.

1.3.C. Active Elements: These are elements which require a dedicated power supply before they become functional. For instance, transistors and operational amplifiers are active elements. In a consequential sense these elements are also called dependent sources (i.e., whose values are functions of other circuit parameters), because when they become functional they act as dependent sources. We only study operational amplifiers briefly in Section 11.7.

1.3.C. Dependent Elements [including sources]: These are elements whose values are changing as a function of other circuit variables [parameters]. For instance, as is described in Section 8.2 a mutual inductance is one such dependent element. Generally, a dependent element is a subset of variable elements whose values change as a function of other circuit's parameters.

1.4. CHARACTERISTIC CURVES

In general, the relationship between the **voltage across** and the **current through** an element is called the characteristic curve for that element. For instance, a battery is shown in Fig. 1.4-1a. Note that the direction of the current through is always as shown for any voltage source, i.e., the current is coming out of the positive polarity of the source as shown in Fig. 1.4-1a. This source has an ideal characteristic curve which is shown in Fig. 1.4-1b as a straight line. It is to be noted that only a finite amount of current can be drawn from a voltage source.

In reality, however, as we draw more currents from a source, the characteristic curve becomes less ideal but more practical than before as shown in Fig. 1.4-2 by a bended line that shows the source's power is decaying as we use it regularly. The non-idealness of a source is due to its internal circuitry which absorbs energy. For instance, we always assume that an ideal battery has zero internal resistance – if so it must work indefinitely. While we know that is not the case and after some finite number of hours of continuous usage the battery dies out partially because of its imperfection or non-idealness – a situation that we have experienced with batteries.

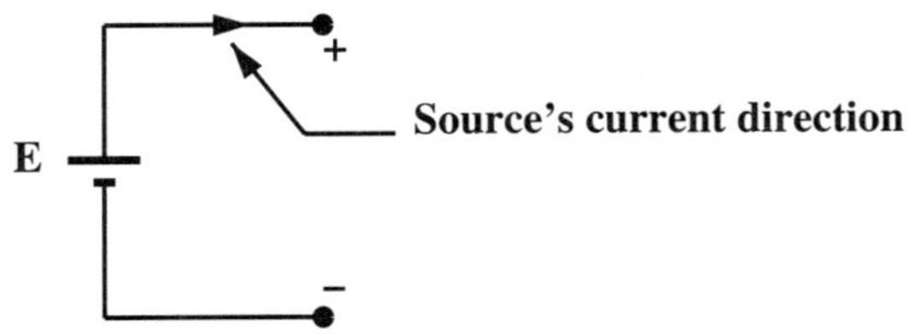

Fig. 1.4-1a. An ideal constant (DC) voltage source in a standard configuration.

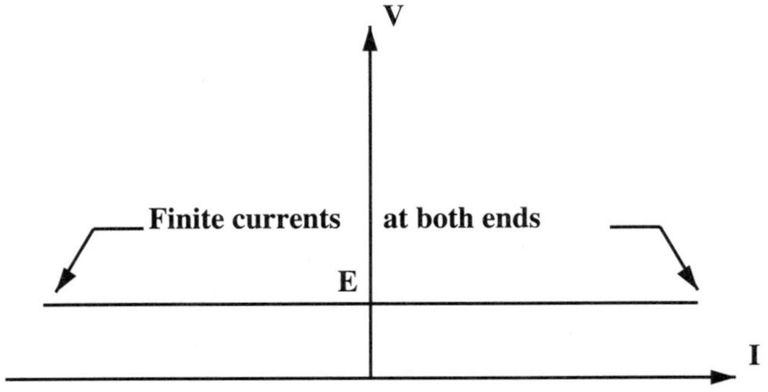

Fig. 1.4-1b. Characteristic curve of an ideal constant (DC) voltage source.

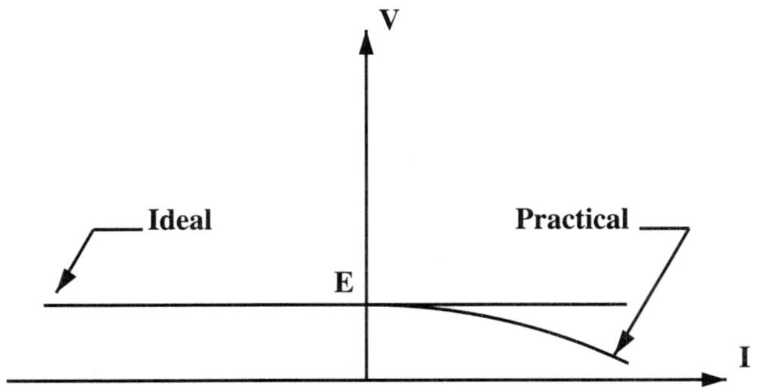

Fig. 1.4-2. Characteristic curves of an ideal (straight line) versus a practical (bended line) voltage sources.

In general, the characteristic curves of most circuit elements are not so straightforward as those for constant voltage (similarly current) sources. In order to generalize these concepts to sinks or loads we need to define the voltage across and the current through these elements. This process is called labeling and is very important in the sense that if we do not follow the particular convention to be defined next, then our analysis will not be consistent with the standard in the field.

For each sink in a circuit, we define the positive polarity of its voltage initially as we please. When our choice is made, however, the direction of its corresponding current through is such that it must enter the positive polarity of this choice. For instance, when we have a resistor R_k connecting node i and node j, we choose the voltage across this element to be as shown in Fig. 1.4-3. Based on this choice, the current through this element must be in the direction from node i to node j, as shown in Fig. 1.4-4.

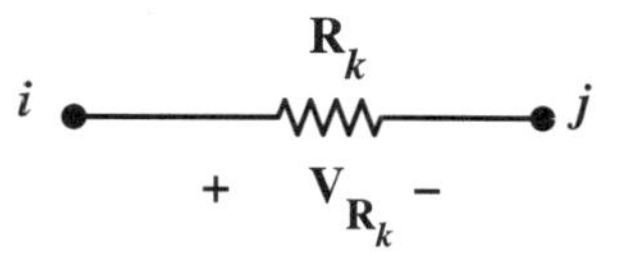

Fig. 1.4-3.

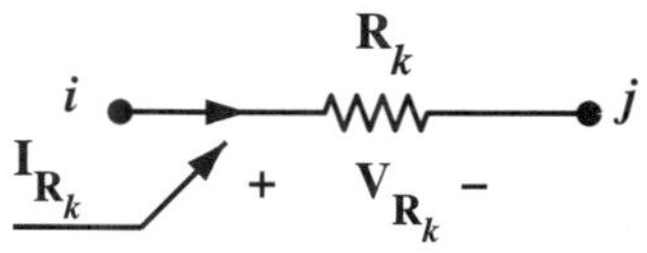

Fig. 1.4-4.

What if we have made a mistake and the actual direction of the positive polarity is the reverse of our choice? Then what will happen? The beauty of the physics behind the circuit is that, when the analysis is completed, it would correct our mistake by resulting in a number that is the negative of our initial choice flagging our mistake.

> Thus, all we need to do first is to label each and every element particularly every load with a set of possible voltage and the corresponding current values which are consistent in the above sense. If we set up correct equations and solve those equations correctly, then the physics of the circuit will correct any possible mistake.

1.5. CHARACTERISTIC CURVE of RESISTORS

The characteristic curves of all ideal resistive elements are linear, i.e.,

$$V = R\,I. \qquad (1.5\text{-}1)$$

This is also called **Ohm's law,** which reads as follows. The voltage across and the current through each ideal resistor is linearly proportion as shown in Fig. 1.5-1. By this definition, we also say that if such a linear relationship does not exist, then the element is not linear and in most cases is nonlinear. However, before closing this section, we note that characteristic curves are generally drawn as I versus V for most elements.

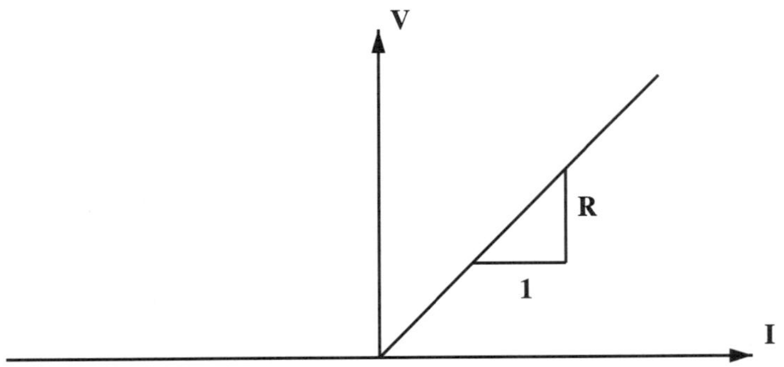

Fig. 1.5-1. Characteristic curve of a resistor.

We may use characteristic curves similar to that in Fig. 1.5-1 for elements which are not linear. Because, in most realistic cases such curves are not linear functions, resulting in nonlinear circuits. Such topics are beyond the scope of our current study, nevertheless we maintain our options open should we face a special problem with such a nonlinear characteristic curve (cf., Section 11.5).

For the time being, we are very content with simple linear circuits involving only linear resistors and ideal sources to describe various methodologies which exist in circuit analysis. When we have mastered the many facets of circuit theory for this class of problems, we can explore the many possibilities of extending our specific knowledge gained in this process to include other types of circuits. In the context of Table 1.1-1, we expand our view both horizontally and vertically. However, there are plenty of ideas in circuits with simple real elements to keep us busy for a while.

1.6. CONCLUDING REMARKS

It is clear that we need to start at some point to describe the very issues which form the basics and must be developed in a typical introductory circuit course. To that end in this chapter we have given some overviews of matters which we have found useful. We hope to cover this body of materials in the balance of the textbook in a one-semester course. However each instructor has his/her way of presenting these topics as he/she feels they may fit in a given program. Chapters one to nine, in our opinion, are required for a first course, while chapters 10 to 12 are optional. We do feel strongly that chapters 10 to 12 should be covered in the first course as well, although conceptually. Students are ready to be familiar with the essential aspects of these topics, although some details, for instance in the feedback amplifiers section, maybe slightly advanced for an early course. We have found students like to be challenged about these matters, and in fact if we capture their attention with some ideas in chapters 10 to 12 they feel more at home with this course. Because, many of them have experienced for instance using their stereo systems, they want to know how the amplifiers work *per se*. So if we do not get into too much detail, then they will concentrate and master the essential theory behind it quickly. This sort of review will prepare students for their electronics and other circuit-related courses.

Thus, in short we hope the arrangement of the covered theory is found helpful and sufficient for everyone concerned. We strongly suggest that the entire textbook be taught in a one-semester course.

Remark 1.6-1: Throughout this textbook we use symbols such as 1^{Ω}, or 2^{A}, or 3^{V}, where the superscripts display the physical dimension of the underlying quantity in the standard MKS system (length "L" in meters; mass "M" in kilograms; time "T" in seconds) as in Table 1.6-1, followed by the list of prefixes in Table 1.6-2.

Table 1.6-1. The essential MKS units for electrical quantities.

Item	*Unit*	*MKS dimensional symbol*
Energy^{J}	Joule (one Newton-meter)	ML^2T^{-2}
Power^{W}	Watt (one Joule per second)	ML^2T^{-3}
Charge^{C}	Coulomb	Q
Current^{A}	Ampere (one Coulomb per second)	QT^{-1}
Voltage^{V}	Volt (one Joule per Coulomb)	$ML^2T^{-2}Q^{-1}$
$\text{Resistance}^{\Omega}$	Ohm	$ML^2T^{-1}Q^{-2}$
Inductance^{H}	Henry	ML^2Q^{-2}
Capacitance^{F}	Farad	$M^{-1}L^{-2}T^2Q^2$

Table 1.6-2. The list of prefixes as used in the MKS system.

$\text{T} - \text{tetra} = 10^{12}$	$\text{G} - \text{giga} = 10^{9}$	$\text{M} - \text{mega} = 10^{6}$	$\text{K} - \text{kilo} = 10^{3}$
$\text{h} - \text{hecto} = 10^{2}$	$\text{dk} - \text{deka} = 10$	$\text{d} - \text{deci} = 10^{-1}$	$\text{c} - \text{centi} = 10^{-2}$
$\text{m} - \text{milli} = 10^{-3}$	$\mu - \text{micro} = 10^{-6}$	$\text{n} - \text{nano} = 10^{-9}$	$\text{p} - \text{pico} = 10^{-12}$

1.7. PROBLEM SET

From time to time and in each problem set, we try to categorize problems according to the covered subject matters and as best as we can. This effort is made only to assist our readers partially. However, the boundaries among questions may not be very finely defined. In fact, in some cases we may not want to be very specific, because choosing the proper methodology to solve a problem by itself may be a part of the question. Thus, try to get whatever information you can from our categorizations.

Definitions

1.7.1. For the circuit in Fig. 1.2-3, find all possible paths between node m and node n.

1.7.2. For the circuit in Fig. 1.2-3, find all possible closed paths.

1.7.3. In the preceding two problems, we refer to Fig. 1.2-3 as a circuit. Is this description acceptable to you and why?

Comment: If your answer is yes, and you realize that irrespective of the nature of elements connecting, say node m to node i, you may extract some generic properties of this circuit. In other words, the structure of this circuit is such that certain paths exist between two of its nodes, or it has so many closed paths, of which these properties are independent from the types of elements connecting the nodes. Then you have a new way of looking at a circuit which is called a topological representation. Another name for Fig. 1.2-3 is the graph for the corresponding circuit. We do not explore these issues any further, but we find similar figures make our explanations easy and we intend to use them in our subsequent presentations.

1.7.4. For the circuit in Fig. 1.2-4, find all possible closed paths.

Comment: Here, we have an interesting circuit which you must redraw slightly in order to answer the above question. Referring to Table 1.1-1, this is called a planar network and we discuss that *subsequently.* Normally, our style is to be specific and give the section number, but for the sake of promoting independent thinking we are not specific now.

Conceptual

1.7.5. Referring to Table 1.1-1, which names or descriptions have not yet been explored? For instance, we have not yet mentioned anything about one-port networks. Have we? Please make a tally of all the outstanding issues and keep that list for crossing as we proceed in this course.

1.7.6. In Table 1.1-1, arrows are used when external sources are applied, but on each level, no arrow is used between different entities. For instance, on Level 1, we have a box with the word "passive" inside, but no arrow goes out of it to the box with resistive elements inside. The reason is that here some of the orders among various requirements are not so critical as the actual requirements themselves. We could have said *one-port, planar, resistive networks* which in turn covers passivity. In your own words, describe how we go from one level to another in this Table. Can you develop an alternative way of looking at these issues?

2 ELEMENTS OF NETWORK ANALYSIS

2.1. INTRODUCTION

Our primary objective is network analysis, and to that end the following materials are prepared. But first we define what we mean by network analysis.

Definition 2.1-1: Analyzing a network means to *set up* a number of independent equations that are to be solved numerically, in order to yield the voltage across and the current through each and every element of the underlying network.

This course is **not** about number crunching, nor is about using a computer to set up the corresponding set of equations to analyze a network, but rather is about learning the process of generating a correct set of equations systematically, whose numerical solutions will complete the network analysis in the sense of Definition 2.1-1.

Here we want to gain experience about methodology and algorithmic way of looking at network analysis. When we master this aspect of the course, we concentrate on numerical solutions and related issues with expansion toward advanced topics in network analysis. We start our approach by considering a few simple though instructive examples first.

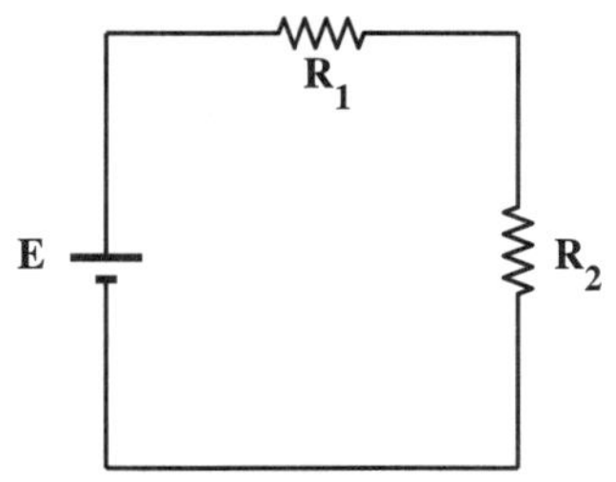

Fig. 2.1-1.

Example 2.1-1: Analyze the circuit in Fig. 2.1-1.

Solution: We can analyze this network in the sense of the **Definition 2.1-1,** from what we have seen so far, i.e., KCL and KVL, and of course, some explicit knowledge about the typical elements for this circuit as shown in Fig. 2.1-1. This approach is refined subsequently in this chapter. Here, we have R_1 and R_2, each satisfies Ohm's law, and E is an ideal voltage source with a known value. For the record and throughout this textbook we assume again that all circuit elements are lumped and not distributed. Clearly, we always assume that all elements are ideal according to their own definitions. In the following we study the above example and give its solution through a deliberate process which we call Method #Zero.

2.2. METHOD #ZERO – SOLUTION by the PRIMITIVE APPROACH

Invoking the following steps forms an approach called Method #Zero herein, and is described as follows.

STEP Zero: To analyze any circuit we must start with this step, which refers to characterizing or labeling each and every element with an appropriate set of variables. For instance, in Fig. 2.1-1, we have three elements and five unknowns as shown in Fig. 2.2-1 for convenience, and those are: (V_{R_1}, I_{R_1}), (V_{R_2}, I_{R_2}), and I_s the current through the voltage source.

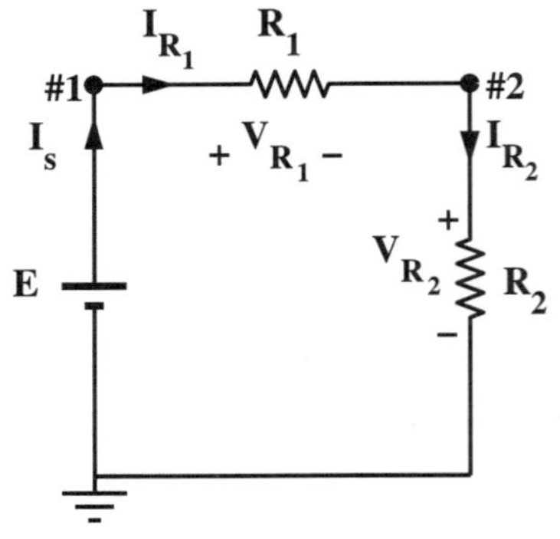

Fig. 2.2-1.

STEP One: Using KVL, we set up as many independent equations as we can. In Fig. 2.2-2, we have only one closed path, and tracing all voltages along this path and according to our **arbitrary** direction of clockwise rotation yields

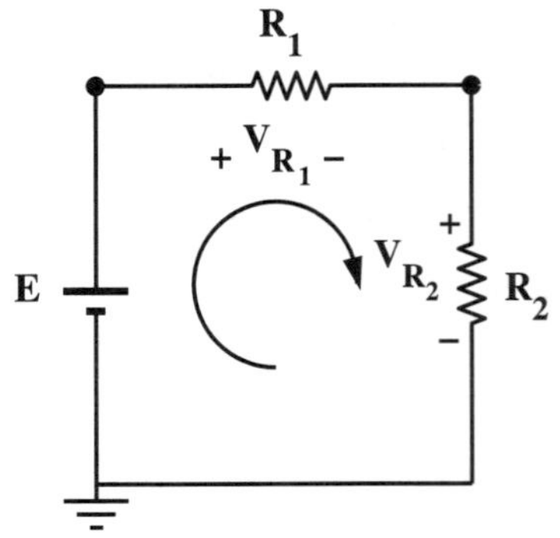

Fig. 2.2-2.

$$E - V_{R_1} - V_{R_2} = 0, \qquad (2.2\text{-}1a)$$

which is the conventional way of writing KVL and this is the same as

$$-E + V_{R_1} + V_{R_2} = 0. \qquad (2.2\text{-}1b)$$

There is no more independent KVL equation and we have exhausted Step One.

Remark 2.2-1. The arrow in Fig. 2.2-2 enters the negative polarity of the source and leaves at its positive polarity, while it enters the positive polarities of both resistors (loads), as we have initially selected so. This procedure is the convention which we must always follow. Had we made a mistake and the actual current through an element were in the reverse direction relative to this choice, then as stated before the physics of the circuit will take over and the final result will be the negative of our initial choice, i.e., we get a "minus" sign in front of this choice to flag our mistake.

STEP Two: Using KCL and also referring to Fig. 2.2-2, we have

$$I_s = I_{R_1}, \quad \text{at node \#1}, \qquad (2.2\text{-}2)$$

$$I_{R_1} = I_{R_2}, \quad \text{at node \#2}. \qquad (2.2\text{-}3)$$

STEP Three: In general, using characteristic curves for elements, we can set up hybrid relationships among the above KVL and KCL equations. Here, we have two resistors R_1 and R_2, thus we have

$$V_{R_1} = R_1 I_{R_1}, \qquad (2.2\text{-}4)$$

$$V_{R_2} = R_2 I_{R_2}. \qquad (2.2\text{-}5)$$

Our solution is completed in that we have five independent equations (2.2-1) to (2.2-5) with five unknowns. The numerical solution for this set of equations will complete the analysis of the circuit that is shown in Fig. 2.2-1.

STOP

Comment 2.2-1: In future, we continue this process by actually solving for a numerical solution, but for now we concentrate only on setting up the relevant set of equations. Certainly, anyone who writes these equations correctly can actually solve them numerically. Therefore, for the time being we concentrate only on developing different methods which yield similar set of equations instead of actual numerical solutions for the underlying equations.

Example 2.2-1: Analyze the circuit in Fig. 2.2-3.

Fig. 2.2-3.

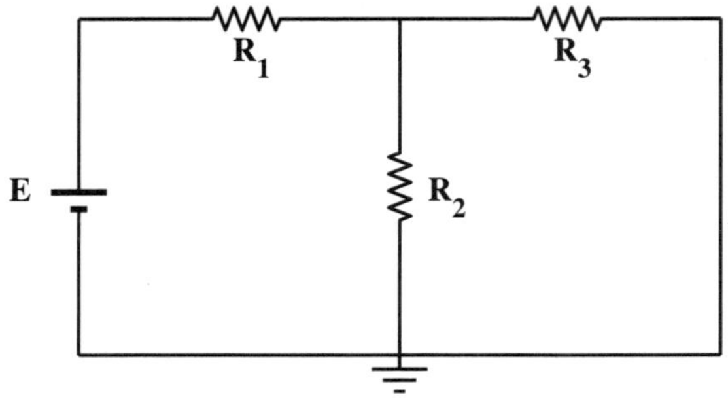

Solution: We invoke all the steps in the preceding example to solve this problem.

STEP Zero: To label or identify all pertinent variables as shown in Fig. 2.2-4.

Fig. 2.2-4.

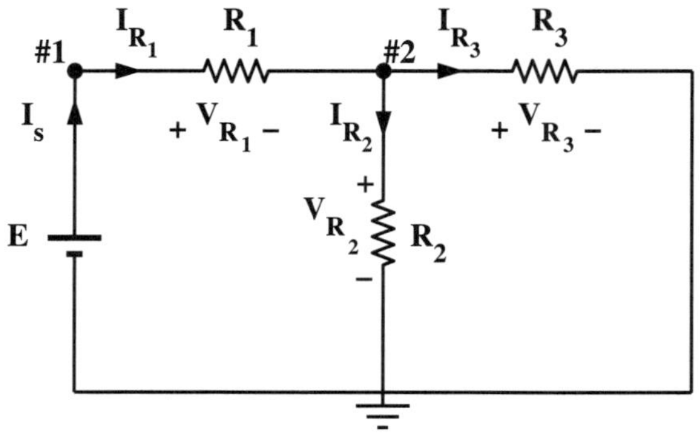

STEP One: Using KVL, we set up as many independent equations as we can. Here, we need to choose directions for two arrows. Each arrow corresponds to one closed path as shown in Fig. 2.2-5, or we could have done so in Fig. 2.2-4.

Fig. 2.2-5.

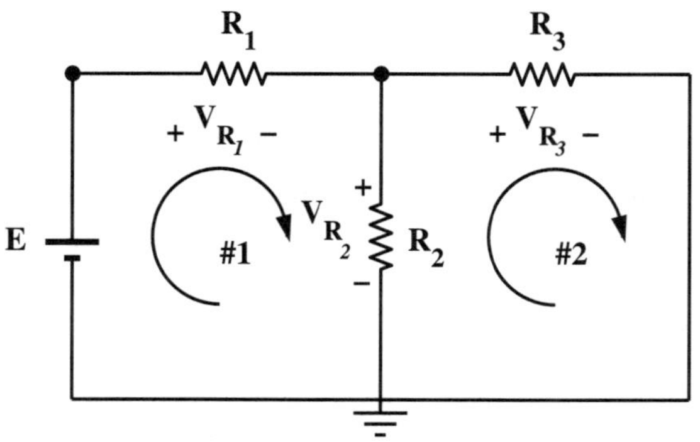

$$-E + V_{R_1} + V_{R_2} = 0, \quad \text{in loop \#1,} \tag{2.2-6}$$

$$-V_{R_2} + V_{R_3} = 0, \quad \text{in loop \#2.} \tag{2.2-7}$$

These equations are written by tracing the arrow and choosing the polarities as we face. For instance, in loop #1, the arrow is entering into the negative edge of the source, while it is entering into the positive polarities of the two resistors. Similarly, in loop #2, there is no independent source, and the corresponding arrow enters in the negative polarity of R_2 while it enters in the positive polarity of R_3. As stated before, if we had made a mistake, then the physics of the circuit would have reversed our choices by generating a negative value for the corresponding loop current.

STEP Two: Using KCL, we have

$$I_s = I_{R_1}, \qquad \text{at node \#1}, \tag{2.2-8}$$

$$I_{R_1} = I_{R_2} + I_{R_3}, \qquad \text{at node \#2}. \tag{2.2-9}$$

STEP Three: Using the characteristic curves of elements we have

$$V_{R_1} = R_1 I_{R_1}, \tag{2.2-10}$$

$$V_{R_2} = R_2 I_{R_2}, \tag{2.2-11}$$

$$V_{R_3} = R_3 I_{R_3}. \tag{2.2-12}$$

We have seven independent equations and seven unknowns: (V_{R_1}, I_{R_1}), (V_{R_2}, I_{R_2}), (V_{R_3}, I_{R_3}) and I_s. Thus, we can solve these equations numerically. For instance, let us simplify them into a set of more manageable forms. Straightforward substitution of (2.2-10 & 11) into (2.2-6) yields

$$\begin{aligned} E = V_{R_1} + V_{R_2} &= R_1 I_{R_1} + R_2 I_{R_2} \\ &= R_1(I_{R_2} + I_{R_3}) + R_2 I_{R_2} \\ &= (R_1 + R_2) I_{R_2} + R_1 I_{R_3}. \end{aligned} \tag{2.2-13}$$

Similarly, using (2.2-12) in (2.2-7) yields

$$0 = -V_{R_2} + V_{R_3} = -R_2 I_{R_2} + R_3 I_{R_3}. \tag{2.2-14}$$

Equations (2.2-13 & 14) form a set of two equations in two unknowns I_{R_2} and I_{R_3}. Thus, the problem is manageable.

STOP

Comment 2.2-2: Although, we can analyze any circuit using the above steps, the entire procedure as named is primitive and should not be pursued in any real situation. Partly because there are simpler methods to address even more complicated problems than the above. Also, partly because newer methods are more systematic and amenable with computer simulations than the above method, as we explore some of them in the forthcoming sections of this textbook.

2.3. MESH EQUATIONS

In the process of solving Example 2.2-1, we have introduced two arrows to describe the two paths in the corresponding circuit. Before proceeding further, however, we note that all the circuits we study in this course are the so-called **planar,** as mentioned in Table 1.1-1. For instance, the circuit in Fig. 2.3-1 is a planar circuit, meaning that it can be easily redrawn, as shown in that figure, on a plane by rotating *klj* triangle along the *kj* axis as shown, and that is why this circuit or network is called planar.

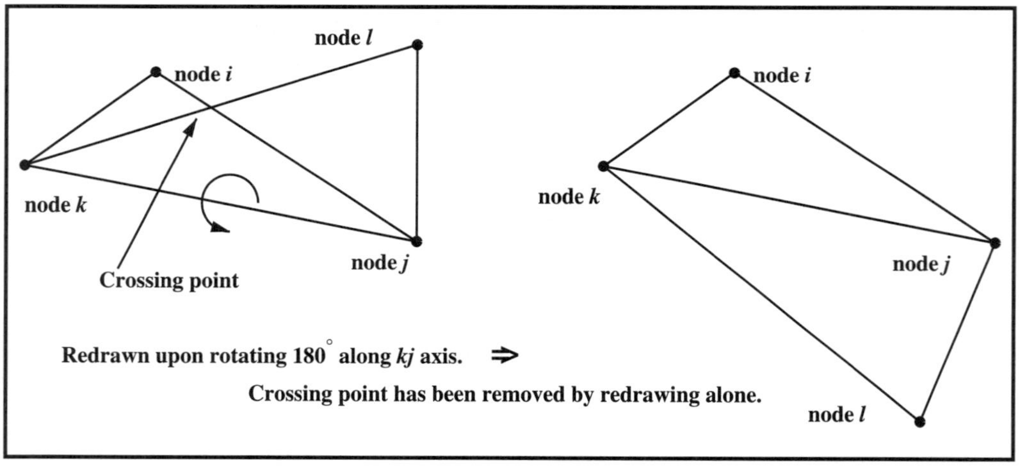

Fig. 2.3-1. A planar network.

On the other hand, a **non-planar** network, as shown in Fig. 2.3-2, cannot be drawn on a plane without some wires crossing each other.

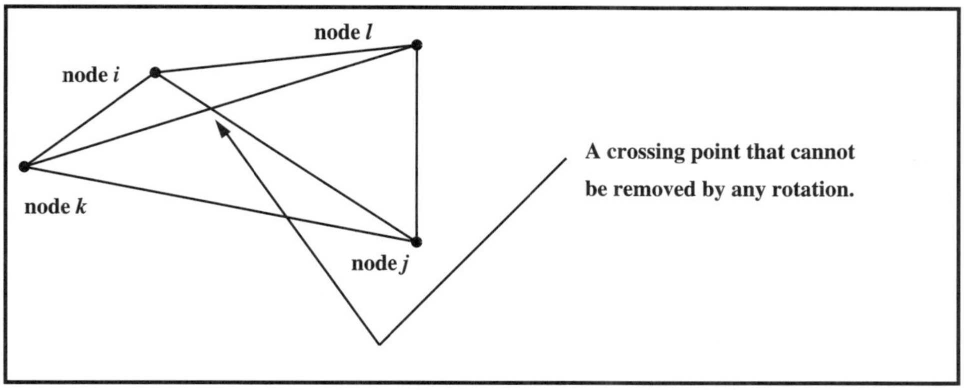

Fig. 2.3-2. A non-planar network with one crossing point.

Definition 2.3-1: In general, given a planar network, we define the shortest closed path (i.e., a closed path with the least number of elements) a mesh.

For instance, in Fig. 2.3-3, each mesh has only four elements (edges). Here, we note that mesh is a property of the network topology and has nothing to do with the actual or physical nature of the circuit elements. That is why we have drawn circuits in their topological forms instead of displaying them with their elements' value. This brief mentioning of the notion of circuit topology (in addition to what we have discussed in Problem 1.7.3) is as far as we plan to use this concept. By referring to network topology we are concerned only with the structure or the interconnection of circuit elements rather than their actual values or types.

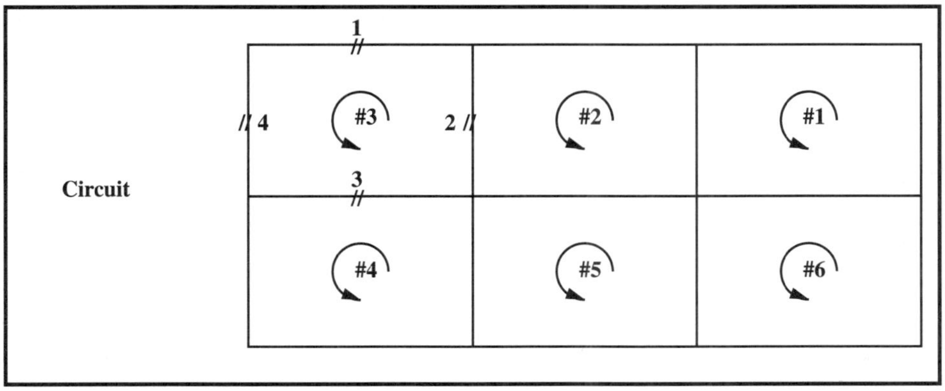

Fig. 2.3-3. A circuit with six meshes. Each mesh has four edges.

Definition 2.3-2: Contrary to Definition 2.3-1, a loop is any closed path in a network with no regard to its number of elements.

For instance, for the circuit in Fig. 2.3-3, we can show two out of its many loops in Figs. 2.3-4 & 5.

Referring to Figs. 2.3-4 & 5, we state the following definition of the so-called **super loop** for our subsequent applications.

Definition 2.3-3: In general, a *super loop* contains two or more adjacently connected meshes.

Perhaps the connotation of "super" needs clarification. For instance, in Figs. 2.3-4 & 5, both loops can also be called super loops, however, for a more relevant application of this concept we refer to Fig. 3.6-3.

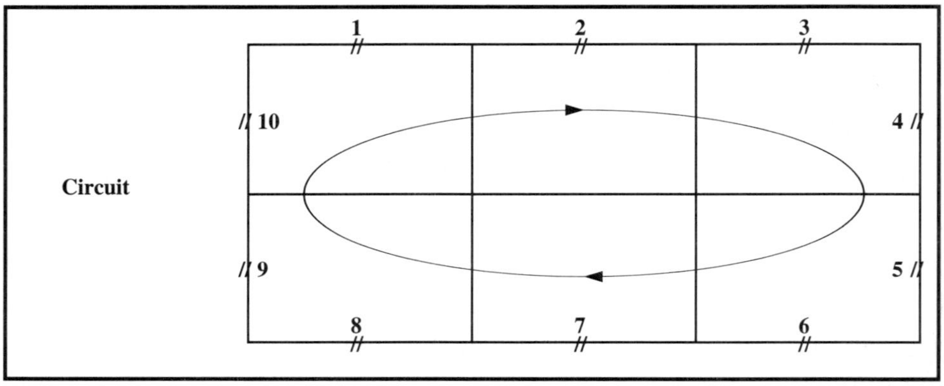

Fig. 2.3-4. A loop with ten edges.

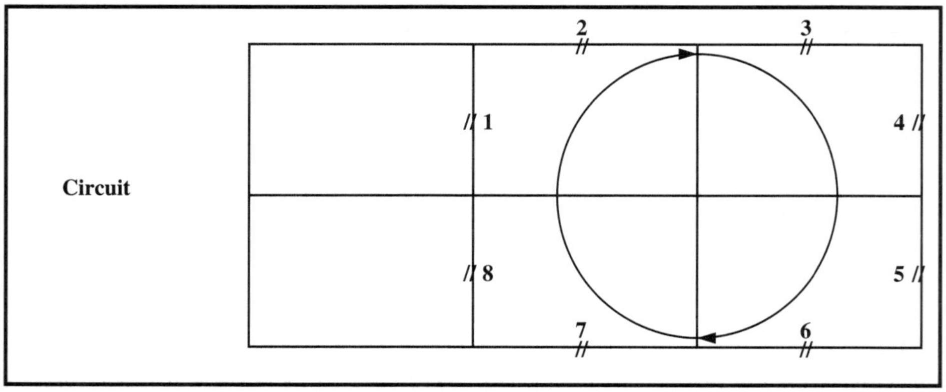

Fig. 2.3-5. A loop with eight edges.

Now, going back to Example 2.2-1, we introduce a new set of variables. Namely, let I_1 and I_2 be the so-called mesh currents corresponding to mesh #1, and mesh #2. If we somehow determine these two variables, then we have

$$I_s = I_{R_1} = I_1 \quad \Rightarrow \quad V_{R_1} = R_1 I_1, \tag{2.3-1a}$$

$$I_{R_2} = I_1 - I_2 \quad \Rightarrow \quad V_{R_2} = R_2(I_1 - I_2), \tag{2.3-1b}$$

$$I_{R_3} = I_2 \quad \Rightarrow \quad V_{R_3} = R_3 I_2. \tag{2.3-1c}$$

What we have discovered herein is remarkable in the sense that if we can determine I_1 and I_2, then all the other variables can be expressed in terms of these two quantities straightforwardly. So let us go back and study the same example using the above variables in order to explain the new approach.

Example 2.3-1: Analyze the circuit of Example 2.2-2, using mesh currents as variables.

Fig. 2.3-6.

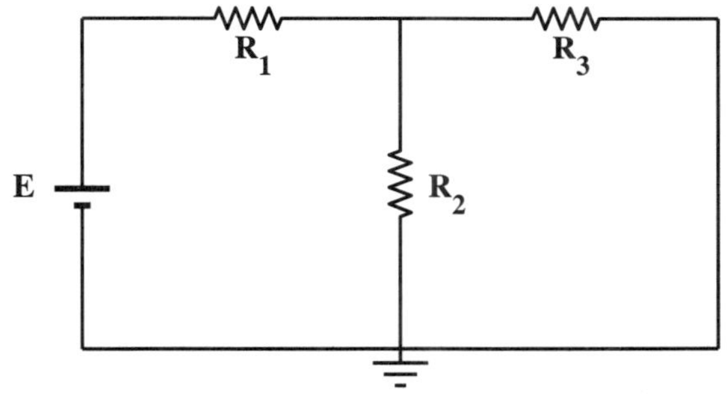

Solution: Again, we invoke the previous steps to solve this problem, but we interpret the new equations in terms of I_1 and I_2.

STEP Zero: To label and/or identify all variables including mesh variables.

Fig. 2.3-7.

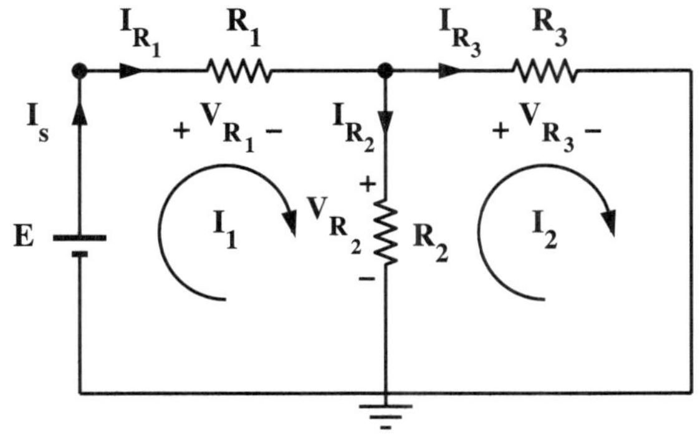

STEP One: Using KVL, we have

$$-E + V_{R_1} + V_{R_2} = 0, \quad \text{in loop \#1}, \tag{2.3-2a}$$

$$-V_{R_2} + V_{R_3} = 0, \quad \text{in loop \#2.} \tag{2.3-2b}$$

In terms of mesh variables, we have the following two corresponding equations.

$$E = R_1 I_{R_1} + R_2 I_{R_2}$$

$$= R_1 I_1 + R_2(I_1 - I_2)$$

$$= (R_1 + R_2)I_1 - R_2 I_2. \tag{2.3-3a}$$

$$0 = -R_2 I_1 + (R_2 + R_3)I_2. \tag{2.3-3b}$$

STEP Two: Using KCL, we have

$$I_s = I_{R_1} = I_1, \tag{2.3-4a}$$

$$I_{R_2} = I_1 - I_2, \tag{2.3-4b}$$

$$I_{R_3} = I_2, \tag{2.3-4c}$$

which have already been incorporated in Step One. Thus, if we establish that step correctly, then we have the right number of independent equations in terms of mesh variables I_1 and I_2. In other words, solving

$$(R_1 + R_2)I_1 - R_2 I_2 = E, \tag{2.3-5a}$$

$$-R_2 I_1 + (R_2 + R_3)I_2 = 0, \tag{2.3-5b}$$

for I_1 and I_2, that is quite manageable, enables us to complete setting up equations in this step, and in turn those equations which we develop in the next step.

STEP Three: Using characteristic curves (here, Ohm's law) associated with the resistive elements, we have:

$$V_{R_1} = R_1 I_{R_1} = R_1 I_1, \tag{2.3-6a}$$

$$V_{R_2} = R_2 I_{R_2} = R_2(I_1 - I_2), \tag{2.3-6b}$$

$$V_{R_3} = R_3 I_{R_3} = R_3 I_2, \tag{2.3-6c}$$

which all the above quantities have already been determined.

Thus, the new approach, i.e., introducing a new set of variables called mesh variables, has a great potential to simplify our computations. But that is not all, if we look at the two **essential equations** in Steps One & Two, repeated below for convenience

$$(R_1 + R_2)I_1 - R_2I_2 = E, \tag{2.3-7a}$$

$$-R_2I_1 + (R_2 + R_3)I_2 = 0. \tag{2.3-7b}$$

We note that we can generate these equations by inspection as shown subsequently. Having generated I_1 and I_2 in this manner, i.e., by solving these equations, say (2.3-7a & b), the rest of the analysis seems straightforward.

But what do we see in (2.3-7a & b), which make them so attractive?

What we see in these equations is an **evolving pattern** which we can generalize to other cases. For instance, looking at the first term in (2.3-7a), i.e., $(R_1 + R_2)I_1$, it is clear that $(R_1 + R_2)$ is the **total resistance** in mesh #1, which is multiplied by the current of that mesh.

The next term $-R_2I_2$ is the **total resistance** between mesh #1 and mesh #2, which is multiplied by the current of mesh #2, i.e., the other mesh. Why the minus sigh? Because the current of mesh #2 (the other mesh) enters into R_2 in the opposite direction with respect to the current of the mesh #1.

Where does E come from? This is the *net* voltage source as we trace in mesh #1, whose polarity coincides with the current of mesh #1, in the sense that $I_s \equiv I_1$.

We can say similar things about the terms in (2.3-7b). In this equation, $(R_2 + R_3)I_2$ is the total resistance of mesh #2 multiplied by the current of that mesh.

The $-R_2I_1$ is the total resistance between mesh #2 and mesh #1. The minus sign is because I_2 and I_1 are opposing each other as they go through R_2.

There is no independent nor any voltage source in mesh #2 and that is why we have the net zero on the right-hand side of (2.3-7b).

Putting together these facts, we can present a new algorithm to analyze networks that consist of only voltage sources, based on the mesh equations concept. For networks with mixed (current and voltage) sources this approach must be modified.

2.4. METHOD #ONE – SOLUTION by the MESH EQUATIONS

To analyze a network with only voltage sources, we invoke the following steps. Again, we emphasize that this method cannot be used directly in a network with mixed sources.

STEP Zero: To label or identify all pertinent variables including the mesh current variables. *Comment:* It is always advisable to select the direction for mesh currents in accordance with the net voltage source in each mesh as much as feasible.

STEP One: Using the mesh currents, and KVL, set up all the independent equations needed in the underlying problem. *Comment:* This step can be set up in the following manner by inspection. In each independent mesh, for instance in mesh #1, we write:

the **net voltage source in mesh #1** =

+ (the total sum of the resistance in mesh #1) × I_1

± (the total sum of resistances between mesh #1 and mesh #2) × I_2

± · · · ± (the total sum of resistances between mesh #1 and the mesh #last) I_{last}.

The ± sign in the above is selected based on the relative direction of various mesh currents with respect to the current in mesh #1, as these currents enter the same common element(s) between these meshes.

STEP Two: For mesh #2 to mesh #last, the above procedures are repeated to establish similar equations in terms of all mesh currents.

STEP Final: Upon solving the above set of equations in terms of each mesh current I_1 to I_{last}, we have all the necessary information to complete the analysis.

STOP

Example 2.4-1: For the circuit in Fig. 2.4-1, *set up only* the corresponding mesh equations.

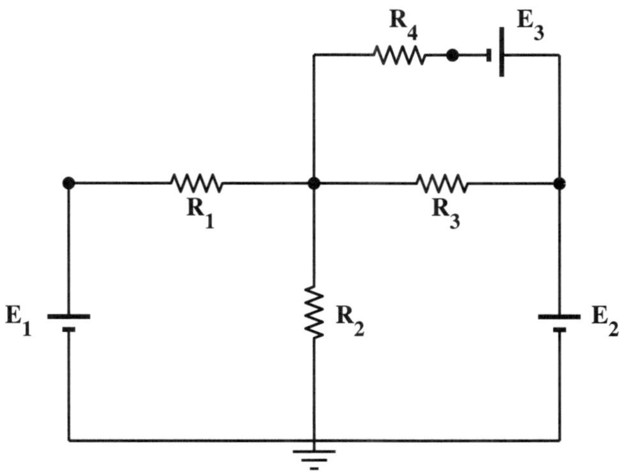

Fig. 2.4-1.

Solution: We invoke the following steps to complete this task.

STEP Zero: To label and/or identify all variables.

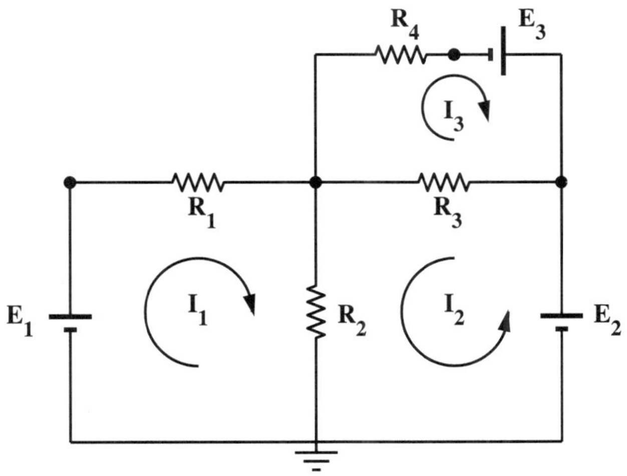

Fig. 2.4-2.

STEP One: Now, the essential mesh equations for this circuit can be set up by inspection. Recall that we are only establishing the essential equations.

$$E_1 = (R_1 + R_2)I_1 + R_2I_2, \tag{2.4-1}$$

$$E_2 = R_2I_1 + (R_2 + R_3)I_2 + R_3I_3, \tag{2.4-2}$$

$$E_3 = R_3I_2 + (R_3 + R_4)I_3. \tag{2.4-3}$$

STOP

Example 2.4-2: Analyze the circuit in Fig. 2.4-3.

Fig. 2.4-3.

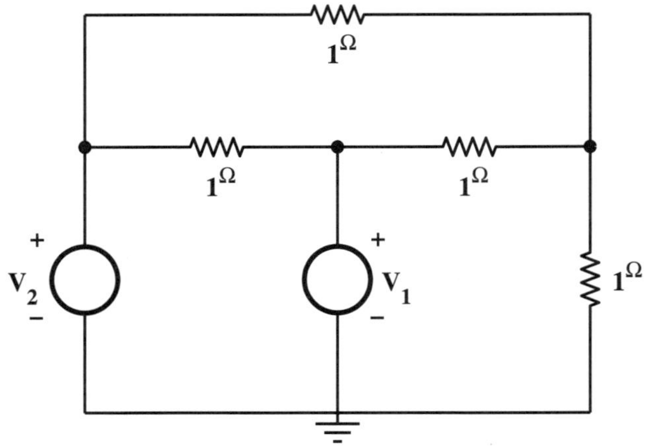

Solution: The circuit in Fig. 2.4-3 consists of only voltage sources. Thus, the Method #One is a right approach to tackle this problem.

STEP Zero: To label and/or identify all variables for this method accordingly.

Fig. 2.4-4.

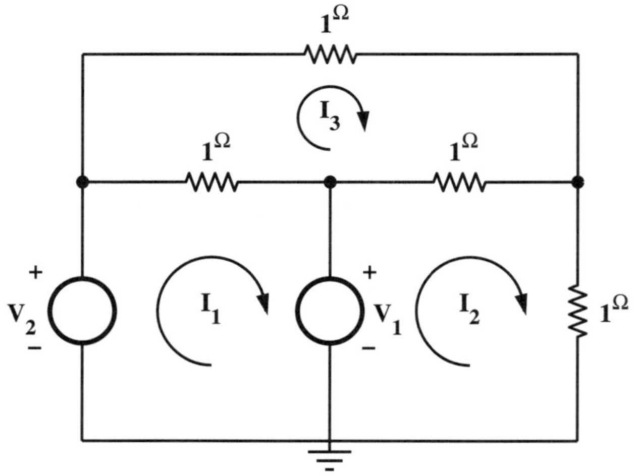

STEP One: The mesh equations for this circuit are as follows.

$$V_2 - V_1 = (1^{\Omega})I_1 - (1^{\Omega})I_3, \tag{2.4-4}$$

$$V_1 = (1^{\Omega} + 1^{\Omega})I_2 - (1^{\Omega})I_3, \tag{2.4-5}$$

$$0 = (1^{\Omega} + 1^{\Omega} + 1^{\Omega})I_3 - (1^{\Omega})I_2 - (1^{\Omega})I_1. \tag{2.4-6}$$

STOP

Comment 2.4-1: The preceding approach is inherently based on having only voltage sources in the circuit. In practice, that is hardly the case. Not only do we have current sources in the network, but also we have another class of sources, which is called **dependent (voltage or current) sources.** These dependent sources are shown by 45° rotated squares instead of circles as shown in Fig. 2.4-5. Furthermore, we seldom have one source in a mesh, and for that reason we need to look into different types of sources with their corresponding interconnections.

Starting with a dependent source, consider the following circuit, which is an equivalent model for a transistor (there are many other similar equivalent configurations, which are not studied herein).

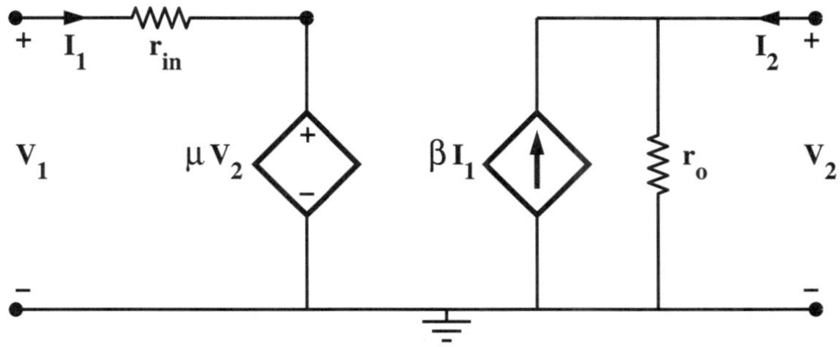

Fig. 2.4-5.

In Fig. 2.4-5, the μV_2 is a dependent voltage source, whose value depends on the output voltage V_2. The βI_1 is a dependent current source, whose value depends on the input current I_1. Note that for current sources (dependent or independent) we use an arrow inside the box to show its direction.

2.5. SOURCE TRANSFORMATION

In most cases we have both current and voltage sources in a given circuit. In these situations, the preceding method of mesh equations cannot be applied directly, and we need an alternative approach to analyze the circuit. One such approach is called source transformation, which is an advanced topic in its own right, but we next describe a simplified version of this transformation, and its expanded version in Section 5.3.

Suppose we have a box with a functional circuit which has a number of elements including sources and correspondingly many nodes. For instance, consider the wirings inside a building with many units. Here, the box is this building and the wirings correspond to the electrical contents of its various units, and sources are provided by the electrical company. Now, suppose we select any two nodes from its many, and pull them out of the box for further study. This process of choosing a particular *entrance* into a given circuit is called selecting a "port", Fig. 2.5-1. Clearly, if we had selected any other two nodes, then the electrical behavior of the same circuit would have been different at the new port. Thus selection of a port is important for the final outcome of a circuit study. Often, we may be interested in looking into only one specific port, out of possibly many, for a given circuit. That is why in Table 1.1-1 we have emphasized "one-port networks" for more than three quarters of this course, since we need to know how to initiate such a study and how to develop the final answer in a given problem for its only one port.

In Fig. 2.5-1, we have a box containing a circuit with many elements, and we have selected port ab. Inside this box there is possibly a complicated circuit, but we assume this circuit can be represented by a simple one, consisting of only a voltage source and a series resistor. In other words, when we look inside a box of an unknown circuit, we can model its entire circuitry with the above R_s in series with the V_s. It is important to understand this presentation, because this box may contain a complicated circuit involving many elements, and yet its electrical behavior at the specific port ab can be described in exactly the same manner as the above circuit shown in Fig. 2.5-1.

Now, consider the same box but select for now another port called $a'b'$. Furthermore, suppose that at this new port the same circuit can be ’ represented with a current source and a parallel resistor, Fig. 2.5-2. That means the electrical behavior of the circuit in the box at port $a'b'$ is identical to that of one current source in parallel with one resistor.

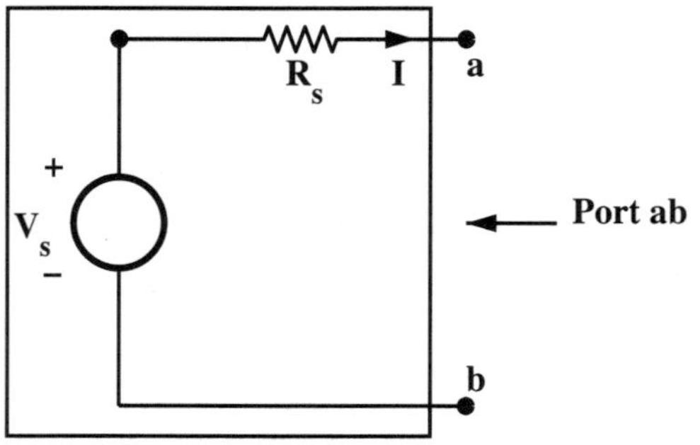

Fig. 2.5-1.

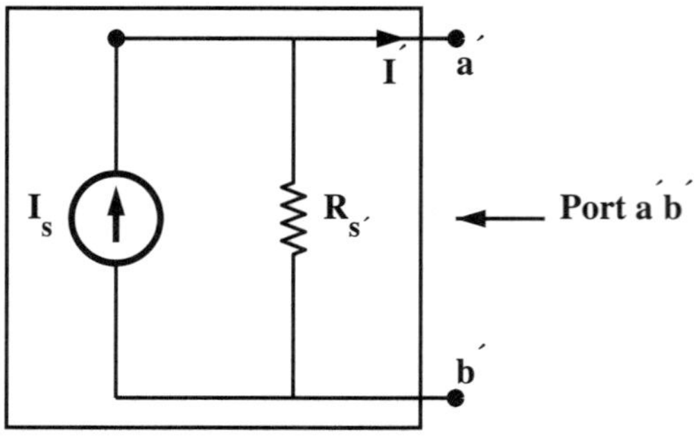

Fig. 2.5-2.

Clearly, we realize that the port behavior of any circuit changes as the underlying port changes. However, if we allow ourselves to be imaginative and ask what if port $a'b'$ becomes the same as port ab? Then under what conditions the two representative circuits in Figs. 2.5-1 & 2 become equivalent? The equivalency means for someone looking inside port ab in Fig. 2.5-1 sees exactly the **same electrical properties** as is seen in port $a'b'$ in Fig. 2.5-2. In other words, when each circuit in Figs. 2.5-1 & 2 is connected to an external circuit, each of these two circuits sends out the **same current** into the applied external circuit, and the **same voltage** is maintained at port ab or port $a'b'$. For instance, consider the circuit in Fig. 2.5-3, with I the current that is entering the external circuit on the right-hand side. As shown in this figure, we have

$$V_{ab} = V_s - R_sI. \tag{2.5-1}$$

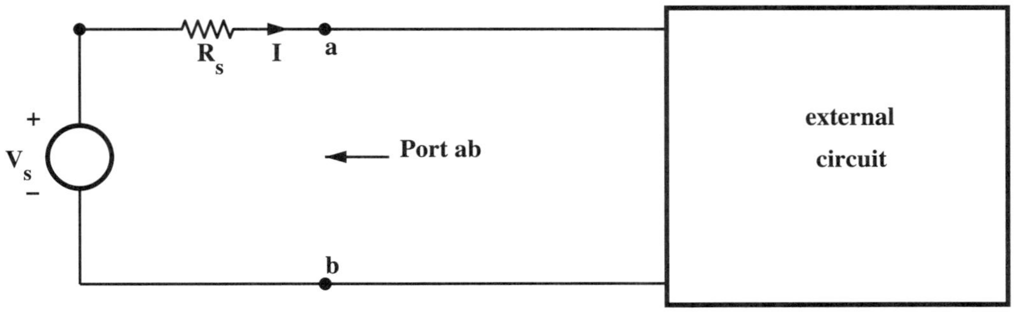

Fig. 2.5-3.

Meanwhile, consider the same external circuit as applied to the circuit in Fig. 2.5-2, and shown in Fig. 2.5-4. Here, we have

$$V_{a'b'} = R_{s'}(I_s - I') = R_{s'}I_s - R_{s'}I', \tag{2.5-2}$$

where again I' is the same current that is entering the external circuit.

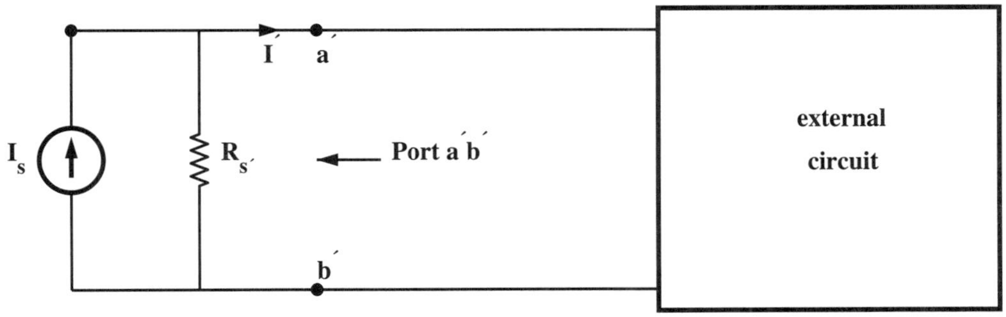

Fig. 2.5-4.

Now, under what requirements do we have equivalency between the above two circuits in Figs. 2.5-1 & 2? That means, if we want these two circuits behave exactly the same electrically, then we must have $I \equiv I'$ and $V_{ab} \equiv V_{a'b'}$, i.e., we have

$$V_s - R_s I \equiv R_{s'} I_s - R_{s'} I. \tag{2.5-3}$$

If we let $R_{s'} \equiv R_s$, i.e., we impose this condition which is perfectly reasonable, then that choice results in $V_s = R_s I_s$. Thus, under the above specific conditions ($I = I'$, $V_{ab} = V_{a'b'}$ and $R_s = R_{s'}$) the following two circuits are identical electrically at port ab and that is a very strong result.

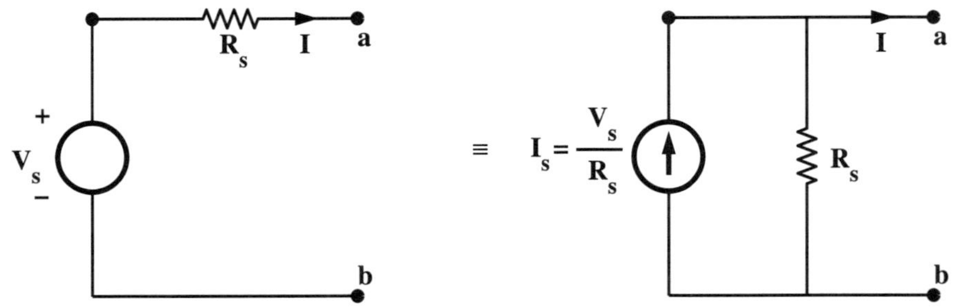

Fig. 2.5-5.

The above transformation makes our future analysis very straightforward. Thus, for now we are able to apply mesh equations approach to solve circuits involving both current and voltage sources. However, the current source must have a resistor in parallel. There are some subtle cases which this transformation alone as it may seem is not so obvious, and we study those cases in Section 3.4.

2.6. PROBLEM SET

Fundamentals

2.6.1. For the circuit in Fig. 2.6-1, find current I and voltage V as indicated.

Fig. 2.6-1.

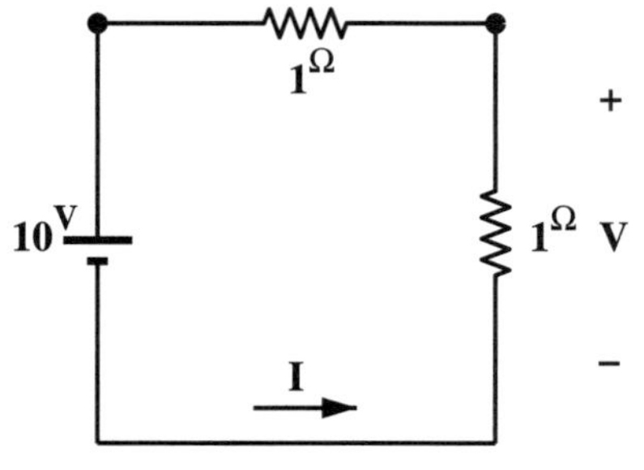

2.6.2. In circuit of Fig. 2.6-2, if $I_1 = 1^A$, then what is the value of current I_2?

Fig. 2.6-2.

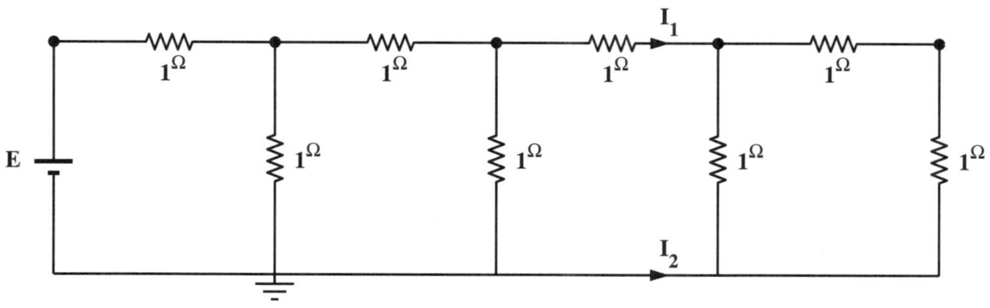

2.6.3. For the circuits in Figs. 2.6-3 to 5, find the indicated V_o numerically.

Fig. 2.6-3.

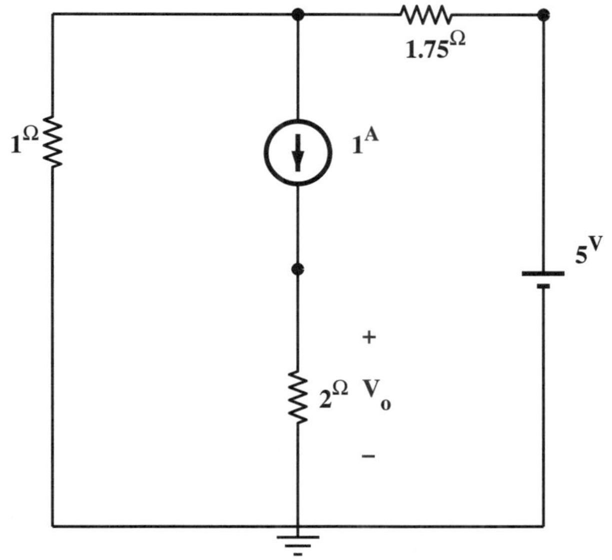

Fig. 2.6-4.

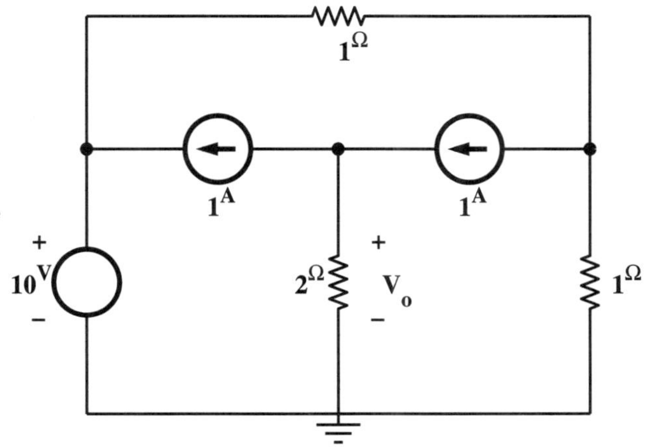

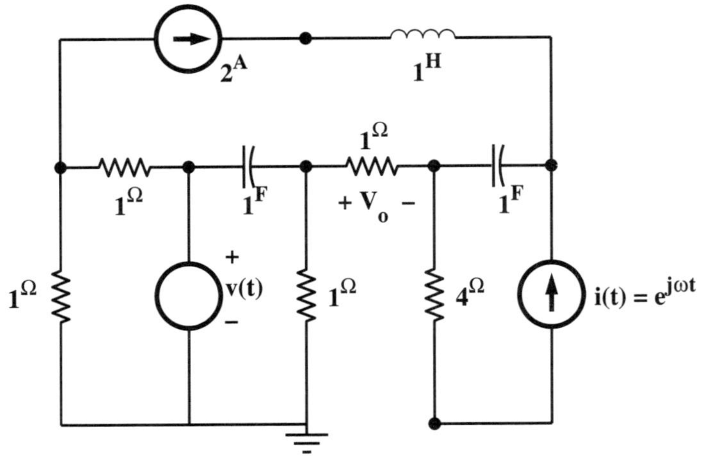

Fig. 2.6-5.

2.6.4. What is the value of I_o in the circuit of Fig. 2.6-6?

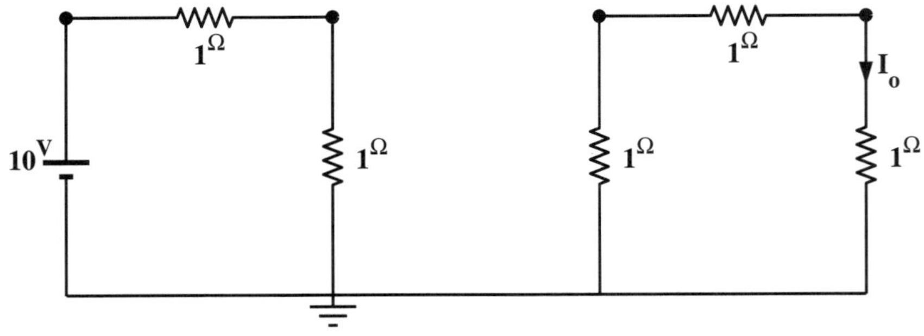

Fig. 2.6-6.

Primitive Analysis

2.6.5. For the circuits in Figs. 2.6-7 to 9, find V_o numerically.

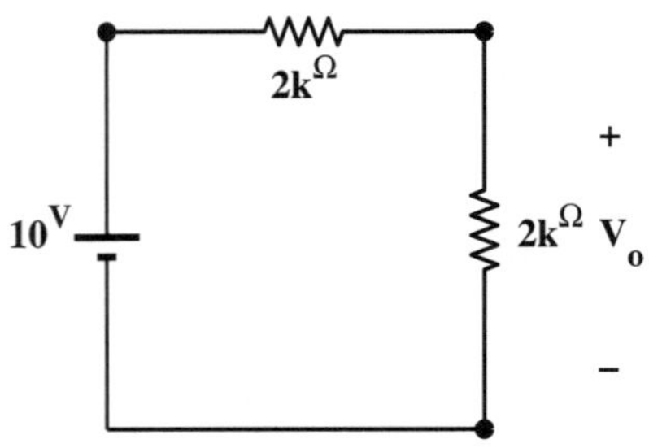

Fig. 2.6-7.

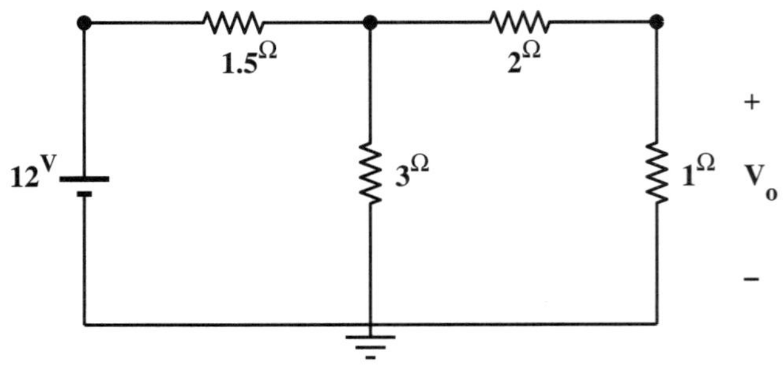

Fig. 2.6-8.

Fig. 2.6-9.

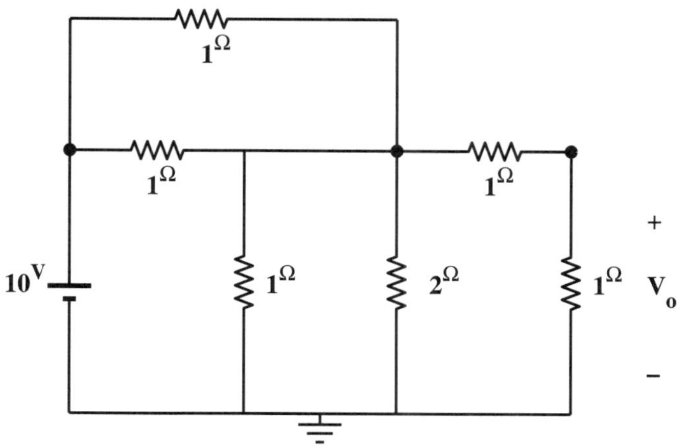

Remark 2.6-1: The circuit in Fig. 2.6-9 may seem complicated, but this circuit is not any different from the one in Fig. 2.6-8. Can you tell us what tool you need to make this problem as simple as that in Fig. 2.6-8.

Mesh Equations

2.6.6. For the circuits in Figs. 2.6-10 to 13 write the corresponding set of mesh equations following similar steps used in this chapter.

Fig. 2.6-10.

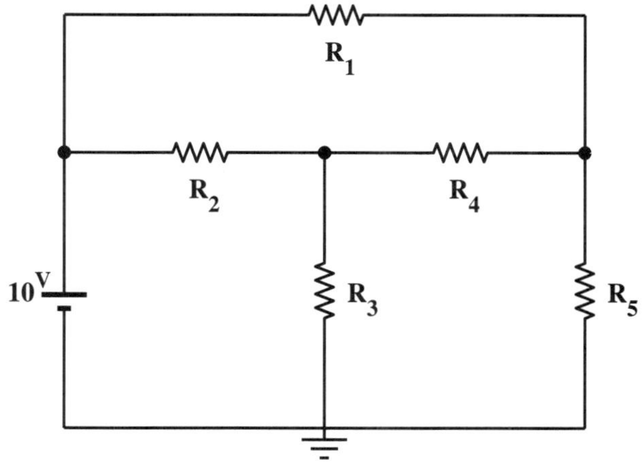

Fig. 2.6-11.

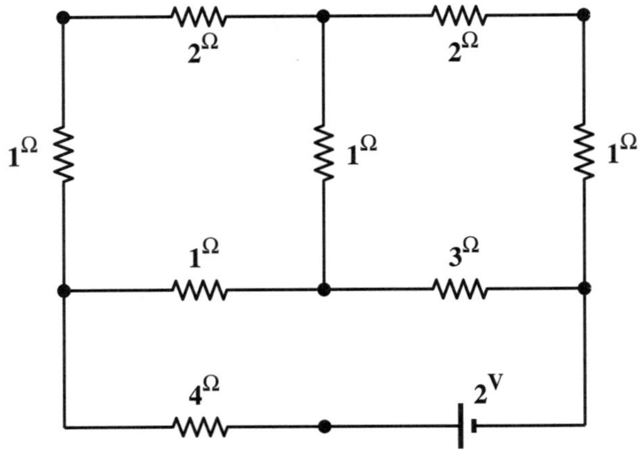

Fig. 2.6-12.

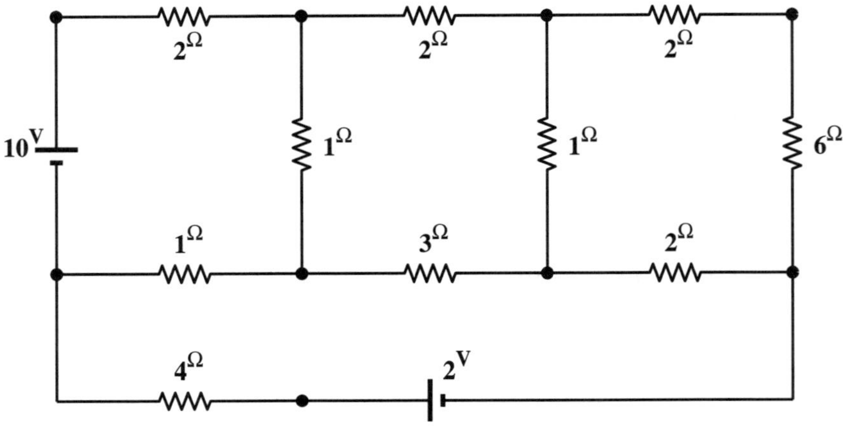

Fig. 2.6-13.

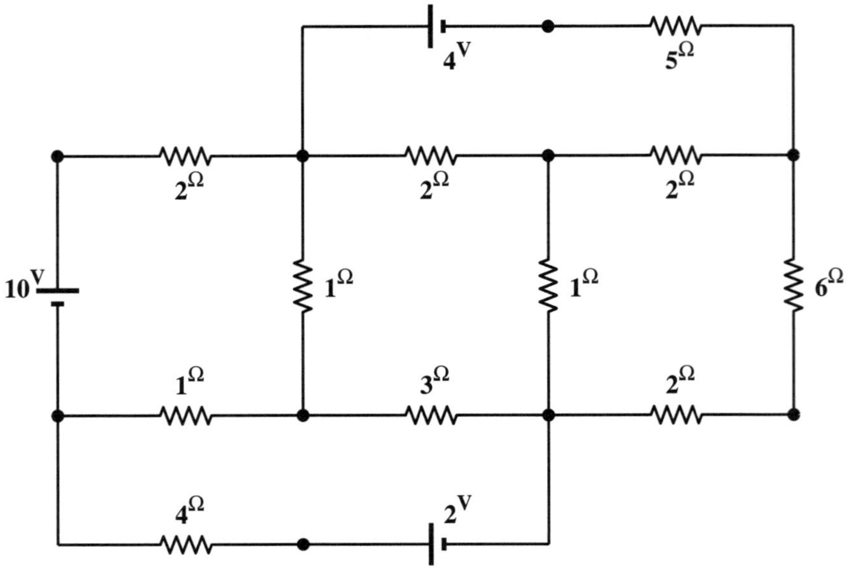

Source Transformation

2.6.7. For the circuits in Figs. 2.6-14 to 16 write the corresponding set of mesh equations in steps by using the source transformation method first.

Fig. 2.6-14.

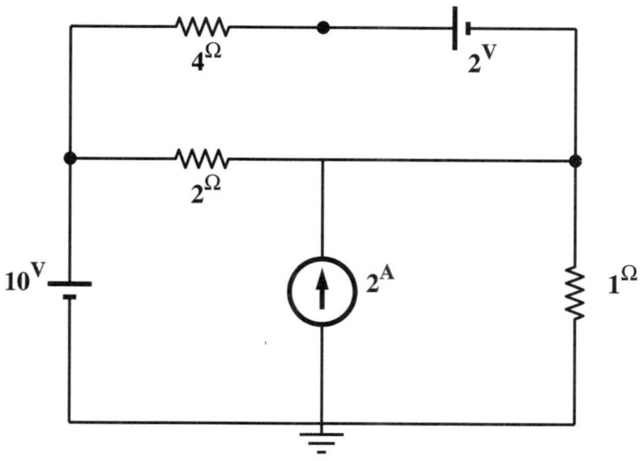

Fig. 2.6-15.

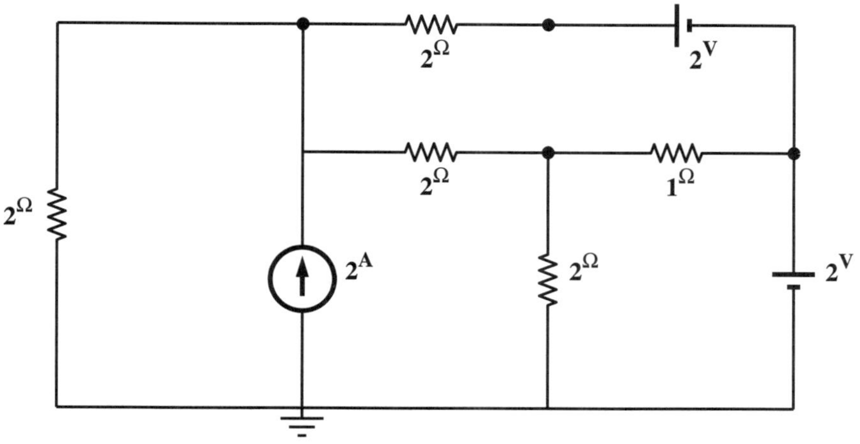

Fig. 2.6-16.

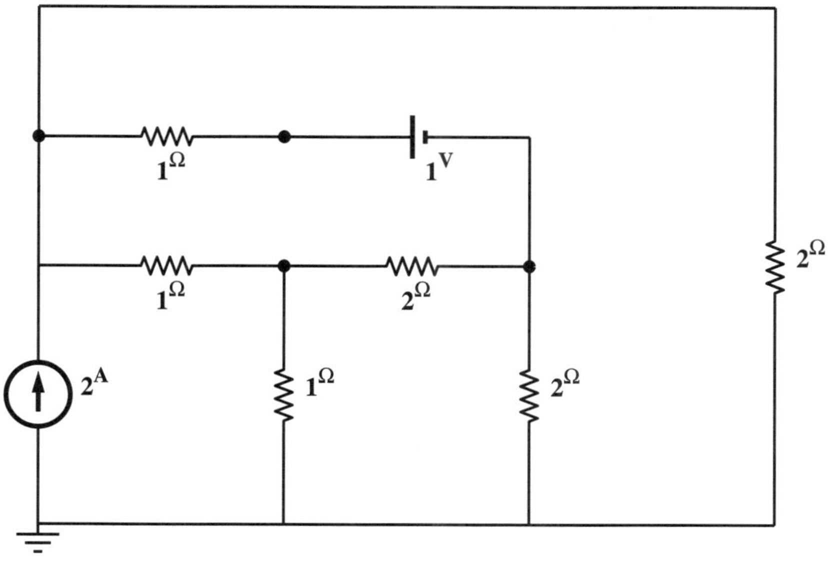

2.6.8. Repeat Problem 2.6.7 for the circuit in Fig. 2.6-17. If you cannot proceed as before, then explain why this circuit is different from those in Problem 2.6.7. Can you come up with a suggestion to tackle the difficulty?

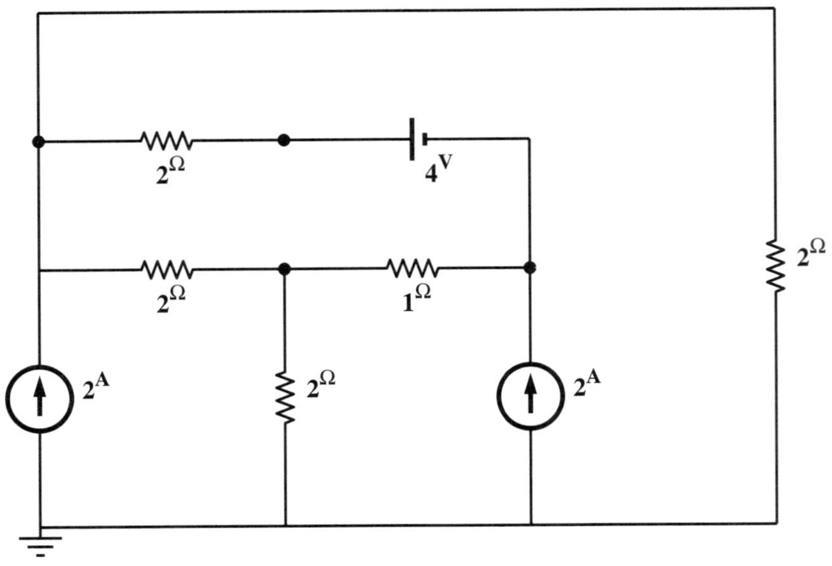

Fig. 2.6-17.

Comment 2.6-1: Regarding this and future problem sets we need to say a few words to clarify our position. Throughout this textbook we have emphasized that our approach has been to build knowledge based on knowledge gained and proceed forward accordingly. However, in Problem 2.6.3 (Fig. 2.6-5) we have used elements such as inductor and capacitor which we have not fully described, nor have we given any explanation for $j = \sqrt{-1}$ used in that problem. Adding to the difficulty we are giving this comment further away from that problem. Similarly in Problem 2.6.8, we have a situation that has not been explicitly covered in the textbook thus far. In all these cases the message to students is that they should pay attention to the covered materials. If so, they can easily see that they can answer Problem 2.6.3 (Fig. 2.6-5), even though we have not described complex elements, in other words, the answer to this problem is independent from these elements. Regarding Problem 2.6.8, some students may solve the problem, but we are more interested to know their responses as why they cannot solve in that configuration. By realizing what is the source of difficulty, or which topic has not been covered the students participate in the learning process. Otherwise, the process becomes only unidirectional, coming from instructor to the students and without any contribution from them.

3 NETWORK ANALYSIS CONTINUED

3.1. INTRODUCTION

In the preceding chapter we concentrate on a method that is primarily based on the KVL. Now, we look at a class of problems for which the KCL approach is the natural way of completing circuit analysis.

Example 3.1-1: Analyze the following circuit.

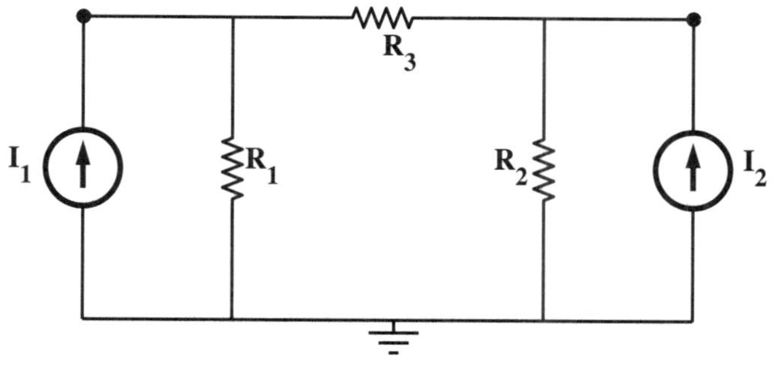

Fig. 3.1-1.

Solution: Here, we can solve this particular circuit using the source transformation technique, because each current source is in parallel with a given resistor. This approach is possible as we have learned in Section 2.5, followed by a mesh equations method. However, any source transformation with all its subsequent extra computations is additional work and must be avoided if possible. Because, any time that we use a source transformation, we must then go back and find the actual current through and voltage across each element **before** the transformation according to Definition 2.1-1. So we need to do something else here to reduce our computation.

STEP Zero: Clearly, we must invoke this step to label various variables and/or unknowns before proceeding further. But we have not yet defined the proper set of variables for circuits of this type, i.e., when we have only current sources. We may stipulate that if we somehow find all node voltages (this circuit has three nodes as depicted in Fig. 3.1-2), then we have enough information to finish the analysis. We also recall that it is perfectly acceptable to choose one node voltage as the reference node, which typically is the node which has most connected elements. For instance, in this circuit four elements are connected to node #3 that is shown by the ground symbol in Fig. 3.1-2. We select the voltages of nodes #1 & 2 with respect to ground, i.e., V_1, V_2, as our set of node-voltage variables. If we can determine these unknowns, then we can complete the network analysis, since knowing these values (with respect to the ground or other reference) enables us to compute the currents through R_1, R_2 and R_3. Furthermore, V_1 and V_2 are the voltages across the two current sources, and voltage across R_3 is $(V_1 - V_2)$. Thus, the analysis can be completed by this choice for a set of variables.

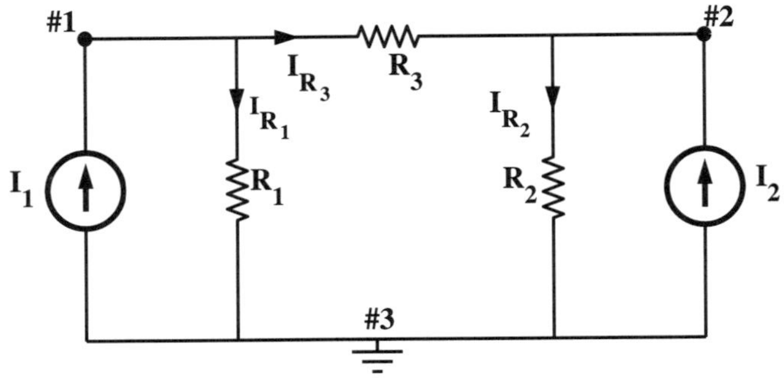

Fig. 3.1-2.

STEP One: Now that we have selected V_1 and V_2 as our unknowns or **essential variables**, we need to write KCL to generate independent equations in terms of these variables. Here, applying KCL at nodes #1 & 2 yields

$$I_1(\text{incoming current into node \#1}) = I_{R_1} + I_{R_3}, \tag{3.1-1a}$$

$$I_2(\text{incoming current into node \#2}) = I_{R_2} - I_{R_3}. \tag{3.1-1b}$$

There is no more node and we cannot write any additional independent equation, thus we have exhausted this step.

STEP Two: Let us write the above current values in terms of V_1 and V_2 our essential variables, i.e., using KVL, we have

$$I_{R_1} = V_1/R_1,\ I_{R_2} = V_2/R_2,\ I_{R_3} = (V_1 - V_2)/R_3. \tag{3.1-2}$$

Substituting these values in (3.1-1) yields the following in terms of V_1 and V_2.

$$I_1 = \frac{V_1}{R_1} + \frac{V_1 - V_2}{R_3} = (\frac{1}{R_1} + \frac{1}{R_3})V_1 - (\frac{1}{R_3})V_2, \quad (3.1\text{-}3a)$$

$$I_2 = \frac{V_2}{R_2} - \frac{V_1 - V_2}{R_3} = -(\frac{1}{R_3})V_1 + (\frac{1}{R_2} + \frac{1}{R_3})V_2. \quad (3.1\text{-}3b)$$

Looking at the above two equations with two unknowns (V_1, V_2), we realize that a *pattern* is evolving in generating these equations. We note that these equations can be written by inspection. Namely,

$I_1 \triangleq$ (the total currents coming into node #1) =

+ (the total conductances connected at node #1) $\times V_1$

− (the total conductances between node #1 and node #2) $\times V_2$. (3.1-4a)

Similarly,

$I_2 \triangleq$ (the total currents coming into node #2) =

− (the total conductances between node #2 and node #1) $\times V_1$

+ (the total conductances connected at node #2) $\times V_2$. (3.1-4b)

In passing, recall that conductance is shown by G and is measured in mhos.

STEP Final: We have sufficient number of independent equations whose solutions yield (V_1, V_2), from which the analysis of the network can be completed.

STOP

Putting all these thoughts together, we have our next methodology.

3.2. METHOD #TWO – SOLUTION by the NODE EQUATIONS

To analyze a network which consists of only current sources, we invoke the following steps.

STEP Zero: To label and/or identify all pertinent variables particularly the essential node voltages.

Remark 3.2-1: Normally, the node that has the most elements connected to is called the reference node and its voltage is chosen at zero volt. This node is also called the ground and can be any node in the network. In other words, if someone chooses a different reference point all node voltages will be shifted upward or downward depending on the actual voltage value of that reference node with respect to the global voltage reference.

STEP One: Using essential node voltages and KCL, set up all independent equations needed to complete the analysis in the underlying problem. *Comment:* When all independent variables which are the node voltages are specified, we can write the corresponding equations by inspection as follows.

$I_1 \triangleq$ (the total currents coming into node #1) =

$$+\,(\text{the total conductances connected at node \#1}) \times V_1$$

$$-\,(\text{the total conductances between node \#1 and node \#2}) \times V_2 - (\cdots) \times V_{(\ldots)}$$

$$-\,(\text{the total conductances between node \#1 and node \#last}) \times V_{last}. \qquad (3.2\text{-}1)$$

Similarly, this pattern is repeated for the other nodes accordingly, until we have all equations generated in this manner.

STEP Final: Upon solving the above set of independent equations and generating $\{V_i, i=1, \ldots, \text{last}\}$, then we complete the analysis by finding numerically the current through and the voltage across each remaining element.

STOP

Example 3.2-1: Analyze the circuit in Fig. 3.2-1, using an appropriate approach.

Solution: The circuit in Fig. 3.2-1 contains only current sources, therefore the approach stemming from the node equations must be used herein.

STEP Zero: To label and/or identify all pertinent variables (here, V_1, V_2, V_3) for a node equation approach as depicted in Fig. 3.2-2.

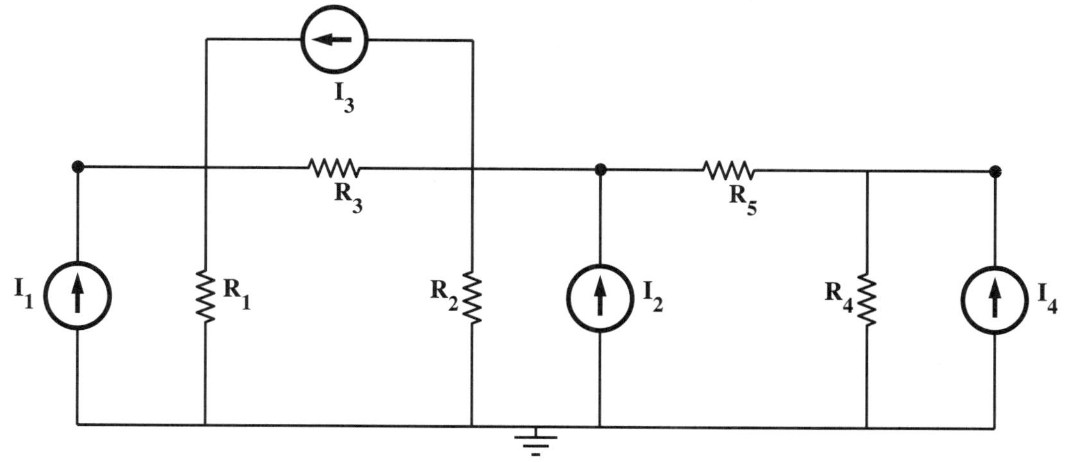

Fig. 3.2-1.

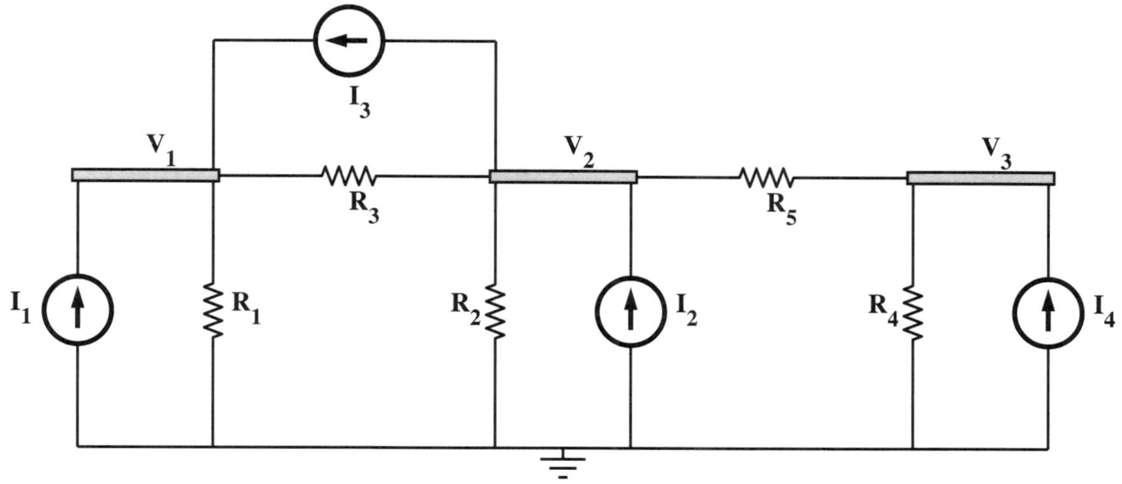

Fig. 3.2-2.

STEP One: Based on the set of pertinent variables as selected in Step Zero, we write three node equations in terms of V_1, V_2 and V_3 by inspection.

$$I_1 + I_3 = (\frac{1}{R_1} + \frac{1}{R_3})V_1 - (\frac{1}{R_3})V_2, \tag{3.2-2}$$

$$I_2 - I_3 = -(\frac{1}{R_3})V_1 + (\frac{1}{R_2} + \frac{1}{R_3} + \frac{1}{R_5})V_2 - (\frac{1}{R_5})V_3, \tag{3.2-3}$$

$$I_4 = -(\frac{1}{R_5})V_2 + (\frac{1}{R_4} + \frac{1}{R_5})V_3. \tag{3.2-4}$$

STEP Final: We have three equations with three unknowns and setting up these equations completes this phase of our analysis.

STOP

Comment 3.2-1: In the above analysis we use the conductance of various elements including current sources. Perhaps we should remind ourselves that the conductance of an ideal current source is zero. Because, by definition an ideal current source has an infinite internal resistance, thus its conductance is zero.

Example 3.2-2: Analyze the circuit in Fig. 3.2-3, using an appropriate approach.

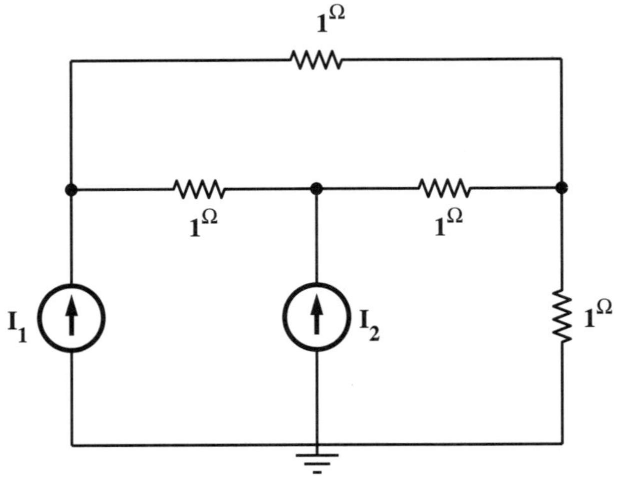

Fig. 3.2-3.

Solution: Here, we have all current sources and the Method #Two is the right candidate to tackle this problem.

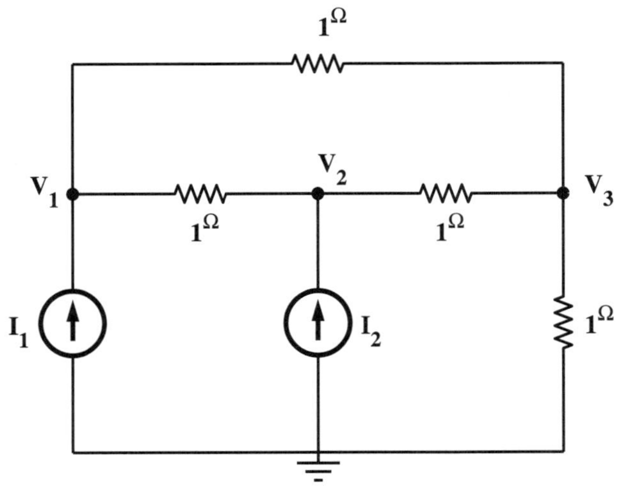

Fig. 3.2-4.

STEP Zero: To label and/or identify all pertinent node voltages as shown in Fig. 3.2-4.

STEP One: By inspection, we have the following set of three equations with three unknowns whose solution yields (V_1, V_2, V_3).

$$I_1 = 2V_1 - V_2 - V_3, \tag{3.2-5}$$

$$I_2 = -V_1 + 2V_2 - V_3, \tag{3.2-6}$$

$$0 = -V_1 - V_2 + 3V_3. \tag{3.2-7}$$

STOP

3.3. SOURCE TRANSFORMATION CONTINUED

Regarding the voltage sources: What if we had more than one voltage source, which are connected in the manners of Figs. 3.3-1 or 2. The series connection in Fig. 3.3-1 is straightforward, and although we have used two sources, the concept can be easily generalized to many such sources connected in that manner. The net voltage source at the port ab can be easily computed, which is one equivalent voltage source, whose net value is the net algebraic summation of V_1 and V_2. For instance, if $|V_1| > |V_2|$, then $V_{ab} = V_1 - V_2$ with the positive polarity at node a. Similarly, if $|V_1| < |V_2|$, then $V_{ba} = V_2 - V_1$, with the positive polarity at node b.

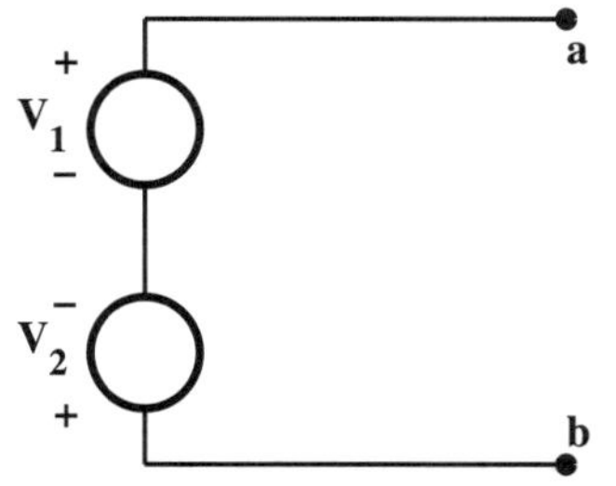

Fig. 3.3-1. Voltage sources: Series connection.

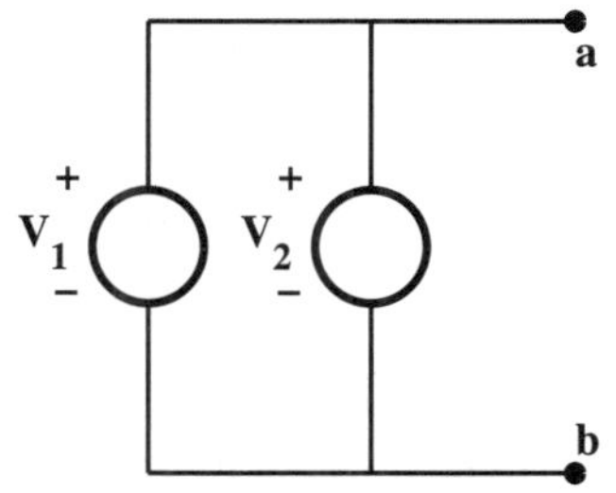

Fig. 3.3-2. Voltage sources: Parallel connection.

The parallel connection in Fig. 3.3-2 is, however, problematic, because we cannot have two different voltage values, for instance V_1 and V_2, for the same port ab. Therefore, if we put two non-ideal voltage sources in parallel, they will charge and discharge each other (i.e., one becomes like a load for the other) until they reach one practical value. Thus, the configuration depicted in Fig. 3.3-2 is not meaningful and is not practiced as far as ideal voltage sources are concerned..

What about the current sources? Here, the series connection of current sources is not practiced, but the parallel connection of two or more current sources between two nodes yields a net equivalent current source, whose value is the algebraic combinations of all these current sources between the two nodes.

Example 3.3-1: Analyze the circuit in Fig. 3.3-3 by mesh equations.

Fig. 3.3-3.

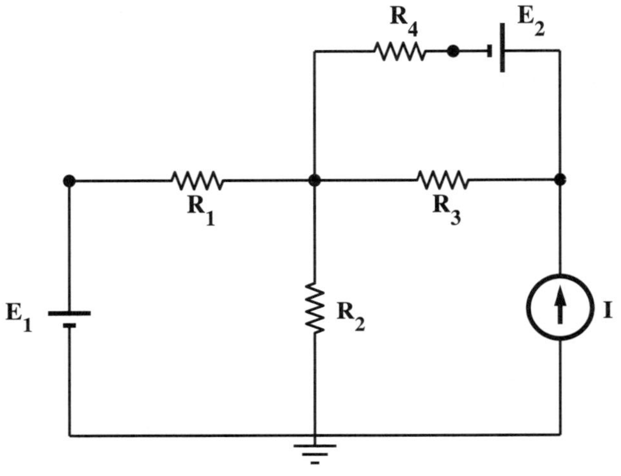

Solution: Here we have a circuit with mixed (both voltage and current) sources. To invoke steps of Method #One to this problem directly, we note that because we have a current source I in the circuit whose voltage is unknown, some of the steps in that approach must be modified. We also note that the current source is not directly connected to a resistor, i.e., the conventional source transformation is not possible herein. However, we proceed accordingly though cautiously expecting difficulty.

STEP Zero: To label or identify all variables for a mesh analysis as depicted in Fig. 3.3-4. Here, we push ourselves into a box. First of all, I_2 must be identically equal to I, in other words, the current in mesh #2 must be equal to the current source in that mesh. There is no way out of this **constraint**. Second, the voltage across this current source, shown by V_x is unknown and its value must be computed as a part of the analysis. We still can continue this approach, but we must be careful to interpret the next steps correctly.

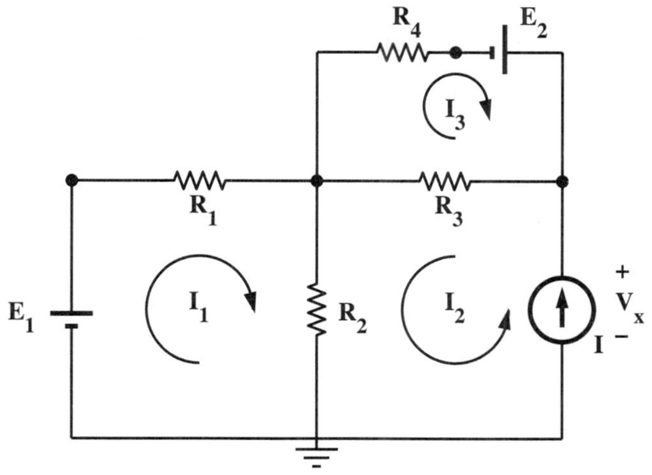

Fig. 3.3-4.

STEP One: We can write a few equations by inspection. For instance,

$$E_1 = (R_1 + R_2)I_1 + R_2I_2, \tag{3.3-1}$$

$$E_2 = R_3I_2 + (R_3 + R_4)I_3. \tag{3.3-2}$$

Now, we need one more independent mesh equation, which is for mesh #2 containing the independent current source. How do we handle this situation? For the time being we write

$$V_x = R_3(I_2 + I_3) + R_2(I_1 + I_2), \tag{3.3-3}$$

as though V_x, an unknown voltage quantity, is the voltage source in mesh #2. However, we also know that

$$I_2 \equiv I, \tag{3.3-4}$$

because the value of the current in mesh #2 is constrained by the value of the current source in that mesh. The physics of the system does not allow any other value for this current. This type equation is called the **constraint equation.** With the straw of good luck and by letting $I_2 \equiv I$, we can simplify (3.3-1) and (3.3-2) for I_1 and I_3 in terms of known quantities as follows.

$$I_1 = \frac{E_1 - R_2I}{R_1 + R_2}, \tag{3.3-5}$$

$$I_3 = \frac{E_2 - R_3I}{R_3 + R_4}. \tag{3.3-6}$$

Substituting these values in (3.3-3) yields V_x. But sooner or later we run into a situation which is not so straightforward as the above case and we must find a more systematic way of handling the situation. For instance, the constraint equations may not be so obvious. The remedy is, when using the mesh equations approach, we must transform all current sources into voltage sources before proceeding further. But can we transform the current source in circuit of Fig. 3.3-3, which is not in a conventional form, into a voltage source with reasonable effort? We address that question next.

STOP

3.4. SOURCE TRANSFORMATION in NON-CONVENTIONAL SETTINGS

What we have studied so far is the following equivalency that shows a voltage source with a series resistance R_s is equivalent to a current source with a parallel resistor R_s as depicted in Fig. 3.4-1.

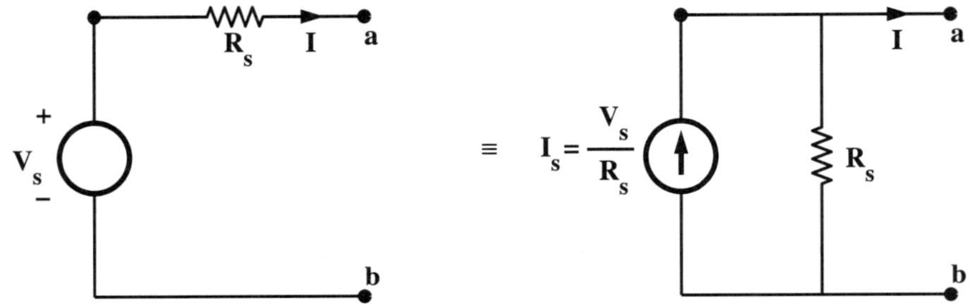

Fig. 3.4-1. Conventional equivalent circuits.

However, in many instances, we may not have any resistance (load) connected directly in series with a voltage source, nor in parallel with a current source as shown in Fig. 3.4-1. For instance, a voltage source may have two or more (instead of one) resistors (loads) connected to it, as shown in Fig. 3.4-2. Similarly a current source may not be directly connected to a parallel resistor as shown in Fig. 3.4-3.

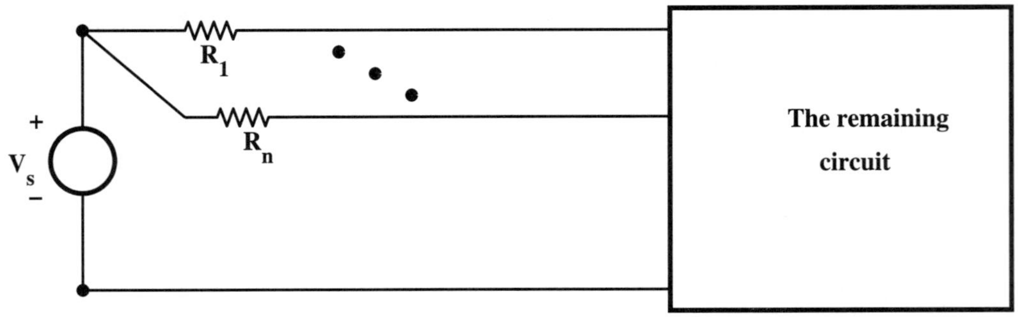

Fig. 3.4-2. One voltage source connected to several resistors.

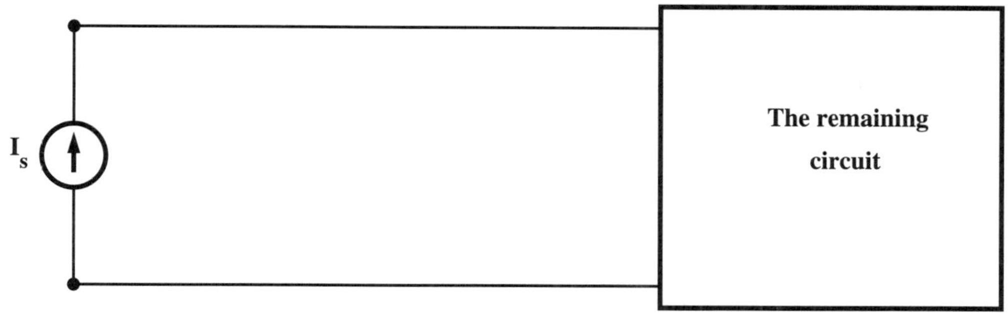

Fig. 3.4-3. One current source not connected to a parallel resistor.

How do we handle these non-conventional settings?

Here, we must **parameterize** the problem by adding appropriately a resistor, say R_s, to the circuit as follows. In the case of the voltage source we add this resistor in series with V_s and complete the analysis. Then the results of our circuit analysis contains R_s as a parameter. Letting $R_s \to 0$ in the corresponding analytical results of the circuit yields the actual final results for the original circuit. This method always works, but it may become cumbersome in certain cases as we show in Sections 3.8 to 3.11, subsequently. Similarly, for a current source without a parallel resistor, as is required in the conventional sense, we can put an R_s in parallel with the source I_s, then let $R_s \to \infty$ in the final parameterized results. Although this method always works, it may become cumbersome in some instances.

To summarize the above, suppose in the circuit of Example 3.3-1, we put R_s in parallel with the current source as shown in Fig. 3.4-4.

Fig. 3.4-4.

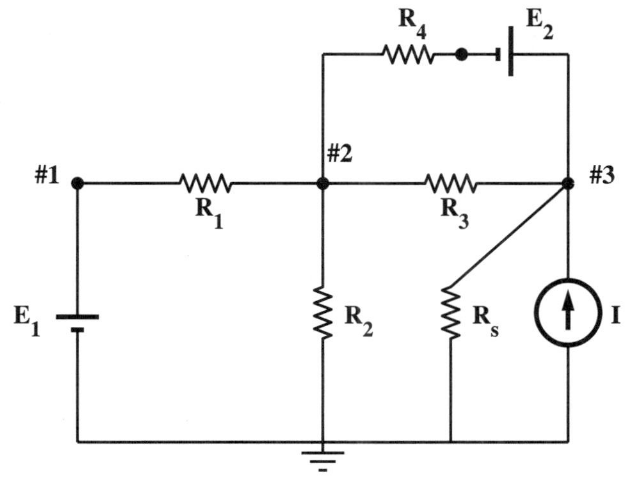

When $R_s \to \infty$, i.e., we open or remove this element, the circuit in Fig. 3.4-4

becomes exactly the same as our original circuit in Fig. 3.3-3. But the circuit in Fig. 3.4-4 can be transformed into a circuit with only voltage sources, Fig. 3.4-5, and thus easily amenable with mesh equations approach, with one parameter R_S.

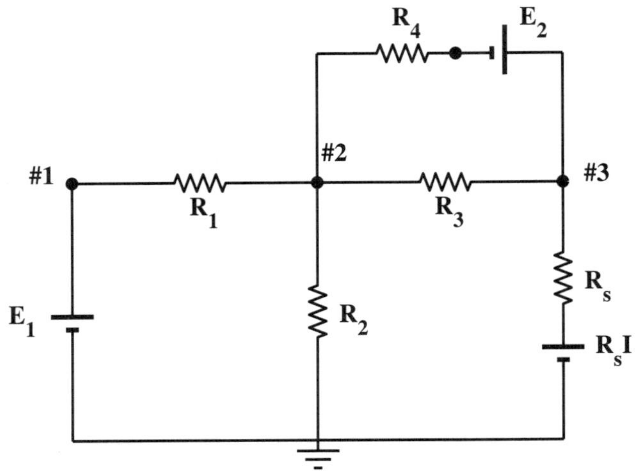

Fig. 3.4-5.

An alternative to the above parameterization approach is called **reconfiguration method** herein.

For instance, for the voltage source connected to many resistors as shown in Fig. 3.4-2, an alternative approach is to increase the number of loops as shown in Fig. 3.4-6. Here, the voltage source is connected to **"n"** resistors as shown, and each pair can be transformed into one current source in the conventional sense, in order to proceed with the analysis.

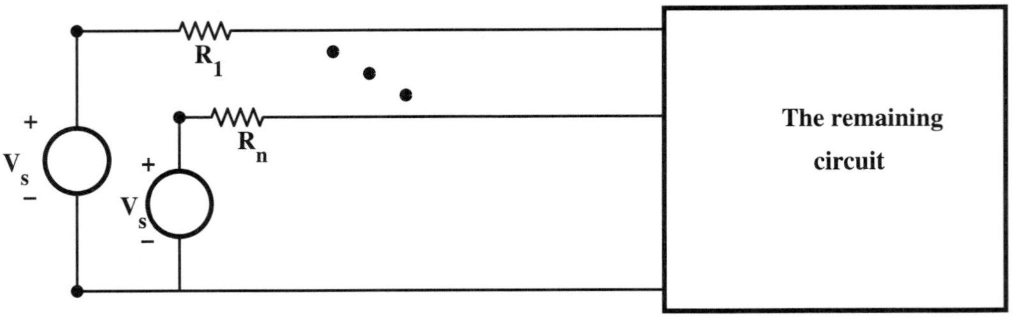

Fig. 3.4-6. Reconfigured circuit of Fig. 3.4-2.

Just as the above case for voltage sources, we may reconfigure the circuit by switching around the current source that is not in parallel with a load such that it becomes parallel with one. For instance, in the case of the circuit in Example 3.3-1, we can redraw that circuit by switching around its current source as shown in Fig. 3.4-7.

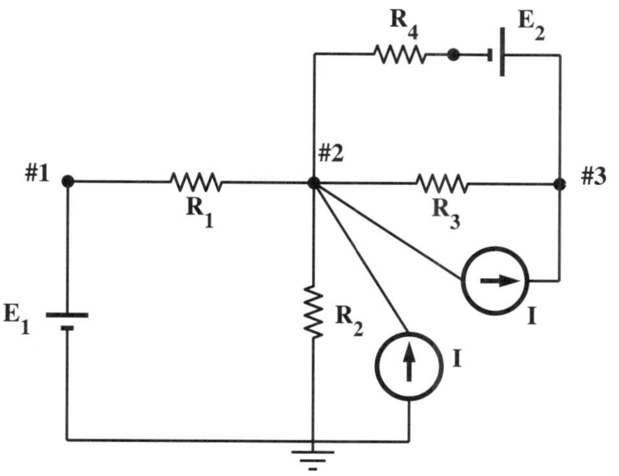

Fig. 3.4-7. Reconfigured circuit of Fig. 3.3-3.

Note that the current source which was pumping into node #3 in Fig. 3.3-3 or Fig. 3.4-4, is now pumping into node #2 first, then leaves that node to pump into node #3 second. The net effect for node #2 is zero, i.e., no changes are made in the circuit as far as the currents coming into all nodes are concerned. However, this reconfiguration puts both sources in parallel with some resistors. As the result of this reconfiguration the circuit in Example 3.3-1 is now amenable with the conventional source transformation. We elaborate further on these reconfiguration schemes in our subsequent numerical examples in Chapter Four.

3.5. METHOD #THREE – SOLUTION by the CONVENTIONAL SOURCE TRANSFORMATION

In the preceding two methods, we have either all voltage sources or all current sources. As Example 3.3-1 reveals, a circuit with both sources, brings us to the present approach.

To analyze a network with mixed sources, we use source transformation to bring the problem into one-type source problem amenable with either Method #One or Method #Two. Here, we have to make a sound judgment that the selected transformation makes the problem easier than before. Otherwise, we must seek alternative approaches. In certain instances, we may solve the problem by some *ad hoc* methods, but the conventional source transformations, if possible, always work and must be explored.

3.6. CIRCUITS with MIXED SOURCES

What we have discovered so far is that when we face a circuit involving both current and voltage sources, i.e., circuit with mixed sources, we need to reexamine the previous two powerful methods of mesh and node equations before applying them directly to the underlying problem. In other words, these two methods work perfectly when the circuits have only one type of sources which is often a very special case. So the challenge is to come up with procedures which can be easily extended to problems involving mixed sources. To that end, we introduce the next example which is used extensively to show various techniques and compare those techniques when applied to the same problem.

Example 3.6-1: Analyze the circuit in Fig. 3.6-1.

Solution: Here, we have both current and voltage sources which are not in conventional settings, i.e., we cannot directly use either Method #One, or Method #Two, or for that matter Method #Three. What should we do here? We start with Method #One, then proceed to see where do we face difficulty and how can we address that situation.

STEP Zero: To label and/or identify all *pertinent variables* for this method as depicted in Fig. 3.6-2.

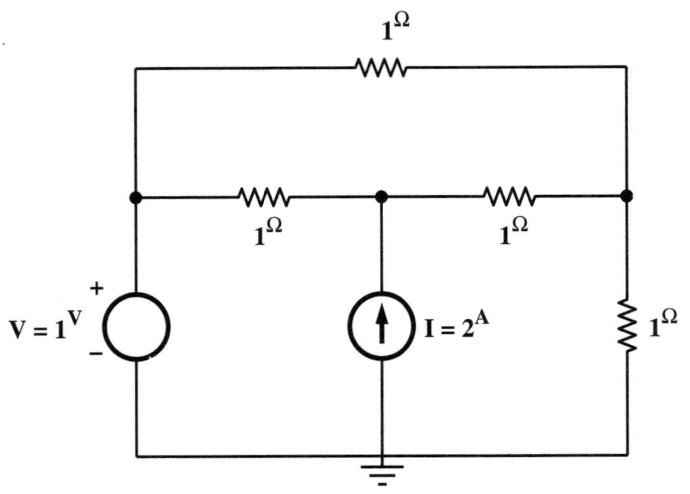

Fig. 3.6-1.

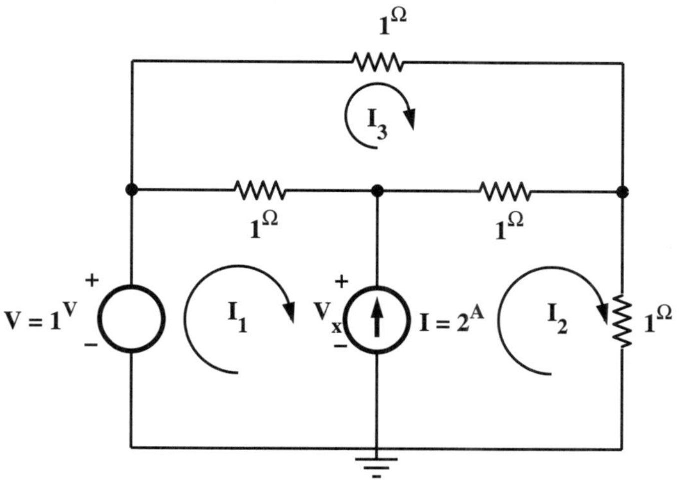

Fig. 3.6-2.

Here,

$$I_2 - I_1 = 2^A, \tag{3.6-1}$$

is the so-called **constraint equation,** which is a relationship that must be incorporated in the overall analysis. Also V_x in Fig. 3.6-2 refers to the voltage across the current source, which for now is an unknown quantity.

STEP One: Based on the above configuration, we can write

$$V - V_x = (1^{\Omega})I_1 - (1^{\Omega})I_3, \tag{3.6-2}$$

$$V_x = (1^{\Omega} + 1^{\Omega})I_2 - (1^{\Omega})I_3, \tag{3.6-3}$$

$$0 = -(1^{\Omega})I_1 - (1^{\Omega})I_2 + (1^{\Omega} + 1^{\Omega} + 1^{\Omega})I_3. \tag{3.6-4}$$

Clearly, the above equations with the constraint equation (3.6-1) form a basis to complete the analysis for the above circuit. However, we note that if we add (3.6-2) and (3.6-3) we get i.sp

$$V = 1^V = I_1 + 2I_2 - 2I_3. \tag{3.6-5}$$

Looking at (3.6-5) we realize that this equation is independent from V_x and is the loop equation for what is known as a **super loop** and is shown in Fig. 3.6-3.

It seems that had we chosen this super loop to begin with, we could have bypassed completely the current source, which is a troublesome element when using a mesh equations approach. If we look at the above super loop, and trace the voltage across each element around the direction in that loop, then we see that indeed (3.6-5) can be constructed directly by this approach. This important observation leads us to the following methodology for a circuit that has mixed sources and is being analyzed by applying the mesh equations technique.

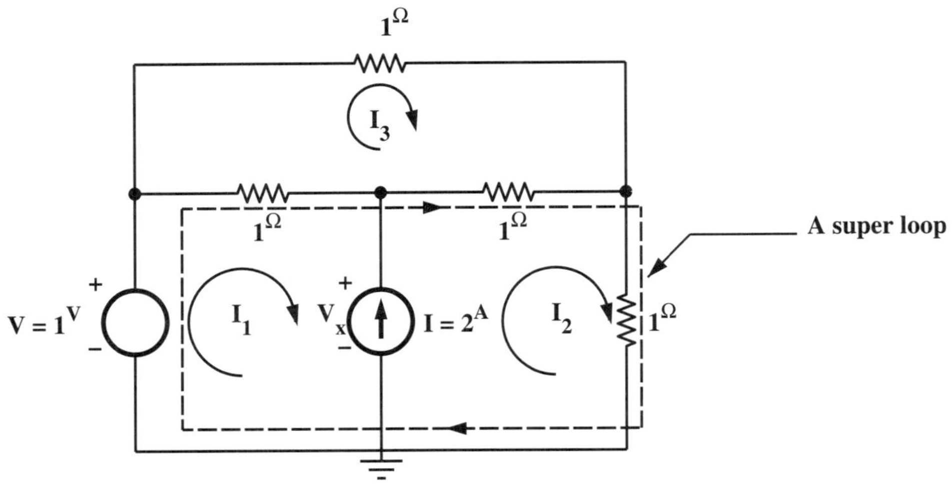

Fig. 3.6-3. A super loop.

3.7. METHOD #FOUR – SOLUTION by the UTILIZATION of SUPER LOOP

When circuits contain mixed sources, we may apply Method #One together with one proper or several proper super loops in order to bypass every current source. The loop equations generated from these super loops with mesh equations involving only meshes with voltage sources together with constraint equations will form the needed set of independent equations to analyze the network by a generalized version of Method #One, which is called Method #Four. Here, (3.6-2) and (3.6-3) yields (3.6-5), which together with (3.6-1) and (3.6-4) are three independent equations for the three unknown mesh variables. It is to be noted that often super loop is not unique, in fact in the circuit of Fig. 3.6-3, we could have chosen the circumference as a super loop.

Before closing this section we also note that we may call a similar approach using what is known as a **super node** approach, and is described in the next example.

Example 3.7-1: Analyze the circuit shown in Fig. 3.7-1.

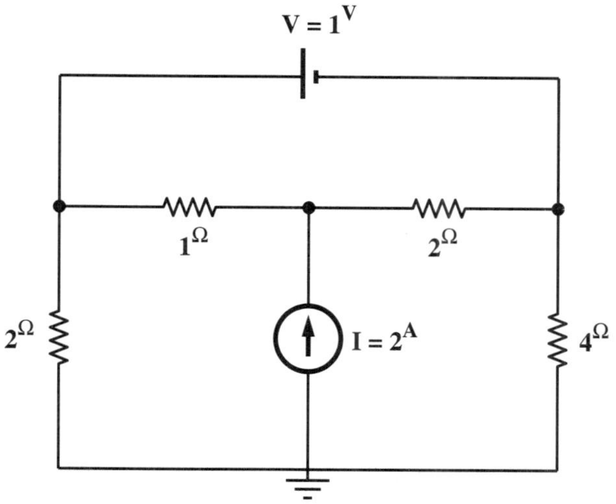

Fig. 3.7-1.

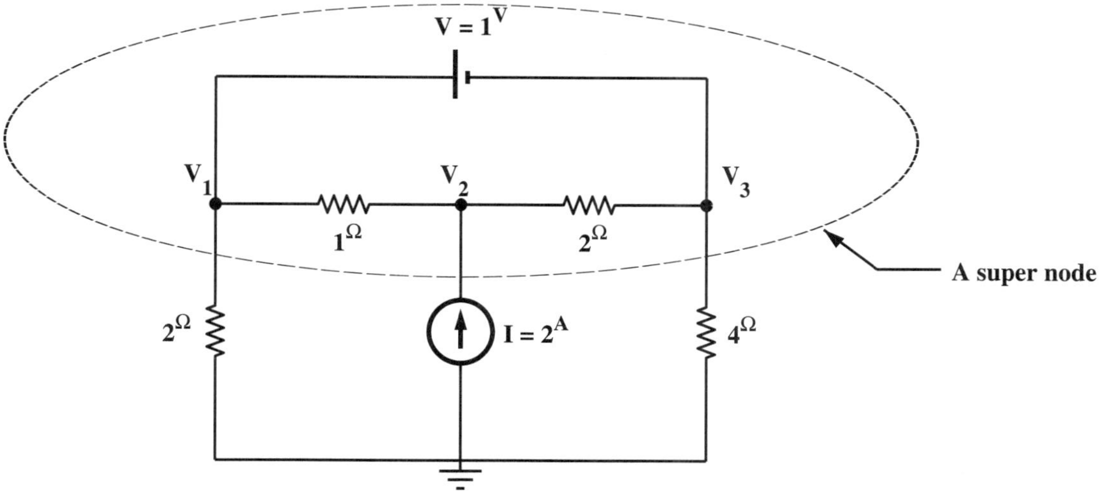

Fig. 3.7-2. A super node.

Solution: Here, we study this problem using the concept of super node as shown in Fig. 3.7-2. Normally, a super node is chosen such that the voltage source that is now a troublesome element is contained. Clearly, for the circuit in Fig. 3.7-1, where we have only one voltage source, the choice of super node is straightforward, though this node is not unique either.

STEP Zero: To label and/or identify all pertinent variables as depicted in Fig. 3.7-2.

STEP One: Here, the node equations are as follows.

$$2^A = \frac{V_1}{2} + \frac{V_3}{4} \quad \text{at the chosen super node,} \tag{3.7-1}$$

$$1^V = V_1 - V_3 \quad \text{constraint equation,} \tag{3.7-2}$$

$$2^A = -V_1 + (1 + \frac{1}{2})V_2 - \frac{1}{2}V_3 \quad \text{at node \#2.} \tag{3.7-3}$$

Resulting in

$$V_1 = 3^V, \quad V_2 = 4^V, \quad V_3 = 2^V. \tag{3.7-4}$$

STOP

We call the above approach pertaining to the super node as **Method #Four Prime.**

3.8. EXAMPLE 3.6-1 CONTINUED

Going back to Example 3.6-1, we may have started the analysis of that circuit using Method #Two, the node equations approach, as follows.

STEP Zero: To label and/or identify all pertinent variables, which in this approach are node voltages, as shown in the circuit of Fig. 3.8-1.

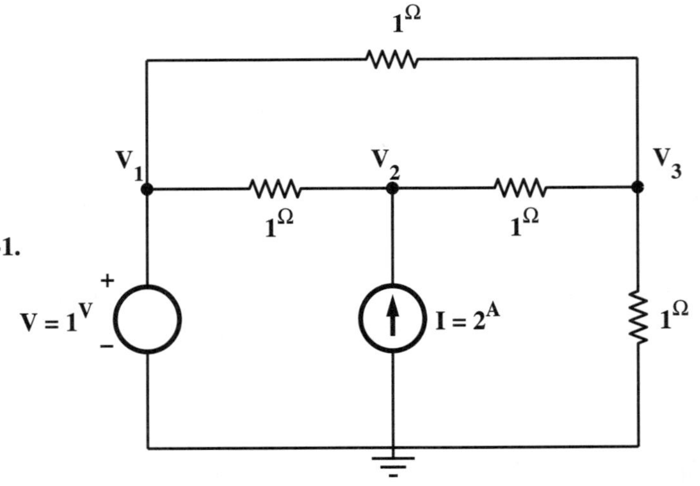

Fig. 3.8-1.

Here,

$$V_1 \equiv V = 1^V, \tag{3.8-1}$$

is the **constraint equation.** Proceeding accordingly yields the necessary equations to analyze the circuit as follows.

STEP One: The two independent node equations are as follows.

$$2^{A} = -(\frac{1}{1})V_1 + (\frac{1}{1} + \frac{1}{1})V_2 - (\frac{1}{1})V_3, \qquad \text{at node \#2,} \tag{3.8-2}$$

$$0 = -(\frac{1}{1})V_1 - (\frac{1}{1})V_2 + (\frac{1}{1} + \frac{1}{1} + \frac{1}{1})V_3, \quad \text{at node \#3.} \tag{3.8-3}$$

Using the constraint equation $V_1 \equiv 1$, we have

$$3 = 2V_2 - V_3, \tag{3.8-4}$$

$$1 = -V_2 + 3V_3. \tag{3.8-5}$$

Resulting in $V_2 = 2^V$ and $V_3 = 1^V$ as expected.

STOP

Comment 3.8-1: Let us go back and use source transformation approach in Example 3.6-1. Starting with the 2^A current source, let us assume that this source is in parallel with an R_s , Fig. 3.8-2. In other words, we parameterize the problem in terms of R_s. When $R_s \to \infty$, the circuit in Fig. 3.8-2 becomes the same as the initial circuit in Example 3.6-1.

Fig. 3.8-2.

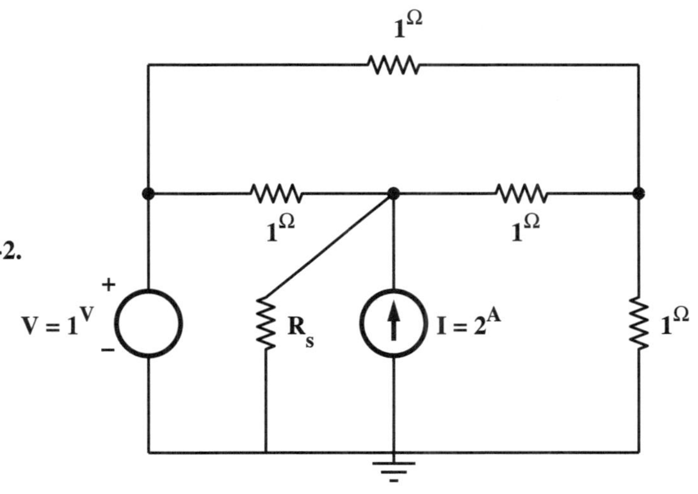

STEP Zero: To transform the current source into a voltage source and label all pertinent parameters for a mesh equations approach, Fig. 3.8-3.

STEP One: Now, using mesh equations yields

$$V - 2R_s = (1^{\Omega} + R_s)I_1 - R_sI_2 - (1^{\Omega})I_3, \tag{3.8-6}$$

$$2R_s = -R_sI_1 + (2^\Omega + R_s)I_2 - (1^\Omega)I_3, \tag{3.8-7}$$

$$0 = -I_1 - I_2 + 3I_3. \tag{3.8-8}$$

Fig. 3.8-3.

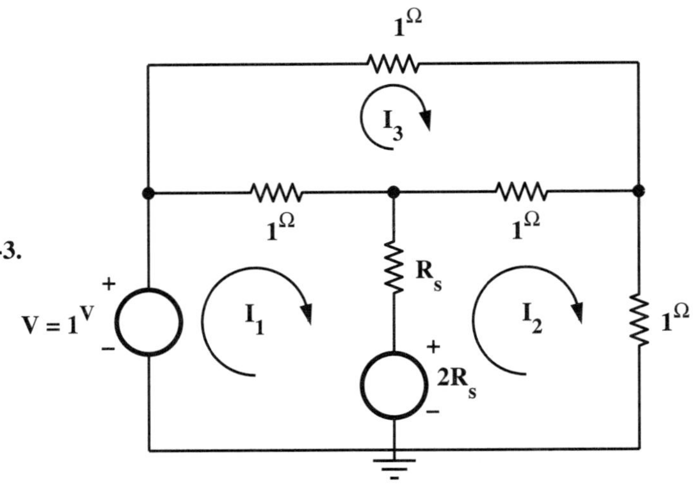

From (3.8-8), we have

$$I_3 = \frac{I_1 + I_2}{3}. \tag{3.8-9}$$

Substituting this I_3 in (3.8-6) and (3.8-7) yields

$$1 - 2R_s = (\frac{2}{3} + R_s)I_1 - (\frac{1}{3} + R_s)I_2, \tag{3.8-10}$$

$$2R_s = -(\frac{1}{3} + R_s)I_1 + (\frac{5}{3} + R_s)I_2. \tag{3.8-11}$$

Adding the above two equations yields

$$I_1 + 4I_2 = 3. \tag{3.8-12}$$

Using (3.8-12) in (3.8-10) yields

$$I_2 = \frac{1 + 5R_s}{3 + 5R_s}. \tag{3.8-13}$$

As expected this current is parameterized in terms of R_s. Thus,

$$I_2 = \lim_{R_s \to \infty} \frac{1 + 5R_s}{3 + 5R_s} = 1^A. \tag{3.8-14}$$

Substituting this limiting value in (3.8-12) results in

$$I_1 = -1^A. \tag{3.8-15}$$

Using (3.8-9) and the above values for I_1 and I_2 yields $I_3 = 0$. Resulting in,

$$V_1 - V_2 = (1^{\Omega})(I_1 - I_3) = -1, \qquad (3.8\text{-}16)$$

where $V_1 = 1^V$ volt, and therefore $V_2 = 2^V$. Similarly,

$$V_2 - V_3 = (1^{\Omega})(I_2 - I_3) = 1^V, \qquad (3.8\text{-}17)$$

where $V_2 = 2^V$, and therefore $V_3 = 1^V$ as expected.

So, with two pages of algebra we get the same answer, of course, as before for the above node voltages. However, the real message is that the proper choice of methodology is very important. Parameterization always works, but it is involved.

3.9. METHOD #THREE CONTINUED

What about transforming the voltage source in Example 3.6-1 into a current source? Here, because we do not have a series resistor with that source, we need to parameterize the problem by using R_s in series with the voltage source and letting $R_s \to 0$ in the final result. On that note, we rearrange the initial circuit into the following form, Fig. 3.9-1.

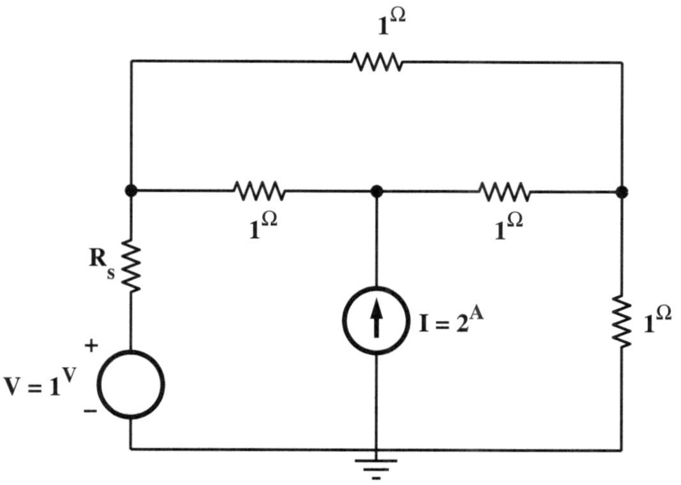

Fig. 3.9-1.

Step Zero: To transform the voltage source into a current source and label all pertinent variables for a node equation analysis as depicted in Fig. 3.9-2.

STEP One: Now, using node equations approach yields

$$\frac{1}{R_s} = \left(\frac{1}{R_s} + 2\right)V_1 - V_2 - V_3, \quad \text{at node \#1,} \tag{3.9-1}$$

$$2 = -V_1 + 2V_2 - V_3, \quad \text{at node \#2,} \tag{3.9-2}$$

$$0 = -V_1 - V_2 + 3V_3, \quad \text{at node \#3.} \tag{3.9-3}$$

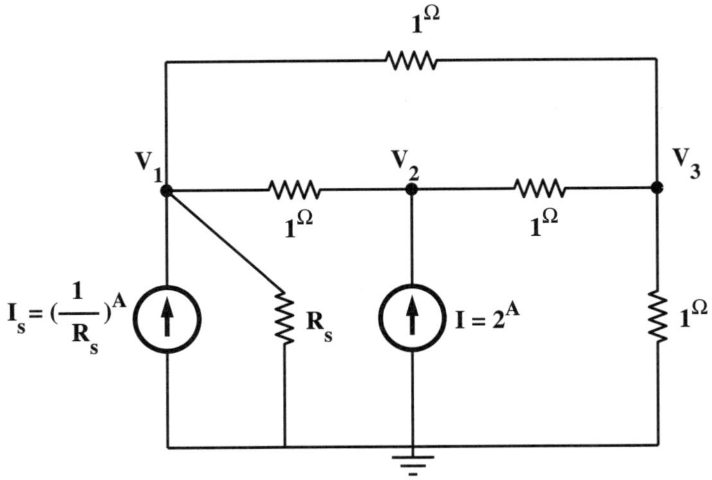

Fig. 3.9-2.

We have three equations and three unknowns with one parameter R_s. Always, we can find a way to sort these equations in a manner that makes the analysis manageable. For instance, we solve (3.9-2) and (3.9-3), in terms of V_1, the voltage of the node connected to R_s. This choice yields

$$2V_2 - V_3 = 2 + V_1, \tag{3.9-4}$$

$$-V_2 + 3V_3 = V_1. \tag{3.9-5}$$

From which we have

$$V_2 = \frac{6 + 4V_1}{5}, \tag{3.9-6}$$

$$V_3 = \frac{2 + 3V_1}{5}. \tag{3.9-7}$$

Substituting these values in (3.9-1) and solving for V_1 yields

$$V_1 = \frac{\frac{1}{R_s} + \frac{8}{5}}{\frac{1}{R_s} + \frac{2}{5}}, \tag{3.9-8}$$

where as expected in the limiting case we have

$$V_1 = \lim_{R_s \to 0} \frac{\frac{1}{R_s} + \frac{8}{5}}{\frac{1}{R_s} + \frac{2}{5}} = 1^V. \tag{3.9-9}$$

Similarly, $V_2 = 2^V$ and $V_3 = 1^V$, which are, of course, the same as before. We could have solved V_2 and V_3 in terms of R_s then let $R_s \to 0$.

STOP

Looking back at the preceding two approaches, we propose to classify the above approach under Method #Three as a new category called **parameterization.** As an engineer we constantly try to seek solutions for as a general class of problems as we can in order to answer simpler or possibly special situations. For instance, in the preceding approaches when we let R_s reach a limiting value, that is only one solution out of many possible solutions for all values of the corresponding R_s. We may use different R_s for different design objectives, but the fact remains that with this resistor in the circuit we have a larger class of circuits (i.e., a class with more members) than before. This way of thinking is a natural way of looking at engineering problems. However, it is not always simple to parameterize a problem in a meaningful way, nevertheless when is done correctly the wealth of information stemming from that parameterized solution is worth all the efforts. Thus, we categorize the above two approaches for source transformations under the following method.

3.10. METHOD #FIVE – SOLUTION by the PARAMETERIZATION

When a current source and/or a voltage source is not directly associated with a resistor in the conventional sense, we can parameterize the problem by introducing a resistor R_s appropriately, i.e., either in parallel or in series, depending on the corresponding source. This resistor acts as a catalyst. In the case of a current source, we put the R_s in parallel with the initial source and let $R_s \to \infty$ in the final values of all parameterized quantities associated with the circuit analysis. In the case of a voltage source, however, we put this R_s in series with the initial source and let $R_s \to 0$ in the final values of all parameterized quantities associated with the circuit. For those cases that we have more than one R_s, the procedure is the same, but with substantially more computations.

3.11. METHOD #SIX – SOLUTION by the RECONFIGURATION

In many situations the Method #Five is not the optimal approach, since a simple reconfiguration of the circuit may result in a more manageable network than before. For instance, in Example 3.6-1, we may redraw the circuit by redirecting the current source to node #3 first, then from that node to node #2 as shown in Fig. 3.11-1.

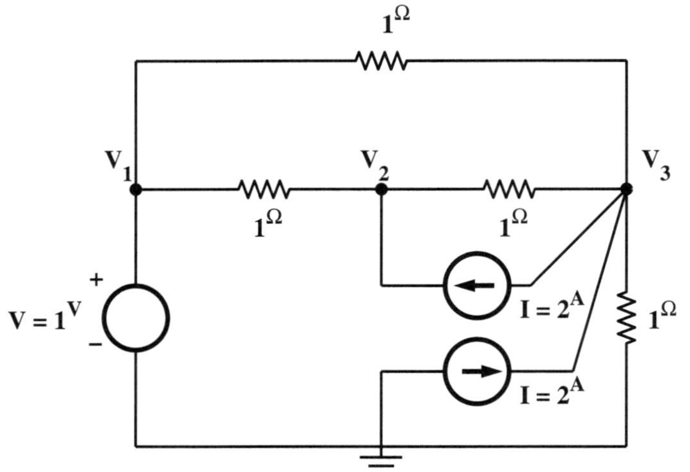

Fig. 3.11-1.

In Fig. 3.11-1, the current which enters node #3, leaves that node in exactly the same amount and the net effect for node #3 is zero. For node #2, the entering current is the same 2^A as before. However, both of these two current sources are now in conventional settings, each having a 1^Ω resistor in parallel with the source. Thus, the problem is now manageable under Method #Three, followed by Method #One.

We could have also reconfigured the circuit in Example 3.6-1 by working on the voltage source as follows.

Here, the 1^V voltage source is drawn as shown in Fig. 3.11-2. We note that $V_1 = V_1' = 1^V$ as before. We can show that at nodes #2 & 3

$$2 = \frac{V_2 - V_1'}{1} + \frac{V_2 - V_3}{1}, \tag{3.11-1}$$

$$0 = \frac{V_3 - V_1}{1} + \frac{V_3 - V_2}{1} + \frac{V_3}{1}. \tag{3.11-2}$$

Using $V_1 = V_1' = 1^V$ in the above yields

$$2V_2 - V_3 = 3, \tag{3.11-3}$$

$$-V_2 + 3V_3 = 1. \tag{3.11-4}$$

Solving for V_2 and V_3 yields the same results ($V_2 = 2^V$ and $V_3 = 1^V$) as before. The key issue to realize here is that we need not take the actual source transformation of the voltage source to a current source, since the voltage sources are the constraint equations and this fact is always the case with similar voltage source configurations.

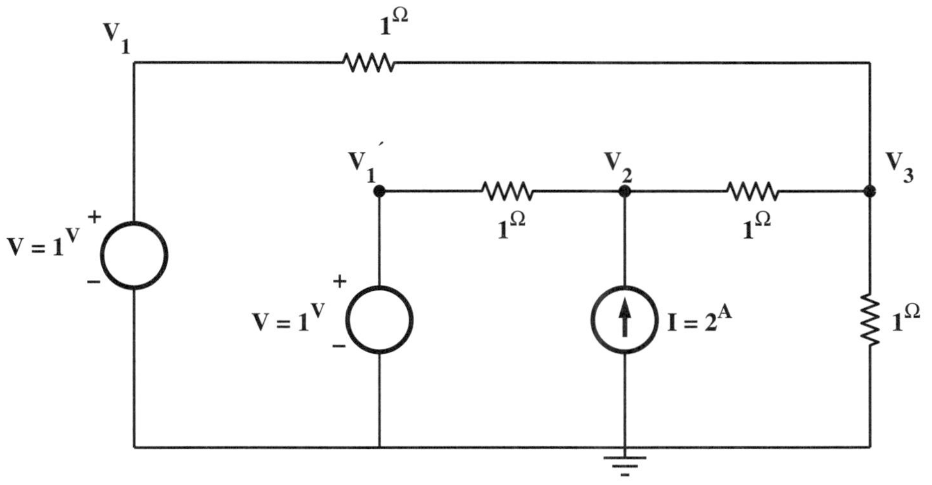

Fig. 3.11-2.

STOP

Comment 3.11-1: We could have also solved Example 3.6-1 by another powerful method called the superposition rule as is described next. In general, a dynamic system is described by a nonlinear differential equation as follows.

$$\dot{x}(t) = f(x, t, u), \qquad x(0) = x_o, \tag{3.11-5}$$

where $x(t) \in R^n(t)$ is the system state vector, $u(t) \in R^m(t)$ is the system input vector, and t is the real variable, which here is time. When u(t) changes, x(t) changes and often in no particular or easily detectable manner with respect to changes in u(t). However, for a large class of systems, which are of interest the above dynamics can be described by a linear differential equation. In fact, the entire class of problems that we are studying in this course are associated with linear systems. In these situations we have an interesting property which is different from the above general dynamic case. Namely, when the system under consideration is linear and is described by a set of linear differential equations, such as $\dot{x}(t) = Ax(t) + Bu(t)$, as its input u(t) changes, its state x(t) changes proportionally. In other words, if we have two inputs u_1 and u_2, each resulting in x_1 and x_2 correspondingly. Then if we apply any linear combination of these two inputs, say, $k_1u_1(t) + k_2u_2(t)$, the corresponding state becomes $x(t) = k_1x_1(t) + k_2x_2(t)$ and that is a remarkable property.

Although we have not shown the linearity of the network in Example 3.6-1, and that requires some effort, this circuit is linear and the superposition rule applies.

3.12. METHOD #SEVEN – SOLUTION by the SUPERPOSITION RULE

To analyze a network with mixed sources, we may break the main problem into two sub-problems each containing one type of source only. For instance, in the Example 3.6-1, we break the problem into one sub-problem with only current source and another one with only voltage source as shown in Figs. 3.12-1a & b. Two rules must be observed when solving any circuit in this manner.

Rule #One: Here, we note that one circuit contains only voltage source(s), while the current source(s) in the original circuit are **opened**. The other circuit contains only current source(s), while the voltage source(s) in the original circuit are **short**.

Rule #Two: Each of the above two sub-problems must be solved in its entirety. That means all currents through each and every element must be specified. Then these two sets of current values must be added algebraically, element by element, to get the current values in every element of the original circuit.

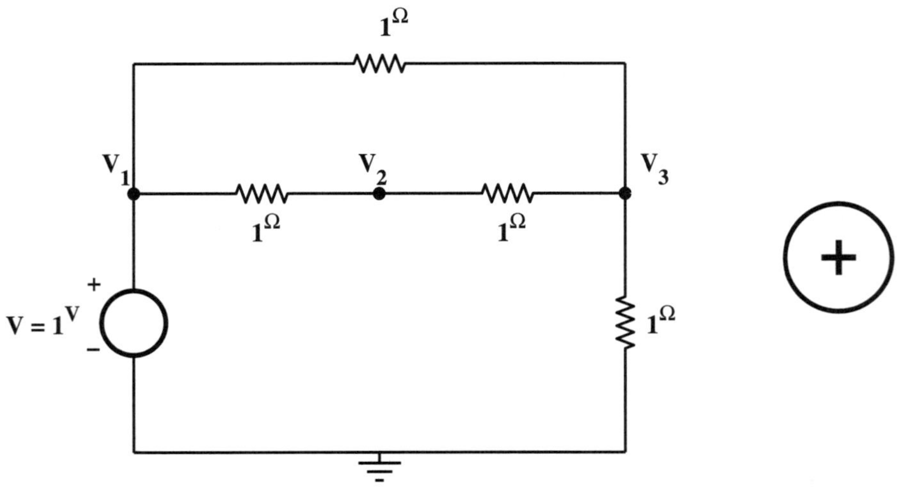

Fig. 3.12-1a. Open all current sources.

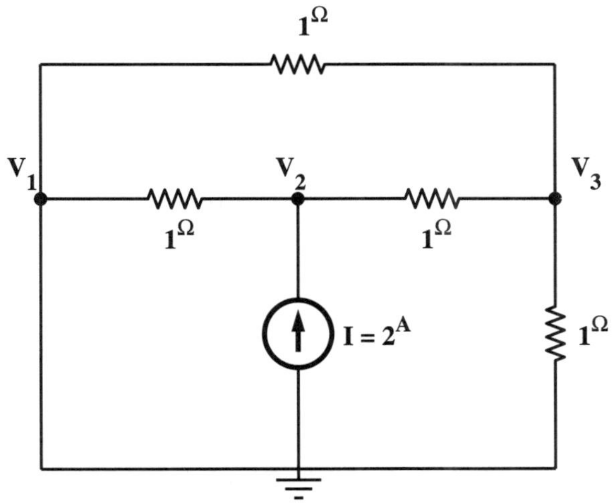

Fig. 3.12-1b. Short all voltage sources.

For instance, in circuits with only voltage sources (cf., Fig. 3.12-2) we have the following set of mesh equations.

$$1 = 3I_1 - 2I_2, \tag{3.12-1}$$

$$0 = -2I_1 + 3I_2. \tag{3.12-2}$$

Resulting in, $I_1 = (3/5)^A$ and $I_2 = (2/5)^A$. The task ahead is to figure out each and every current through the elements of this circuit as generated based on the mesh currents and shown in Fig. 3.12-3.

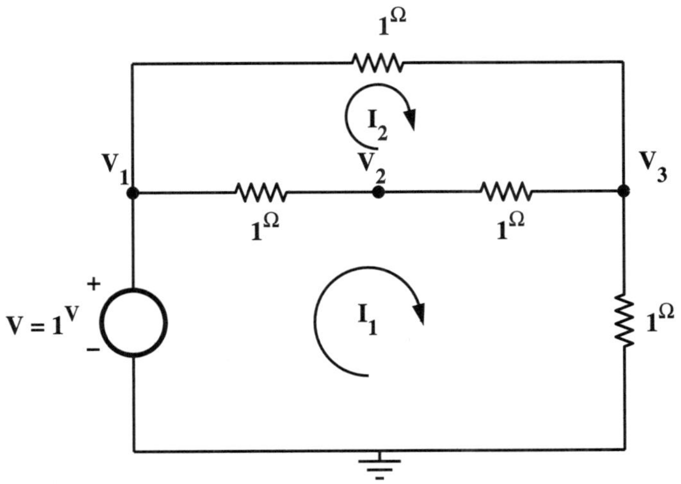

Fig. 3.12-2.

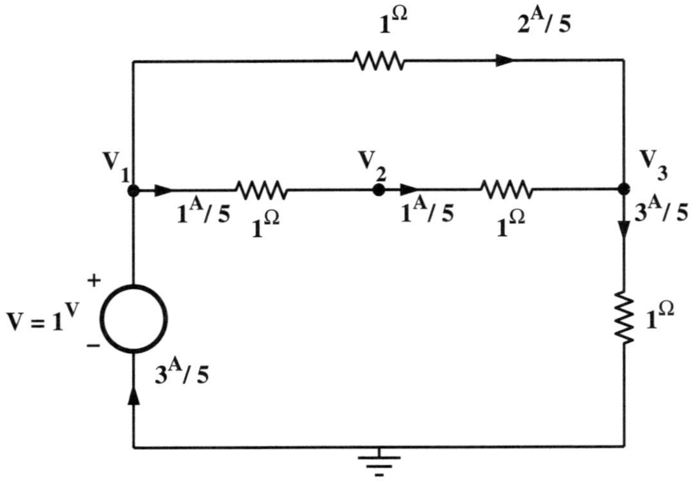

Fig. 3.12-3.

In circuits with only current sources (cf., Fig. 3.12-1b), we note that $V_1 \equiv 0$, while

$$2 = (1 + 1)V_2 - V_3, \tag{3.12-3}$$

$$0 = -V_2 + (1 + 1 + 1)V_3. \tag{3.12-4}$$

Resulting in, $V_2 = (6/5)^V$ and $V_3 = (2/5)^V$. The corresponding current through each element is computed from the voltage across each element as shown in Fig. 3.12-4. In certain cases we have made choices for the current signs which turned out to be reversed. For instance, in Fig. 3.12-4, we have assumed that the current is going from node #3 to node #2, while this choice is perfectly valid, but the resulting minus sign says that the actual or physical direction of this current is the reverse of this choice.

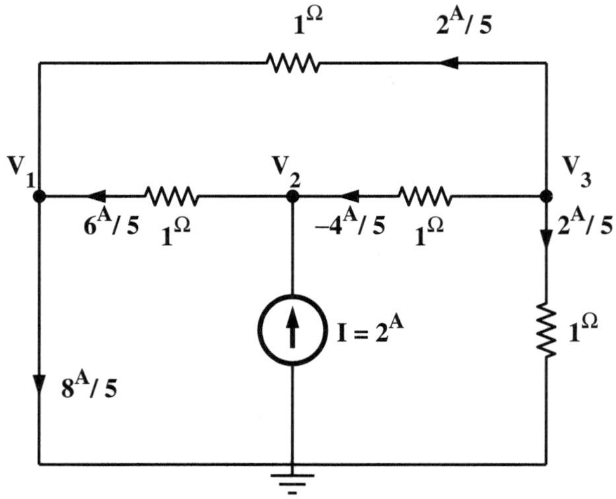

Fig. 3.12-4.

Having generated all currents through each and every element as shown in Figs. 3.12-3 & 4, the net final currents are now calculated by just adding these current values algebraically, element by element, as shown in Fig. 3.12-5.

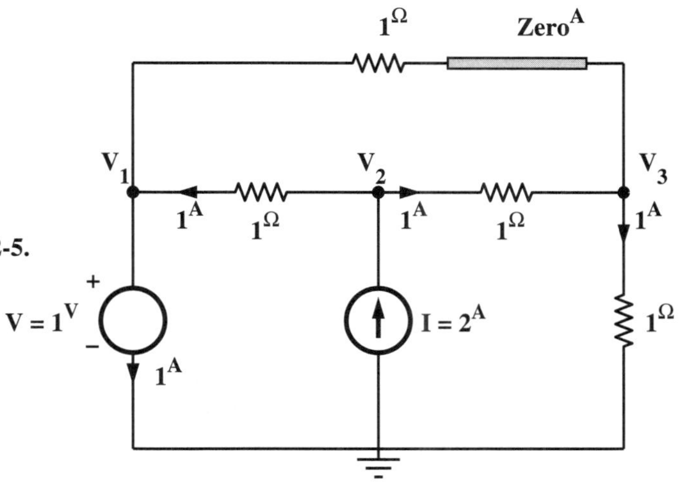

Fig. 3.12-5.

Also, we note that the the 1^{A} is now going into the voltage source, rather than coming out of that source as we often suggest. The reason is that in this circuit the 2^{A} current source is the more dominant independent source than is the 1^{V} voltage source. From the above circuit we get

$$V_3 = V_1 = 1^{V}, \quad \text{and} \quad V_2 - V_3 = 1 \;\Rightarrow V_2 = 2^{V}. \qquad (3.12\text{-}5)$$

Conclusion: All in all, no matter which approach we take to analyze a network, the final result is the same. We need to know, however, which approach is more suitable in a given problem and the answer to that comes with experience and that comes with practice, practice and more practice.

3.13. PROBLEM SET

Node Equations

3.13.1. Write a set of node equations to analyze the following circuits. In each case follow the procedure according to the steps shown in this chapter.

Fig. 3.13-1.

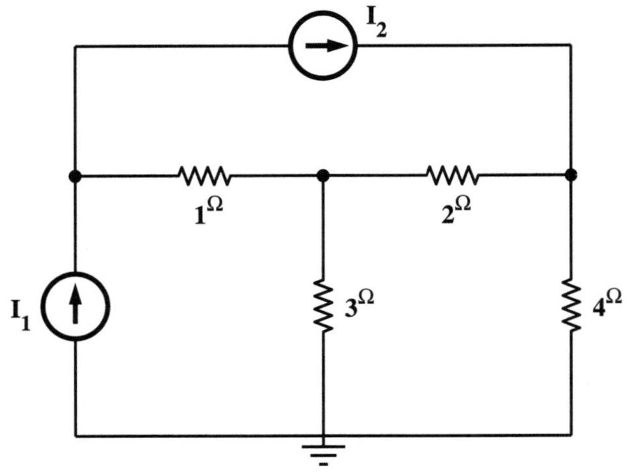

Fig. 3.13-2.

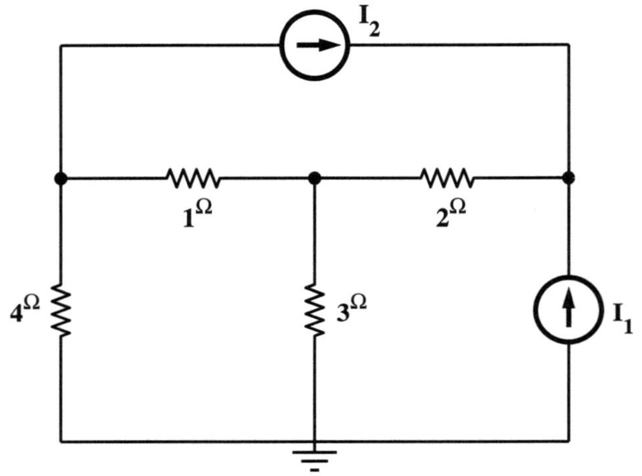

Fig. 3.13-3.

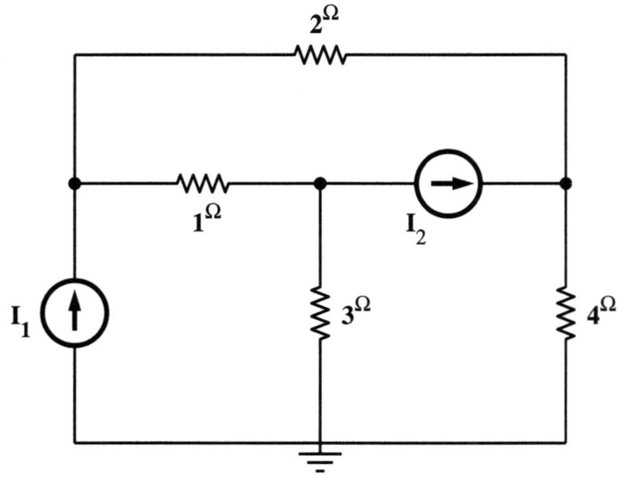

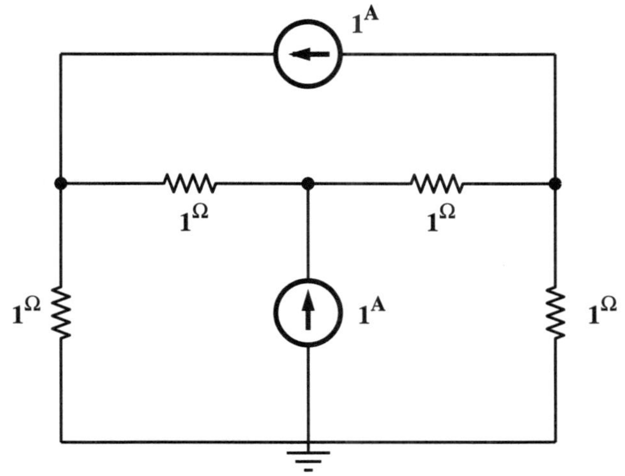

Fig. 3.13-4.

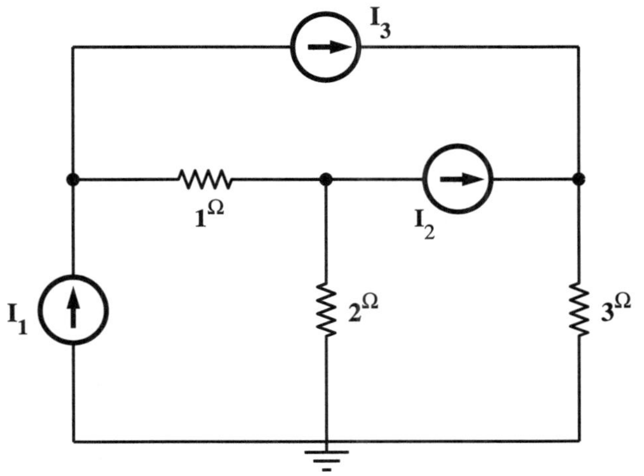

Fig. 3.13-5.

Mesh Equations (Mixed Sources)

3.13.2. Write a set of mesh equations to analyze the following circuits. In each case follow the procedure according to the steps shown in this chapter.

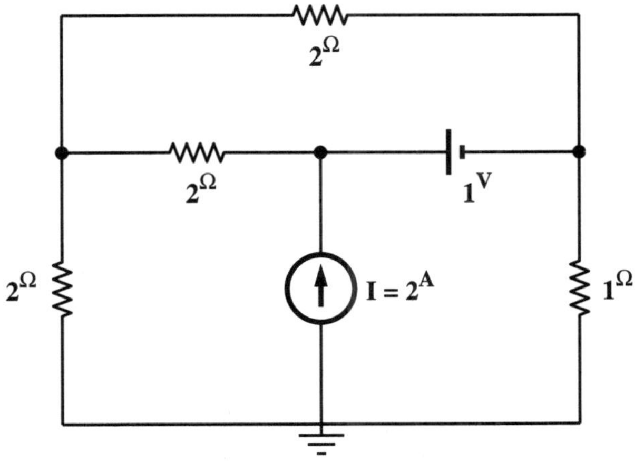

Fig. 3.13-6.

Fig. 3.13-7.

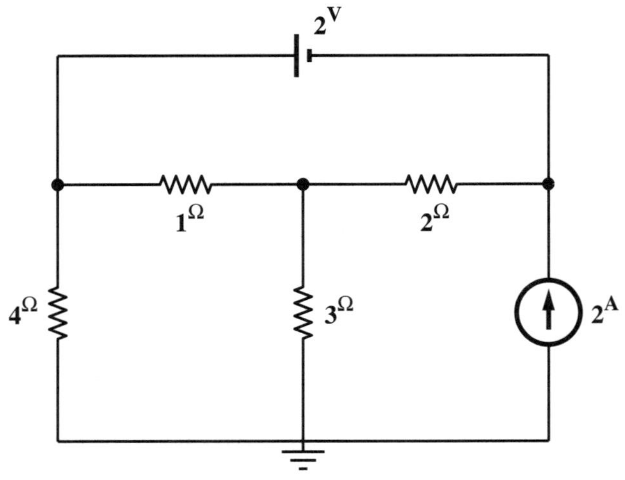

Fig. 3.13-8.

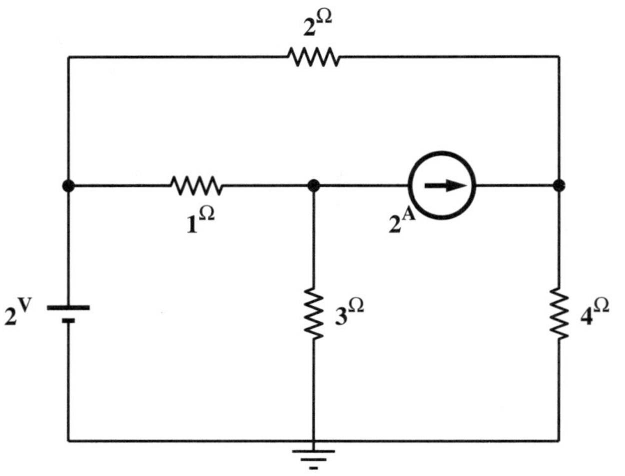

Fig. 3.13-9.

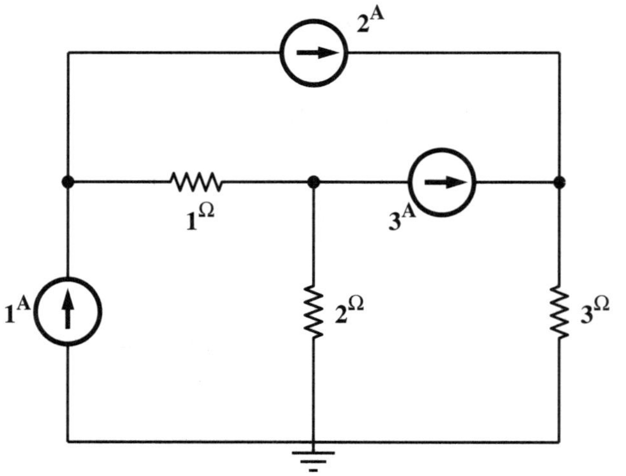

General Methods

3.13.3. Write a set of sufficient equations in order to analyze the following circuits. In each case specify the procedure selected step by step.

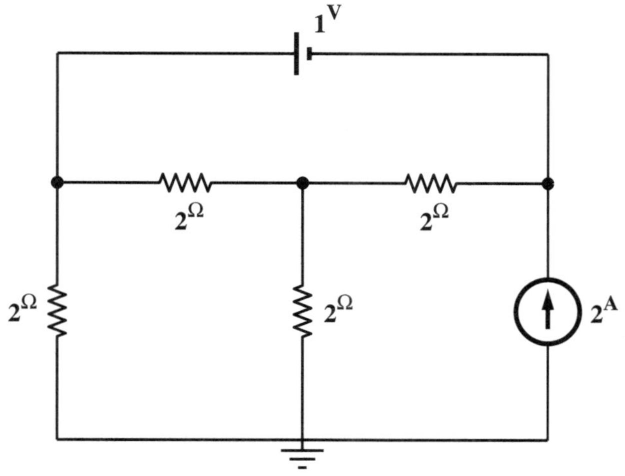

Fig. 3.13-10.

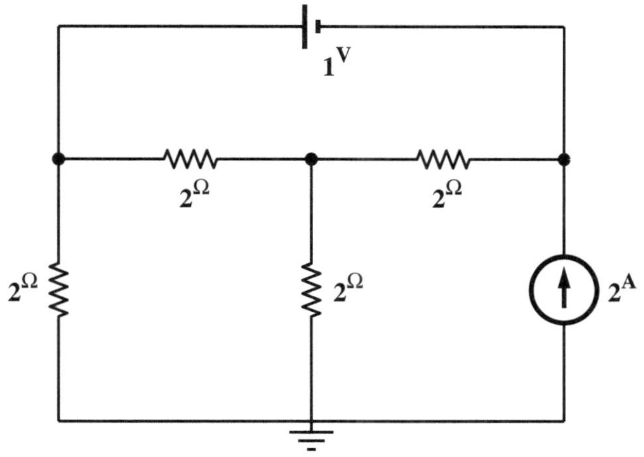

Fig. 3.13-11.

Superposition Rule

3.13.4. Using superposition rule analyze the following circuits.

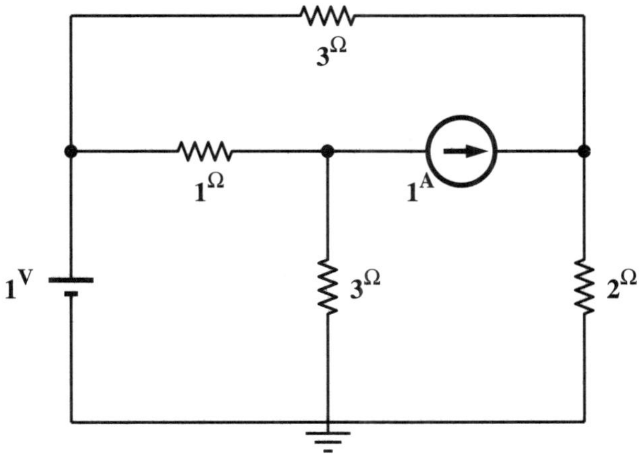

Fig. 3.13-12.

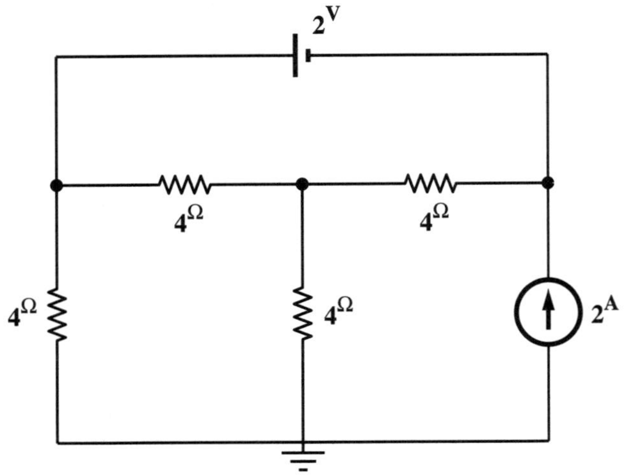

Fig. 3.13-13.

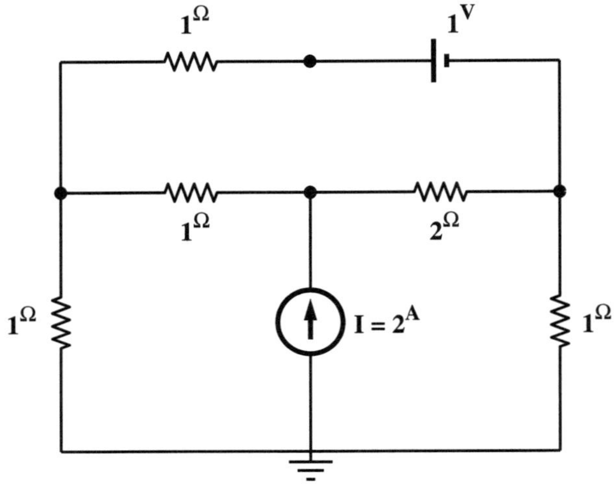

Fig. 3.13-14.

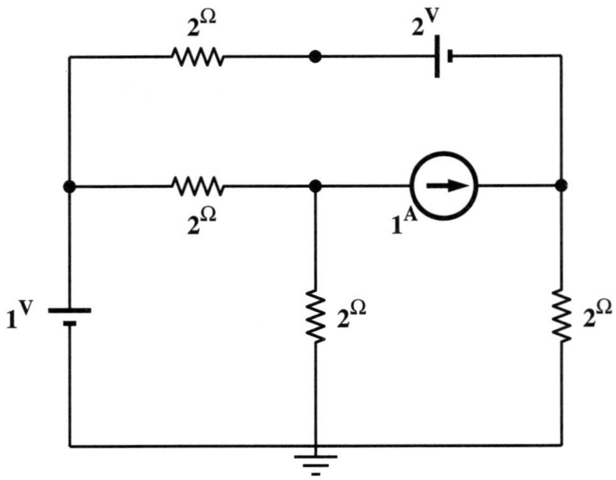

Fig. 3.13-15.

4 NUMERICAL EXAMPLES AND PROBLEMS

4.1. INTRODUCTION

In the preceding three chapters the essentials of linear network analysis which take a long time to develop in the traditional sense are presented. We realize that in many instances we need to use several more examples and drill problems to elucidate the underlying issues. However, we also believe that one effective way for students to remember all those different methods is through a parallel teaching of these methods, in order that students be constantly confronted with these issues and will be asked to distinguish among various methods. Now, we introduce and solve a few examples to explain these cases further. This approach will help students to master all essential aspects of circuit analysis quickly.

4.2. NUMERICAL EXAMPLES

In the following we present two examples that each is solved in many different ways to elucidate various methods discussed in this textbook so far.

Example 4.2-1: Analyze the circuit in Fig. 4.2-1 in as many different ways as you know how.

Solution: The first thing that we do is to pause and study the circuit elements and its configuration. For instance, this is a circuit with mixed sources, where neither of the two independent sources are applied in the conventional settings, i.e., there is no resistor in series with the voltage source, nor is there one in parallel with the current source. Thus, all the methods we have studied must be applied cautiously in this example to avoid mistakes. Nevertheless we attempt to apply those methods one by one to this problem and see where we face difficulty.

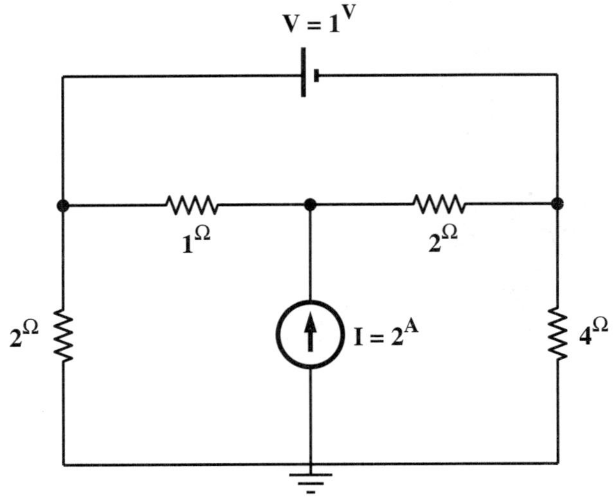

Fig. 4.2-1.

Method #One – Solution by the Mesh Equations

STEP Zero: Clearly, we must invoke this step to label variables and/or unknowns according to this method before proceeding further. The essential variables are mesh currents. We **interpret** V_x, the unknown voltage across the current source with its own polarity selected in accordance with the direction of that source (cf., Fig. 4.2-2), as a voltage source with zero internal resistance. (This interpretation is based on a slightly more advanced topic than what has been discussed in this textbook.)

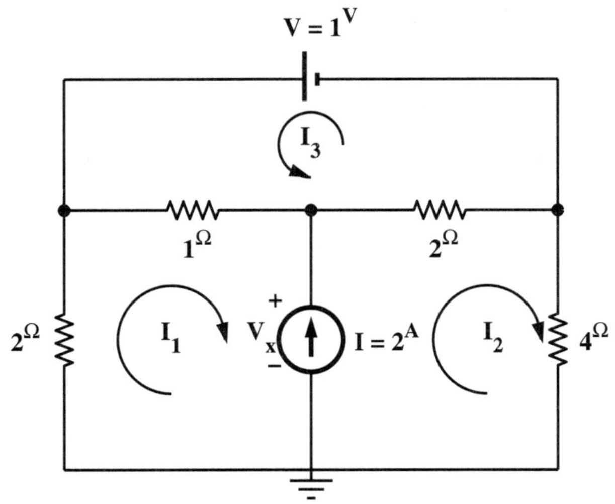

Fig. 4.2-2.

STEP One: Considering the above variables, the mesh equations for this circuit and the corresponding constraint equation are developed as follows.

$$-V_x = 3I_1 + I_3, \quad \text{mesh \#1,} \tag{4.2-1}$$

$$V_x = 6I_2 + 2I_3, \quad \text{mesh \#2,} \tag{4.2-2}$$

$$1^V = I_1 + 2I_2 + 3I_3, \quad \text{mesh \#3,} \tag{4.2-3}$$

$$2^A = I_2 - I_1, \quad \text{constraint equation.} \tag{4.2-4}$$

STEP Final: Clearly, we have generated all equations we can using this approach and we can solve these equations numerically in a number of ways. Generating that numerical solution is the last phase of our study.

STOP

Method #Two – Solution by the Node Equations

STEP Zero: Here we have an independent voltage source in the circuit, which makes this approach challenging. Nevertheless we label all pertinent essential variables as shown in Fig. 4.2-3. Here, I_x refers to the current of the independent voltage source which is an unknown quantity that is **interpreted** as a current source.

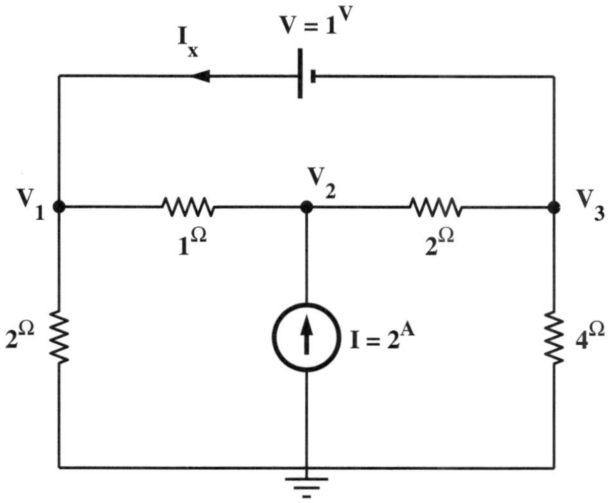

Fig. 4.2-3.

STEP One: Using the above essential node voltages, we have the following set of equations.

$$I_x = (1 + ½)V_1 - V_2, \quad \text{at node \#1,} \tag{4.2-5}$$

$$2^A = -V_1 + (1 + ½)V_2 - (½)V_3, \qquad \text{at node \#2,} \qquad (4.2\text{-}6)$$

$$-I_x = -(½)V_2 + (½ + ¼)V_3, \qquad \text{at node \#3,} \qquad (4.2\text{-}7)$$

$$1^V = V_1 - V_3, \qquad \text{constraint equation.} \qquad (4.2\text{-}8)$$

STEP Final: Upon solving the above set of independent equations numerically, the analysis of this network is completed.

STOP

Method #Three – Solution by the Conventional Source Transformation

Because neither the 1^V voltage source, nor the 2^A current source are in the conventional settings, i.e., there is no series resistor with the voltage source, nor any parallel resistor with the current source, this method is not applicable to this problem directly. However, Methods #Five and #Six are applicable and we explore those subsequently.

Method #Four – Solution by the Utilization of Super Loop

STEP Zero: In Method #One, if we use the dotted line as a super loop, instead of choosing V_x, as shown in Fig. 4.2-4, then we have two mesh/loop equations plus one constraint equation. This is the same as adding (4.2-1) and (4.2-2).

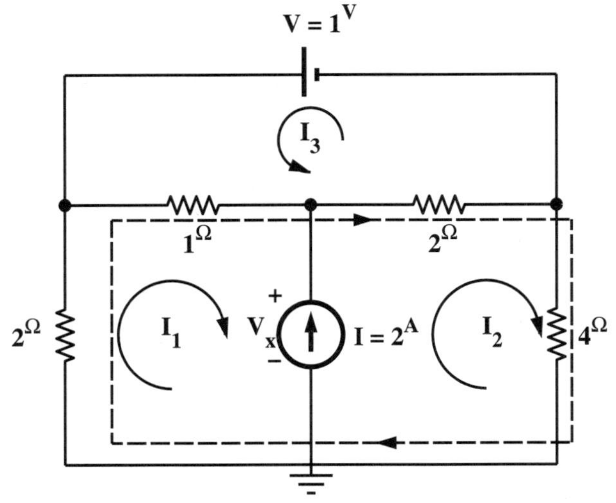

Fig. 4.2-4.

STEP One: Here, we have

$$1^V = I_1 + 2I_2 + 3I_3, \qquad \text{for mesh \#3,} \qquad (4.2\text{-}9)$$

$$0^V = 1^\Omega(I_1 + I_3) + 2^\Omega(I_2 + I_3) + 4^\Omega I_2 + 2^\Omega I_1$$

$$= 3I_1 + 6I_2 + 3I_3, \qquad \text{for super loop,} \qquad (4.2\text{-}10)$$

$$2^A = I_2 - I_1, \qquad \text{constraint equation.} \qquad (4.2\text{-}11)$$

STEP Final: Again, upon solving the above set of independent equations numerically the analysis of this network is completed.

STOP

Method #Four Prime – Solution by the Utilization of Super Node

This case has already been solved in Example 3.7-1, which we can use as a point of reference. Recall that $V_1 = 3^V$, $V_2 = 4^V$ and $V_3 = 2^V$.

Method #Five – Solution by the Parameterization

First we look at the current source. We inject a resistor as the parameter in parallel with the current source as shown in Fig. 4.2-5a.

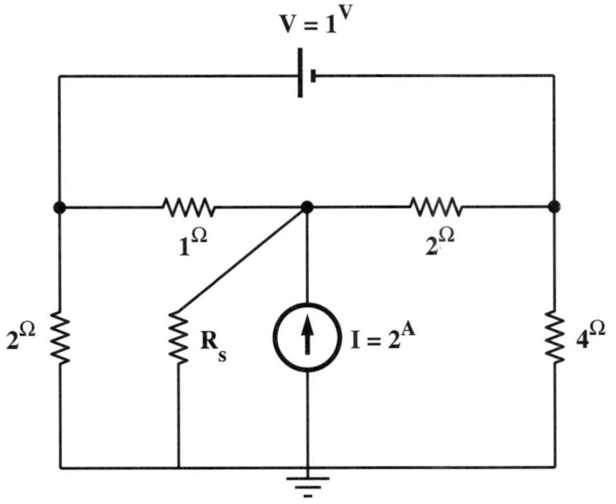

Fig. 4.2-5a.

STEP Zero: Here, we have transformed the current source into a voltage source whose value is $(2R_s)^V$, and label all pertinent variables.

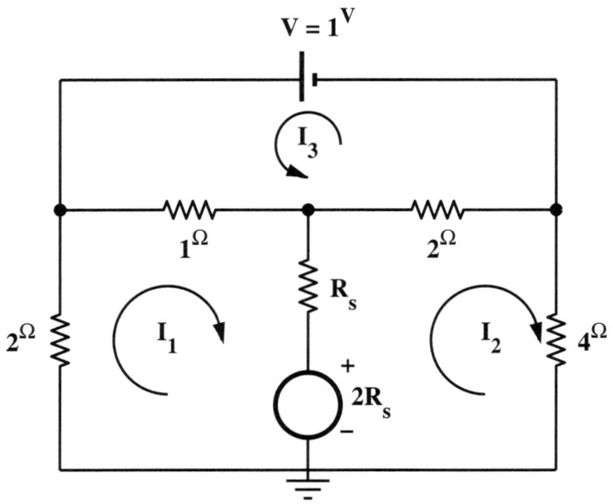

Fig. 4.2-5b.

STEP One: The corresponding mesh equations are as follows.

$$-2R_s = (1^{\Omega} + 2^{\Omega} + R_s)I_1 - R_sI_2 + (1^{\Omega})I_3, \tag{4.2-12}$$

$$2R_s = -R_sI_1 + (2^{\Omega} + 4^{\Omega} + R_s)I_2 + (2^{\Omega})I_3, \tag{4.2-13}$$

$$1^V = (1^{\Omega})I_1 + (2^{\Omega})I_2 + (3^{\Omega})I_3. \tag{4.2-14}$$

STEP Final: We solve these equations for both pedagogical reason and the fact that we need some numerical results to compare values with those generated in Example 3.7-1.

Adding (4.2-12) and (4.2-13) results in

$$0 = I_1 + 2I_2 + I_3. \tag{4.2-15}$$

Substituting (4.2-15) in (4.2-14) yields

$$1 = I_1 + 2I_2 + I_3 + 2I_3 \;\Rightarrow\; I_3 = (½)^A. \tag{4.2-16}$$

Using this value for I_3 in (4.2-14) yields

$$I_1 = -2I_2 - ½. \tag{4.2-17}$$

Substituting for I_1 and I_3 in (4.2-12) yields

$$I_2 = \frac{3R_s - 2}{6R_s + 12}, \tag{4.2-18}$$

where the limiting numerical value as $R_s \to \infty$ is

$$I_2 = \lim_{R_s \to \infty} \frac{3R_s - 2}{6R_s + 12} = (½)^A. \qquad (4.2\text{-}19)$$

Thus, the corresponding limiting value for I_1 is $I_1 = -2I_2 - ½ = -1.5^A$.

STOP

Second, we look at the voltage source. Here, we put an R_s (the parameter) in series with the voltage source and transform the circuit into a circuit with only current sources as shown in Fig. 4.2-6.

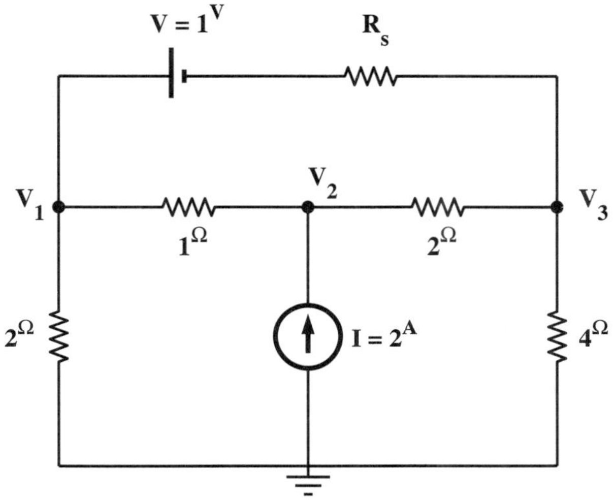

Fig. 4.2-6a.

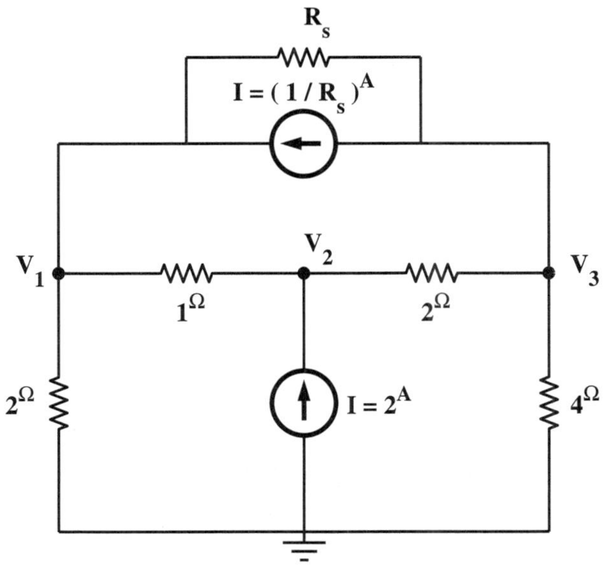

Fig. 4.2-6b.

STEP Zero: The circuit in Fig. 4.2-6 is labeled with the appropriate set of variables to perform a node analysis.

STEP One: The equations for the above circuit can be generated by inspection as follows.

$$(\frac{1}{R_s})^A = (\frac{1}{R_s} + 1 + ½)V_1 - V_2 - (\frac{1}{R_s})V_3, \tag{4.2-20}$$

$$2^A = -V_1 + (1 + ½)V_2 - (½)V_3, \tag{4.2-21}$$

$$(-\frac{1}{R_s})^A = -\frac{1}{R_s}V_1 - (½)V_2 + (½ + ¼ + \frac{1}{R_s})V_3. \tag{4.2-22}$$

STEP Final: Adding (4.2-20) and (4.2-22) yields

$$2V_1 - 2V_2 + V_3 = 0 \Rightarrow V_3 = 2(V_2 - V_1). \tag{4.2-23}$$

Substituting this V_3 in (4.2-21) yields

$$2^A = -V_1 + 1.5V_2 - (½)(2V_2 - 2V_1) \Rightarrow V_2 = 4^V. \tag{4.2-24}$$

Substituting $V_2 = 4^V$, and $V_3 = 2(4 - V_1)$ in (4.2-20) yields

$$V_1 = \frac{\frac{9}{R_s} + 4}{\frac{3}{R_s} + \frac{3}{2}}, \tag{4.2-25}$$

where the limiting numerical value as $R_s \to 0$ is

$$V_1 = \lim_{R_s \to 0} \frac{\frac{9}{R_s} + 4}{\frac{3}{R_s} + \frac{3}{2}} = 3^V, \Rightarrow V_3 = 2^V. \tag{4.2-26}$$

These values as expected are the same as those presented in Example 3.7-1.

STOP

Method #Six – Solution by the Reconfiguration

First the current source. Here we may reconfigure the circuit of Fig. 4.2-1 into either the circuit in Fig. 4.2-7, or that in Fig.4.2-8.

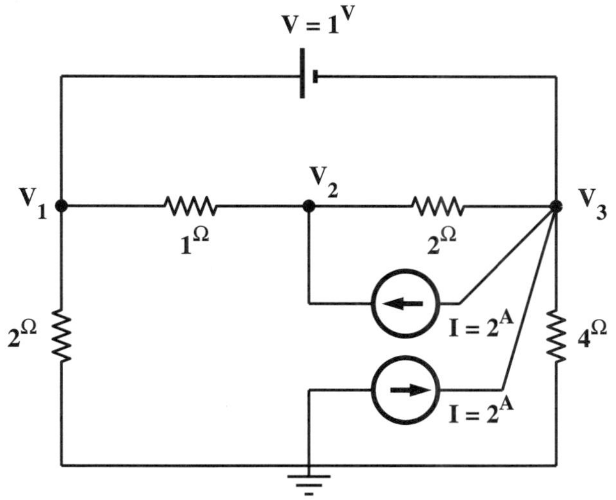

Fig. 4.2-7.

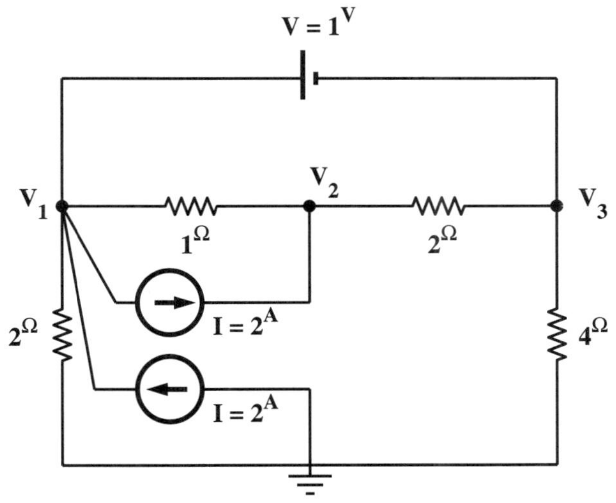

Fig. 4.2-8.

Then either of these two circuits can be transformed into circuits with only voltage sources, because each current source is in parallel with a corresponding resistor, i.e., both current sources are in the conventional settings. We may then apply the Method #One – the mesh equations – to the resulting circuits. There are major efforts involved in this process. We leave carrying out the remaining analysis as an exercise for the reader. (Problem 4.3.1).

Second, we consider the voltage source. There are choices to reconfigure the initial circuit in Fig. 4.2-1, such that its only voltage source becomes in series with a resistor and thus the conventional or normal source transformation can be applied as shown in Fig. 4.2-9 and/or Fig. 4.2-10. In each of these two circuits the ground has been changed, and the choice of the new ground makes computation easy since we have only two nodes.

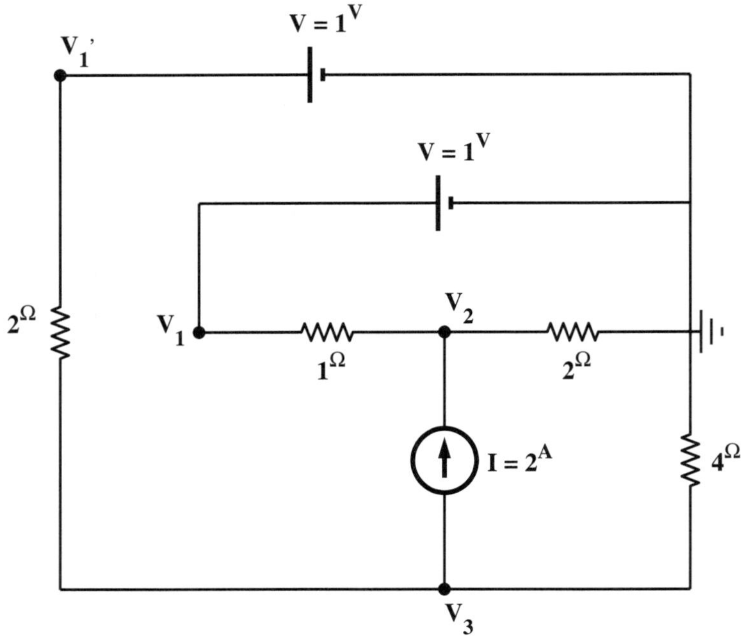

Fig. 4.2-9.

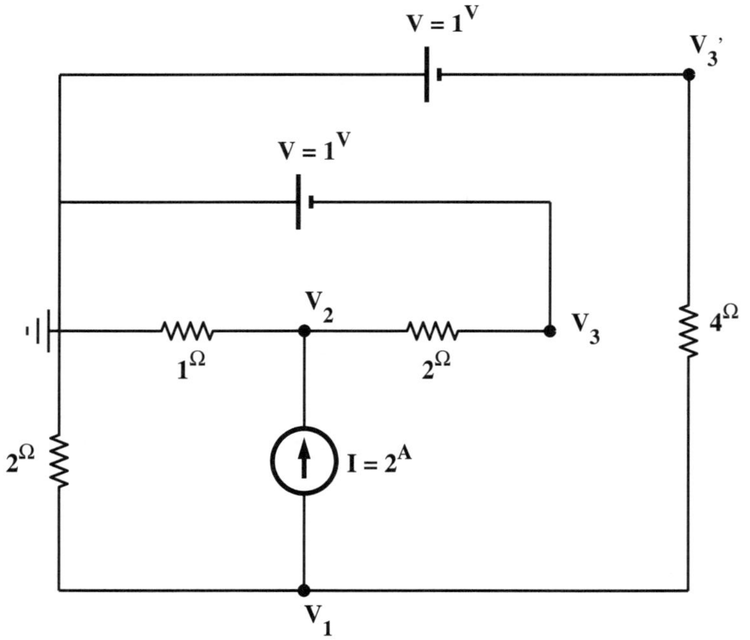

Fig. 4.2-10.

Although, each of the above circuit seems straightforward and the Method #Two is easily applicable to the corresponding configuration, more work is left to complete the analysis. Furthermore, the change of reference or ground voltage does not complicate our problem, but we must be very careful to reinterpret the new voltage values in terms of the old reference or ground. This interpretation is necessary in order to compare the final result of this approach with the results developed in the preceding approach. Therefore, our work has been expanded to cover all these intermediate cases and we leave the details for exercise. (Problem 4.3.1).

Method #Seven – Solution by the Superposition Rule

The solution to our problem can also be generated by this procedure. As stated Section 3.12, the **currents** in each element of the initial circuit is equivalent to the algebraic summation of currents in each and every element of the following two circuits.

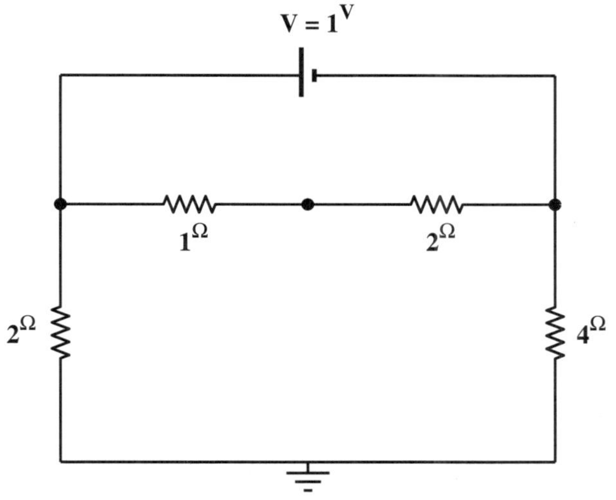

Fig. 4.2-11. Open all current sources.

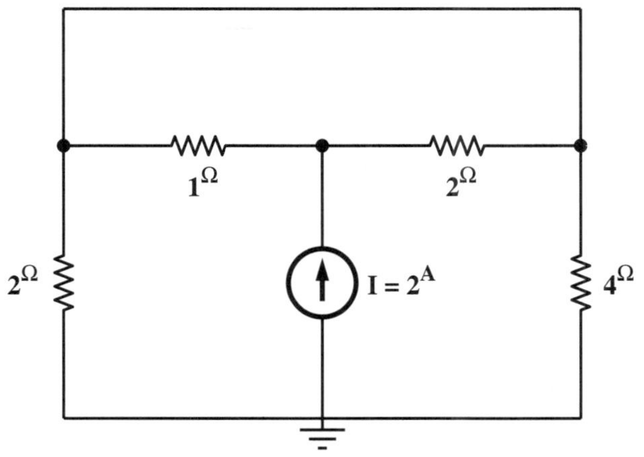

Fig. 4.2-12. Short all voltage sources.

Either of the above two circuits (Figs. 4.2-11 & 12) can be solved and all corresponding **currents** through each and every element can be calculated rather easily. Then we must add algebraically all these current values.

For the circuit in Fig. 4.2-11, we use a mesh analysis approach with its variables and the corresponding circuit as shown in Fig. 4.2-13.

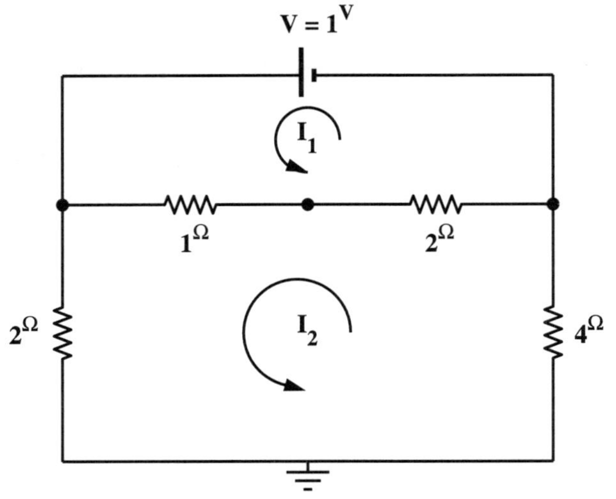

Fig. 4.2-13.

Here,

$$1^{V} = (3^{\Omega})I_1 - (3^{\Omega})I_2, \tag{4.2-27}$$

$$0 = -(3^{\Omega})I_1 + (9^{\Omega})I_2. \tag{4.2-28}$$

Resulting in $I_2 = (1/6)^{A}$, and $I_1 = (1/2)^{A}$. Thus, for this circuit the currents through each and every element are those shown in Fig. 4.2-14.

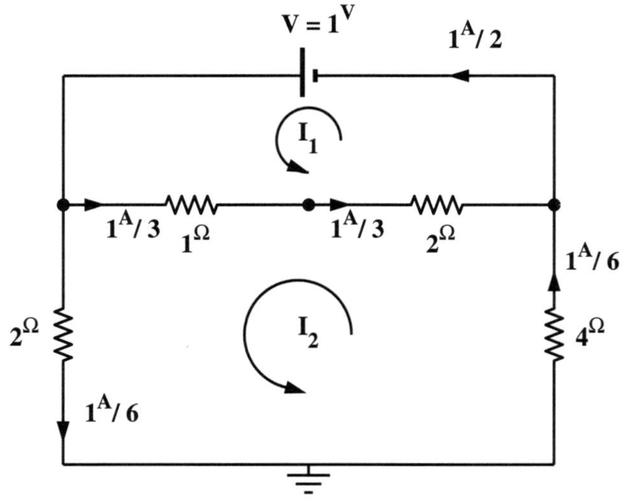

Fig. 4.2-14.

For the circuit in Fig. 4.2-12, repeated in Figs. 4.2-15 & 16 for convenience, we have $V_1 \equiv V_3$. Also from Fig. 4.2-16, which is more instructive than Fig. 4.2-15, we have

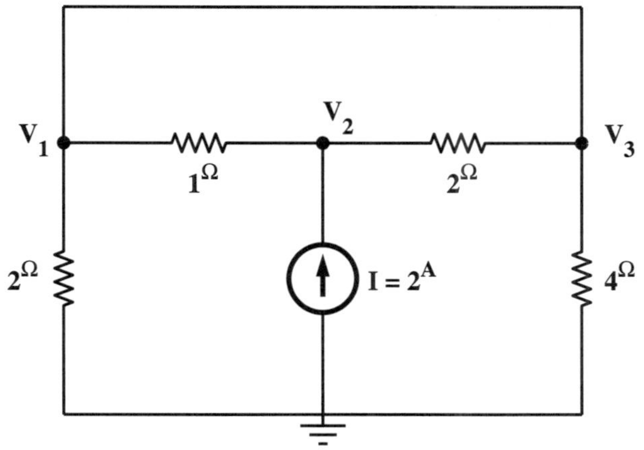

Fig. 4.2-15.

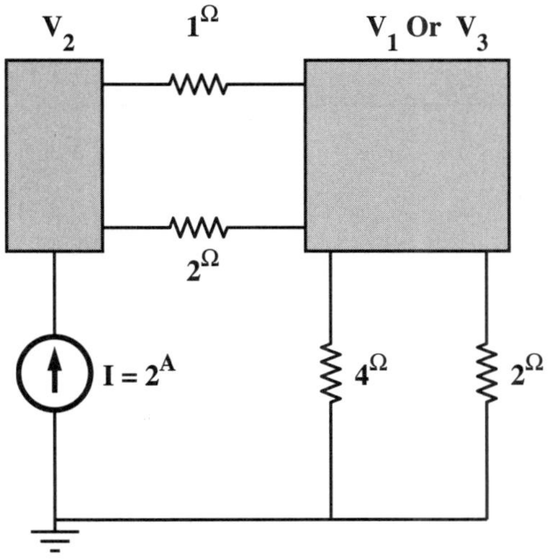

Fig. 4.2-16.

$$2^A = (1 + ½)V_2 - (1 + ½)V_1, \tag{4.2-29}$$

$$0 = -(1 + ½)V_2 + [1 + 2(½) + ¼]V_1, \tag{4.2-30}$$

resulting in $V_1 = (8/3)^V$ and $V_2 = 4^V$. Using these values, it is easy to calculate all currents through each and every element of the circuit in Fig. 4.2-12, as depicted in Fig. 4.2-17.

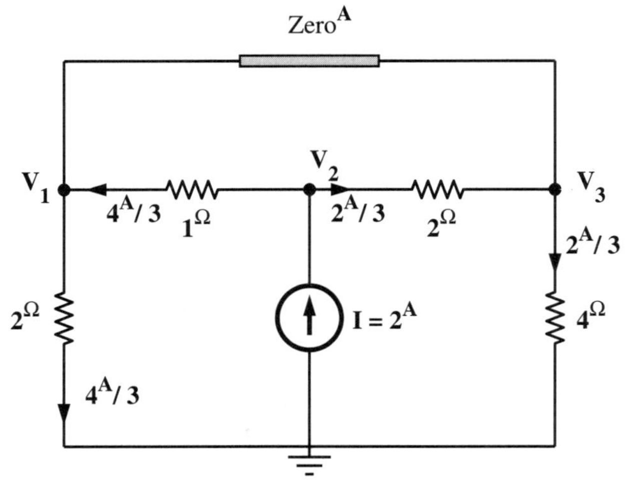

Fig. 4.2-17.

Finally, currents through each and every element of the original circuit are calculated as the algebraic sum of the currents in Figs. 4.2-14 & 17, as shown in Fig. 4.2-18.

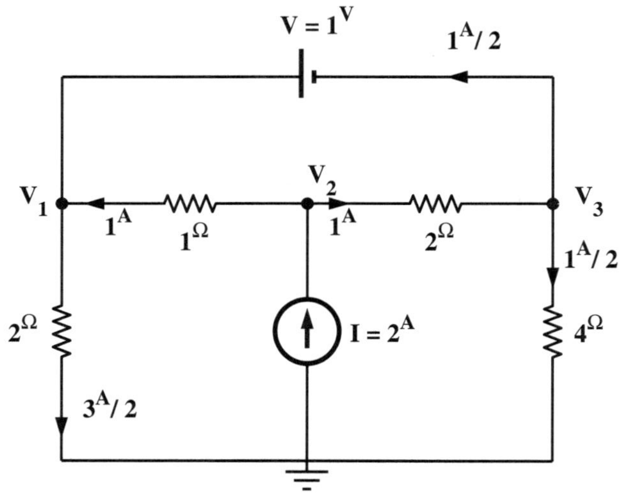

Fig. 4.2-18.

From Fig. 4.2-18 we can calculate the voltages across each and every element as follows.

$$V_1 = (1.5)^A \times (2^\Omega) = 3^V, \tag{4.2-31}$$

$$V_3 = (\tfrac{1}{2})^A \times (4^\Omega) = 2^V, \tag{4.2-32}$$

$$V_2 - V_3 = (1^A) \times (2^\Omega) = 2^V, \tag{4.2-33}$$

$$V_1 - V_2 = (-1)^A \times (1^\Omega) = -1^V, \tag{4.2-34}$$

$$V_2 = V_2 - V_3 + V_3 = 2 + 2 = 4^V. \tag{4.2-35}$$

STOP

Example 4.2-2: Analyze the circuit in Fig. 4.2-19 in as many different ways as you can.

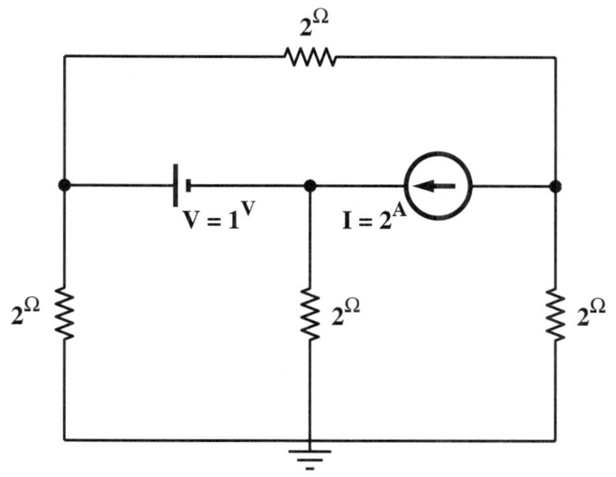

Fig. 4.2-19.

Solution: Here again, the first thing that we do is to pause and study the circuit elements and its configuration. This is also a circuit with mixed sources, where neither of the two independent sources are applied in the conventional settings, i.e., there is no resistor in series with the voltage source, nor is there one in parallel with the current source. We now apply all the preceding methods to this circuit. Clearly, there are a few repetitive lines between this example and the last one.

Method #One – Solution by the Mesh Equations

STEP Zero: We invoke this step to label various variables and/or unknowns before proceeding further according to this methodology. The essential variables are mesh currents, and V_x is the voltage across the current source.

STEP One: Considering the above variables, the mesh equations for this circuit and the corresponding constraint equation are as follows.

$$1^V = (2^\Omega + 2^\Omega)I_1 - (2^\Omega)I_2, \qquad \text{mesh \#1}, \tag{4.2-36}$$

$$V_x = -(2^{\Omega})I_1 + (2^{\Omega} + 2^{\Omega})I_2, \qquad \text{mesh \#2,} \qquad (4.2\text{-}37)$$

$$1^V + V_x = (2^{\Omega})I_3, \qquad \text{mesh \#3,} \qquad (4.2\text{-}38)$$

$$2^A = I_2 + I_3, \qquad \text{constraint equation.} \qquad (4.2\text{-}39)$$

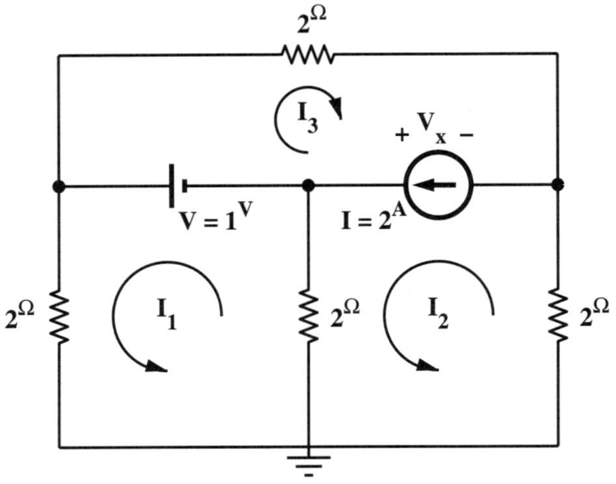

Fig. 4.2-20.

STEP Final: Clearly, we have generated all essential equations, and a numerical solution of these equations completes our study.

STOP

Method #Two – Solution by the Node Equations

STEP Zero: Here also we have an independent voltage source in the circuit, which makes this approach challenging. Nevertheless we label all pertinent essential variables as shown in Fig. 4.2-21, with I_x as the current of the independent voltage source.

STEP One: Using the above essential node voltages, we have the following set of equations.

$$I_x = (½ + ½)V_1 - (½)V_3, \qquad \text{at node \#1,} \qquad (4.2\text{-}40)$$

$$2^A - I_x = (½)V_2, \qquad \text{at node \#2,} \qquad (4.2\text{-}41)$$

$$-2^A = -(½)V_1 + (½ + ½)V_3, \qquad \text{at node \#3,} \qquad (4.2\text{-}42)$$

$1^V = V_1 - V_2$, constraint equation. (4.2-43)

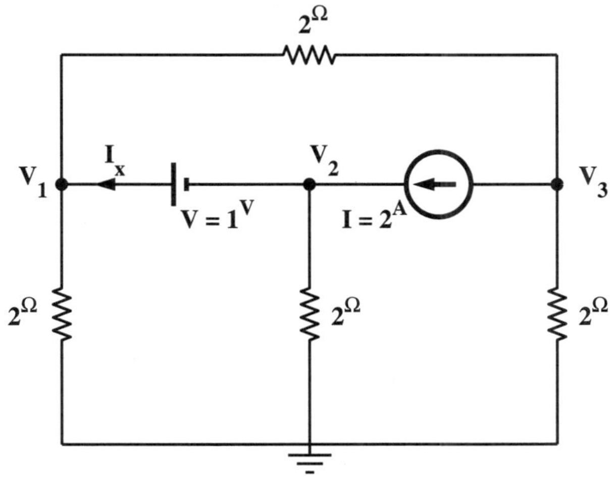

Fig. 4.2-21.

STEP Final: The numerical solution of the above set of independent equations completes the analysis of this network.

STOP

Method #Three – Solution by the Conventional Source Transformation

Because neither the 1^V voltage source, nor the 2^A current source are in the conventional settings, this method is not applicable to this problem.

Method #Four – Solution by the Utilization of Super Loop

STEP Zero: In Method #One and as shown in Fig. 4.2-22, if we use the dotted line as a super loop, instead of choosing V_x, then we have two mesh/loop equations plus one constraint equation. Incidentally, this is not the only super loop in this circuit.

STEP One: Here, we have

$1^V = 4I_1 - 2I_2$, for mesh #1, (4.2-44)

$1^V = 2I_1 - 4I_2 + 2I_3$, for super loop, (4.2-45)

$2^A = I_2 + I_3$, constraint equation. (4.2-46)

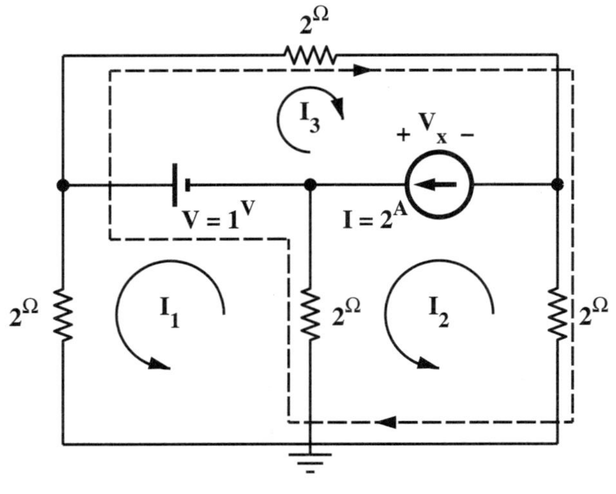

Fig. 4.2-22.

STEP Final: Again, upon solving the above set of independent equations numerically the analysis of this network is completed.

STOP

Method #Four Prime – Solution by the Utilization of Super Node

STEP Zero: The super node is chosen to shield the troublesome voltage source.

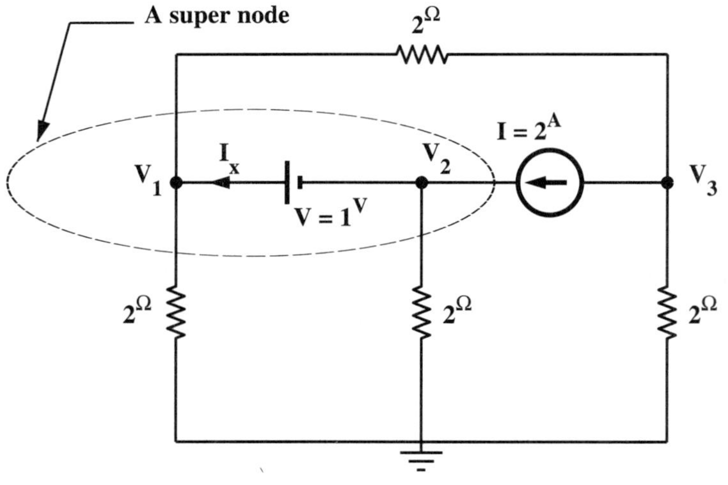

Fig. 4.2-23.

STEP One: Here, we use the super node, node #3 and the constraint equation to generate a proper set of independent equations.

$$2^A + (½)(V_3 - V_1) = (½)V_1 + (½)V_2, \qquad \text{at super node,} \quad (4.2\text{-}47)$$

$$-2^A = -(½)V_1 + (½ + ½)V_3, \qquad \text{at node \#3,} \qquad (4.2\text{-}48)$$

$$1^V = V_1 - V_2, \qquad \text{constraint equation.} \qquad (4.2\text{-}49)$$

STEP Final: The numerical solution of the above set completes the analysis.

STOP

Method #Five – Solution by the Parameterization

First we look at the current source. We inject a resistor as the parameter in parallel with the current source as shown in Fig. 4.2-24.

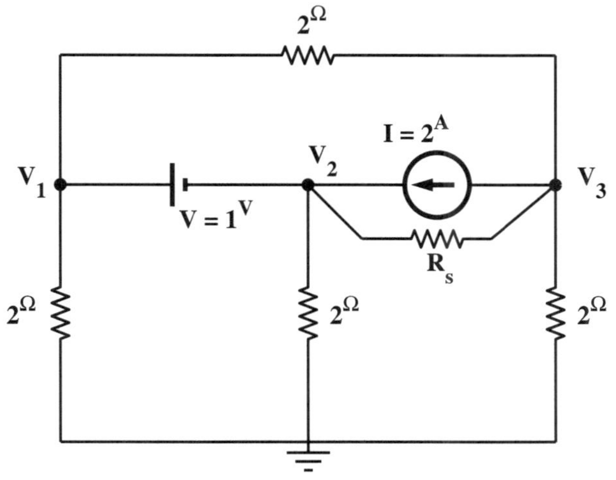

Fig. 4.2-24a.

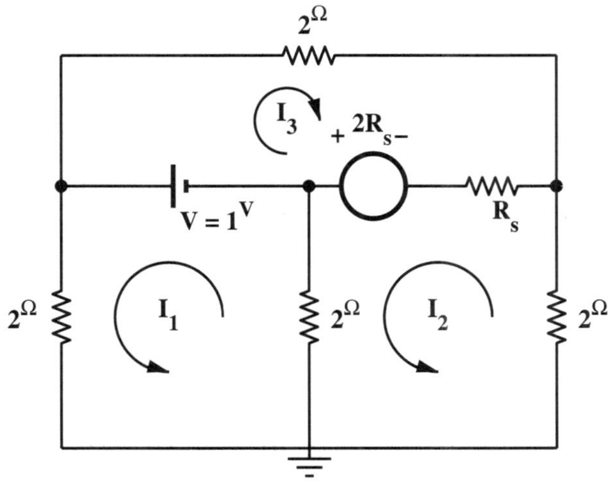

Fig. 4.2-24b.

STEP Zero: All pertinent variables are identified in Fig. 4.2-24.

STEP One: The corresponding mesh equations are as follows.

$$1^V + (2R_s)^V = R_sI_2 + (2^\Omega + R_s)I_3, \tag{4.2-50}$$

$$(2R_s)^V = -(2^\Omega)I_1 + (2^\Omega + 2^\Omega + R_s)I_2 + R_sI_3, \tag{4.2-51}$$

$$1^V = (2^\Omega + 2^\Omega)I_1 - (2^\Omega)I_2. \tag{4.2-52}$$

STEP Final: We may attempt to solve these equations and compute the limiting values as $R_s \to \infty$. However, we leave that as an exercise. (Problem 4.3.2).

STOP

Second, we look at the voltage source.

STEP Zero: Here, we put an R_s (the parameter) in series with the voltage source and transform the circuit into one with only current sources, Fig. 4.2-25.

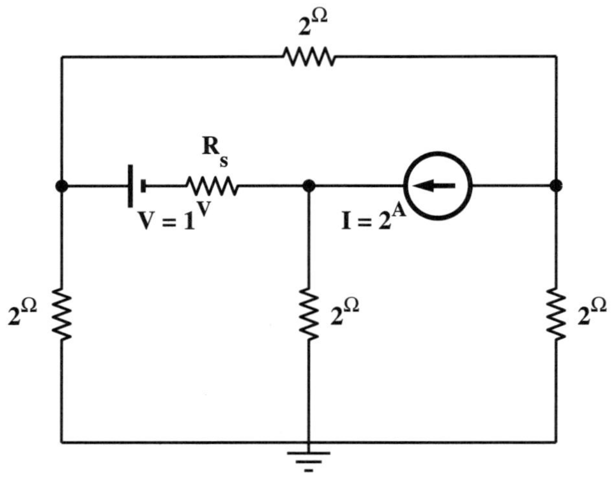

Fig. 4.2-25a.

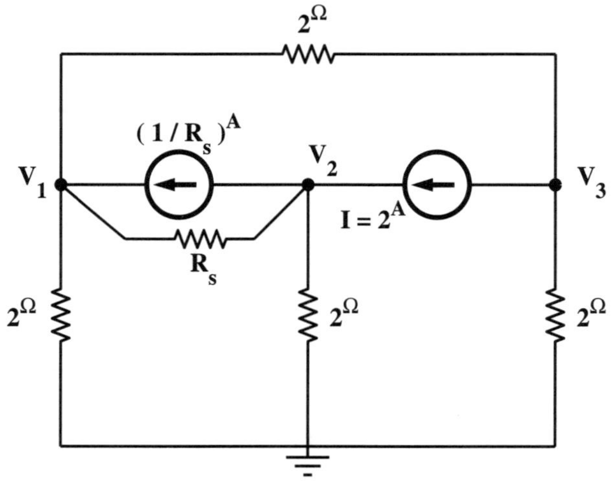

Fig. 4.2-25b.

STEP One: The equations for the above circuit can be generated by inspection as follows.

$$\left(\frac{1}{R_s}\right)^A = \left(\frac{1}{R_s} + ½ + ½\right)V_1 - \left(\frac{1}{R_s}\right)V_2 - (½)V_3, \tag{4.2-53}$$

$$2^A - \left(\frac{1}{R_s}\right)^A = -\left(\frac{1}{R_s}\right)V_1 + \left(\frac{1}{R_s} + ½\right)V_2, \tag{4.2-54}$$

$$-2^A = -(½)V_1 + (½ + ½)V_3. \tag{4.2-55}$$

STEP Final: Again, we may attempt to solve these equations symbolically and let $R_s \to 0$ as the limiting value, in order to complete the numerical solution of these equations. We leave that as an exercise. (Problem 4.3.3).

STOP

Method #Six – Solution by the Reconfiguration

First the current source. Here we may reconfigure the circuit of Fig. 4.2-19 into the circuit of Fig. 4.2-26, which is transformed into the circuit of Fig. 4.2-27.

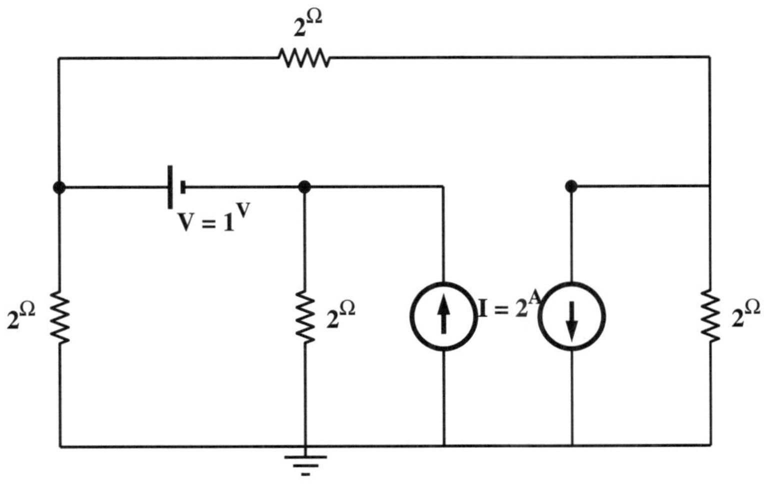

Fig. 4.2-26.

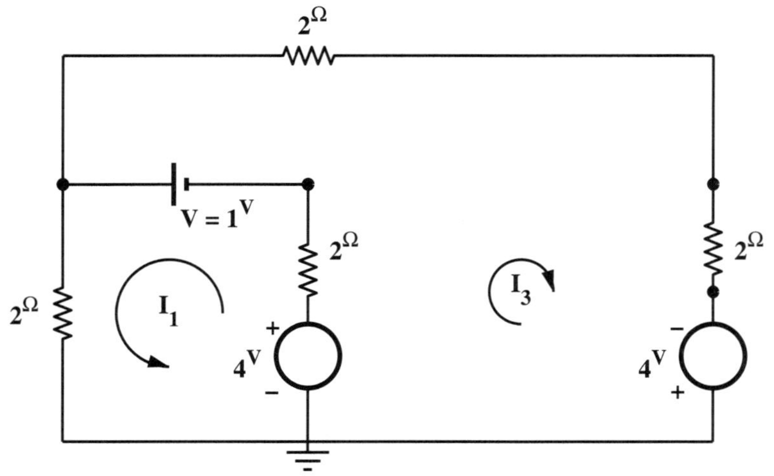

Fig. 4.2-27.

Then the circuit in Fig. 4.2-27 can be analyzed using the following mesh equations.

$$1^V + 4^V = 4I_1 + 2I_3, \tag{4.2-56}$$

$$1^V - 4^V + 4^V = 2I_1 + 6I_3. \tag{4.2-57}$$

The numerical solution of the above set of equations completes the analysis and we leave as an exercise for the reader.

Second, we consider the voltage source. Here, the circuit is not so clear compared with its counterpart in the preceding example. First, we propose to redraw the circuit of Fig. 4.2-19 as that shown in Fig. 4.2-28.

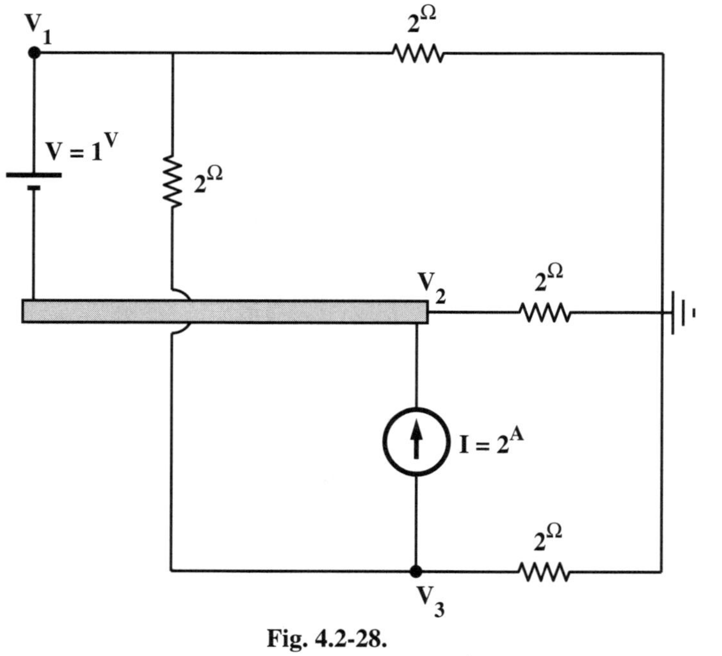

Fig. 4.2-28.

By choosing V_2 as the new ground, we reconfigure the above circuit to that of Fig. 4.2-29, where the old ground is now shown by V_4.

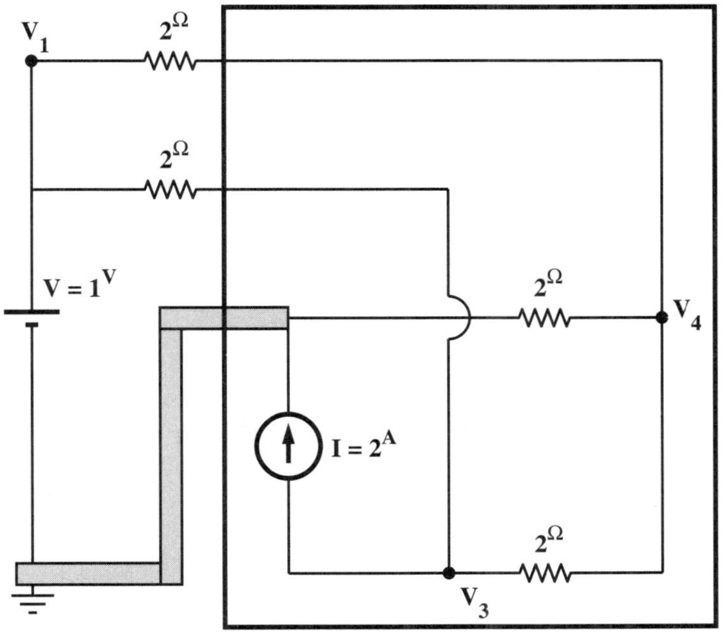

Fig. 4.2-29.

The circuit in Fig. 4.2-29 can be transformed into a circuit with only current sources, as we have described in Sections 3.4 & 3.11. Using a node analysis we can compute V_1, V_3 and V_4 with respect to the new ground. Then we must recalculate these values in term of the old ground by shifting upward or downward all voltage values with respect to the value of V_4. Certainly, there are extra work involved here, but we need to learn the technique just in case we want to use them in a problem that other methods fail. Therefore, our work has clearly been expanded to cover these cases and we leave the details as an exercise. (Problem 4.3.3).

Method #Seven – Solution by the Superposition Rule

The solution can be generated by computing the **currents** in each and every element of the following two circuits, followed by the algebraic summation of these current values in the elements of these two circuits, Figs. 4.2-30 & 31.

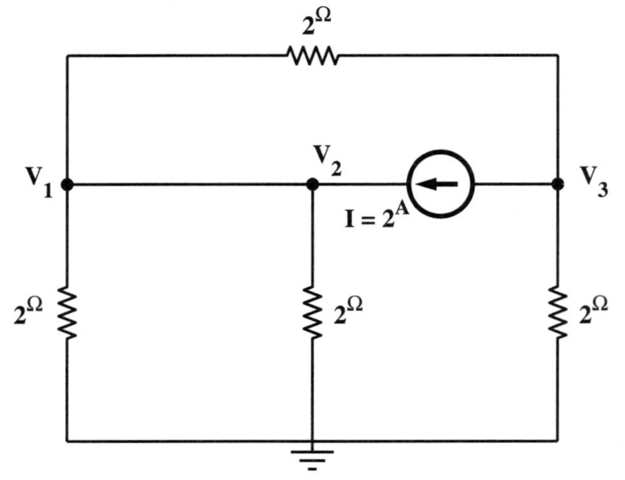

Fig. 4.2-30.

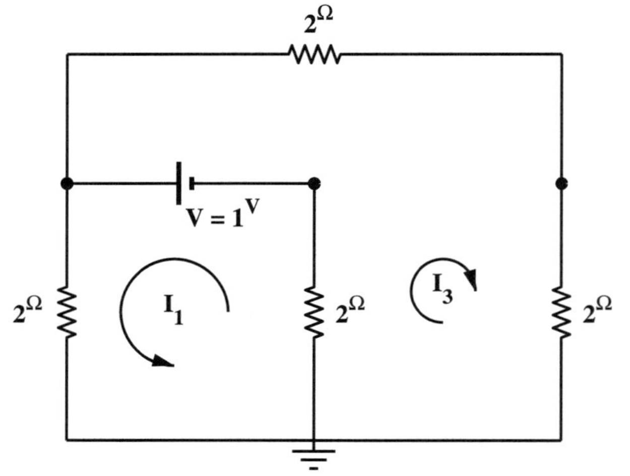

Fig. 4.2-31.

Either of the above two circuits can be easily solved and all corresponding **currents** through each and every element can be calculated. Then we must add algebraically all these current values. For circuit in Fig. 4.2-30 we have

$$V_1 \equiv V_2, \tag{4.2-58}$$

$$2^A = (½ + ½ + ½)V_2 - (½)V_3, \tag{4.2-59}$$

$$-2^A = -(½)V_2 + (½ + ½)V_3. \tag{4.2-60}$$

Resulting in $V_2 = 0.8^V$, $V_3 = -1.6^V$ and $V_1 = 0.8^V$, which in turn resulting in current values for all elements.

For circuit of Fig. 4.2-31, we use a mesh analysis approach and the results are:

$$1^V = 4I_1 + 2I_3, \tag{4.2-61}$$

$$1^{V} = 2I_1 + 6I_3. \tag{4.2-62}$$

Resulting in $I_3 = 0.1^{A}$ and $I_1 = 0.2^{A}$.

Finally, currents through each and every element of the original circuit are calculated as the algebraic sum of the currents in each and every elements of circuits in Figs. 4.2-30 & 31. The details are left as an exercise. (Problem 4.3.4).

STOP

4.3. PROBLEM SET

4.3.1. Complete the analysis for the reconfiguration circuits in Figs. 4.2-7 to 10.

4.3.2. Complete the analysis for the parameterization circuits in Figs. 4.2-24 & 25.

4.3.3. Complete the analysis for the reconfiguration circuit in Fig. 4.2-29.

4.3.4. Complete the analysis for the superposition circuits in Figs. 4.2-30 & 31.

4.3.5. Going back to the problem set in Section 3.13, analyze those circuits with mixed sources in as many different ways as you can. In each case, detail steps of your analysis similar to the steps in the preceding two examples. Clearly, this is a major task, but you have already established the necessary set of equations in each of these circuits for at least one method.

4.3.6. Going back to Problem 2.6.8, is this problem still an outstanding issue? If so, analyze the circuit in Fig. 2.6-17 in as many different ways as you can following the steps shown in this chapter.

4.3.7. For circuits shown in Figs 4.3-1 to 3, compute the desired output voltages as indicated in each circuit.

Comment 4.3-1: We are leaving you with the following thought. In most cases of interest, just as the three circuits in Figs. 4.3-1 to 3 confirm, we are not interested nor do we need to analyze the entire circuit in the sense of Definition 2.1-1. In other words, before embarking a comprehensive circuit analysis according to the Definition 2.1-1, we must ask whether we can simplify the underlying circuit any further as needed in the specific case under consideration. For instance, if we are asked to compute only one specific quantity, say V_o in circuits of Figs. 4.3-1 to 3, then we do not need to develop the entire set of current through and voltage across each and every element of these circuits as required in Definition 2.1-1. To address these issues we study some new techniques in Chapter Five.

Fig. 4.3-1.

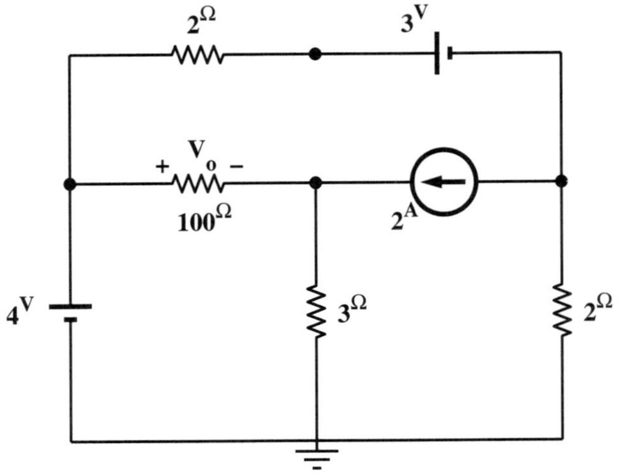

Fig. 4.3-2.

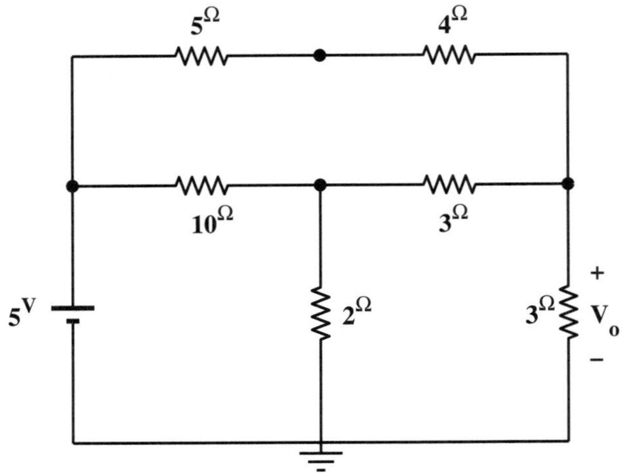

Fig. 4.3-3.

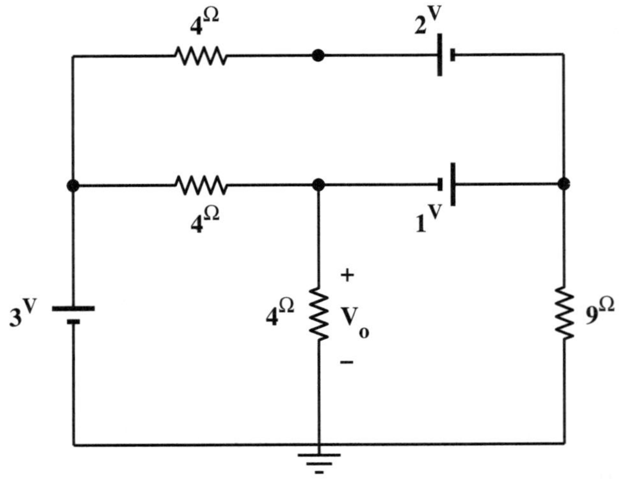

5 ESSENTIAL FORMULAE AND THEOREMS

5.1. INTRODUCTION

Although we have primarily concerned ourselves with networks involving only real elements, the methodologies which have been presented thus far are general. These methodologies enable us to finish the analysis of any complicated circuit, by establishing a sufficient number of equations to analyze a network *completely.* There are, however, many cases in practice for which we do not need the *complete* analysis of a network. For instance, when concerned with only one output, possibly out of many variables in a complicated network, we must seek methods which simplify this complicated network to its "equivalent" version, keeping that particular variable intact, while substantially simplifying the rest of the circuit analysis. By putting the word equivalent in quotation marks, we remind ourselves that this concept must be defined rigorously.

The reduction approach which we have mentioned above uses inspection to simplify a given network into a more manageable form by simplifying the non-essential parts of the initial problem substantially, rendering it almost to a network no more complicated than what we have seen thus far. As a point of view, consider the next example.

Example 5.1-1: In the circuit of Fig. 5.1-1 compute the output voltage V_o.

Solution: Here, we note that there are 14 nodes, plus the ground, thus there are potentially 14 equations to be solved completely to finish analyzing this problem. To solve this set of equations numerically is involved and we must avoid this seemingly straightforward approach if it is possible at all.

What we have emphasized thus far is to *study* the configuration of a given circuit first, before embarking any approach to analyze it completely. For instance, the circuit in Fig. 5.1-1 looks giant, but let us see how we can solve it systematically. Certainly, we can solve the circuit in Fig. 5.1-1 using one of the several methods discussed in the preceding chapters, but that is not the whole story. What we really want to do is to solve this, or for that matter any circuit problem, with the least computations. For instance, in this case, if we choose Method #Seven, then the two corresponding sub-problems are shown in Figs. 5.1-2 & 3. These circuits can be simplified substantially by some sort of source transformations followed by combining series and parallel elements, as we show subsequently. Therefore, what we are studying in this chapter is to use new procedures to simplify a circuit such that all non-essential variables, as far as a particular application is concerned, are set aside and we get directly to the desired variable or solution with the least computations. We continue this solution in § 5.4.3.

Thus, in the following we develop new rules or formulae to calculate a host of important equivalent circuits which complement the preceding methodologies in this textbook. We develop formulae to calculate parallel and series connections of elements, albeit real elements. We expand our horizon by introducing a new concept of dissipated power in a network subsequently. In a due course, we generalize these ideas to cover general circuits with different complex elements.

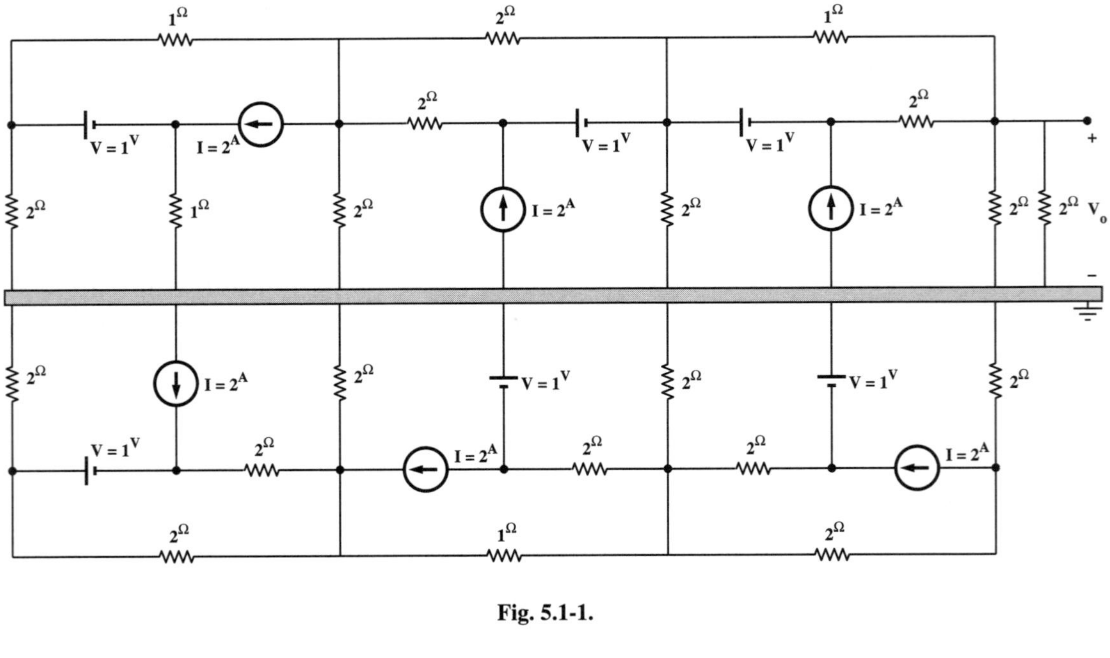

Fig. 5.1-1.

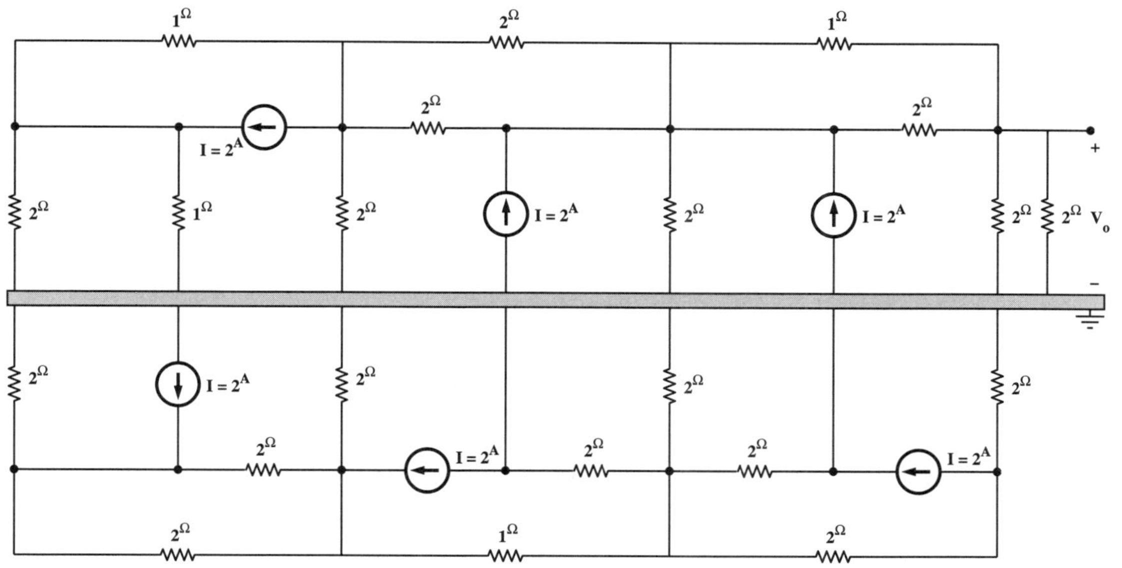

Fig. 5.1-2. Voltage sources are short.

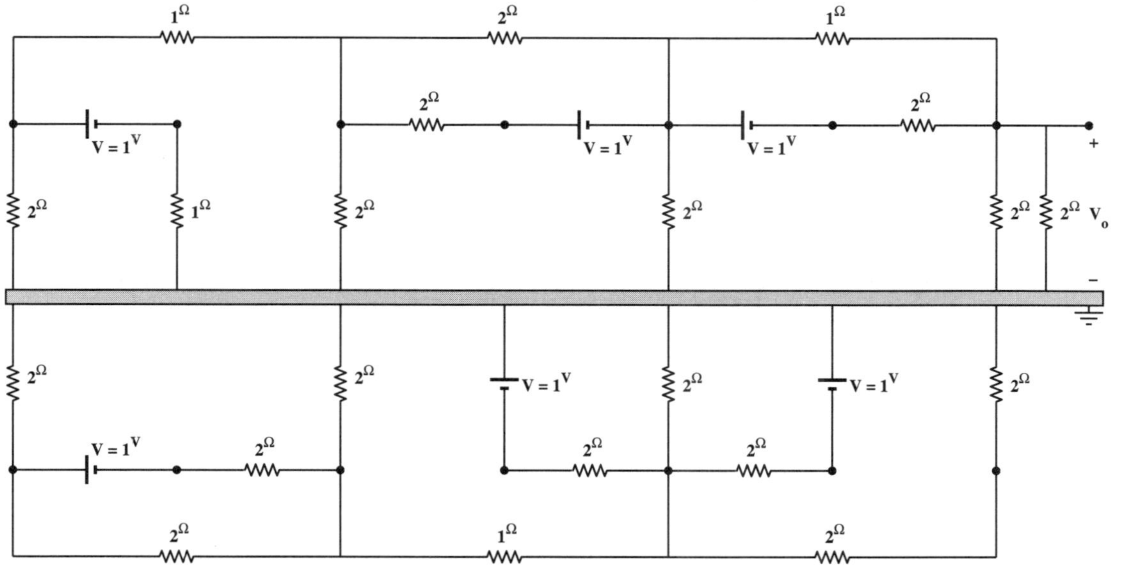

Fig. 5.1-3. Current sources are opened.

5.2. PARALLEL and SERIES CONNECTIONS

Starting with parallel connections, consider the circuit in Fig. 5.2-1. Here, R_1 and R_2 are in parallel and this is shown as $R_1 \| R_2$. Parallel means the elements are connected end to end and tail to tail. For someone who is *looking into* the port ab, what matters is that this circuit can be replaced by a so-called equivalent circuit consisting of only one resistor called R_{eq}. The question is how do we determine this R_{eq}?

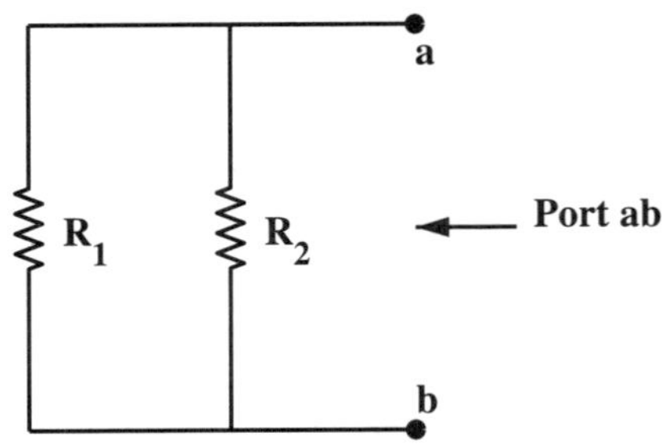

Fig. 5.2-1.

But first, by *looking into* a port we mean that if we replace the network on the left-hand side of port ab, *per se,* by an equivalent one, then its electrical effect in the circuit on the right-hand side of port ab will remain the same as shown in Fig. 5.2-2. This idea is the same as saying that the voltage across port ab and the current flowing between the two sub-networks on the right- and the left-hand sides of port ab remain unchanged. Thus, the new and the old sub-networks are equivalent in the sense of port ab, i.e., they both have the identical electrical effect on the remaining circuit on the right-hand side.

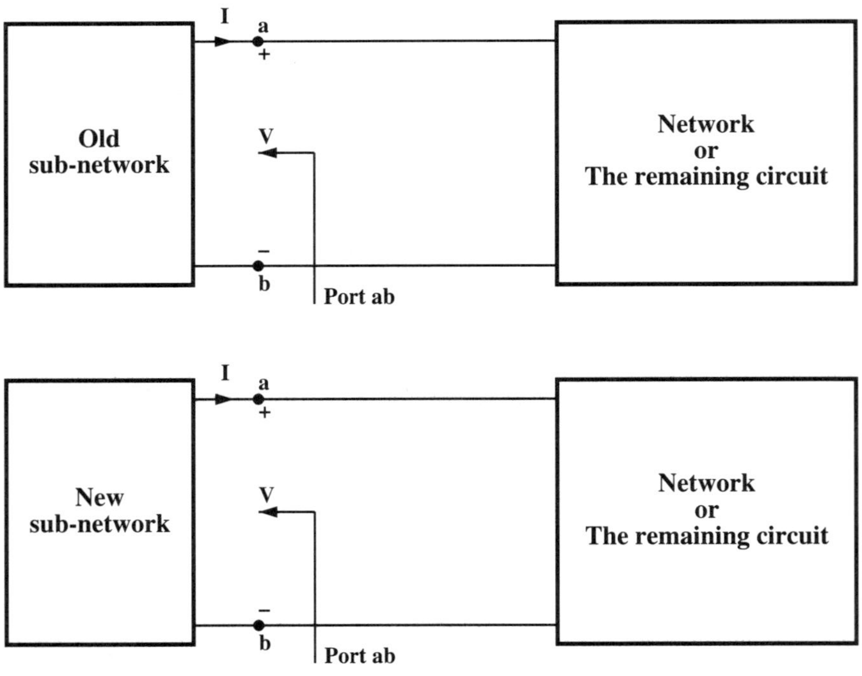

Fig. 5.2-2.

Now, going back to $R_1 \| R_2$, we note that if the current I is injected into the network, then that current is distributed as shown in Fig. 5.2-3.

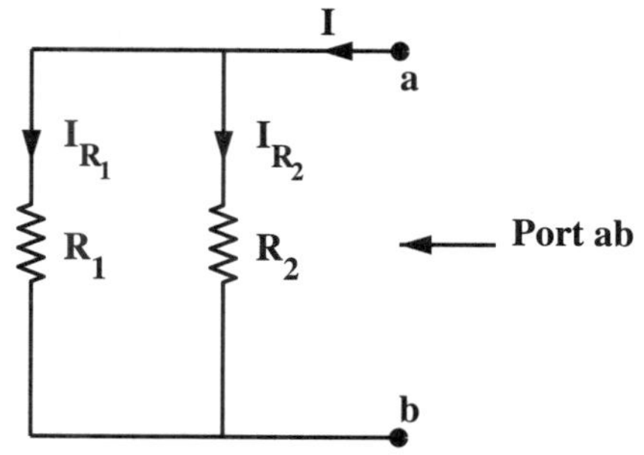

Fig. 5.2-3.

Here,

$$V_{ab} = R_1 I_{R_1} = R_2 I_{R_2}, \qquad (5.2\text{-}1)$$

$$I = I_{R_1} + I_{R_2}. \qquad (5.2\text{-}2)$$

On the other hand, if we set this circuit equivalent to that in Fig. 5.2-4, where the same current I is injected, then for some yet unknown equivalent resistor R_{eq} we have $V_{ab} = R_{eq} I$.

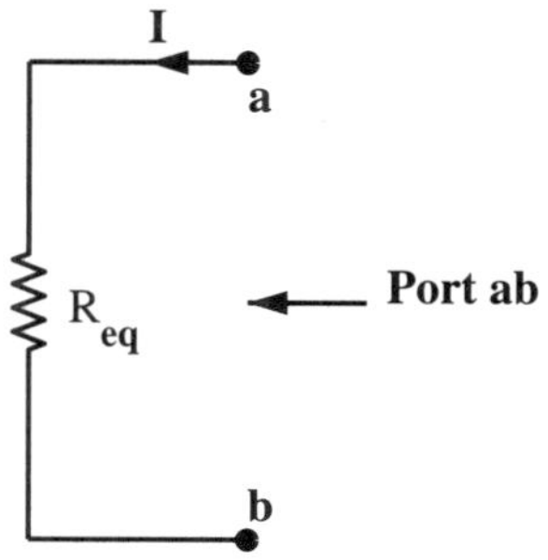

Fig. 5.2-4.

Substituting for I from (5.2-2) in V_{ab}, with each current given by (5.2-1), results in

$$V_{ab} = R_{eq}\left(\frac{V_{ab}}{R_1} + \frac{V_{ab}}{R_2}\right) \Rightarrow R_{eq} = \frac{R_1 R_2}{R_1 + R_2}. \qquad (5.2\text{-}3)$$

In terms of the conductance $G = (1/R)$, we have $G_{eq} = G_1 + G_2$, which is more appealing for parallel resistors than the previous expression.

In the case of having many resistors in parallel we have the circuit in Fig. 5.2-5.

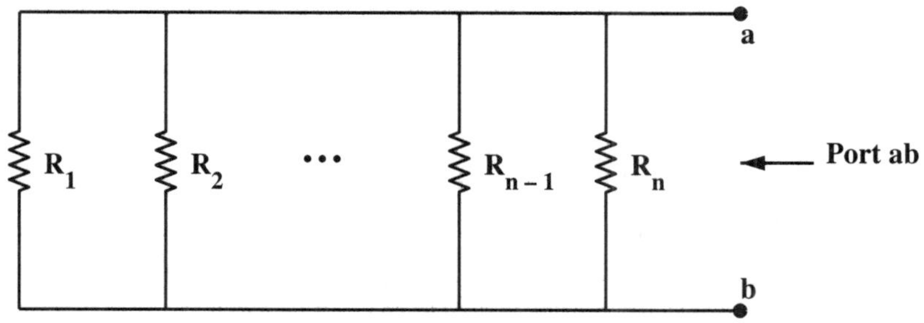

Fig. 5.2-5.

Here, we group together the first two resistors which are in parallel, i.e., $R_1 \| R_2$, then we group this resultant with the next parallel one. This is the same as saying $(R_1 \| R_2) \| R_3, \ldots,$ until we have exhausted the circuit. This expression in terms of conductances can be written as

$$G_{eq}(\text{parallel connections}) = \sum_{i=1}^{n} G_i. \tag{5.2-4}$$

Series connection is also widely used and the equivalent circuit for that case can be found similarly. For instance, consider the circuit in Fig. 5.2-6, where $R_1, \ldots, R_n$ are connected in series, i.e., the connections are end to tail and vice versa for two consecutive elements. Here, we note that $V_{ab} = V_{R_1} + \cdots + V_{R_n}$, with each $V_{R_i} = R_i I$. Thus, $R_{eq} I = (R_1 + \cdots + R_n) I$, or

$$R_{eq}(\text{series connections}) = \sum_{i=1}^{n} R_i. \tag{5.2-5}$$

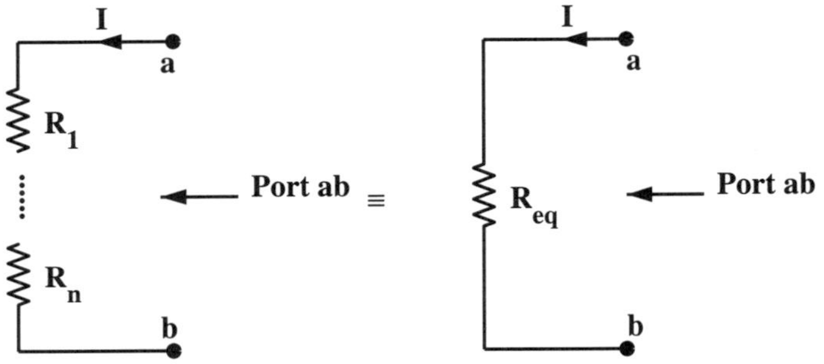

Fig. 5.2-6.

5.3. EQUIVALENT NETWORK THEOREMS

In the context of the preceding equivalent networks, we now present the following important theorems. But first, let us entertain the following example that the author had used repeatedly and found helpful to elucidate the underlying concept. Consider a classroom with several doors. Depending on which door someone enters and looks inside the room, he/she sees a different configuration or arrangement among the attendants, despite the fact that the classroom configuration remains the same. So it is very important to understand that equivalency concept is defined for a selected port.

Now, consider a network with many (real) elements and mixed sources connected as shown in Fig. 5.3-1. We pull out two nodes, say a and b, to ask the following question. Is there a network consisting of only one resistor and one source, which is equivalent to the rest of this network at port ab? In other words, the equivalency consists of only one sink and one source and is defined in terms of the given port.

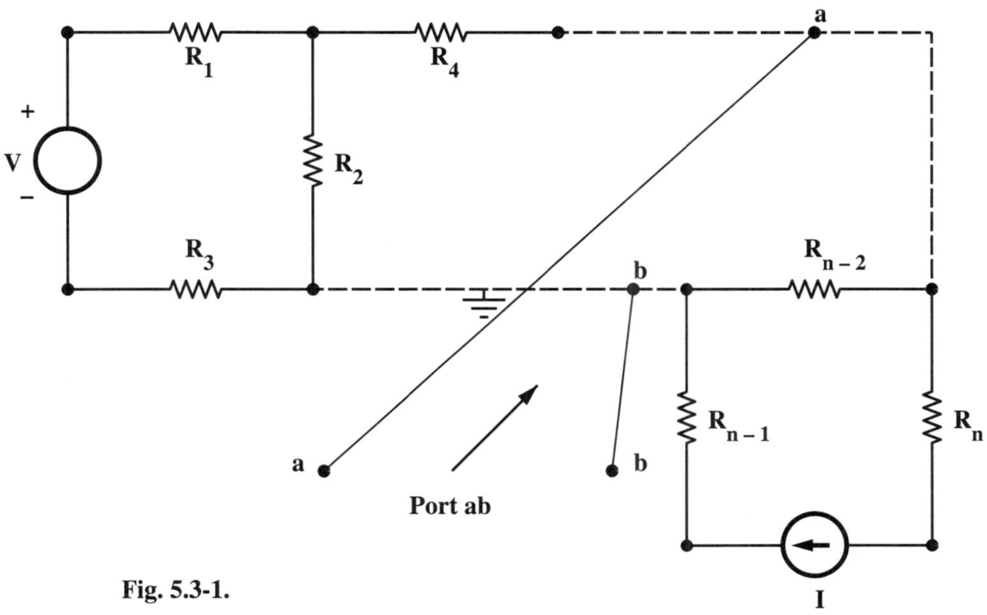

Fig. 5.3-1.

Thus, the question is: Can we replace the entire circuit of Fig. 5.3-1 at port ab, and for the purpose of analysis, with a simpler circuit consisting of one sink (here a resistor) and one source? The answer is yes and the following theorems are the procedures to construct that equivalent circuit.

Theorem 5.3-1 (Thevenin's Equivalent Circuit): Given a circuit consisting of many elements (sinks and sources) (presently, these are real quantities, but in general these elements may be both real and complex quantities), and at any specific port, an equivalent circuit can be found, which contains only one sink and one voltage source.

For instance, at port ab the circuit in Fig. 5.3-1 is equivalent to that in Fig. 5.3-2.

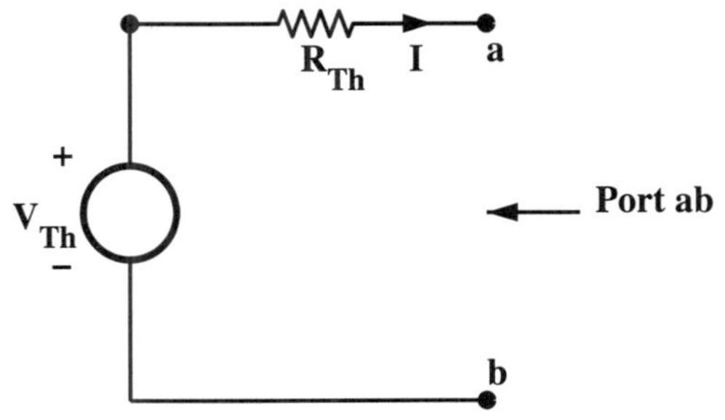

Fig. 5.3-2.

To prove this theorem we offer a constructional procedure. We propose to utilize repeatedly techniques in source transformations (Section 2.5), in order to simplify the initial circuit into the above equivalent Thevenin's circuit. In the final analysis when such a circuit is found, that circuit has the following properties. Namely, if we connect or short circuit port ab, then the open-circuit voltage V_{Th} and the short-circuit current I_{sc} are related through

$$I_{sc} = \frac{V_{Th}}{R_{Th}} \quad \Rightarrow \quad \frac{V_{Th}}{I_{sc}} = R_{Th}. \tag{5.3-1}$$

We call $\{V_{Th}, I_{sc}, R_{Th}\}$ the set of Thevenin's parameters.

This physical property provides us with a measuring technique to construct the Thevenin's equivalent circuit at any given port. There is also another method based on the concept of dissipated power which enables us to find experimentally the Thevenin's equivalent network as is described in Section 5.5.

For now, we entertain the following circuit and seek its Thevenin's equivalent circuit through a constructional procedure as is detailed below.

Example 5.3-1: Find the Thevenin's equivalent circuit at port ab, for the circuit in Fig. 5.3-3.

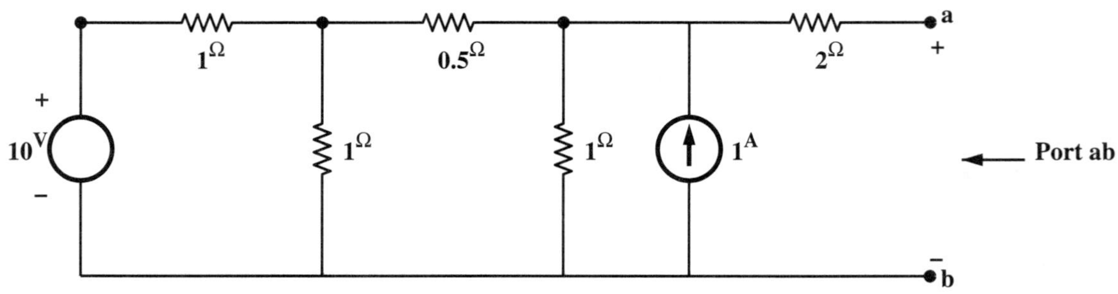

Fig. 5.3-3.

Solution: We offer two separate procedures to answer this question.

Procedure I: We can use the definition of the Thevenin's parameters to construct and solve two circuits shown in Figs. 5.3-4 & 5, in order to compute the Thevenin's parameters, which is rather an involved process.

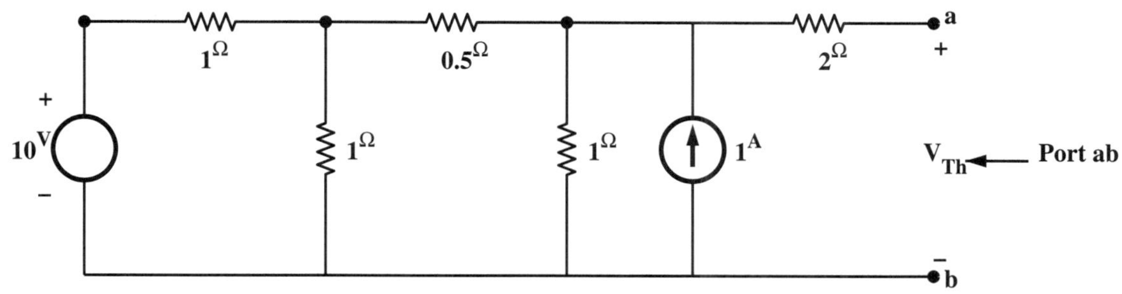

Fig. 5.3-4. This circuit is used to compute (measure) V_{Th}.

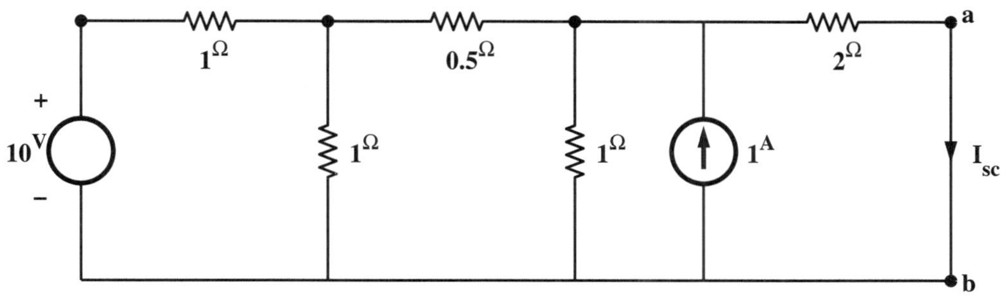

Fig. 5.3-5. This circuit is used to compute (measure) I_{sc}.

Although, we can solve these two circuits in Figs. 5.3-4 & 5 straightforwardly to compute Thevenin's parameters (Problem 5.8.8), we use the next procedure which is perhaps easier than the analysis for these two circuits, and can be carried out almost by inspection. Namely,

Procedure II: We start using repeatedly source transformation to simplify the circuit in Fig. 5.3-3 as follows. We always start from the left-hand side of the circuit and work our way through the right-hand side. For instance, the circuit in Fig. 5.3-3 is equivalent to that in Fig. 5.3-6.

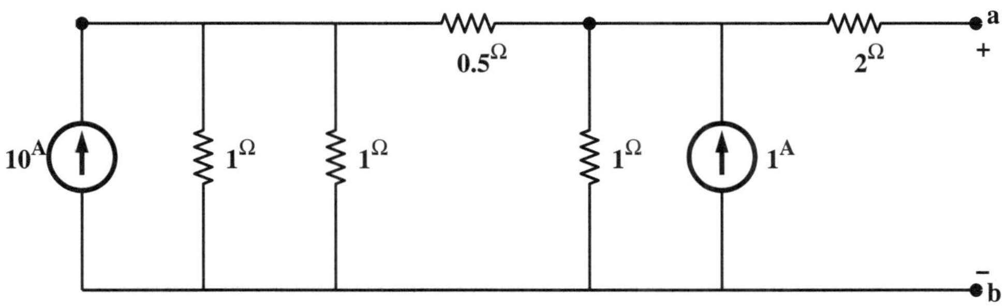

Fig. 5.3-6.

Here, the 10^V voltage source with its 1^Ω series resistor is transformed into a 10^A current source with the same resistor in parallel. Now, the source resistor and the other 1^Ω circuit resistor are in parallel, which we may combine to get an equivalent resistor that is $[1 \times 1/(1 + 1) = ½]^\Omega$. Then we may regroup this equivalent resistor with the previous current source into a 5^V voltage source with its $(½)^\Omega$ series resistor as shown in Fig. 5.3-7. This circuit is then transformed back into the one shown in Fig. 5.3-8.

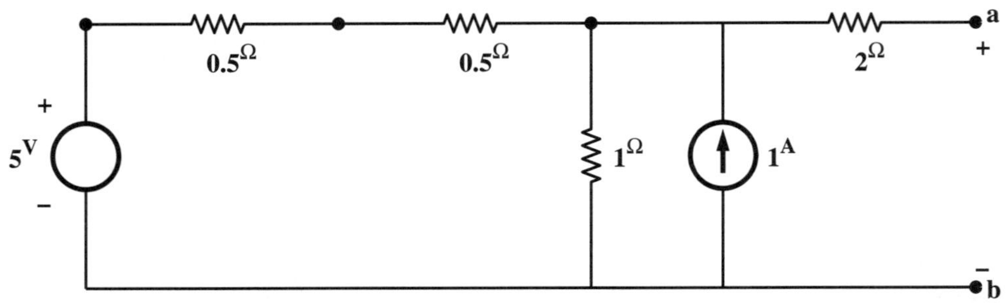

Fig. 5.3-7.

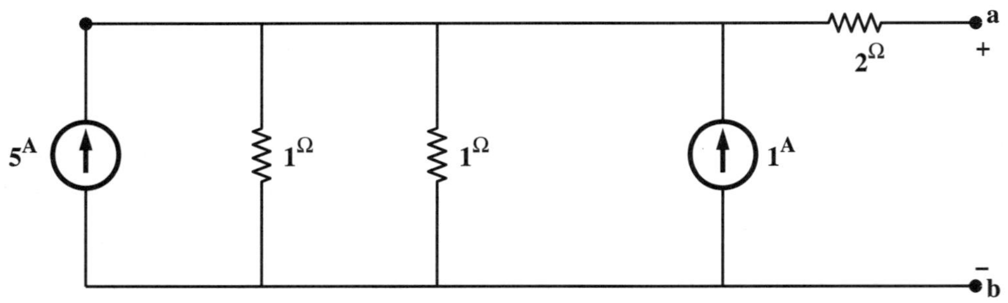

Fig. 5.3-8.

At this stage we note that we have two current sources and two 1^Ω resistors, which are all in parallel, resulting in;

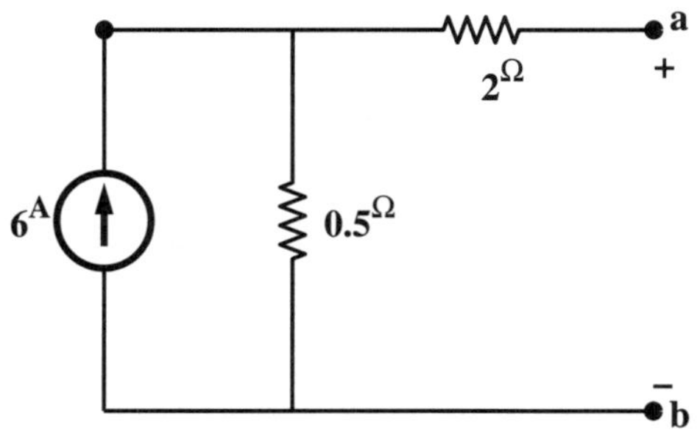

Fig. 5.3-9.

The circuit in Fig. 5.3-9 is identical to the final result of Fig. 5.3-10.

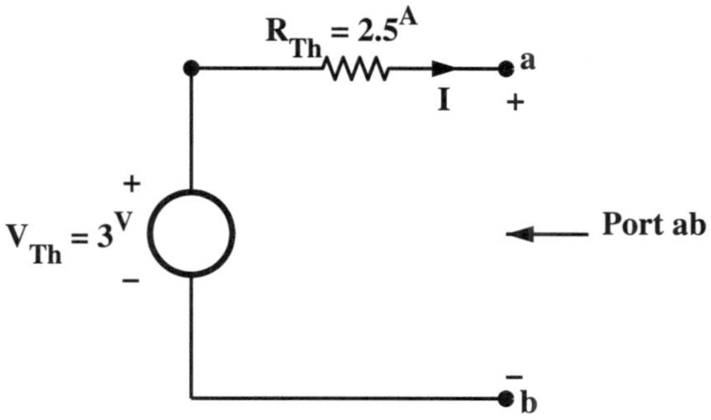

Fig. 5.3-10.

A reason that we often prefer Procedure II is that with experience most of the above computations can be done by inspection. However, Procedure I always works and is the method which we should be familiar with as well. We invite the reader to try again the above example using Procedure I. (Problem 5.8.8).

We now introduce the counterpart of the previous theorem using a current source.

Theorem 5.3-2 (Norton's Equivalent Circuit): Given a circuit consisting of many elements (sinks and sources) (presently, these are real quantities, but in general these elements may be both real and complex quantities), and given a port, an equivalent circuit can be found consisting of only one sink and one current source.

This result is shown in the circuit of Fig. 5.3-11.

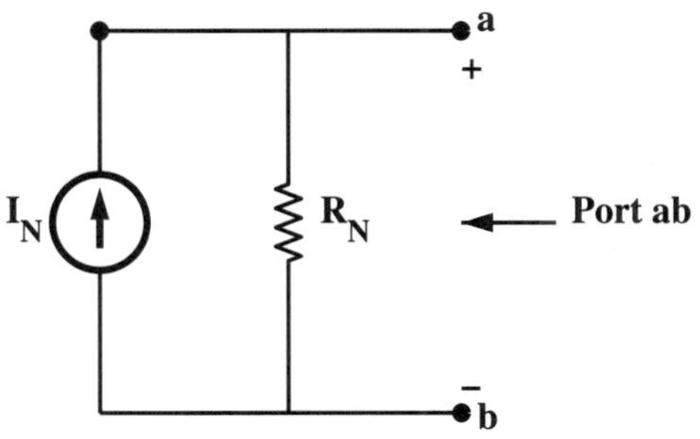

Fig. 5.3-11.

Because we have already established the source transformation procedure, the proof uses the fact that the following two circuits are equivalent.

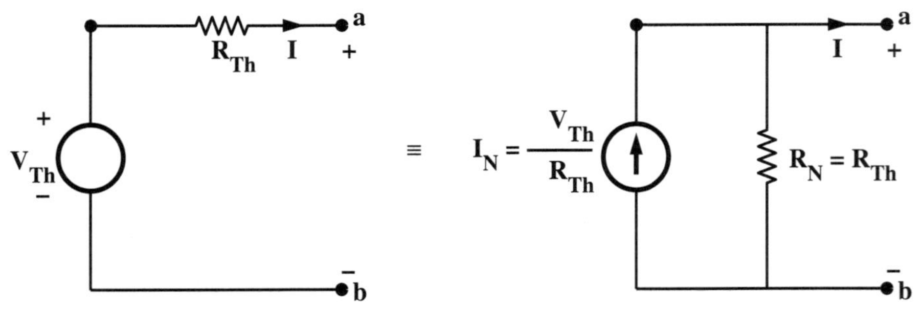

Fig. 5.3-12.

Thus, the previous $R_N \equiv R_{Th}$ and $I_N \equiv V_{Th}/R_{Th}$. Generally, these two circuits are used interchangeably.

The Norton's equivalent circuit in Example 5.3-1 is developed as follows.

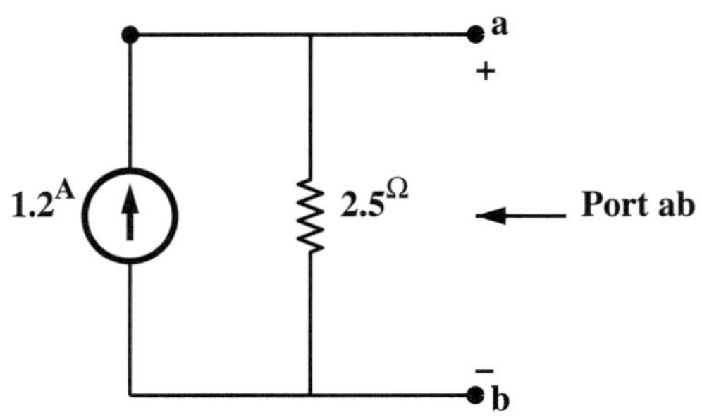

Fig. 5.3-13. Referring to Example 5.3-1.

In closing, we remind the reader that there is another procedure for finding the above equivalent circuit which is based on the concept of power dissipated and is described in Section 5.5.

5.4. USEFUL FORMULAE

There are many other circuits of interest, which we cannot review them all, but we study a few here. For instance, consider the circuit in Fig. 5.4-1, which is called a voltage divider. Clearly, V_2 (the voltage across R_2) is only a portion of V_1 (the voltage across $R_1 + R_2$). So the name comes from the fact that the original source is divided into two portions and we are taking one. Some textbooks call this circuit attenuator. If we write the mesh equation, we note that $V_1 = (R_1 + R_2)I$ and $V_2 = R_2I$. Thus the ratio of V_2/V_1 is

$$\frac{V_2}{V_1} = \frac{R_2}{R_1 + R_2} < 1. \tag{5.4-1}$$

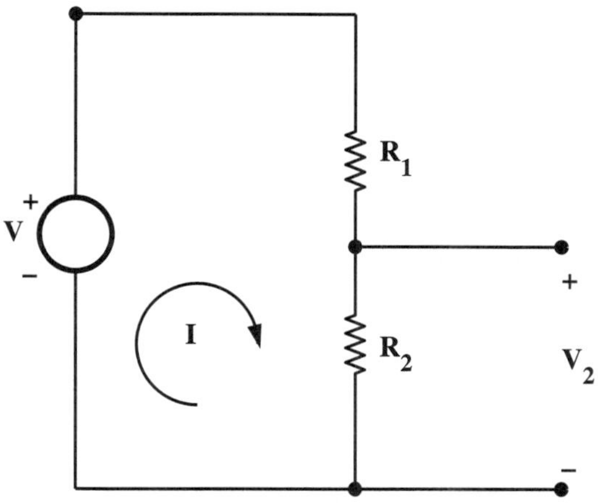

Fig. 5.4-1. A voltage divider.

Similarly, the following circuit is called a current divider.

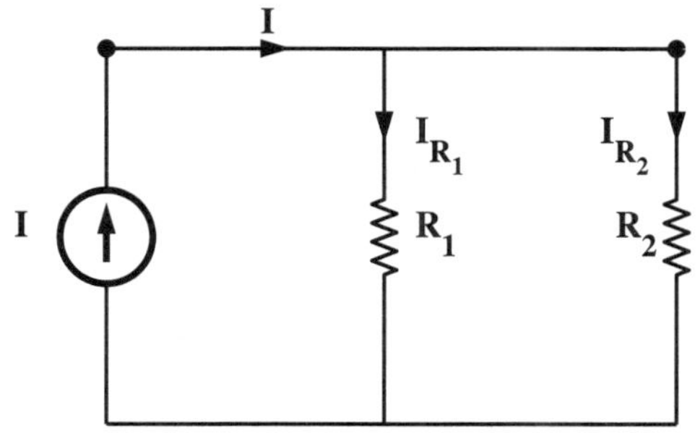

Fig. 5.4-2. A current divider.

Here, $I = I_{R_1} + I_{R_2}$ and $R_1 I_{R_1} = R_2 I_{R_2}$. Thus,

$$\frac{I_{R_2}}{I} = \frac{R_1}{R_1 + R_2} < 1. \tag{5.4-2}$$

5.4.1. The Wheatstone Bridge

This historic circuit is used to measure the resistance of an unknown resistor R_x as shown in Fig. 5.4-3. The main idea centers on the fact that when the bridge is in the so-called balanced mode, the galvanometer G (which is sensitive within a micro-ampere) shows zero. In that case, we have $i_g \equiv 0$. We can show that under this balanced condition, we have

$$R_x = \frac{R_2}{R_1} \times R_3, \tag{5.4-3}$$

where R_3 is a variable resistor (the arrow is used for this purpose). By using the values of R_1 and R_2, and adjusting the value of R_3 we can impose the balanced condition and establish (5.4-3).

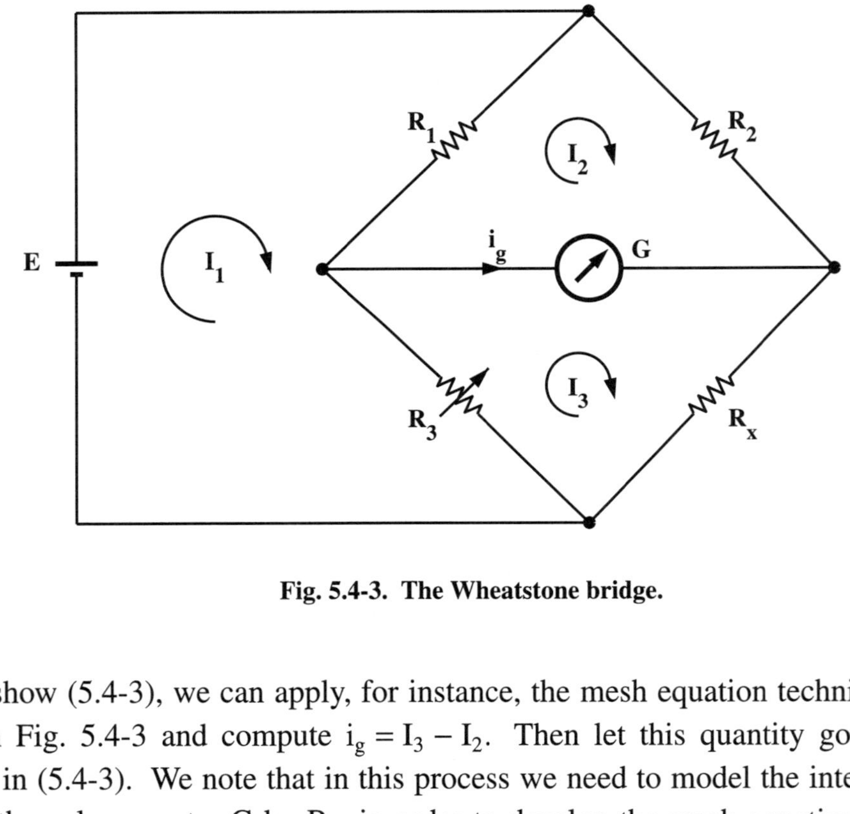

Fig. 5.4-3. The Wheatstone bridge.

To show (5.4-3), we can apply, for instance, the mesh equation technique to the circuit in Fig. 5.4-3 and compute $i_g = I_3 - I_2$. Then let this quantity goes to zero resulting in (5.4-3). We note that in this process we need to model the internal resistance of the galvanometer G by R_g, in order to develop the mesh equations, but this parameter does not play a role in the outcome. To show this claim, we set up the mesh equations for this circuit in the following matrix form (cf., Section 5.7).

$$\begin{Bmatrix} E \\ 0 \\ 0 \end{Bmatrix} = \begin{pmatrix} (R_1 + R_3) & -R_1 & -R_3 \\ -R_1 & (R_1 + R_2 + R_g) & -R_g \\ -R_3 & -R_g & (R_x + R_3 + R_g) \end{pmatrix} \times \begin{Bmatrix} I_1 \\ I_2 \\ I_3 \end{Bmatrix}. \tag{5.4-4}$$

From this equation we have

$$I_3 - I_2 = \frac{E}{\Delta}[R_3R_2 - R_1R_x], \tag{5.4-5}$$

where Δ is the determinant of the coefficient matrix in (5.4-4). As $I_3 - I_2 \to 0$, $R_3R_2 - R_1R_x \to 0$ resulting in (5.4-3), and that completes the proof.

5.4.2. Delta to Star and Star to Delta Transformations

There are also a few circuit configurations as shown in Figs 5.4-4 & 5, which have many applications in circuits, especially in polyphase electrical networks.

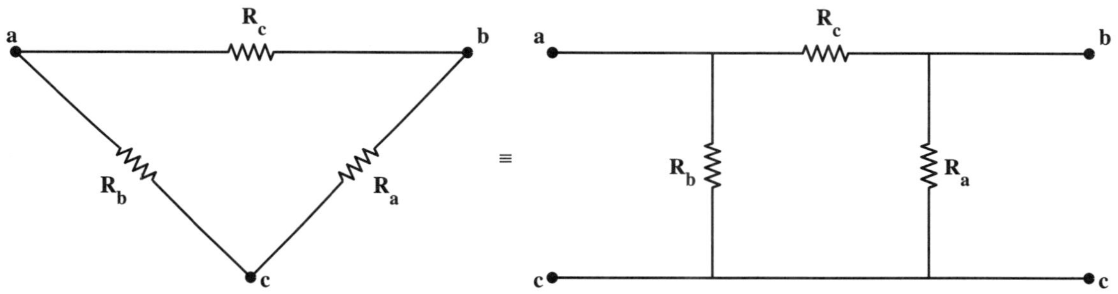

Fig. 5.4-4. Δ configuration.

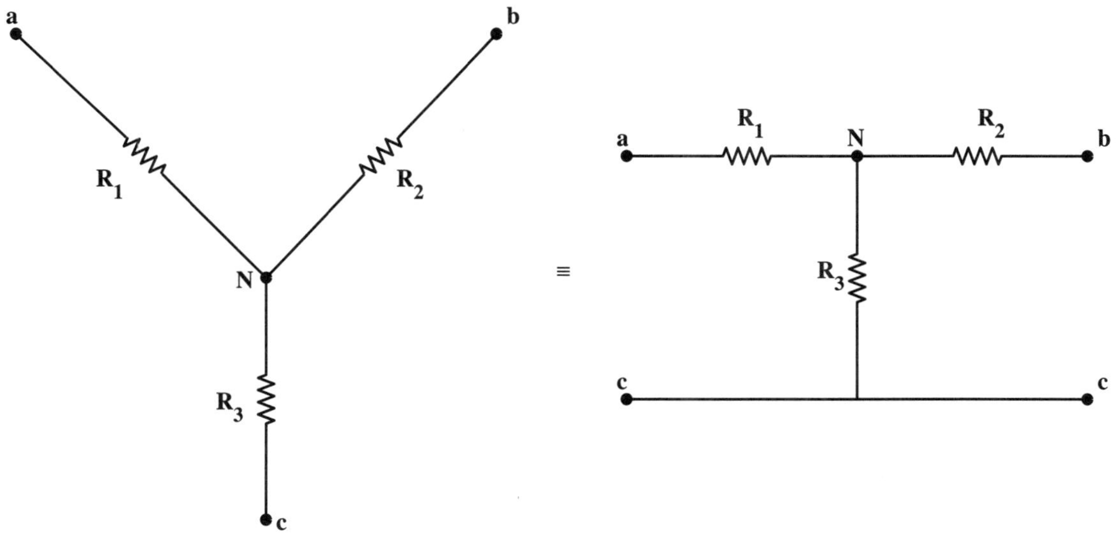

Fig. 5.4-5. Y/Star configuration.

Clearly, an inquisitive mind always searches for ways of relating these two configurations, by asking how do we transform $\Delta \rightarrow Y$ or $Y \rightarrow \Delta$?

To that end, we offer the following formulae, which are developed in Section 12.4.

$$\Delta \rightarrow Y: \qquad R_1 = \frac{R_b R_c}{R_a + R_b + R_c}, \tag{5.4-6a}$$

$$R_2 = \frac{R_c R_a}{R_a + R_b + R_c}, \tag{5.4-6b}$$

$$R_3 = \frac{R_a R_b}{R_a + R_b + R_c}. \tag{5.4-6c}$$

$$Y \rightarrow \Delta: \qquad R_a = \frac{R_1 R_2 + R_2 R_3 + R_3 R_1}{R_1}, \tag{5.4-7a}$$

$$R_b = \frac{R_1 R_2 + R_2 R_3 + R_3 R_1}{R_2}, \tag{5.4-7b}$$

$$R_c = \frac{R_1 R_2 + R_2 R_3 + R_3 R_1}{R_3}. \tag{5.4-7c}$$

5.4.3. Solution of Example 5.1-1 Continued

Going back to Figs. 5.1-2 & 3, we note that these circuits can be analyzed quite easily. Starting with Fig. 5.1-2, where all current sources are kept, we mark those elements which are in parallel for regrouping as shown in Fig. 5.4-6.

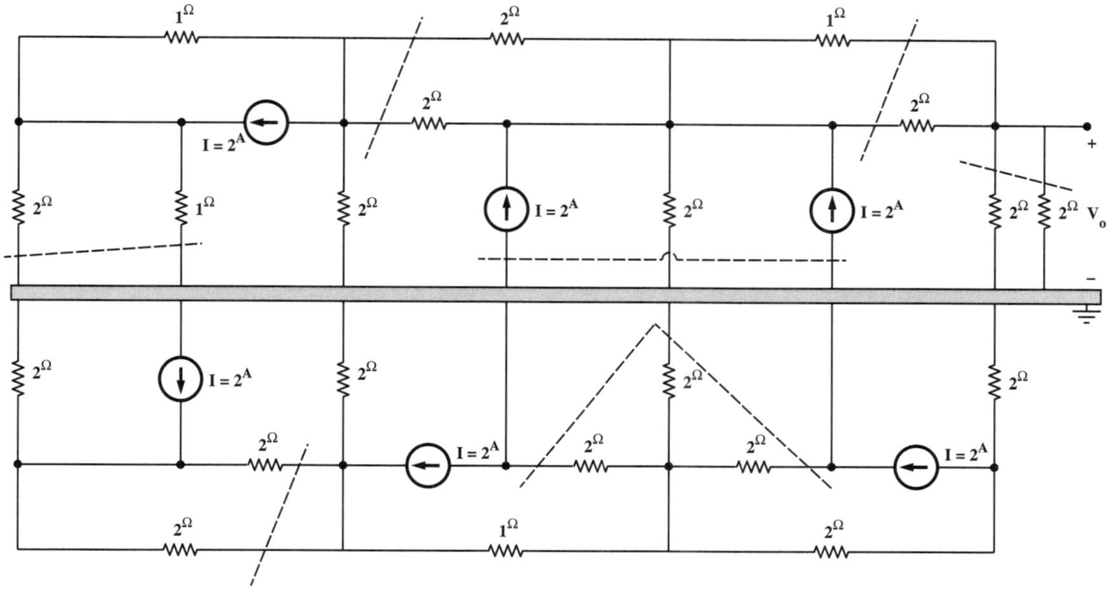

Fig. 5.4-6. Crossing lines indicate parallel elements.

Upon regrouping the marked elements, using (5.2-4) for parallel resistors, the circuit in Fig. 5.4-6 becomes that in Fig. 5.4-7.

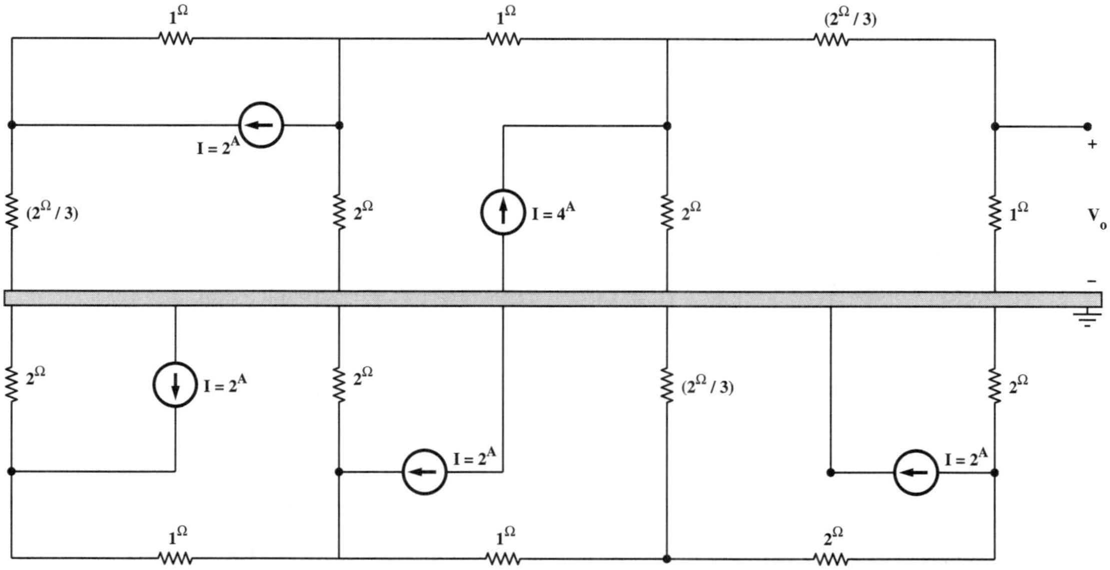

Fig. 5.4-7.

Here, we may be tempted to use a few $Y \rightarrow \Delta$ and $\Delta \rightarrow Y$ transformations, but those are not going to help. Instead, we use the conventional current to voltage source transformations to modify Fig. 5.4-7 to that in Fig. 5.4-8.

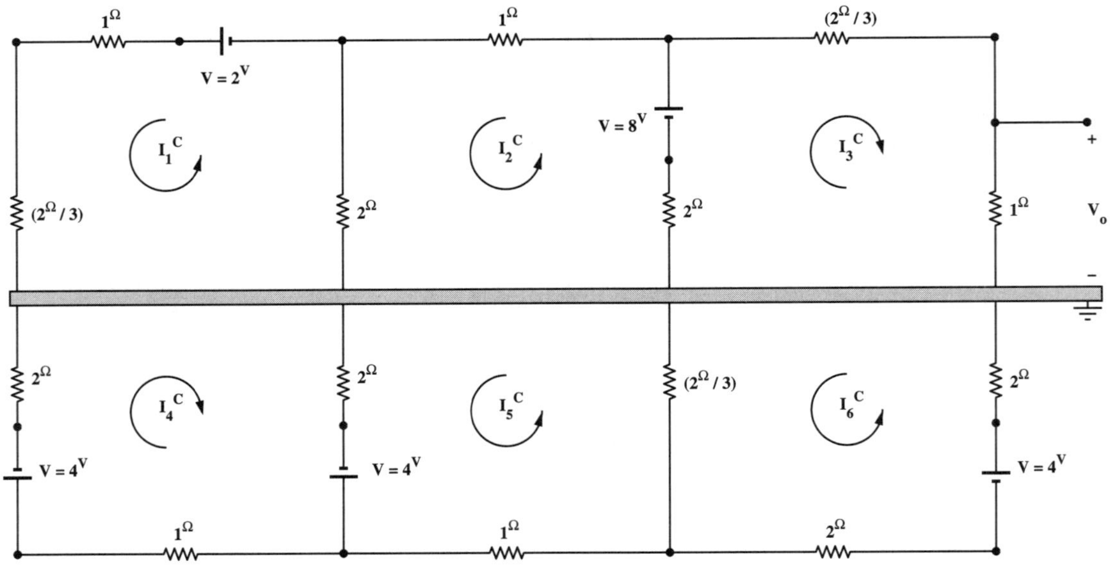

Fig. 5.4-8. The superscript C stands for circuit with initial current sources.

The above circuit can be analyzed using mesh equations and the resulting set of equations is as follows.

$$\begin{Bmatrix} 2 \\ 8 \\ 8 \\ 0 \\ 4 \\ 4 \end{Bmatrix} = \begin{pmatrix} 11/3 & -2 & 0 & 0 & 0 & 0 \\ -2 & 5 & 2 & 0 & 0 & 0 \\ 0 & 2 & 11/3 & 0 & 0 & 0 \\ 0 & 0 & 0 & 5 & 2 & 0 \\ 0 & 0 & 0 & 2 & 11/3 & -2/3 \\ 0 & 0 & 0 & 0 & -2/3 & 14/3 \end{pmatrix} \times \begin{Bmatrix} I_1^C \\ I_2^C \\ I_3^C \\ I_4^C \\ I_5^C \\ I_6^C \end{Bmatrix}. \quad (5.4\text{-}8)$$

From this equation we begin to see an interesting property, which we could not have explained it before. We note that although we are interested in V_o and for that we need to compute I_3^C, this current is completely decoupled from I_4^C, I_5^C and I_6^C. Because the circuit in Example 5.1-1 consists of two circuits connected by a node as shown in Fig. 5.4-9, and there is no dependent source as in circuit of Fig. 8.2-1, to couple these two parts. We may cut the circuit in Fig. 5.4-9 into two pieces at the common node and that operation does not affect the other piece of this circuit at all. At the moment this observation may seem unusual, thus, we continue the analysis of the circuit in Fig. 5.1-3, where all voltage sources are retained, to complete the superposition rule. There we have, similar elements which we group them together as indicated by circles in Fig. 5.4-10.

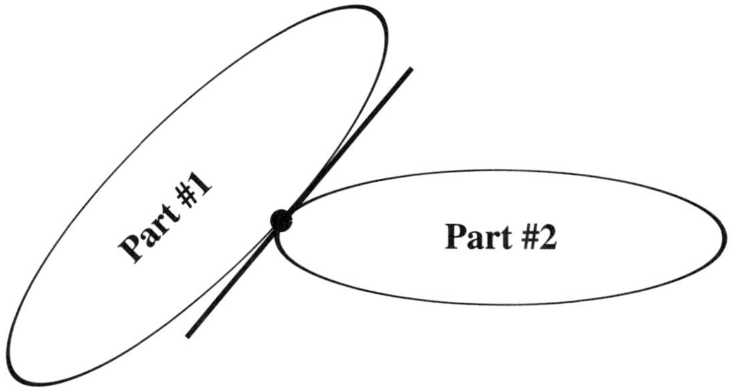

Fig. 5.4-9. A circuit with two separate parts connected at one node.

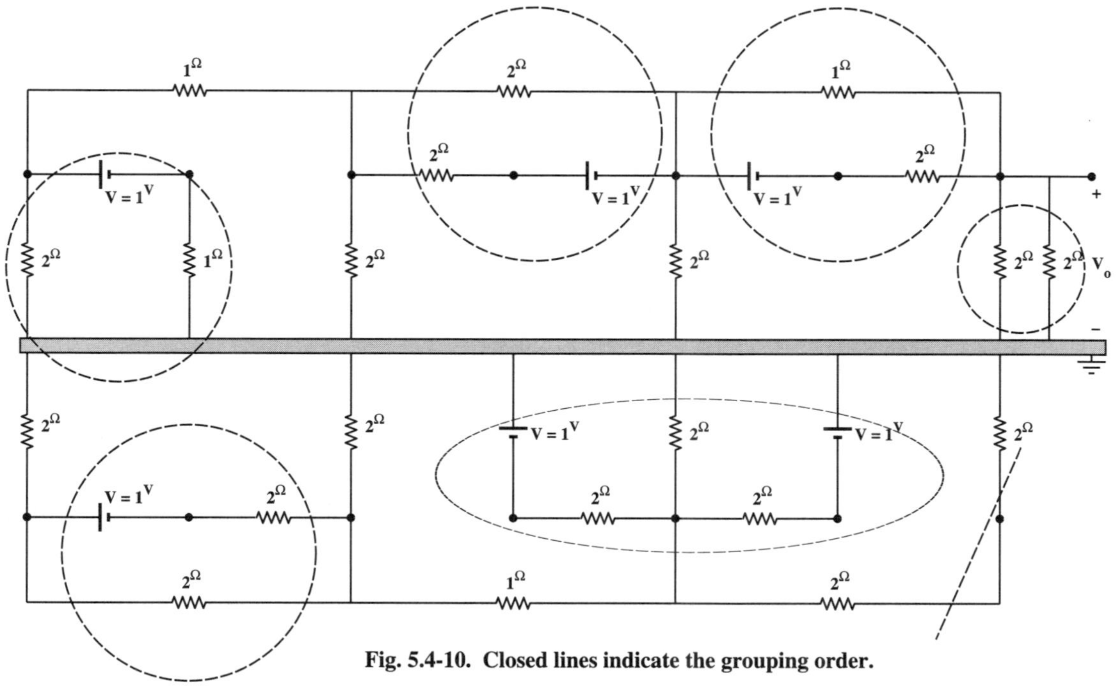

Fig. 5.4-10. Closed lines indicate the grouping order.

Upon a voltage to current then back to voltage transformations of segments for the circuit in Fig. 5.4-10, we modify that circuit into the one shown in Fig. 5.4-11, which is easily amenable with mesh equations.

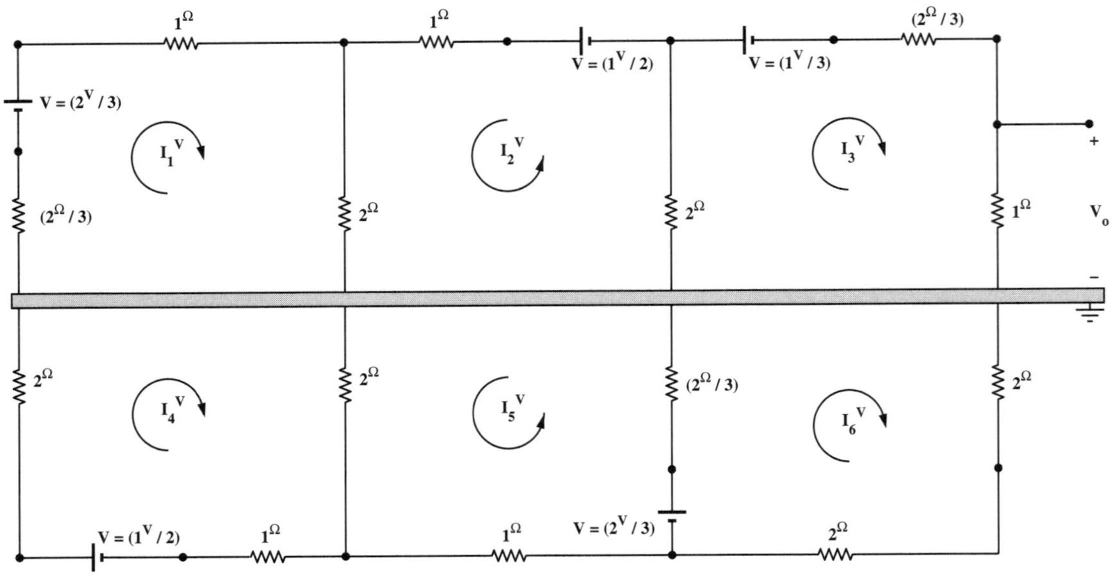

Fig. 5.4-11. The superscript V stands for circuit with initial voltage sources.

The mesh equations for the above circuit results in

$$\begin{Bmatrix} 2/3 \\ 1/2 \\ -1/3 \\ 1/2 \\ 2/3 \\ 2/3 \end{Bmatrix} = \begin{pmatrix} 11/3 & 2 & 0 & 0 & 0 & 0 \\ 2 & 5 & 2 & 0 & 0 & 0 \\ 0 & 2 & 11/3 & 0 & 0 & 0 \\ 0 & 0 & 0 & 5 & 2 & 0 \\ 0 & 0 & 0 & 2 & 11/3 & 2/3 \\ 0 & 0 & 0 & 0 & 2/3 & 14/3 \end{pmatrix} \times \begin{Bmatrix} I_1^V \\ I_2^V \\ I_3^V \\ I_4^V \\ I_5^V \\ I_6^V \end{Bmatrix}. \qquad (5.4\text{-}9)$$

Clearly, as we expect this set of equations is also decoupled and we are left with the first three equations to calculate I_1^V, I_2^V and I_3^V. The results of these two sets of equations (5.4-8 & 9), using the matrix inversion rule (cf., Section 5.7), are as follows.

$$\begin{Bmatrix} I_1^C \\ I_2^C \\ I_3^C \end{Bmatrix} = \begin{pmatrix} 11/3 & -2 & 0 \\ -2 & 5 & 2 \\ 0 & 2 & 11/3 \end{pmatrix}^{-1} \times \begin{Bmatrix} 2 \\ 8 \\ 8 \end{Bmatrix}, \qquad (5.4\text{-}10)$$

and

$$\begin{Bmatrix} I_1^V \\ I_2^V \\ I_3^V \end{Bmatrix} = \begin{pmatrix} 11/3 & 2 & 0 \\ 2 & 5 & 2 \\ 0 & 2 & 11/3 \end{pmatrix}^{-1} \times \begin{Bmatrix} 2/3 \\ 1/2 \\ -1/3 \end{Bmatrix}. \tag{5.4-11}$$

From (5.4-10) and following the standard procedure to invert a matrix (cf., Example 5.7-1), we have

$$I_3^C = \frac{2\Delta_{31} + 8\Delta_{32} + 8\Delta_{33}}{\Delta}. \tag{5.4-12}$$

Here, Δ is the determinant of the coefficient matrix in (5.4-10) and is equal to $11 \times 31/9$. Similarly, each Δ_{ij} is the ij-cofactor of the above coefficient matrix as follows. $\Delta_{31} = -4$, $\Delta_{32} = -22/3$ and $\Delta_{33} = 43/3$. Thus,

$$I_3^C = \frac{2(-4) - 8(22/3) + 8(43/3)}{11 \times 31/9} = 1.267^A. \tag{5.4-13}$$

Similarly from (5.4-11) we have

$$I_3^V = \frac{(2/3)\Delta_{31} + (1/2)\Delta_{32} - (1/3)\Delta_{33}}{\Delta}. \tag{5.4-14}$$

Here, Δ is the determinant of the coefficient matrix in (5.4-11), which is $11 \times 31/9$, and similarly, $\Delta_{31} = 4$, $\Delta_{32} = -22/3$ and $\Delta_{33} = 43/3$. Resulting in,

$$I_3^V = \frac{(2/3)(4) - (1/2)(22/3) - (1/3)(43/3)}{11 \times 31/9} = -0.152^A. \tag{5.4-15}$$

Thus, $I_3 = I_3^C + I_3^V = 1.115^A$, and $V_o = 1.115^V$.

The education we earn in this example is two-fold. (1) We should carefully study the circuit configuration before embarking on any analytical procedure. For instance, had we chosen to set up all 14 node equations initially, then it would have taken us a long time to finish the corresponding computations, resulting in many additional pieces of information which we are not asked to develop. (2) On a positive note, even if we do not see any special property, and conduct a network analysis systematically by using our knowledge gained thus far, then we can finish the task, because there really is not any circuit that we cannot analyze completely.

5.5. The BASIC CONCEPTS in ELECTRIC POWER

The concept of dissipated power in an electric circuit is an important part of the overall circuit analysis. Without getting deeply into theoretical evaluation of this quantity, we present the formula for dissipated power in its most basic form as follows.

$$P = V\,I, \tag{5.5-1}$$

where V is measured in volts, I is measured in amperes, and P, the power, is measured in watts. To apply this formula to a resistive element R, we note that because $V = RI$, we have $P = RI^2$.

For the circuit in Fig. 5.5-1, that is modeled by its Thevenin's equivalent network, and a load R_l, the dissipated power in this load is calculated as follows.

$$P_l = R_l\,I_l^2, \tag{5.5-2}$$

where

$$I_l = \frac{V_{Th}}{R_{Th} + R_l}, \tag{5.5-3}$$

and therefore the dissipated power becomes

$$P_l = \frac{R_l V_{Th}^2}{(R_{Th} + R_l)^2}. \tag{5.5-4}$$

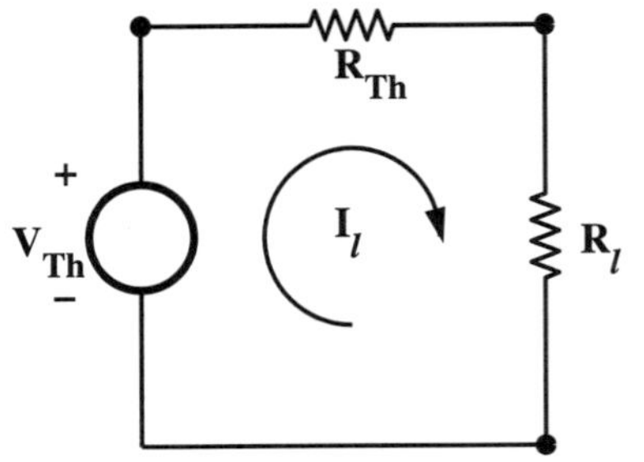

Fig. 5.5-1.

Theorem 5.5-1 (The maximum power transfer): The dissipated power transfer P_l in (5.5-4) is maximum when $R_l = R_{Th}$.

Proof: Clearly the dissipated power in (5.5-4) is a function of R_l, and the optimal R_l is generated by letting $dP_l/dR_l = 0$ as a necessary condition. Here,

$$\frac{dP_l}{dR_l} = V_{Th}^2 \frac{(R_{Th} + R_l)^2 - 2R_l(R_{Th} + R_l)}{(R_{Th} + R_l)^4}$$

$$= V_{Th}^2 \frac{R_{Th} - R_l}{(R_{Th} + R_l)^3} \,. \tag{5.5-5}$$

Letting $dP_l/dR_l = 0$ yields $R_l = R_{Th}$. To show that this solution yields the maximum dissipated power, we need to show that at $R_l = R_{Th}$ the $d^2P_l/dR_l^2 < 0$, which its verification is straightforward. Thus this solution corresponds to the maximum.

That means if we select R_l exactly the same as R_{Th}, then the dissipated power becomes maximum as follows.

$$P_l^{max} = \frac{V_{Th}^2 R_{Th}}{(R_{Th} + R_{Th})^2} = \frac{V_{Th}^2}{4R_{Th}} \,. \tag{5.5-6}$$

This result is interesting, which gives us an alternative way of identifying R_{Th}. This experimental approach is shown as follows [2].

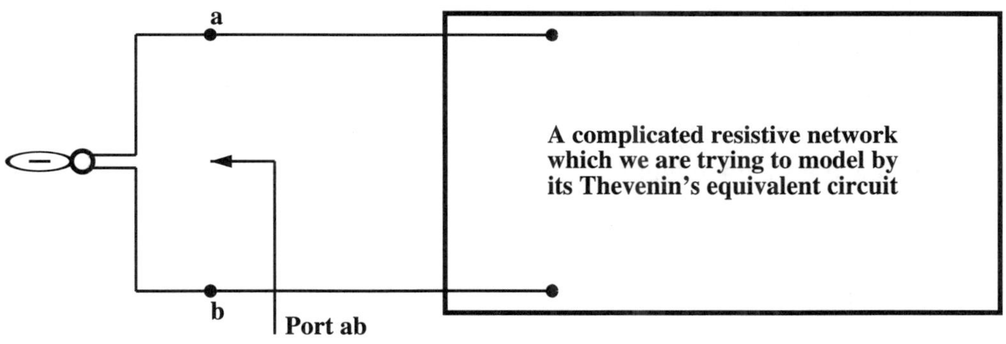

Fig. 5.5-2.

In the above experiment, we choose a series of lamps with different intensities, i.e., their internal filaments are different and thus each provides a different level of light intensity. When the light is the brightest, the resistance of the corresponding filament is equal to the R_{Th}, within our experimental range of accuracy. As far as V_{Th} is concerned, we can measure that with a simple voltmeter. Thus we can construct the Thevenin's equivalent network at port ab experimentally [2].

5.6. CIRCUIT ANALYSIS – A REFLECTION

We now pause to reflect and ask the following question.

What have we accomplished in circuit analysis thus far?

Clearly, we are able to solve any circuit containing resistive elements and constant sources. How? We can establish independent group of equations whose solution yields the necessary information to complete the analysis, i.e., to find the current through and the voltage across each and every element of the underlying circuit. Perhaps, the following issue is not formally discussed. How do we know *a priori* the total number of independent equations in a given circuit?

We try to answer this question with our current knowledge in circuit theory. To that end, we appeal to our intuition in setting up the equations in the preceding chapters. But first of all, we note that all circuits which we have studied, or will be studied in the near future, are the so-called **connected,** i.e., they are in one piece. Also, we recall that we are only concerned with the planar networks. Second, and by the law of physics, no matter which approach we select to analyze a network, the final result is the same. However, when we use the Method #Two, the node equations approach, and apply KCL to all nodes except the reference node or ground whose voltage is zero by definition, we identify all independent equations. So for a circuit with $n+1$ nodes, we select one node as the reference or ground and n nodes as variables, which means we have n independent equations. This observation is the most straightforward approach to identify the number of independent node equations to be established, in order to analyze the network. On the other hand, if we have a total of b *branches* (or elements including sinks and sources), then we must use sufficient number of KVL equations to complete the analysis, which are no more than $(b - n)$ equations, and that is not very specific. We may make a similar argument regarding the total number of independent mesh equations as the above node equations.

Thus in short, the total number of independent equations depends on the number of nodes and meshes in the circuit. We can determine these two numbers by inspection in all connected networks, and that is as far as we take the preceding question in this elementary course.

Having said all the above, an inquisitive mind wants to know where can we go from here? In the context of Table 1.1-1, we are still on Level 1.

If our earlier claim is valid and we are able to analyze all sorts of networks consisting of real elements, then the question would be, where is the natural expansion path from real elements to the more advanced circuits consisting all sorts of elements?

Perhaps the first thing that comes to mind is the following. What if we replace real elements with complex elements, then what are the implications of that replacement? Certainly, if these complex elements are physically realizable, that means they can be built and tested in the laboratory, then we expect to get a response or answer. Because we are working with physical systems that can be built, there will be ways to excite them and make the necessary measurements. We just have to find the laws which govern these elements. For instance, if we use capacitors and inductors which are physically realizable, well-known elements and are studied in the prerequisite physics course, then we can find solutions when we apply *proper sources in proper manners.* In other words, when we have circuits with inductors and capacitors, we need to have other sources beside batteries to excite them. Even batteries or constant current sources must be applied in ways which are different from what we have been discussing so far. Thus, when we advance from circuits with real elements (Level 1 in Table 1.1-1) to circuits with complex elements, we must expect to use time-varying complex sources, in order to analyze and experiment with the corresponding circuits.

To summarize, we begin looking into circuits with complex elements and complex sources to analyze, but the underlying assumption is that these circuits are physically realizable. That means we can build and test them in a laboratory, and therefore there are solutions for these networks.

In the context of Table 1.1-1, we are about to move vertically up to Level 2.

Now, we raise another important question. For the purpose of the analysis, do KCL and KVL work in such a complex setting network?

The answer is yes. These laws just as any law of physics are "independent from the methods of measurements." In § **1.2.A & B,** no mention of how currents and voltages are measured is made. We now explicitly say that these physical laws work in a *point setting* (i.e., resistive elements with constant sources) as well as in the *time-* and *frequency-domain* settings (i.e., all elements and sources including complex ones).

In other words, the KCL reads as follows. The total sum of all currents as functions of either time or frequency entering a node is equal to the total sum of similarly measured currents leaving that node.

Similarly, the KVL reads as follows. The total sum of all elements' voltages as functions of either time or frequency around any closed path is equal to zero.

Thus, we now claim that we can analyze any physically realizable network with any set of elements and sources, in either time- or frequency-domain setting. To demonstrate our claim we need practice, and that we do in the remaining parts of this textbook.

5.7. MATRIX ALGEBRA

To avoid confusion, we review the most basic aspects of matrix algebra in the following. We have no intention of using this concept rigorously, and in fact there are advanced circuit courses solely based on methodologies that use matrix algebra extensively. We find matrix notation as used in § 5.4.1 very convenient and helpful for our presentation. Those who are not familiar with this topic may choose to write the corresponding equations line by line.

A matrix is an "$n \times m$" array of real or complex numbers (and possibly time and frequency dependent quantities) shown in an upper case with "n" rows and "m" columns as follows.

$$A \triangleq \begin{pmatrix} a_{11} & a_{12} & \cdots & a_{1m} \\ a_{21} & a_{22} & \cdots & a_{2m} \\ \cdot & \cdot & \cdots & \cdot \\ \cdot & \cdot & \cdots & \cdot \\ \cdot & \cdot & \cdots & \cdot \\ a_{n1} & a_{n2} & \cdots & a_{nm} \end{pmatrix} \in R^{n \times m}. \tag{5.7-1}$$

Here, R means all entries are real, if those were complex, then we use C. When "m" or "n" becomes "1", the matrix is called vector, and is shown by a lower case and "one" index array as follows. For $m = 1$, a **column** vector, we have

$$b \triangleq \begin{Bmatrix} b_1 \\ \cdot \\ \cdot \\ \cdot \\ b_n \end{Bmatrix} \in R^{n \times 1}, \text{ or } R^n. \tag{5.7-2}$$

For $n = 1$, a **row** vector, we have

$$c \triangleq [c_1, c_2, \ldots, c_m] \in R^{1 \times m}. \tag{5.7-3}$$

Matrix Addition: For two matrices A and B with identical dimensions

$$C = (A + B) = (B + A) \in R^{n \times m}, \tag{5.7-4}$$

where each entry $c_{ij} = a_{ij} + b_{ij}$ for all i and j.

Matrix Multiplication: For two *compatible* matrices $A \in R^{n \times m}$ and $B \in R^{m \times k}$ this operation is performed as follows.

$$BA \neq C = AB \in R^{n \times k}, \text{ where } c_{ij} = \sum_{p=1}^{m} a_{ip} b_{pj}, \; i = 1, \ldots, n, \text{ and } j = 1, \ldots, k. \tag{5.7-5}$$

As noted in (5.7-5), in general $BA \neq AB$.

To solve a linear matrix equation $y = Ax$, for $x, y \in R^n$ and $A \in R^{n \times n}$ (when $m = n$, A is called a square matrix), we define A^{-1} as the inverse of A. If we can calculate A^{-1}, then $x = A^{-1}y$ is the solution of the linear matrix equation. To calculate A^{-1}, if exits, we offer the following steps. But first let us review a determinant.

Determinant: Refers to an array of numbers corresponding to a square matrix, say $n \times n$, and is shown similar to (5.7-1), but with $m = n$ and an addition of "det" as shown in (5.7-6). There is a number associated with each determinant which is calculated by *expanding* the determinant along its rows or columns. For instance, for the following determinant

$$\Delta = \det \begin{pmatrix} a_{11} & a_{12} & \cdots & a_{1n} \\ a_{21} & a_{22} & \cdots & a_{2n} \\ \cdot & \cdot & \cdots & \cdot \\ \cdot & \cdot & \cdots & \cdot \\ \cdot & \cdot & \cdots & \cdot \\ a_{n1} & a_{n2} & \cdots & a_{nn} \end{pmatrix}, \qquad (5.7\text{-}6)$$

and upon expansion along the first row, we have

$$\Delta = a_{11}\Delta_{11} + a_{12}\Delta_{12} + \cdots + a_{1n}\Delta_{1n}. \qquad (5.7\text{-}7)$$

Here, each Δ_{ij} is called the ij^{th}–*cofactor* of Δ, which is constructed by deleting the ith row and the jth column of Δ. The actual value of $\Delta_{ij} = (-1)^{i+j} m_{ij}$, where m_{ij} is called the ij^{th}–*minor* of the Δ, which is the value of the determinant corresponding to the ij^{th}–cofactor. By this process of starting from either a row or a column, we can compute the actual value of a determinant.

Having stated the above, we now present a procedure to compute A^{-1} for a square matrix A whose determinant $\Delta \neq 0$.

STEP One: Construct the *transpose* of A shown by A^T, where each ij entry of A^T is the ji entry of A. Simply stated, rotate each row by its corresponding column.

STEP Two: Construct the *adjoint* matrix, which is generated from A^T as follows. We look at each entry (i,j) of A^T then construct its corresponding cofactor and compute its determinant. We recall that each $\Delta_{ij} = (-1)^{i+j} m_{ij}$, where m_{ij} is the corresponding minor, which is the value of its determinant.

STEP Three: The inverse of A is then becomes

$$A^{-1} = \frac{1}{\Delta} \times \mathrm{Adj}A = \frac{1}{\Delta} \times \begin{pmatrix} \Delta_{11} & \Delta_{12} & \cdots & \Delta_{1n} \\ \Delta_{21} & \Delta_{22} & \cdots & \Delta_{2n} \\ \cdot & \cdot & \cdots & \cdot \\ \cdot & \cdot & \cdots & \cdot \\ \cdot & \cdot & \cdots & \cdot \\ \Delta_{n1} & \Delta_{n2} & \cdots & \Delta_{nn} \end{pmatrix}. \tag{5.7-8}$$

Example 5.7-1: Compute the inverse of the following matrix whose determinant $\Delta \triangleq \det A$ is assumed to be nonzero.

$$A \triangleq \begin{pmatrix} a_{11} & a_{12} & a_{13} \\ a_{21} & a_{22} & a_{23} \\ a_{31} & a_{32} & a_{33} \end{pmatrix}. \tag{5.7-9}$$

STEP One: The transpose matrix is

$$A^T \triangleq \begin{pmatrix} a_{11} & a_{21} & a_{31} \\ a_{12} & a_{22} & a_{32} \\ a_{13} & a_{23} & a_{33} \end{pmatrix}. \tag{5.7-10}$$

STEP Two: The adjoint matrix is

$$\mathrm{Adj}\,A \triangleq \begin{pmatrix} \Delta_{11} & \Delta_{12} & \Delta_{13} \\ \Delta_{21} & \Delta_{22} & \Delta_{23} \\ \Delta_{31} & \Delta_{32} & \Delta_{33} \end{pmatrix}. \tag{5.7-11}$$

Here, each Δ_{ij} is as follows.

$$\Delta_{11} = (-1)^{1+1}\begin{pmatrix} a_{22} & a_{32} \\ a_{23} & a_{33} \end{pmatrix}, \Delta_{12} = (-1)^{1+2}\begin{pmatrix} a_{12} & a_{32} \\ a_{13} & a_{33} \end{pmatrix}, \Delta_{13} = (-1)^{1+3}\begin{pmatrix} a_{12} & a_{22} \\ a_{13} & a_{23} \end{pmatrix},$$

$$\Delta_{21} = (-1)^{2+1}\begin{pmatrix} a_{21} & a_{31} \\ a_{23} & a_{33} \end{pmatrix}, \Delta_{22} = (-1)^{2+2}\begin{pmatrix} a_{11} & a_{31} \\ a_{13} & a_{33} \end{pmatrix}, \Delta_{23} = (-1)^{2+3}\begin{pmatrix} a_{11} & a_{21} \\ a_{13} & a_{23} \end{pmatrix},$$

$$\Delta_{31} = (-1)^{3+1}\begin{pmatrix} a_{21} & a_{31} \\ a_{22} & a_{32} \end{pmatrix}, \Delta_{32} = (-1)^{3+2}\begin{pmatrix} a_{11} & a_{31} \\ a_{12} & a_{32} \end{pmatrix}, \Delta_{33} = (-1)^{3+3}\begin{pmatrix} a_{11} & a_{21} \\ a_{12} & a_{22} \end{pmatrix}. \tag{5.7-12}$$

STEP Three: The inverse matrix is

$$A^{-1} = \frac{1}{\Delta}\,\mathrm{Adj}\,A. \tag{5.7-13}$$

5.8. PROBLEM SET

Analysis

5.8.1. For the circuit in Fig. 5.8-1, compute V_o.

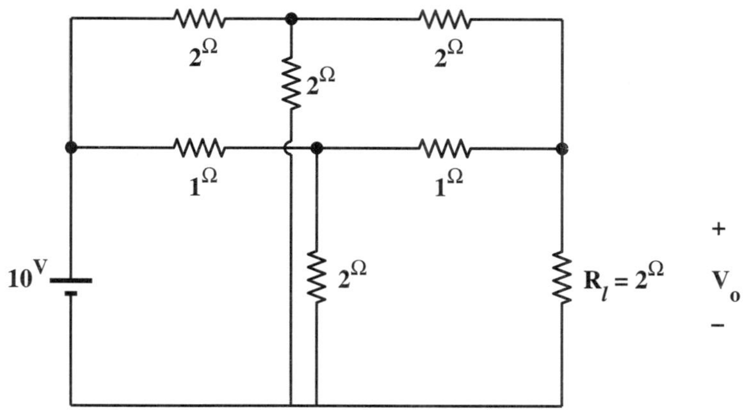

Fig. 5.8-1.

5.8.2. Assuming that all nodes (connection points) are shown by "dots", is the circuit in Fig. 5.8-2 whose elements are arranged on the edges of a cube planar or not and why?

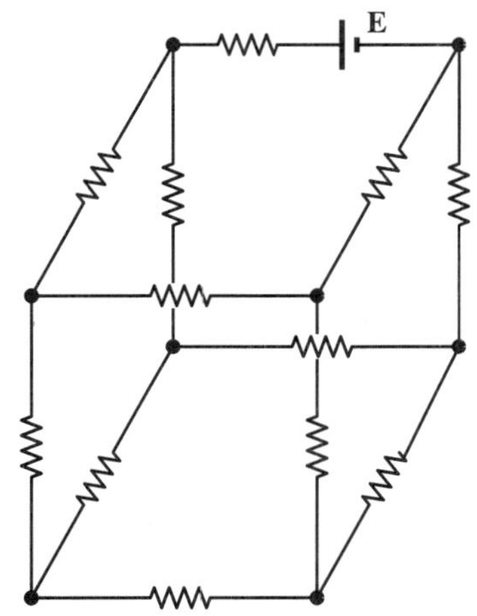

Fig. 5.8-2.

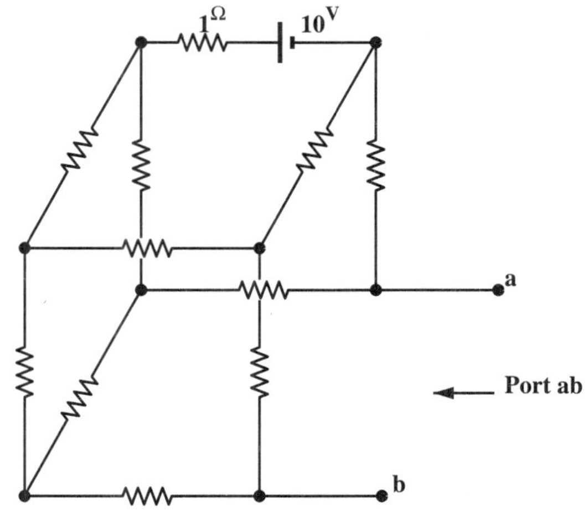

Fig. 5.8-3.

5.8.3. Find the input resistance at port ab for the circuit in Fig. 5.8-3, where all resistors are 1^Ω, and nodes (connection points) are shown by "dots."

5.8.4. For the circuit in Fig. 5.8-4, where all resistors are 1^Ω, and the first row has four resistors, while the last row has one, find R_{eq}. *Hint:* Use $R_{eq} = V_{in}/I_{in}$.

Fig. 5.8-4.

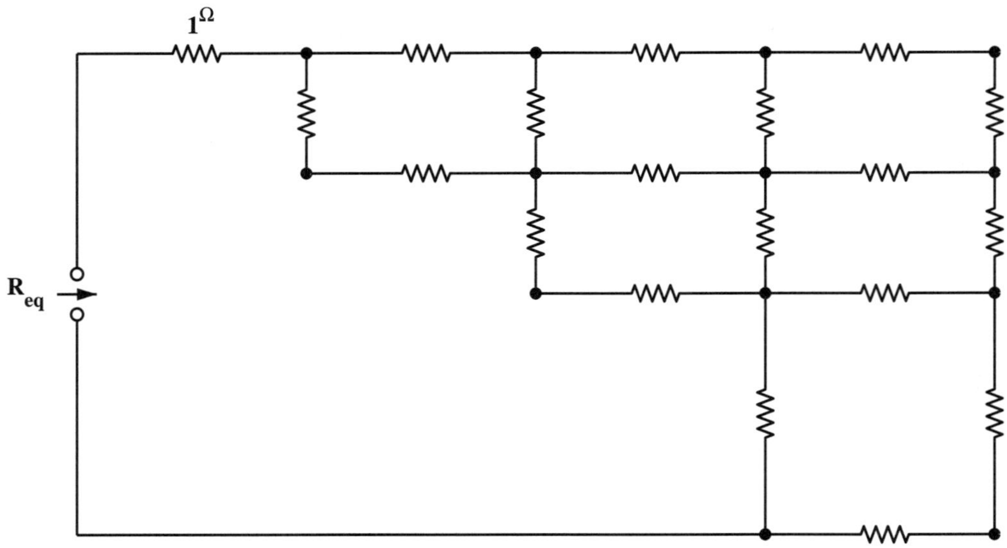

Remark 5.8-1: In general, the circuit in Fig. 5.8-4 has n resistors in the first row, while the nth row has one. Generate R_{eq} in terms of n, and let n becomes large, say $n = 100$.

5.8.5. What is the R_{eq} in the circuit of Fig. 5.8-5 which consists of n blocks of resistive circuit. Compute R_{eq} for $n = 100$.

Fig. 5.8-5.

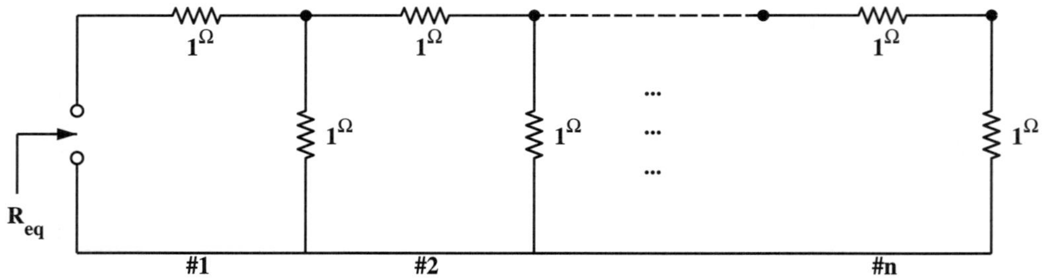

Thevenin's and Norton's Theorems

5.8.6. Determine the Thevenin's and Norton's equivalent circuits at port ab for the circuits in Figs. 5.8-6 to 11.

Fig. 5.8-6.

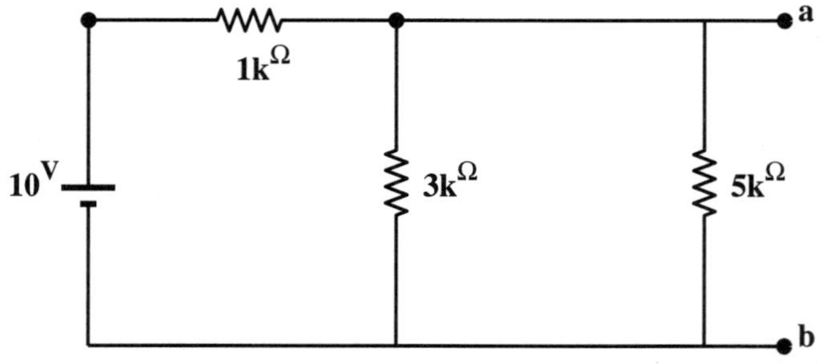

Fig. 5.8-7.

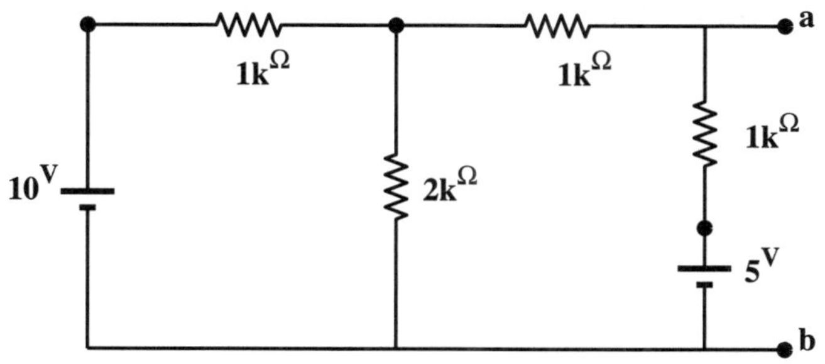

Fig. 5.8-8.

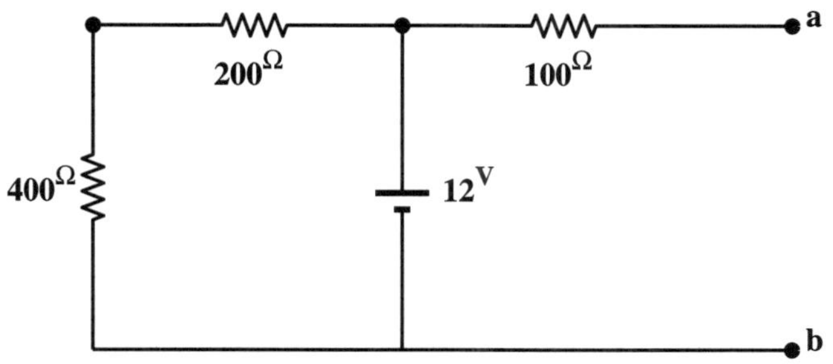

Fig. 5.8-9.

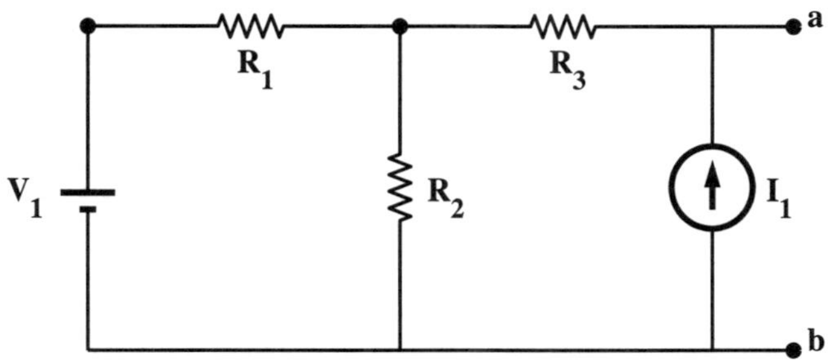

Fig. 5.8-10.

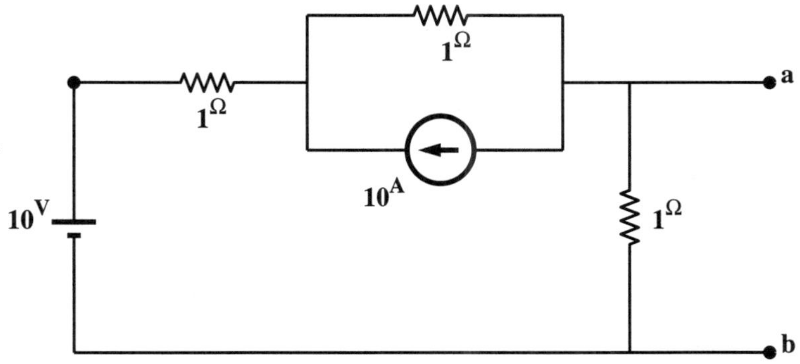

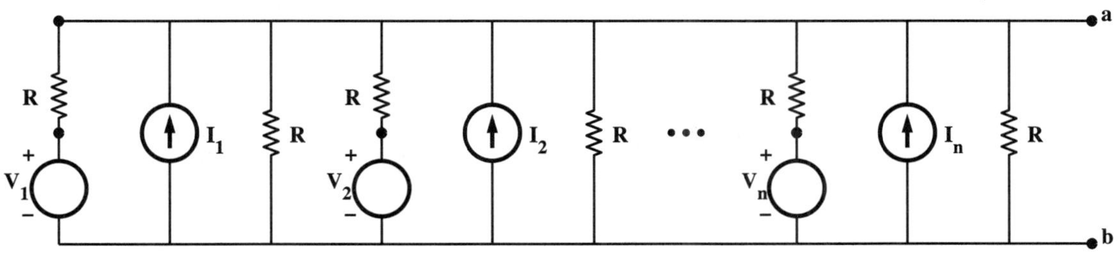

Fig. 5.8-11.

5.8.7. The characteristic curve for a particular two-terminal device is shown in Fig. 5.8-12.

Fig. 5.8-12.

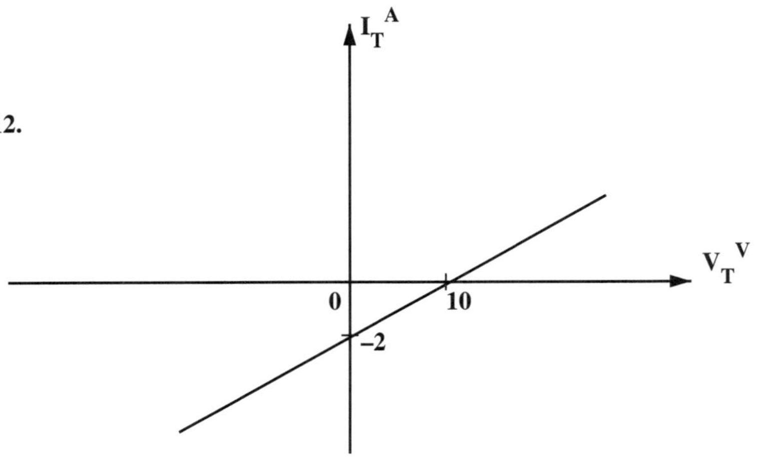

What is the Thevenin's equivalent circuit for this device as depicted in Fig. 5.8-13?

Fig. 5.8-13.

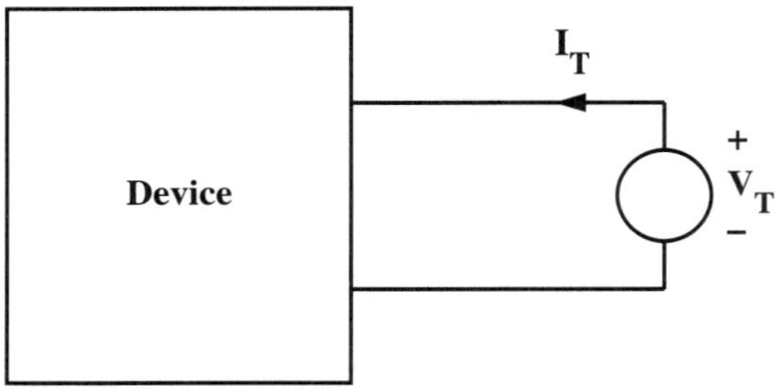

5.8.8. Redo Example 5.3-1 using Procedure I.

Power

5.8.9. Write an expression for power dissipated in element R of the circuit in Fig. 5.8-14 in terms of R_1, R_2 and E.

Fig. 5.8-14.

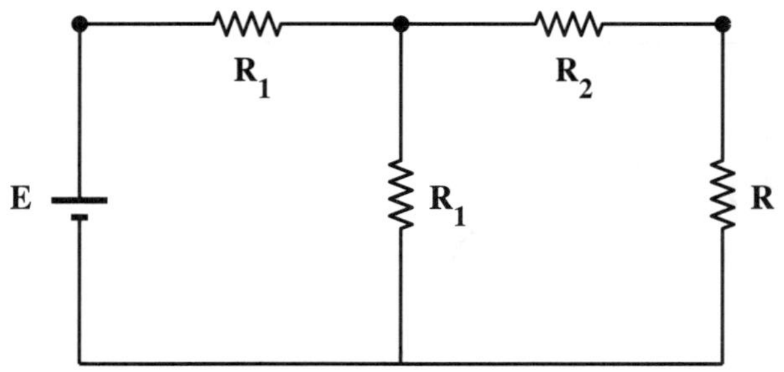

5.8.10. Compute R_l in the circuit of Fig. 5.8-15 for the maximum power transfer.

Fig. 5.8-15.

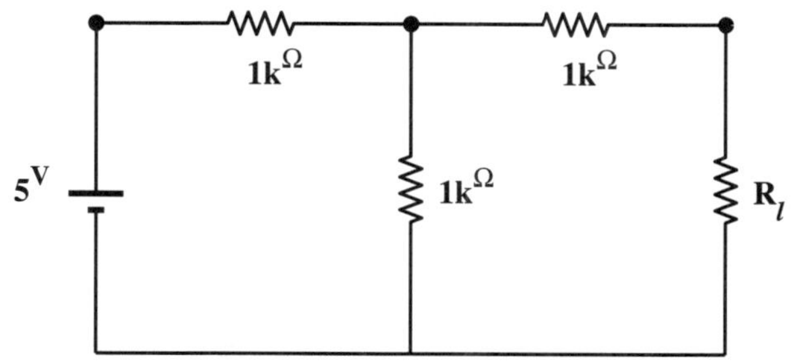

5.8.11. Compute the dissipated power in every resistive element in the circuit of Fig. 5.8-16, and find its total. Compare that value with the total power delivered by the two independent sources.

Fig. 5.8-16.

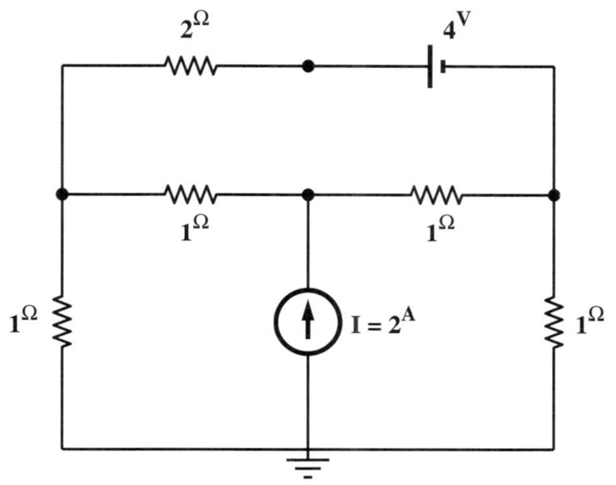

5.8.12. For the circuit in Fig. 5.8-17, choose a proper voltage source V, such that the dissipated power in R_l becomes 10^W.

Fig. 5.8-17.

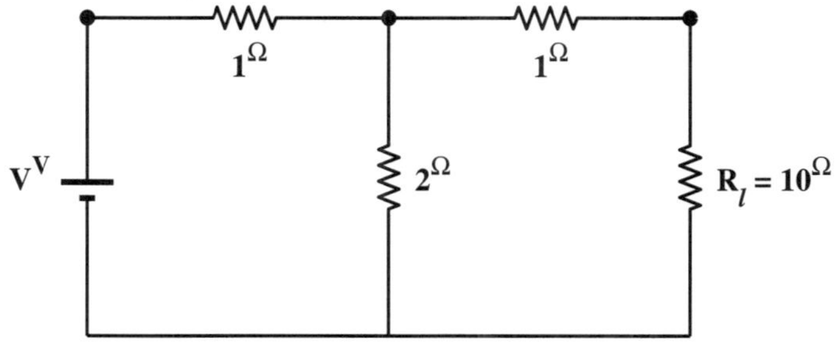

5.8.13. Repeat Problem 5.8.12 for the circuit in Fig. 5.8-18.

Fig. 5.8-18.

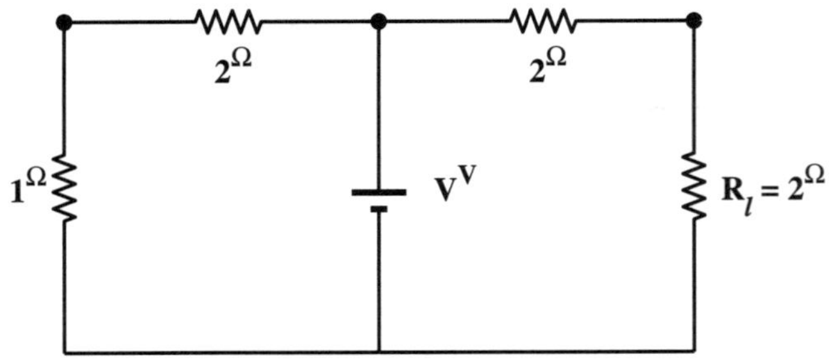

5.8.14. Find R_x in the circuit of Fig. 5.8-19 such that the dissipated output power as shown by an element whose voltage is V_o becomes 1^W.

Fig. 5.8-19.

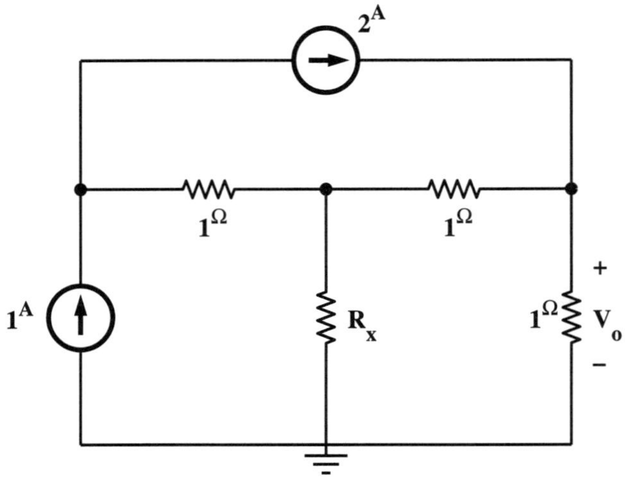

Input-Output Ratio

5.8.15. For the circuit in Fig. 5.8-20, compute the ratio of V_o/V_{in}.

Fig. 5.8-20.

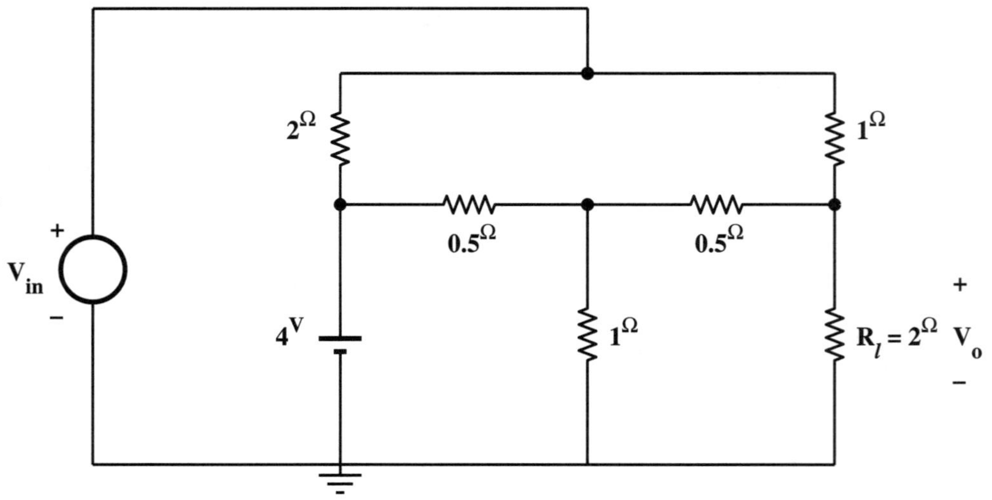

5.8.16. For the circuit in Fig. 5.8-21, compute the ratio of V_o/I_{in}. (Problem 11.8.8).

Fig. 5.8-21.

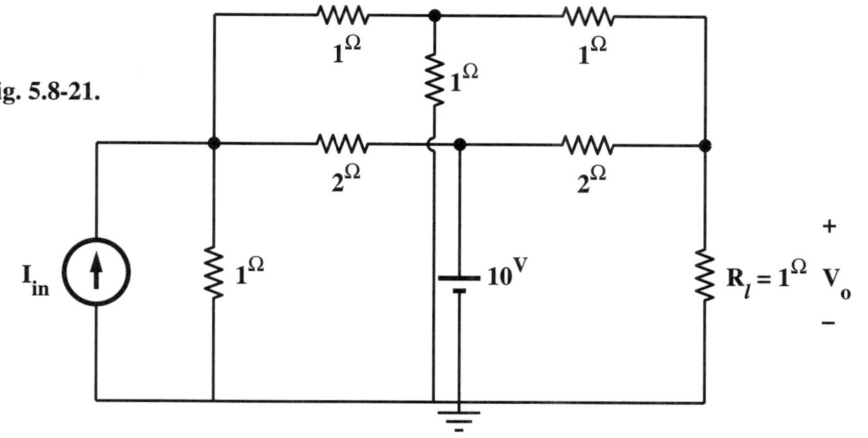

5.8.17. For the circuit in Fig. 5.8-22, compute the ratio of V_o/V_{in}.

Fig. 5.8-22.

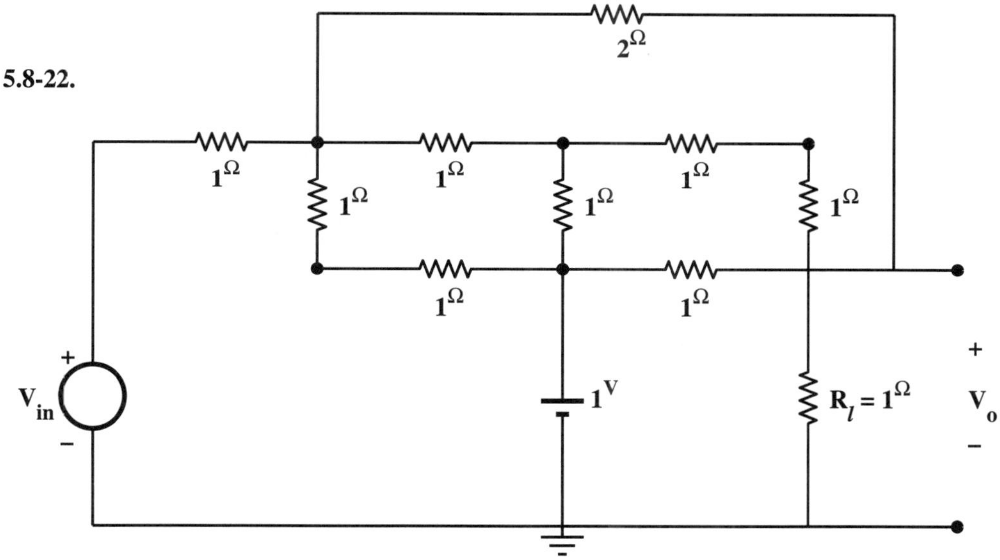

Transformational Analysis

5.8.18. In the circuit of Fig. 5.8-23, compute I_x as shown.

Fig. 5.8-23.

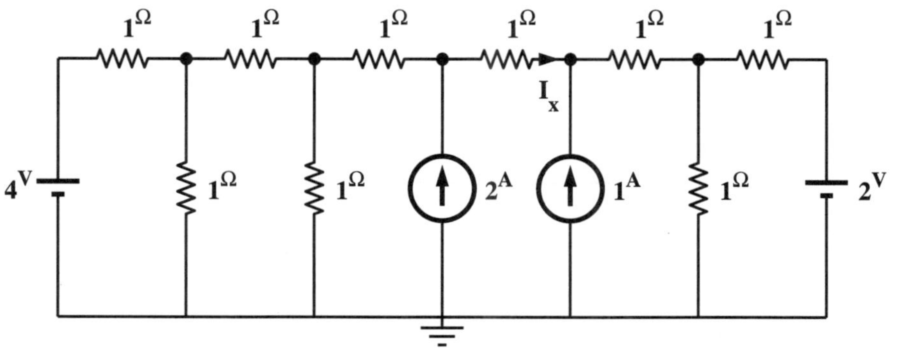

5.8.19. Compute V_o in the circuit of Fig. 5.8-24.

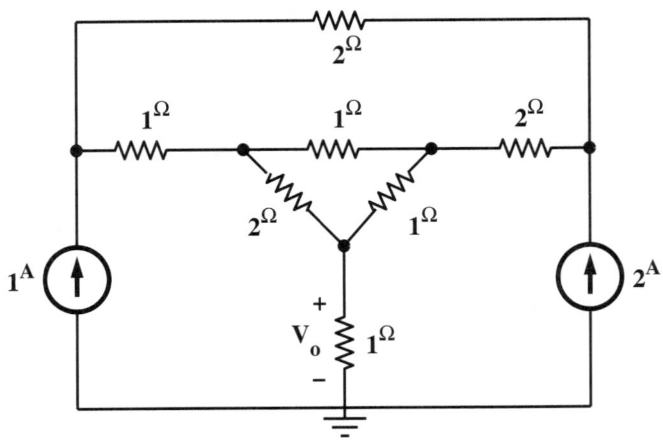

Fig. 5.8-24.

5.8.20. Find I_o in the circuit of Fig. 5.8-25.

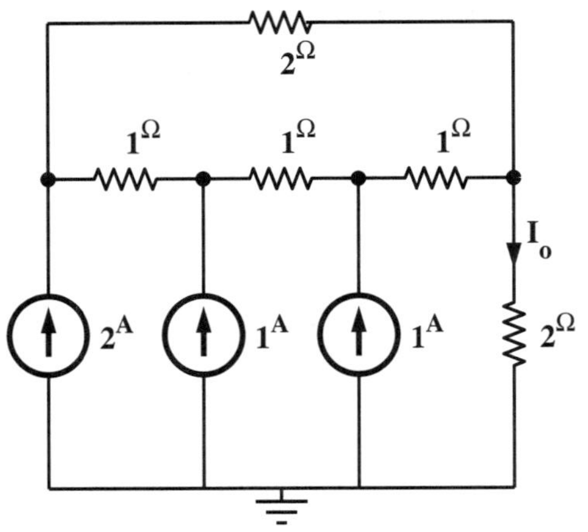

Fig. 5.8-25.

5.8.21. In the circuit of Fig. 5.8-26, find I_o.

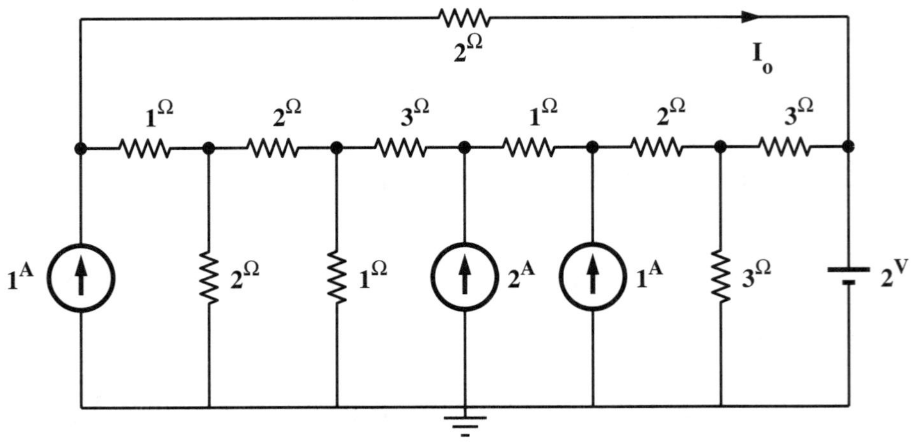

Fig. 5.8-26.

General Circuits

5.8.22. Consider the circuit in Fig. 5.8-26, with a nonlinear resistive element R_N. We assume the characteristic curve for this element in a given range of applied voltage is as follows. For I_N as its current through, we have $I_N = I_0(e^{20V_N} - 1)$, where I_0^A is a small constant parameter, V_N is the applied voltage across. Analyze this circuit in the sense of developing a proper set of equations for its essential variables. Suggest a numerical approach to simplify this set numerically?

Fig. 5.8-27.

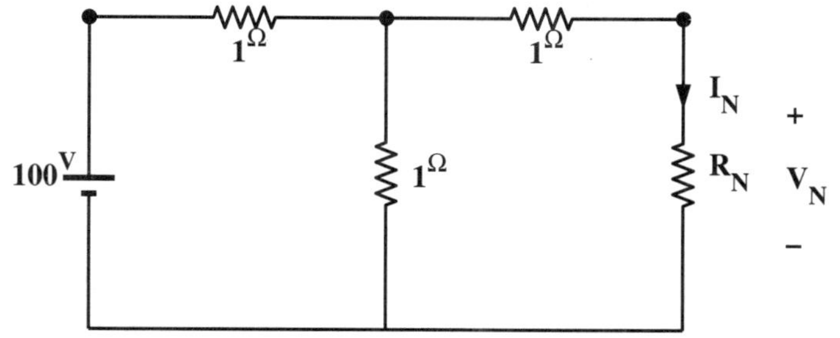

5.8.23. Repeat Problem 5.8.22 for the circuit in Fig. 5.8-28 with two nonlinear resistive elements. Here, the characteristic curves corresponding to each element is as follows. $I_{N_1} = I_0(e^{20V_{N_1}} - 1)$, $I_{N_2} = I_0(e^{20V_{N_2}} - 1)$. Assume an appropriate range for the applied operating voltages in order to make sense out of the circuit equations.

Fig. 5.8-28.

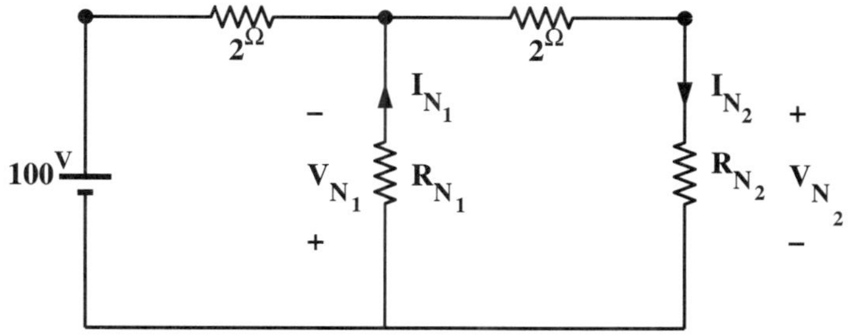

Comment 5.8-1: Our intention has always been to work on circuits which help to clarify the underlying topic and with the minimum concern for its actual physical limitations. For instance, in the preceding two problems we have used 100^V constant sources in circuits with two small resistors. Clearly, these resistors must tolerate high currents in order that we can build these circuits. Also, in Problem 5.8.23, we have not specified the operating range for applied voltages upon which these devices are dependent. Sometimes numbers [or lack of them] are selected so awkwardly that they leave no room for guessing and we must complete the analysis, and that selection is by choice.

5.8.24. Given a set of linear algebraic equations $y = Ax$, where $y \in R^n$ is the input sources, $x \in R^n$ represents the essential variables of a network, and $A \in R^{n \times n}$ is the coefficient matrix constructed from the elements of the network. For instance, both node and mesh equations are of this type. Now, suppose we have

$$y \triangleq \begin{Bmatrix} 1 \\ 0 \\ 0 \end{Bmatrix} = Ax \triangleq \begin{pmatrix} 3 & -1 & 0 \\ -1 & 4 & -1 \\ 0 & -1 & 5 \end{pmatrix} \begin{Bmatrix} x_1 \\ x_2 \\ x_3 \end{Bmatrix}. \tag{5.8-1}$$

If y is the input current sources and x is the node voltages, then design a linear circuit whose node equations is the same as (5.8-1). Now, if y is the input voltage sources and x is the mesh currents, then design a linear circuit whose mesh equations is the same as (5.8-1).

Comment 5.8-2: Clearly, we have not discussed any design methodology which provides procedures to construct the above cases, although we can find an answer for the above example quickly. However, the lesson behind this problem is that many seemingly different circuits have similar representations. Thus, we may get many responses when we meet our specifications, and that situation is at the heart of any engineering design practice. Also, the different circuits which are generated for the same set of linear equations are called the *dual* of each other.

6 CIRCUITS WITH COMPLEX ELEMENTS

6.1. INTRODUCTION

We begin this chapter with a brief review of complex numbers, followed by analysis of a few samples of circuits with complex elements.

Each real number is shown by one parameter which is also called magnitude. But each complex number is shown by two parameters called real and imaginary parts. For instance, for a complex number z, its *rectangular form* representation is

$$z = x + j\,y, \quad \text{with} \quad j = \sqrt{-1}. \tag{6.1-1}$$

where x is the "real part" and y is the "imaginary part" of z.

The complex number z is also shown equivalently in its *polar form* as

$$z = |z| \angle z, \tag{6.1-2}$$

where |z| is called the amplitude or magnitude of z, and $\angle z$ represents its angle or phase. This presentation with its two components of magnitude and angle (also called argument) is shown pictorially in Fig. 6.1-1.

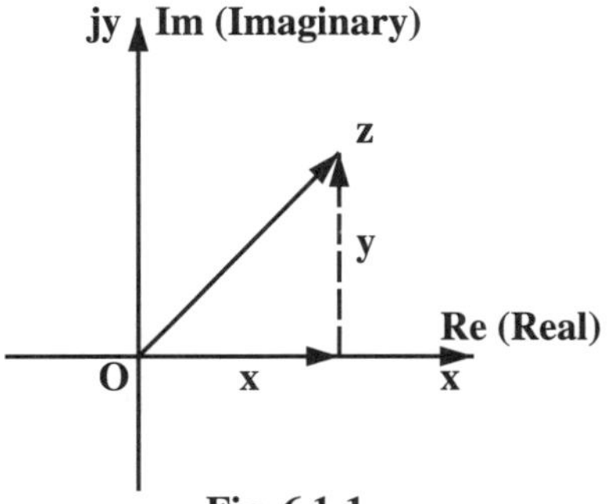

Fig. 6.1-1.

$$z \triangleq \sqrt{x^2 + y^2} \angle \operatorname{tg}^{-1} \frac{y}{x}$$

Using the Euler formula ($e^{jx} = \cos x + j \sin x$), we can show a complex number by its *exponential form* as follows.

$$z = |z| e^{j\theta} = |z|(\cos\theta + j \sin \theta), \tag{6.1-3}$$

where θ is the angle or phase of z. All the above three forms are equivalent.

Rules of complex numbers: Given two complex numbers $z_1 = x_1 + j y_1$ and $z_2 = x_2 + j y_2$, we have the following with proposed forms for their utilization.

Addition [use rectangular form] $z_1 + z_2 = (x_1 + x_2) + j(y_1 + y_2) = z_2 + z_1$.

Subtraction [use rectangular form] $z_1 - z_2 = (x_1 - x_2) + j(y_1 - y_2) \neq z_2 - z_1$.

Multiplication [use polar or exponential form] $z_1 z_2 = (x_1 + j y_1)(x_2 + j y_2) = z_2 z_1 = |z_1||z_2| \angle(\angle z_1 + \angle z_2)$.

Division [use polar or exponential form] $\dfrac{z_1}{z_2} = \dfrac{|z_1| \angle z_1}{|z_2| \angle z_2} = \dfrac{|z_1|}{|z_2|} \angle(\angle z_1 - \angle z_2) \neq \dfrac{z_2}{z_1}$.

6.2. CIRCUITS with COMPLEX ELEMENTS

Suppose we have a circuit with complex elements shown in rectangles as that in Fig. 6.2-1. We claim that we have all the necessary background to analyze such a circuit. We do realize that in many traditional settings this class of circuits is reviewed after such topics as sinusoidal waveforms are presented. However, we believe our approach will resonate better with students and they quickly absorb the essential points and become more confident in their learning experience than otherwise. In other words, contrary to the traditional way of teaching circuits with complex elements in another semester, we can analyze such circuits now with our current knowledge. But we must be cautious to establish the pertinent equations correctly, and the rest is a pure algebra. We elaborate this point in two examples below first, before we explain the meaning of complex elements afterward.

Example 6.2-1: Analyze the circuit in Fig. 6.2-1 containing complex elements shown by rectangles with their complex values, and the corresponding complex sources are also indicated by their complex values.

Solution: Looking at this circuit and studying its configuration, we note that all pertinent variables are defined and we can solve this circuit by mesh equations in a manner as straightforward as that for cases with real elements. Therefore we invoke similar steps as those in the preceding chapters for circuits with real elements to find mesh currents, and do not expect to run into any difficulty.

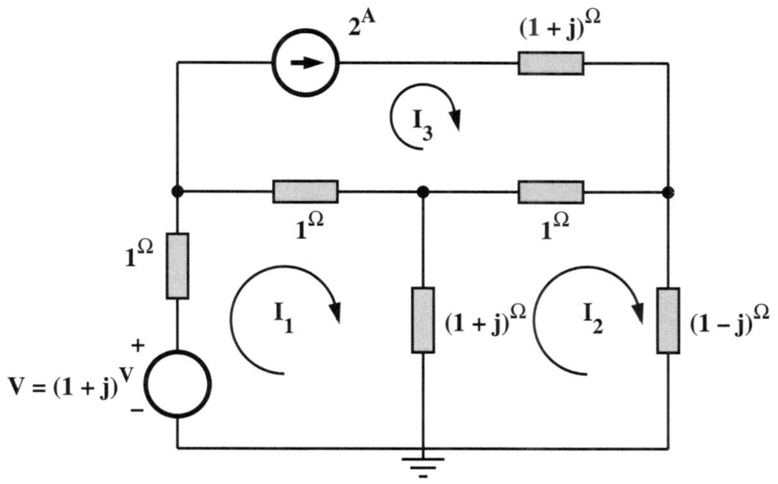

Fig. 6.2-1.

STEP Zero: Three mesh currents I_1, I_2 and I_3 are shown in Fig. 6.2-1.

STEP One: We can write by inspection the corresponding mesh equations for this circuit as follows.

$$(1+j)^V = [1^\Omega + 1^\Omega + (1+j)^\Omega]I_1 - (1+j)^\Omega I_2 - (1^\Omega)I_3, \tag{6.2-1}$$

$$0^V = -(1+j)^\Omega I_1 + [1^\Omega + (1-j)^\Omega + (1+j)^\Omega]I_2 - (1^\Omega)I_3, \tag{6.2-2}$$

$$I_3 \equiv 2^A \angle 0^\circ, \qquad \text{the constraint equation.} \tag{6.2-3}$$

STEP Final: Substituting (6.2-3) in (6.2-1) and (6.2-2) yields

$$(3+j)^V = (3+j)^\Omega I_1 - (1+j)^\Omega I_2, \tag{6.2-4}$$

$$2^V = -(1+j)^\Omega I_1 + 3^\Omega I_2. \tag{6.2-5}$$

Resulting in,

$$I_1 = \frac{11+5j}{9+j}, \tag{6.2-6}$$

$$I_2 = \frac{8+6j}{9+j}. \tag{6.2-7}$$

Example 6.2-2: Analyze the circuit in Fig. 6.2-2.

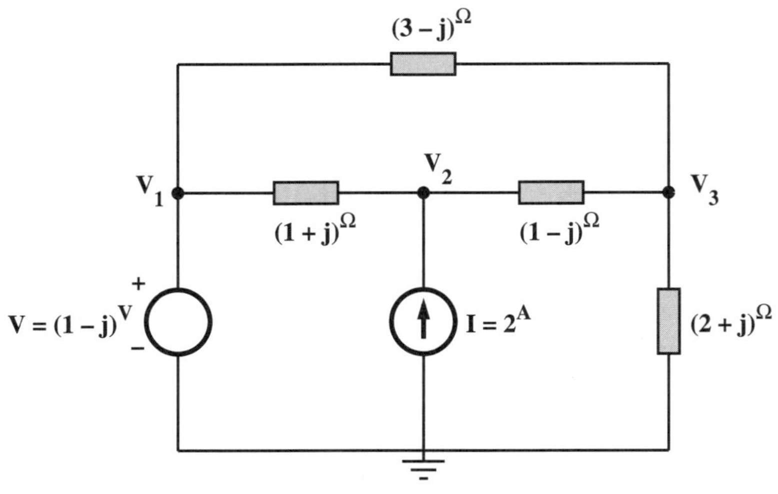

Fig. 6.2-2.

Solution: Here again, looking at this circuit and studying its configuration reveals that the node equations approach is the right choice for analyzing the network. To that end, all pertinent variables are defined and we invoke similar steps as those in the preceding chapters to establish essential equations for the complete analysis.

STEP Zero: We label V_1, V_2 and V_3 as three node voltages, Fig. 6.2-2.

STEP One: We write the corresponding node equations by inspection.

$$V_1 = (1 - j)^V, \qquad \text{the constraint equation,} \tag{6.2-8}$$

$$2^A = -\frac{1}{1+j}V_1 + \left(\frac{1}{1+j} + \frac{1}{1-j}\right)V_2 - \frac{1}{1-j}V_3, \tag{6.2-9}$$

$$0^A = -\frac{1}{3-j}V_1 - \frac{1}{1-j}V_2 + \left(\frac{1}{3-j} + \frac{1}{1-j} + \frac{1}{2+j}\right)V_3. \tag{6.2-10}$$

STOP

Comment 6.2-1: It is clear that we have made our point, since the above two examples can be solved with the same methods used in the preceding chapters for circuits with real elements, for instance, Example 3.6-1. Thus, we can solve a new class of problems with complex elements. Now, the issue that needs clarification is the physical meaning of a complex element or a complex source as we elaborate next.

6.3. COMPLEX ELEMENTS REVISITED

In the preceding two examples, we have used complex elements such as $(2+j)^{\Omega}$. Now, what is the meaning of a complex element? Addressing this issue, we reintroduce the following three physical elements in the context of a complex setting.

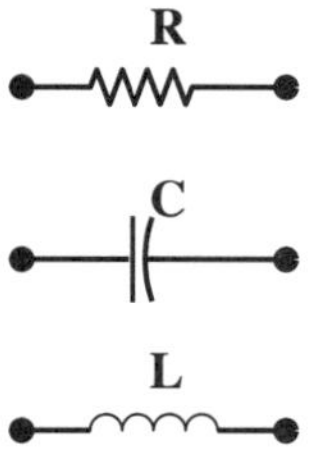

Resistor, its *resistance (R)* is measured in Ohms (Ω).

Capacitor, its *capacitance (C)* is measured in Farads (F).

Inductor, its *inductance (L)* is measured in Henrys (H).

Fig. 6.3-1. The Fig. 1.3-3 revisited.

(a) **Inductor:** A physical element as shown in Fig 6.3-1, which is symbolized by L and is measured in Henrys (H) that has an *impedance* given by $Z_L = j\omega L$ is called an inductor. Here, $j = \sqrt{-1}$, the *angular frequency* $\omega = 2\pi f$ (measured in radian per second – rad/sec), and f is the frequency in hertz (Hz). An impedance such as Z_L is the counterpart of the resistance for a complex element and is also measured in Ohms. For instance, $(j5)^{\Omega}$ is an inductor whose impedance is j5 Ohms. In general, $Z = R + jX$, where R is the resistive (real) part, and X corresponds to the reactive (imaginary) part of an impedance which is called the *reactance.* For an inductor alone $X_L = \omega L$ is its reactance. In practice, each complex circuit is specified with its operational frequency under which the circuit is functioning. When this frequency is selected, we can specify each complex element accordingly. For instance, for the above case and at, say, $\omega = 1$, $j5 = j\omega L = jL$, we have $L = 5^H$ as its inductance, which is the element's specific value.

(b) **Capacitor:** Similarly, a physical element as shown in Fig. 6.3-1, which is symbolized by C and is measured in Farads (F) is called a capacitor. The inverse of C is called *elastance.* This element has an *admittance* given by $Y_C = j\omega C$. Both C and ω must be specified in order to describe $Y_C = 1/Z_C$. Thus, for $Z_C = -j5$ and at $\omega = 1$ (rad/sec) we have $Z_C = 1/0.2j$, which corresponds to a capacitor whose $C = 0.2^F$. In general, $Y = G + jB$, where G is the conductance and B is the *susceptance.* For a capacitor alone $B_C = \omega C$. Finally, the *immittance* is the generic name for **im**pedance and ad**mittance**.

(c) **Resistor:** The resistance of this physical element is just a real number.

Therefore, all +j's in any network correspond to inductors, while all −j's correspond to capacitors and all real numbers represent resistors.

What about sources? Complex sources are frequency-domain representations of certain physical sources as discussed in Section 6.4.

Comment 6.3-1: To appreciate the meaning of a network with complex elements, recall that in a complex plane we show vectors as in Fig. 6.3-2.

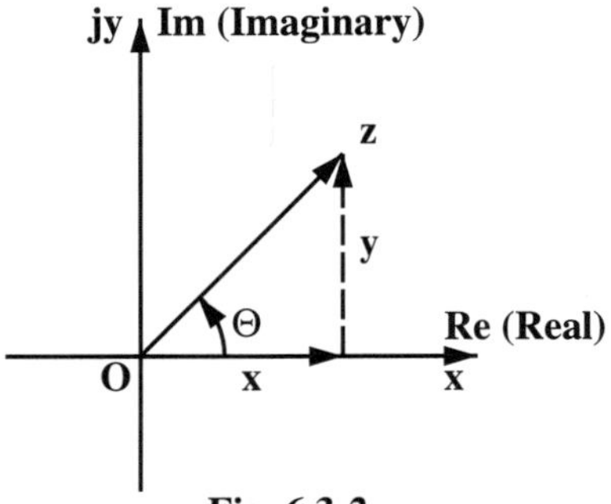

Fig. 6.3-2.

$$z = |z| \angle z = \sqrt{x^2 + y^2} \angle \mathrm{tg}^{-1}\frac{y}{x} \triangleq \rho \angle \theta.$$

Now, imagine this vector is rotating as a function of time and according to ωt, where $\omega = 2\pi f$ rad/sec, and f is the frequency in Hz. That means the above vector at different instances of time may have different magnitudes. As illustrated in Fig. 6.3-3, the vector changes its magnitude and angle as a function of time. Thus, a proper way of showing the above vector is

$$z(t) = |z|(t)\angle z(t) = \sqrt{x^2(t) + y^2(t)} \angle \mathrm{tg}^{-1}\frac{y(t)}{x(t)}. \tag{6.3-1}$$

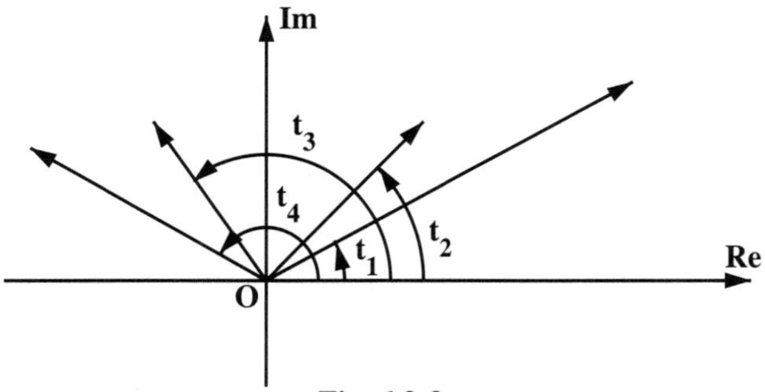

Fig. 6.3-3.

Or in a so-called phasor-form representation $z = \rho\angle(\omega t + \theta)$, where θ is the angle of the original vector z. Thus, to analyze a network with complex elements and sources, not only can we use our earlier methodologies, but also these networks display a well-defined practical circuits as well. We have reasons to establish these networks this way as we elaborate subsequently.

Now, we study a few examples, where each is specified for a particular range of operational frequencies. Generally, this range is specified, and from which one specific ω is chosen to operate the circuit. For instance, when we select an operating frequency of $\omega = 1000$ rad/sec, we set $\omega = 1000$ to compute all relevant complex numbers which are functions of this parameter in order to proceed numerically.

Example 6.3-1: Analyze the circuit in Fig. 6.3-4 for an operating frequency of $\omega = 1000$ rad/sec. (Problem 9.9.4).

Fig. 6.3-4.

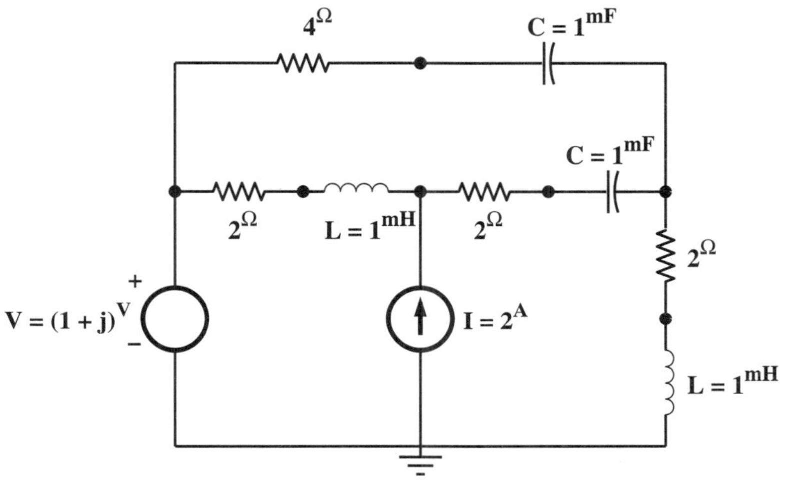

Solution: The key to solving this problem is letting $\omega = 1000$, since without that information we cannot proceed numerically. At this value, we compute all impedances and transform the circuit into its complex-value representation as depicted in Fig. 6.3-5.

Fig. 6.3-5.

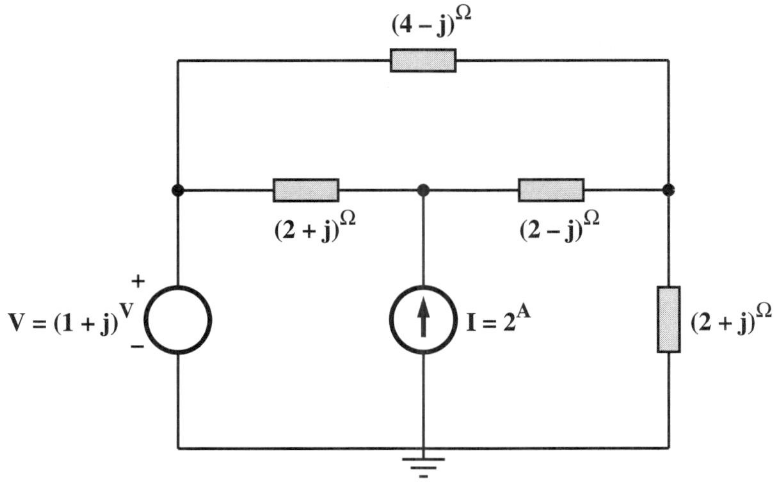

Now, looking at Fig. 6.3-5 and studying its configuration, we note that all pertinent variables are defined and we can solve this circuit in exactly the same manner as we have done for the Example 3.6-1 or in Example 6.2-2. The rest of the analysis is straightforward. We calculate the currents through and the voltages across each and every element in exactly the same manner as in Example 3.6-1. The only difference is that here the circuit is changing in time and all these complex values are indeed multiplied by $e^{j\omega t}$ to display their time variations. (Problem 6.6.8).

6.4. GENERAL CIRCUITS with COMPLEX ELEMENTS

So far, we have studied circuits with resistive elements and constant sources only; as well as circuits with complex elements including complex sources at one given frequency $\omega = 2\pi f$. Now, we study the same class of networks, assuming that all sources are changing as a general function of time. In other words, the constant sources are no longer sufficient to excite complex circuits. We may rephrase this statement by saying that the source must be rich in its frequency contents to excite complex circuits.

Recall that the notation in Fig. 6.4-1 has already been used to display a complex load (element). Here, x is the *net* real part of this load which represents its net resistive portion, if any. On the other hand, the jy is the *net* complex part of the load and depending on its value we may have either an inductor ($Z_L = j\omega L$) or a capacitor ($Z_C = 1/j\omega C$), or a combination of these two elements.

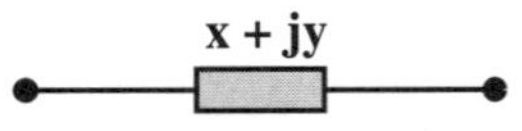

Fig. 6.4-1.

Now, we propose the following alternative representation for the above two complex elements. Namely, from the physics we recall that the inductor is shown as in Fig. 6.4-2.

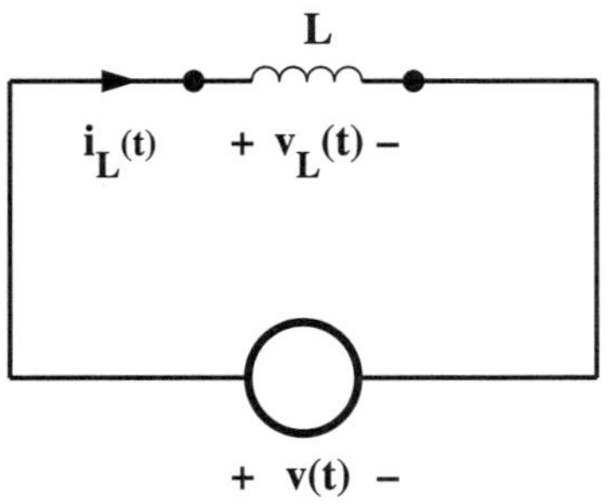

Fig. 6.4-2.

Here,

$$v_L(t) = L\,\frac{di_L(t)}{dt}\,, \tag{6.4-1a}$$

$$i_L(t) = \frac{1}{L}\int_{-\infty}^{t} v_L(\tau)d\tau. \tag{6.4-1b}$$

Clearly, this definition of an inductor, which says the voltage across this element is proportional to the derivative of its current through, confirms the fact that we need time-varying source v(t) to excite this element. The instantaneous energy stored in an inductor is $W_L(t) = (½)Li_L^2(t)$ (in Joules).

Similarly, from our physics we have another complex element called capacitor which is shown in Fig. 6.4-3 and is defined as follows.

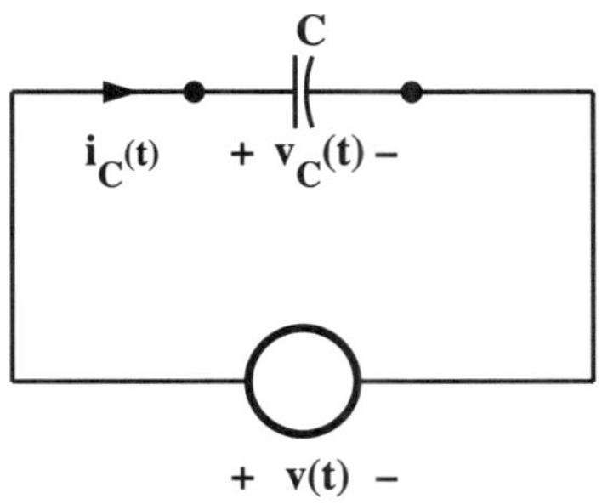

Fig. 6.4-3.

Here,

$$i_C(t) = C\,\frac{dv_C(t)}{dt}\,, \tag{6.4-2a}$$

$$v_C(t) = \frac{1}{C}\int_{-\infty}^{t} i_C(\tau)d\tau. \tag{6.4-2b}$$

Thus, a capacitor is an element whose current is the derivative of its voltage across. The instantaneous energy stored in a capacitor is $W_C(t) = (½)Cv_C^2(t)$ (in Joules).

Looking at (6.4-1) and (6.4-2), we note that there is a similarity in appearance among these equations representing current and voltage of inductor and capacitor. In fact, for general passive elements we have:

$$R\,i \Longleftrightarrow G\,v, \tag{6.4-3a}$$

$$L\frac{di}{at} \Longleftrightarrow C\frac{dv}{dt}\,, \tag{6.4-3b}$$

$$\frac{1}{C}\int i\,dt \Longleftrightarrow \frac{1}{L}\int v\,dt, \tag{6.4-3c}$$

which are similar in appearance. The similarity in appearance with all its implications is referred to as the *principle of duality.* Thus, in this context

$$\text{R and G,} \quad \text{L and C,} \quad \text{loop current and node voltage,} \tag{6.4-4}$$

are dual of each other. In the context of Comment 5.8-2 also (6.4-1) and (6.4-2) are dual. There are procedures to construct the dual of an electrical circuit. Also, there are procedures to construct the mechanical dual for an electrical circuit and vice versa, but we do not pursue these matters.

If we choose the current through an inductor as $i_L(t) = e^{j\omega t}$, which is a special and perhaps the simplest time-varying signal in a phasor form, then

$$v_L(t) = L\frac{de^{j\omega t}}{dt} = j\omega L e^{j\omega t} = j\omega L i_L(t). \tag{6.4-5}$$

This phasor relationship and in comparison with Ohm's law suggests that we may choose $Z_L = j\omega L$, as a generalized Ohm's law in the sense that $v_L = Z_L i_L$.

Similarly, for $v_C = e^{j\omega t}$ we have

$$i_C(t) = j\omega C v_C(t), \tag{6.4-6}$$

with $Z_C = \frac{1}{j\omega C}$, where the generalized Ohm's law holds as well.

With the above preliminaries, we are now ready to move up vertically to Level 2 of Table 1.1-1, and establish integro-differential equations describing any given circuit involving R, L and C elements with complex and time-varying sources.

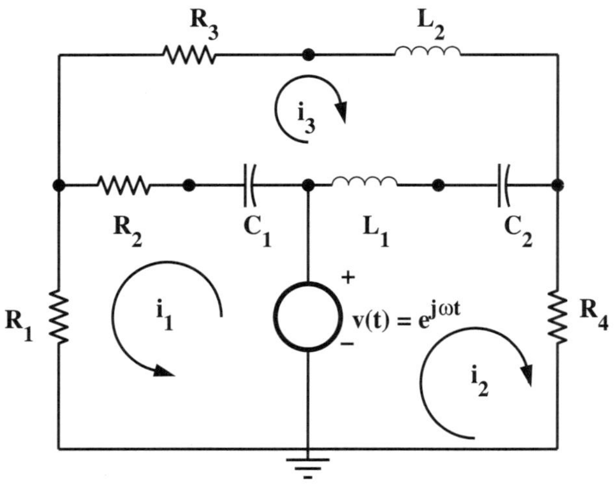

Fig. 6.4-4.

Example 6.4-1: Analyze the circuit in Fig. 6.4-4 by setting up an appropriate group of integro-differential equations whose solution yields the necessary information to compute the current through and the voltage across each and every element of this circuit.

Solution: Looking at the circuit in Fig. 6.4-4 and studying its configuration, we note that this circuit is very much similar and indeed easier than its counterpart resistive circuit in Example 2.4-2, which we have exhaustively analyzed. We note that there is only one voltage source, albeit a complex time-varying source, and the mesh equations are the right choices to complete the analysis.

STEP Zero: We label all pertinent variables as shown in Fig. 6.4-4, and note that the lower case letters are used herein for these variables, because we are in the time-domain setting.

STEP One: By inspection, we can write the corresponding set of mesh equations for this circuit, now represented as a group of integro-differential equations. These equations provide sufficient information to complete the analysis. By sufficient we mean just as any integro-differential equation we also need initial values to get the unique (necessary and sufficient conditions) solution.

$$v(t) = e^{j\omega t} = (R_1 + R_2)i_1(t) + R_2 i_3(t) + \frac{1}{C_1}\int_{-\infty}^{t} [i_1(\tau) + i_3(\tau)]d\tau, \tag{6.4-7}$$

$$v(t) = e^{j\omega t} = R_4 i_2(t) + \frac{1}{C_2}\int_{-\infty}^{t} [i_2(\tau) - i_3(\tau)]d\tau + L_1 \frac{d(i_2 - i_3)}{dt}, \tag{6.4-8}$$

$$\begin{aligned} 0 = R_3 i_3(t) + L_2 \frac{di_3}{dt} + \frac{1}{C_2}\int_{-\infty}^{t} [i_3(\tau) - i_2(\tau)]d\tau \\ + L_1 \frac{d(i_3 - i_2)}{dt} + \frac{1}{C_1}\int_{-\infty}^{t} [i_3(\tau) + i_1(\tau)]d\tau + R_3(i_3 + i_1). \end{aligned} \tag{6.4-9}$$

STOP

The preceding set of equations, when solved numerically (Problem 9.9.5), gives us the information needed to complete the analysis of the associated circuit. To solve in a reasonable manner similar sets of equations as those above is the remaining task in this chapter. To that end we introduce the following important lemma.

Lemma 6.4-1 (Leibnitz): Given

$$\phi(t) = \int_{a(t)}^{b(t)} f(\tau, t)d\tau, \tag{6.4-10}$$

where both limits of integral are functions of t, our parameter, while integration is in term of τ, then

$$\frac{d\phi(t)}{dt} = \int_{a(t)}^{b(t)} \frac{df(\tau,t)}{dt} d\tau + f(\tau,t)\Big|_{\tau=b(t)} \frac{db(t)}{dt} - f(\tau,t)\Big|_{\tau=a(t)} \frac{da(t)}{dt}. \qquad (6.4\text{-}11)$$

□

Using the above formula, we can differentiate the preceding (6.4-7) to (6.4-9), in order to make them differential equations, i.e., to remove the integral parts of these equations.

For instance, in (6.4-7), we have

$$\frac{d}{dt}\left\{\frac{1}{C_1}\int_{-\infty}^{t}[i_1(\tau)+i_3(\tau)]d\tau\right\} = \frac{1}{C_1}\left\{\int_{-\infty}^{t}\frac{d}{dt}[i_1(\tau)+i_3(\tau)]d\tau\right.$$

$$\left. + [i_1(\tau)+i_3(\tau)]\Big|_{\tau=t}\frac{dt}{dt} - [i_1(\tau)+i_3(\tau)]\Big|_{\tau=-\infty}\frac{d(-\infty)}{dt}\right\}. \qquad (6.4\text{-}12)$$

Clearly, the first and the third terms on the right-hand side of (6.4-12) are zero, because $df(\tau)/dt \equiv 0$ (there is no explicit t for differentiating), also $d(-\infty)/dt = 0$. Because $dt/dt \equiv 1$, the right-hand side of (6.4-12) is often written without going through the preceding steps, i.e.,

$$\frac{d}{dt}\left\{\frac{1}{C_1}\int_{-\infty}^{t}[i_1(\tau)+i_3(\tau)]d\tau\right\} = \frac{1}{C_1}[i_1(t)+i_3(t)]. \qquad (6.4\text{-}13)$$

Using Lemma 6.4-1, we differentiate (6.4-7) to (6.4-9) to get a set of ordinary differential equations to complete the analysis for circuit in Example 6.4-1.

$$\frac{dv(t)}{dt} = j\omega v(t) = (R_1+R_2)\frac{di_1(t)}{dt} + R_2\frac{di_3(t)}{dt} + \frac{1}{C_1}[i_1(t)+i_3(t)], \qquad (6.4\text{-}14)$$

$$\frac{dv(t)}{dt} = j\omega v(t) = R_4\frac{di_2(t)}{dt} + \frac{1}{C_2}[i_2(t)-i_3(t)] + L_1\frac{d^2(i_2-i_3)}{dt^2}, \qquad (6.4\text{-}15)$$

$$0 = R_3\frac{di_3(t)}{dt} + L_2\frac{d^2 i_3}{dt^2} + \frac{1}{C_2}[i_3(t)-i_2(t)]$$

$$+ L_1\frac{d^2(i_3-i_2)}{dt^2} + \frac{1}{C_1}[i_3(t)+i_1(t)] + R_2\frac{d(i_3+i_1)}{dt}. \qquad (6.4\text{-}16)$$

Certainly, we need to group these terms before trying to solve them numerically. We note that these are coupled ordinary differential equations, and we need some review of the pertinent mathematics before proceeding further. That review is the subject of Chapter Seven.

Example 6.4-2: Analyze the circuit in Fig. 6.4-5.

Solution: Clearly, this circuit is very similar to that in the last example as far as configuration is concerned, and can be analyzed using node equations. To avoid setting up excessive number of node equations, we combine various elements to have the circuit in Fig. 6.4-6, and that combining of elements brings us some new choices. We continue our analysis using node equations approach.

STEP Zero: We label all pertinent variables as shown in Fig. 6.4-6. Here, we use ω as a parameter versus a fix number in the Example 6.2-1.

Fig. 6.4-5.

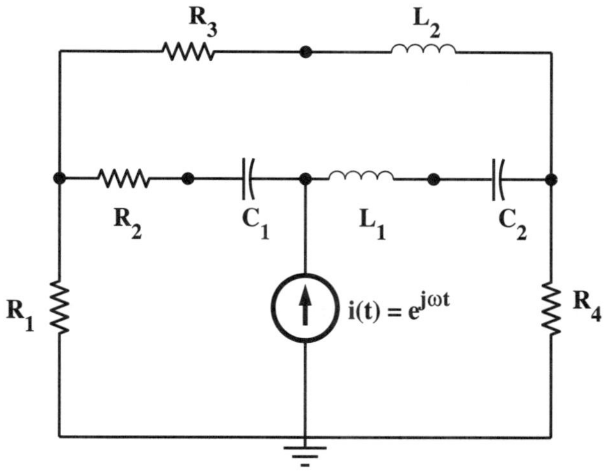

Fig. 6.4-6.

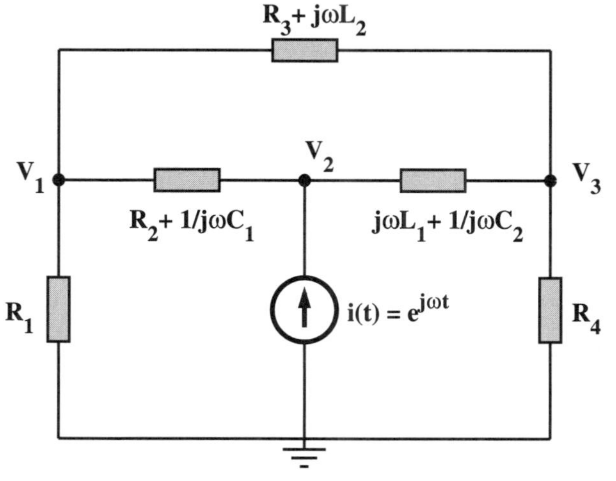

STEP One: We can write by inspection the corresponding node equations for this circuit. These node equations become in the forms of integro-differential equations had we not made the above changes. Further, we have seen that the resulting set of equations is always coupled (i.e., there are crossing terms among unknowns) and we need to use additional algebras to simplify those equations into manageable forms. There are systematic procedures to perform these additional steps under various transformational techniques. For now we introduce a new variable and call that $s \triangleq j\omega$. Using this variable which changes as ω changes, we can set up the node equations for the circuit in Fig. 6.4-6 as follows. This transformation portrays an interesting and powerful direction to analyze a circuit. (Problem 9.9.14).

$$0 = [\frac{1}{R_1} + \frac{1}{R_2 + \frac{1}{sC_1}} + \frac{1}{R_3 + sL_2}]V_1 - [\frac{1}{R_2 + \frac{1}{sC_1}}]V_2 - [\frac{1}{R_3 + sL_2}]V_3, \qquad (6.4\text{-}17)$$

$$e^{j\omega t} = -[\frac{1}{R_2 + \frac{1}{sC_1}}]V_1 + [\frac{1}{R_2 + \frac{1}{sC_1}} + \frac{1}{sL_1 + \frac{1}{sC_2}}]V_2 - [\frac{1}{sL_1 + \frac{1}{sC_2}}]V_3, \qquad (6.4\text{-}18)$$

$$0 = -[\frac{1}{R_3 + sL_2}]V_1 - [\frac{1}{sL_1 + \frac{1}{sC_2}}]V_2 + [\frac{1}{R_4} + \frac{1}{R_3 + sL_2} + \frac{1}{sL_1 + \frac{1}{sC_2}}]V_3. \qquad (6.4\text{-}19)$$

STOP

Comment 6.4-1: In the above process we have mixed ωt (a time-domain quantity) and s (a frequency-domain quantity, also called the Laplace variable) as well as the upper and the lower case letters. Although, this mixing is used here only for the purpose of demonstration and explanation, such a practice is not acceptable and should not be repeated.

Example 6.4-3: Analyze the circuit in Fig. 6.4-7.

Solution: As indicated in Fig. 6.4-7, here we use mesh equations.

STEP Zero: All pertinent variables are shown in Fig. 6.4-7.

STEP One: Here, we only can set up the integro-differential equations.

$$v(t) = (1^{\Omega})(i_1 - i_3) + \frac{1}{C_1}\int_{-\infty}^{t}[i_1(\tau) - i_2(\tau)]d\tau, \qquad (6.4\text{-}20)$$

$$0 = (1^{\Omega})i_2 + \frac{1}{C_1}\int_{-\infty}^{t}[i_2(\tau) - i_1(\tau)]d\tau + \frac{1}{C_2}\int_{-\infty}^{t}[i_2(\tau) - i_3(\tau)]d\tau, \tag{6.4-21}$$

$$0 = (1^{\Omega})i_3 + L_1\frac{di_3}{dt} + \frac{1}{C_2}\int_{-\infty}^{t}[i_3(\tau) - i_2(\tau)]d\tau + (1^{\Omega})(i_3 - i_1). \tag{6.4-22}$$

Fig. 6.4-7.

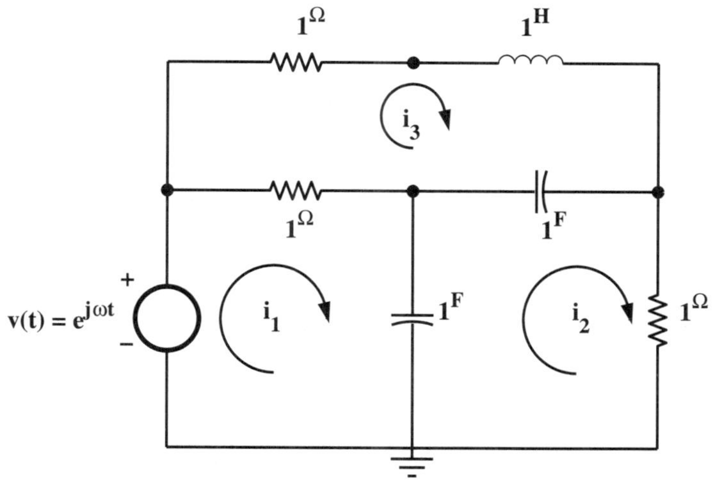

STEP Two: Differentiating the above set of equations and substituting values for various elements results in the following set of ordinary differential equations.

$$\frac{dv(t)}{dt} = j\omega v(t) = \frac{di_1(t)}{dt} - \frac{di_3(t)}{dt} + [i_1(t) - i_2(t)], \tag{6.4-23}$$

$$0 = \frac{di_2(t)}{dt} + [i_2(t) - i_1(t)] + [i_2(t) - i_3(t)], \tag{6.4-24}$$

$$0 = \frac{di_3(t)}{dt} + \frac{d^2i_3}{dt^2} + [i_3(t) - i_2(t)] + \frac{di_3}{dt} - \frac{di_1}{dt}. \tag{6.4-25}$$

STOP

Comment 6.4-2: The preceding set of differential equations must be solved to generate numerical values for $i_1(t)$, $i_2(t)$ and $i_3(t)$. These values are needed, in turn, to complete the analysis, i.e., to find the current through and the voltage across each and every element of the circuit. To do that let $D \triangleq d/dt$ and set the preceding equations in an operator form as follows.

$$j\omega e^{j\omega t} = (D + 1)i_1(t) - i_2(t) - Di_3(t), \tag{6.4-26}$$

$$0 = -i_1(t) + (D + 2)i_2(t) - i_3(t), \tag{6.4-27}$$

$$0 = -Di_1(t) - i_2(t) + (D^2 + 2D + 1)i_3(t). \qquad (6.4\text{-}28)$$

Although at the moment, we are not going to elaborate further regarding this set of equations, we can solve these coupled equations quite easily and development of that solution is one of the topics in the next chapter. (Also, Problem 9.9.15).

6.4.1. Concluding Remarks

We close this section by emphasizing that all analytical methods described in the preceding chapters can be reinterpreted in the context of a complex setting analysis almost line by line. For instance, the formulae for series and parallel of complex elements are exactly the same as their counterparts for real elements. Similarly, the Thevenin's and Norton's theorems hold in a complex setting in exactly the same manners as in the real cases. The maximum power transfer in a complex load changes slightly as is described in Theorem 6.5-1 below. Thus, unless we explicitly state the difference or change, the complex-analysis setting of a circuit is the same as its counterpart in the real setting.

6.5. POWER

In general dissipated power in a complex load is computed in the same way as that for the real load, but with some minor changes. Here,

$$P = VI^* \triangleq R + jQ = |V||I| \angle\phi$$

$$= |V||I|\cos\phi + j|V||I|\sin\phi, \qquad (6.5\text{-}1)$$

where the superscript * stands for complex conjugate.

In (6.5-1), R is the real part of the power also known as *active power,* which is measured in W (Watts). While Q is the imaginary part of the power also known as *reactive power,* and is measured in VAR (volt-ampere reactive). The so-called *apparent power,* is $|P| = \sqrt{R^2 + Q^2}$ and is measured in VA (volt ampere). The angle ϕ is the angle between the voltage and the current in a given load. The $\cos\phi$ is called the *power factor.*

In general, one side of the circuit *delivers* power while the other side *receives*. In the industrial sense, the power companies deliver energy and the consumer receive. Power companies prefer the costumers to minimize their reactive loads, for instance too many motors running in the house will create a reactive load which will generate almost 90° phase shift, i.e., $\cos\phi = 0$, while $\sin\phi = 1$. This is not desirable from the point of view of power company, and that is why we see in streets big capacitors mounted on various distribution lines to "correct" the reactive loads as much as possible.

For the maximum power transfer, the counterpart of Theorem 5.5-1 for complex elements is as follows.

Theorem 6.5-1 (The maximum power transfer): Referring to Fig. 6.5-1, it is easy to show that given $Z_{Th} = X + jY$, the optimal load impedance for the maximum power dissipated is $Z_l = X - jY = Z_{Th}^*$ (i.e., the load's impedance is the complex conjugate of the corresponding Thevenin's impedance).

This theorem for real elements states that given a voltage source V_{Th} and a series resistance R_{Th}, the power dissipated in R_l, which is in series with (V_{Th}, R_{Th}) will be maximum when $R_l = R_{Th}$. To prove Theorem 6.5-1, we follow exactly similar steps as in Section 5.5 for circuit with a real element (cf., Problem 6.6.14).

Fig. 6.5-1.

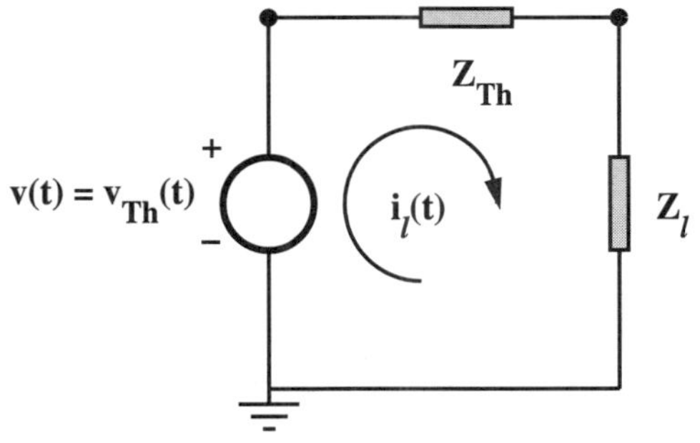

6.6. PROBLEM SET

Fundamentals

6.6.1. Compute voltages across the inductor and capacitor in the circuit of Fig. 6.6-1.

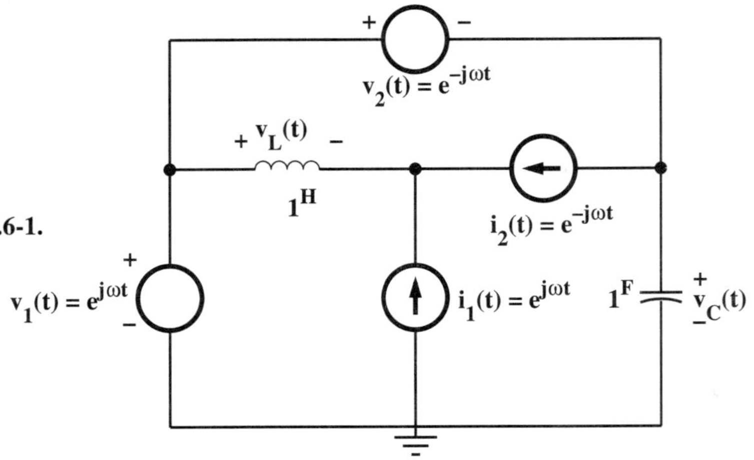

Fig. 6.6-1.

6.6.2. In the circuit of Fig. 6.6-2 with $v_{in}(t) = 50\sin(\omega t + \frac{\pi}{4})$, find $v_o(t)$.

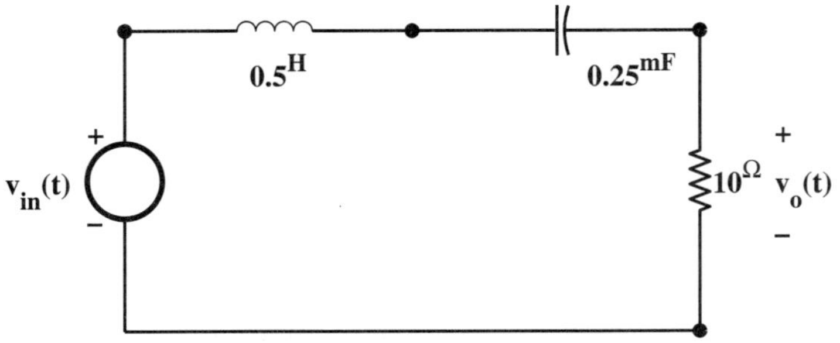

Fig. 6.6-2.

6.6.3. In the circuit of Fig. 6.6-3 with $i(t) = e^{j\omega t}$, find $v_o(t)$.

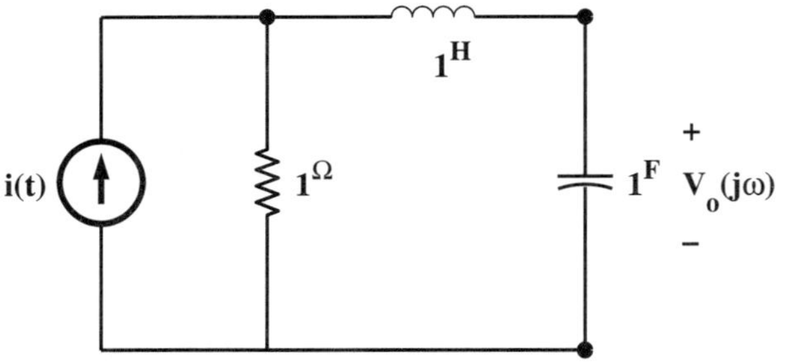

Fig. 6.6-3.

6.6.4. Find Z_{eq} for the circuit in Fig. 6.6-4.

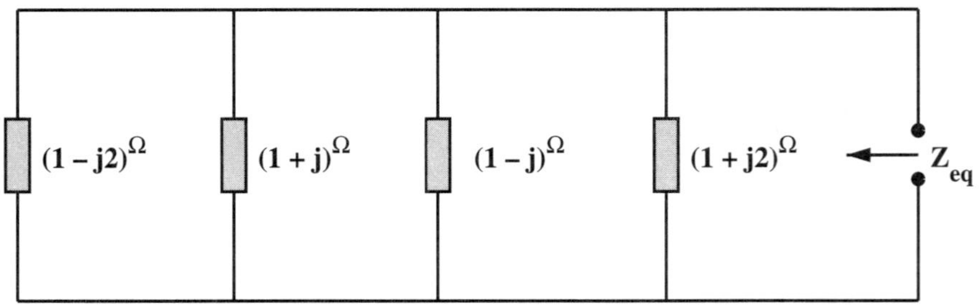

Fig. 6.6-4.

6.6.5. Find Z_{eq} for the circuit in Fig. 6.6-5 at $\omega = 1$ rad/sec.

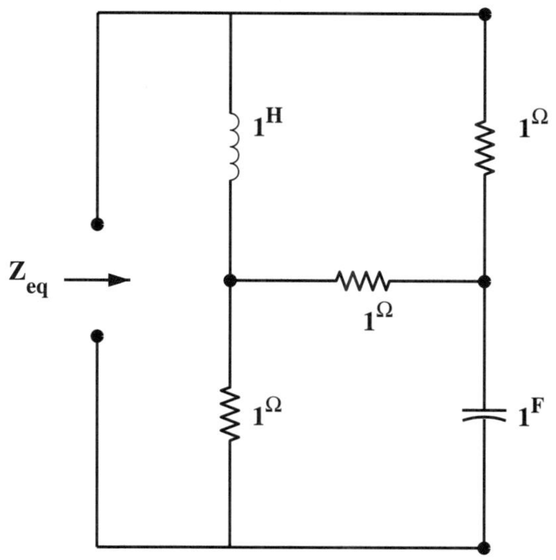

Fig. 6.6-5.

6.6.6. For the circuit in Fig. 6.6-6, find Z_{eq} at $\omega = 1000$ rad/sec.

Fig. 6.6-6.

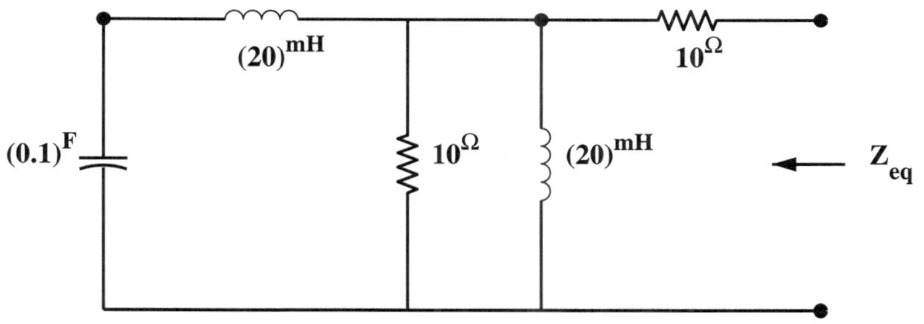

Analysis

6.6.7. For the circuit in Fig. 6.6-7, find I_o.

Fig. 6.6-7.

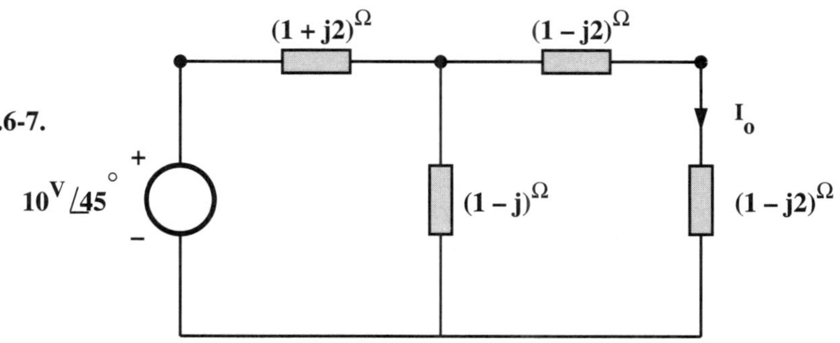

6.6.8. Complete the analysis for the circuit in Fig. 6.3-5.

6.6.9. For circuits of Figs. 6.6-8 & 9, write the necessary set of integro-differential equations, in order to analyze the circuit completely. Use the mesh equations approach with variables as shown.

Fig. 6.6-8.

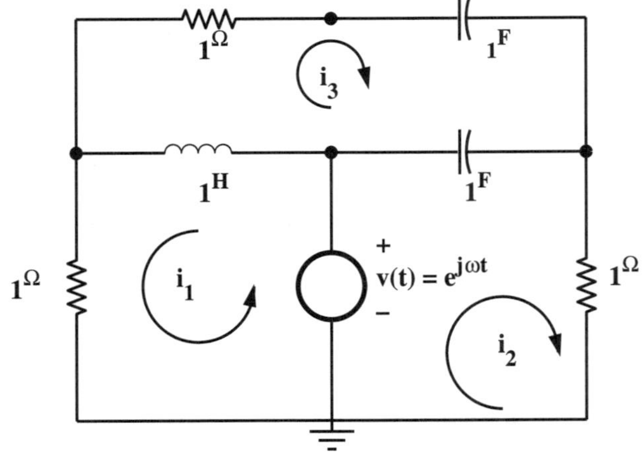

Fig. 6.6-9.

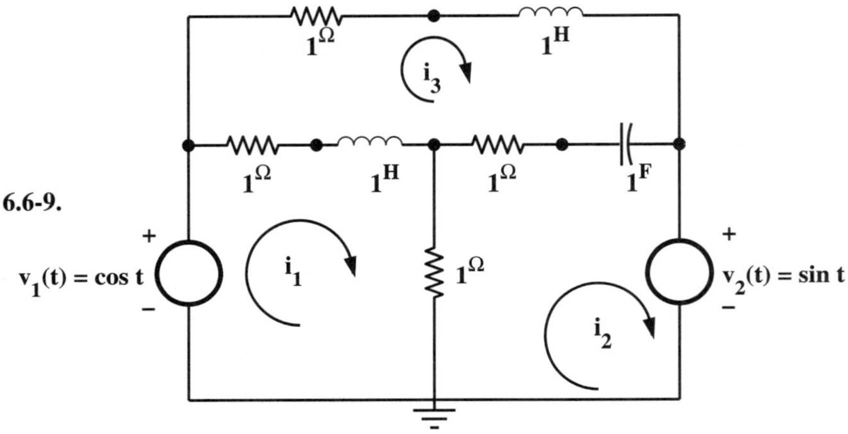

6.6.10. At $\omega = 1$ rad/sec, analyze the circuit in Fig. 6.6-10 with mixed sources using the superposition rule.

Fig. 6.6-10.

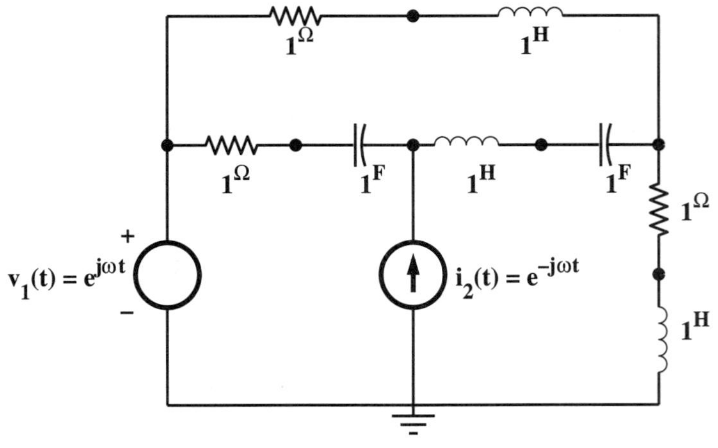

6.6.11. At $\omega = 1$ rad/sec, analyze the circuit in Fig. 6.6-10 with mixed sources using the remaining methods of circuit analysis which we have studied in this textbook.

Reconfiguration approach in its many forms plays a major role in this circuit.

Thevenin's Theorem

6.6.9. Find the Thevenin's equivalent for circuits in Figs. 6.6-11 & 12 at port ab and for $\omega = 1$ rad/sec.

Fig. 6.6-11.

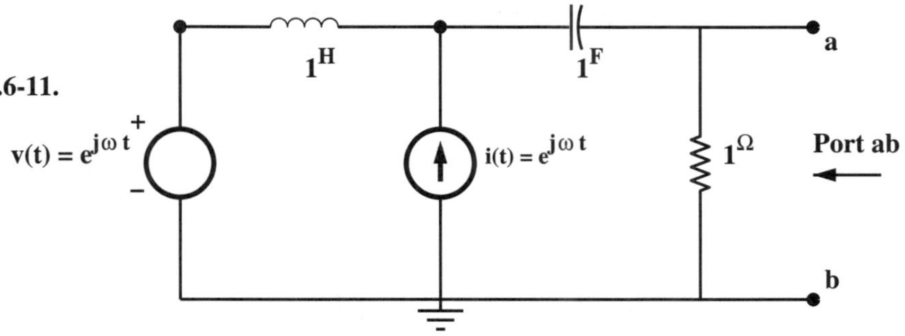

Fig. 6.6-12.

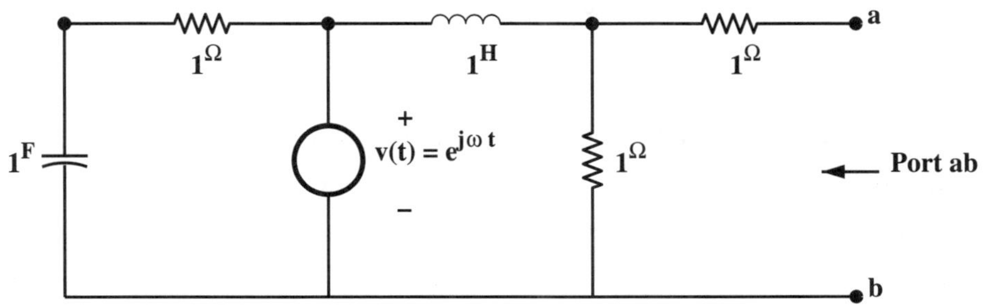

Input-Output Ratio

6.6.13. For the circuit in Fig. 6.6-13, find V_o/V_1 at $\omega = 1$ rad/sec.

Fig. 6.6-13.

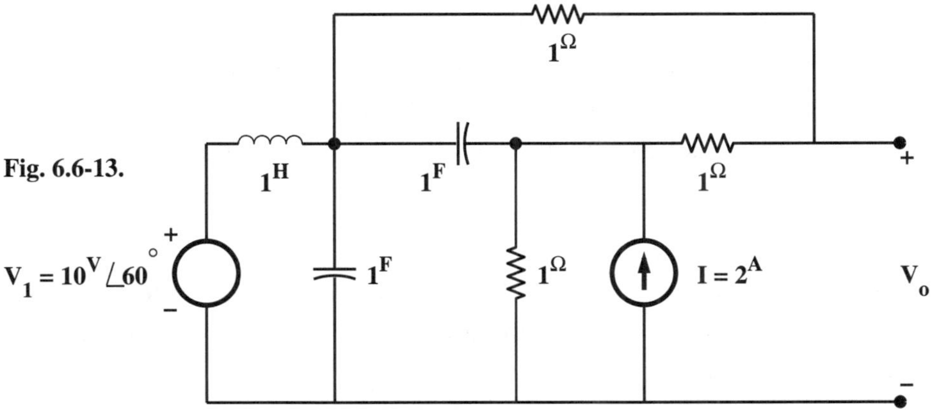

Power

6.6.14. Prove the result stated in Theorem 6.5-1.

6.6.15. For the circuit in Fig. 6.6-14, find Z_l at $\omega = 1$ rad/sec for the maximum power transfer.

Fig. 6.6-14.

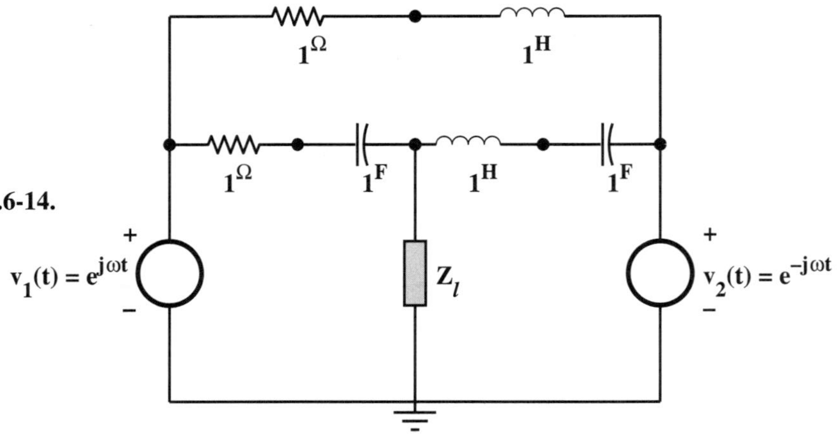

Comment 6.6-1: There are many issues which we have not discussed. For instance, when we have more than one inductor as in the circuit of Fig. 6.6-10, we need to consider another phenomenon called mutual inductance which is discussed in Chapter Eight. Does that mean this circuit is wrong? Of course, not. This means that we have assumed the inductors are isolated from each other. Since we have not defined isolation, we have not used that concept in circuit description yet. There are however some issues which are not so critical at the moment, and if we need to trade them for other topics, we do so. For instance, every time we use energy stored elements such as inductors and capacitors, we need to be concerned about their initial conditions, and we have implicitly considered them at zero in order to proceed. But a complete numerical solution at the moment complicates our efforts to appreciate and distinguish issues which underline and form the essential topics or the core in circuit analysis and not those in mathematics or physics. We consider the full numerical solution of circuits in our subsequent discussions, and we undertake such a task at the right moment. Nevertheless, the message that we want to convey is that by compartmentalizing materials systematically, we can attack them rigorously and orderly.

Comment 6.6-2: Again, it is the right moment to go back to Table 1.1-1 and assess as where are we standing in the course and which directions must we go from here. Reflection is always fruitful.

7 DIFFERENTIAL EQUATIONS AND CIRCUITS WITH SEQUENTIAL SWITCHES

7.1. INTRODUCTION

Our final task is to generate a numerical solution for a practical circuit. To that end, we study all pertinent analytical tools. But first, we must reevaluate certain assumptions which we have made thus far for a better understanding of what we are about to propose. For instance, in any meaningful circuit operation, we do not have, say a constant battery or any source of energy that is sustainable indefinitely. What we have, and in order to make a circuit functional, is a switch that is turned on at a specific instant and possibly for a finite duration. This is the same as saying we need to bring from the shelf all the necessary circuit components together and build the circuit. This process of bringing together all the pieces is the same as saying that at a certain instant we turn on a switch and start a circuit that was at rest prior to that specific moment. Therefore, in reality a battery is only functional from one specific instant then forward. This instant is our reference point. The preceding mechanism is for all sources. Furthermore, circuits with energy stored elements have histories which are carried through their initial conditions. Correspondingly, in this chapter, upon reviewing the essential aspects of generating solution for a differential equation which governs a circuit, we study circuits with one or more sequential switches. This study is the prerequisite for learning the operational and functionality conditions of a circuit with history. We show how to find the initial conditions and their propagations in the circuit for different circuit variables.

7.2. DIFFERENTIAL EQUATIONS

To solve any realistic circuit numerically, we often run into either a set of first-order ordinary differential equations *(ODE's)*, or one n^{th}–order *ODE* with constant coefficients. In the following we review general procedures to tackle such equations.

Consider an nth–order *ODE* with constant coefficients as follows.

$$a_n \frac{d^n x(t)}{dt^n} + a_{n-1} \frac{d^{n-1} x(t)}{dt^{n-1}} + \cdots + a_1 \frac{dx(t)}{dt} + a_0 x(t) = f(t). \tag{7.2-1}$$

Here, x(t) is the unknown variable, which is also a function of the real variable t, herein time, a_i's are all constant quantities, and f(t) is the forcing function or input. In this course, all these quantities are real variables. The general solution for this problem is as follows.

$$x(t) = x_h(t) + x_p(t), \tag{7.2-2}$$

where $x_h(t)$ is the solution to the homogeneous part of (7.2-1), i.e., when

$$a_n \frac{d^n x_h(t)}{dt^n} + a_{n-1} \frac{d^{n-1} x_h(t)}{dt^{n-1}} + \cdots + a_1 \frac{dx_h(t)}{dt} + a_0 x_h(t) \equiv 0. \tag{7.2-3}$$

While $x_p(t)$ is *one* particular solution to (7.2-1) when f(t) is accounted for, i.e., any solution of (7.2-1) which satisfies both sides of this equation.

The above procedure holds globally, i.e., no matter how $x_h(t)$ and $x_p(t)$ are found, the final solution is the summation of $x_h(t)$ and $x_p(t)$. In the following we present one general method to find $x_h(t)$, followed by some *ad hoc* or intuitive methods for finding $x_p(t)$.

7.2.1. Homogeneous Solution

Consider an nth–order *ODE* with constant coefficients as follows.

$$a_n \frac{d^n x_h(t)}{dt^n} + a_{n-1} \frac{d^{n-1} x_h(t)}{dt^{n-1}} + \cdots + a_1 \frac{dx_h(t)}{dt} + a_0 x_h(t) = 0. \tag{7.2-4}$$

Without any loss of generality, we let $a_n \equiv 1$ in the following. To compute $x_h(t)$, we let $D \equiv d/dt$ to serve as an operator, then (7.2-4) becomes

$$(D^n + a_{n-1}D^{n-1} + \cdots + a_1 D + a_0)x_h(t) = 0. \tag{7.2-5}$$

Before proceeding further, we ask the following question. What kind of function or x(t) can possibly satisfy (7.2-5)? In other words, what type of functions can be differentiated and added as in the manner of (7.2-5) to sum to zero? The answer is exponential functions like $e^{\alpha_i t}$, for a proper set of α_i's, will satisfy this type of equations. To verify this claim, we compute various derivatives of $e^{\alpha_i t}$ and substitute those quantities in (7.2-5) to get

$$(\alpha_i^n + a_{n-1}\alpha_i^{n-1} + \cdots + a_1\alpha_i + a_0)e^{\alpha_i t} = 0. \tag{7.2-6}$$

Since $e^{\alpha_i t} \neq 0$, otherwise trivial, the parenthesis in (7.2-6) must be equal to zero. That means α_i's are special values which make the parenthesis in (7.2-6) to vanish. Thus, we look for $e^{\alpha_i t}$ as a candidate for $x_h(t)$ such that the following equation equals zero, and solve for its roots, i.e., we search for α_i's such that

$$\alpha_i^n + a_{n-1}\alpha_i^{n-1} + \cdots + a_1\alpha_i + a_0 \equiv 0. \tag{7.2-7}$$

This n^{th}–order polynomial equation has "n" distinct roots α_1 to α_n (real or pairs of complex conjugate numbers, also called structural modes) and is called the **characteristic equation**. Thus, each exponential $e^{\alpha_i t}$ for $i = 1, \ldots, n$, is a candidate for $x_h(t)$. So is a scalar-multiple of this exponential as $A_i e^{\alpha_i t}$ for some constant A_i's, $i = 1, \ldots, n$, yet to be determined. In other words, if we substitute any one of these terms in (7.2-4), the net effect is that both sides of that equation is multiplied by A_i and that does not change the outcome for the solution of the underlying differential. In fact since this equation represents a linear system, by the superposition rule, $x_h(t)$ is an algebraic combination of all these exponentials as follows.

$$x_h(t) = A_1 e^{\alpha_1 t} + A_2 e^{\alpha_2 t} + \cdots + A_n e^{\alpha_n t} = \sum_{i=1}^{n} A_i e^{\alpha_i t}. \tag{7.2-8}$$

We now face another issue that we have addressed partially, but have not acted upon, which brings us to the following topic. A dynamic system (circuits included) has a memory and remembers its history. For instance, when we charge a capacitor to a certain voltage level, before disconnecting it from the circuit, the capacitor maintains that voltage level, and when it is put back in another circuit, that voltage level acts as an initial condition for this element. Similarly, an inductor remembers its last current value. These values, also known as **initial conditions** must be accounted for in the final solution of a circuit containing these complex elements. That is how we incorporate A_i's in the above solution, such that the overall final solution satisfies all pertinent initial conditions of all complex elements in the circuit.

7.2.2. Particular Solution

The procedure for computing $x_p(t)$, for any given f(t), is not so straightforward as its counterpart $x_h(t)$. Here, we need more experience than before, which is the crux of a special course in differential equations. One such general method is the utilization of the Laplace transformation techniques, which enables us to solve a large class of problems provided that f(t) is Laplace transformable. The procedures to compute $x_p(t)$ for a general f(t) are described in most differential equations textbooks. From our point of view that centers around issues pertaining to basics of circuit analysis, however, we are primarily concerned with a few special f(t)'s. Those special functions are listed below.

(1) The forcing function f(t) is a **constant,** i.e., $f(t) \equiv c$, for $-\infty < t < \infty$. Constant sources fit under this category. Here, we let $a_0 x(t) = f(t) = c$, then $x_p(t) = c/a_0$ satisfies a typical (7.2-1) as this can be shown by straightforward differentiation and substitution of $x_p(t)$ in (7.2-1). Since this $x_p(t)$ satisfies both sides of (7.2-1), this is a particular solution of (7.2-1).

(2) **Step function.** In reality, and from circuit point of view, we never have such a working horizon of $-\infty < t < \infty$, but rather we have $0 < t < \infty$. In other words, each time we insert a "constant" source in the circuit, we do so through a switch, which is open prior to some reference point in time. Thus, in reality we have the following f(t), Fig. 7.2-1, which is zero prior to the reference point of t = 0, and is a constant "c" afterward. This special function is very dear to electrical engineers. In fact, an important assumption that is regularly made regarding circuit analysis is that every forcing function is applied at one specific instant, which for convenience we choose that instant to be t = 0. Without such a formal assumption or incorporation of a switch to turn the circuit on, we cannot solve the many circuit problems. We often take this switching effect for granted.

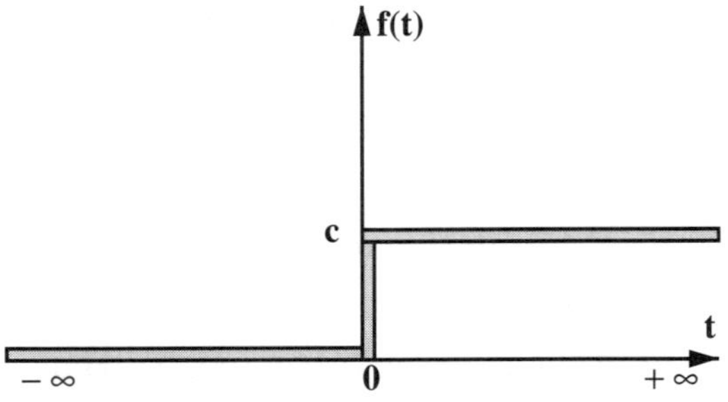

$$f(t) = \begin{cases} c & t > 0 \\ 0 & t \le 0 \end{cases}$$

Fig. 7.2-1. Step function.

(3) **Ramp function.** Here, $f(t) = ct$, for some constant c and $0 < t < \infty$.

(4) **Rational polynomials.** Similarly, $f(t) = \sum_{i=1} c_i t^i$, for some constant c_i's and $0 < t < \infty$.

(5) **Exponentials.** Here, $f(t) = e^{\alpha t}$, which by a Taylor series expansion can be categorized under the above rational polynomials (4), for $0 < t < \infty$.

(6) **Sinusoidals.** Similarly, $f(t) = \sin \omega t$ as well as $f(t) = \cos \omega t$ for some ω can be written in terms of exponentials using the Euler formula, for $0 < t < \infty$.

All the above forcing functions are examples of basic functions which we face in circuit analysis. We study the actual derivations of particular solutions for the above cases in our examples. All in all, we try to build a structure for various forcing functions based on our knowledge gained from the simplest case of a constant input to the advanced cases of sinusoidal inputs. Having mastered these special cases enables us to study the many other input functions, because any other rational function can be described in terms of the above references cases.

7.2.3. Complete or Total Solution

Having generated $x_h(t)$ and $x_p(t)$, then the complete or total solution of (7.2-1) becomes as follows.

$$x(t) = \sum_{i=1}^{n} A_i e^{\alpha_i t} + x_p(t). \tag{7.2-9}$$

Here, A_i's must be computed from the initial conditions of the circuit. To that end, consider the above case with n initial conditions as follows. Suppose at some t_0 we have

$$\left\{\begin{matrix} x(t_0) \\ \left.\dfrac{dx}{dt}\right|_{t=t_0} \\ \cdot \\ \cdot \\ \cdot \\ \left.\dfrac{d^{n-1}x}{dt^{n-1}}\right|_{t=t_0} \end{matrix}\right\} \triangleq \left\{\begin{matrix} x_0 \\ \dot{x}_0 \\ \cdot \\ \cdot \\ \cdot \\ x_0^{(n-1)} \end{matrix}\right\}. \tag{7.2-10}$$

Note that for an nth–order *ODE,* we need the initial condition at t_0 plus its $(n-1)^{st}$ derivatives at the same time t_0. Thus, from (7.2-9) we have:

$$e^{\alpha_1 t_0}A_1 + e^{\alpha_2 t_0}A_2 + \cdots + e^{\alpha_n t_0}A_n = x_0 - x_p(t_0),$$

$$\alpha_1 e^{\alpha_1 t_0}A_1 + \alpha_2 e^{\alpha_2 t_0}A_2 + \cdots + \alpha_n e^{\alpha_n t_0}A_n = \dot{x}_0 - \dot{x}_p(t_0),$$

$$\cdot$$
$$\cdot$$
$$\cdot$$

$$\alpha_1^{n-1} e^{\alpha_1 t_0}A_1 + \alpha_2^{n-1} e^{\alpha_2 t_0}A_2 + \cdots + \alpha_n^{n-1} e^{\alpha_n t_0}A_n = x_0^{(n-1)} - x_p^{(n-1)}(t_0). \tag{7.2-11}$$

Also, note that in (7.2-11), all $x_0^{(i)}$, for $i = 0, \ldots, n-1$, are defined in (7.2-10), while all $x_p^{(i)}(t_0)$, also for $i = 0, \ldots, n-1$, are generated from the known forcing function.

Example 7.2-1: Solve the following ordinary differential equation.

$$\frac{d^3x(t)}{dt^3} + 6\frac{d^2x(t)}{dt^2} + 11\frac{dx(t)}{dt} + 6x(t) = 1, \tag{7.2-12}$$

assuming that

$$x(0) = 0, \quad \frac{dx}{dt}\Big|_{t=0} = 0, \quad \frac{d^2x(t)}{dt^2}\Big|_{t=0} = 0. \tag{7.2-13}$$

Solution: This is a third-order *ODE* with constant coefficients and a forcing function or input of $f(t) \equiv 1$, for all t. The solution consists of two parts $x_h(t)$ and $x_p(t)$ as follows.

STEP Zero: We make sure that all pertinent parameters are given. Here, this is a third-order differential equation and we need to have three initial conditions, which we do, while the forcing function is a constant. In other words, we continue to make a habit of studying the structure of the problem before proceeding further to seek a numerical solution.

STEP One: Solve the homogeneous part by letting $D \equiv d/dt$ as follows.

$$(D^3 + 6D^2 + 11D + 6)x_h(t) = 0. \tag{7.2-14}$$

The roots of the characteristic equation

$$D^3 + 6D^2 + 11D + 6 \equiv 0, \tag{7.2-15}$$

are as follows.

$$D_1 = -1, \quad D_2 = -2, \quad D_3 = -3. \tag{7.2-16}$$

These are the same as α_1, α_2 and α_3. Thus,

$$x_h(t) = A_1e^{-t} + A_2e^{-2t} + A_3e^{-3t}. \tag{7.2-17}$$

STEP Two: Compute the particular solution. Here, we have $f(t) \equiv 1$, and $a_0x(t) = 6x(t) = f(t)$ yields

$$x_p(t) = \frac{1}{6}, \quad \text{for all t.} \tag{7.2-18}$$

STEP Three: Complete the formal solution by adding $x_h(t)$ and $x_p(t)$, and invoke the initial conditions' requirements. Here, the complete solution is

$$x(t) = A_1 e^{-t} + A_2 e^{-2t} + A_3 e^{-3t} + \frac{1}{6}, \tag{7.2-19}$$

and the initial conditions' requirements yield

$$A_1 + A_2 + A_3 = x(0) - x_p(0) = -\frac{1}{6}, \tag{7.2-20}$$

$$-A_1 - 2A_2 - 3A_3 = \dot{x}(0) - \dot{x}_p(0) = 0, \tag{7.2-21}$$

$$A_1 + 4A_2 + 9A_3 = 0. \tag{7.2-22}$$

Adding (7.2-20) and (7.2-21) as well as adding (7.2-21) and (7.2-22) yield

$$-A_2 - 2A_3 = -\frac{1}{6}, \tag{7.2-23}$$

$$2A_2 + 6A_3 = 0. \tag{7.2-24}$$

From the above set we have $A_3 = -1/6$, $A_2 = 1/2$, and finally $A_1 = -1/2$.

STEP Final: The complete or total solution becomes

$$x(t) = -\frac{1}{2}e^{-t} + \frac{1}{2}e^{-2t} - \frac{1}{6}e^{-3t} + \frac{1}{6}. \tag{7.2-25}$$

STOP

Comment 7.2-1: Going back to our fundamental goal in this elementary textbook, namely, to analyze [completely] circuits with all sorts of elements including different sources, we note that the types of differential equations which we are interested to analyze (i.e., those which govern circuits with lumped elements) are such that they enjoy many special properties. For instance, that class of problems is linear and always yields a differential equation with constant coefficients. Thus, we have been reviewing only those aspects of solutions (or methodologies) in differential equations which are of interest in this regard. We further expand our skills using Laplace transformation which changes a differential equation into an algebraic equation to tackle more complicated problems than are possible with the present approach. However, and for now, we study a procedure to solve a class of coupled differential equations which arises naturally in circuit analysis. These coupled equations come from the simplest circuits containing two or more complex elements such as inductors and capacitors.

Example 7.2-2: Solve the following set of two coupled ordinary differential equations with the following initial conditions $x(0) = 0$, $\dot{x}(0) = 0$, $y(0) = 0$ and $\dot{y}(0) = 1/2$, where $D \equiv d/dt$.

$$\begin{cases} (D+1)x(t) + (D+3)y(t) = t, \\ (D+2)x(t) + (3D+9)y(t) = t^2. \end{cases} \tag{7.2-26}$$

Solution: Though (7.2-26) is purely an academic example at the moment, differential equations representing any circuit consisting of capacitors and inductors are at least as complicated as this set. For instance, x(t) may represent the voltage across a capacitor, while y(t) is the current through an inductor. Therefore, we must know how to solve this class of problems for its applications in circuit analysis. Also note that, we now have "t" and "t^2" as the forcing functions, which these functions are rational polynomials of t and as such are new to us.

STEP Zero: We first study the problem. All pertinent parameters are specified in the following sense. Because the equations are both first-order in D, it may seem that only x(0) and y(0) are required. However, the method that we are about to present (known as an operator approach) requires differentiation with respect to time, in order to transform the coupled problem into a set of decoupled, though of higher-order differential equations. Thus, we often need higher-order initial conditions to remove the coupling between the corresponding coupled equations. In a mathematically well-posed problem, these values are specified accordingly, when needed. In a physical circuit, we often can develop methods to calculate higher-order initial conditions from the circuit, as is described in Section 7.4.

STEP One: In order to solve the preceding set of equations, we invoke steps to remove coupling among them. In the same manner as we would have done, if we were solving a set of coupled algebraic equations. Namely, we start with one variable and try to substitute that in the next equation and repeat this process until we have only one equation and one unknown. Here, we use the same concept, except to eliminate a variable we need to differentiate as shown below.

We multiply the first equation in (7.2-26) by (D + 2) and the next equation by (D + 1) to get the following set.

$$\begin{cases} (D+2)(D+1)x(t) + (D+2)(D+3)y(t) = (D+2)t, \\ (D+1)(D+2)x(t) + (D+1)(3D+9)y(t) = (D+1)t^2. \end{cases} \tag{7.2-27}$$

Considering that $(D + 2)(D + 1) = (D + 1)(D + 2)$, the differentiation operator commutes, subtracting the above two equations yields

$$[(D + 1)(3D + 9) - (D + 2)(D + 3)]y(t) = (D + 1)t^2 - (D + 2)t. \tag{7.2-28}$$

Applying the D operation on the right-hand side and simplifying the left-hand side yields

$$(2D^2 + 7D + 3)y(t) = 2t + t^2 - 1 - 2t, \tag{7.2-29}$$

or,

$$(2D + 1)(D + 3)y(t) = t^2 - 1. \tag{7.2-30}$$

To carry on, we note that the complete solution of (7.2-30), just as any such differential equation, consists of two parts as follows.

$$y(t) = y_h(t) + y_p(t), \tag{7.2-31}$$

where $y_h(t)$ is a solution of $(2D + 1)(D + 3)y_h(t) \equiv 0$, i.e.,

$$y_h(t) = A_y e^{-(\frac{1}{2})t} + B_y e^{-3t}. \tag{7.2-32}$$

The particular solution when the forcing function is a rational polynomial of t, is a new case for us. To tackle such problems, the procedure is to let the particular solution be also of the same complexity as the forcing function, but parameterized accordingly in such a manner that our task becomes determination of the corresponding parameters. Thus, we let

$$y_p(t) = a_y t^2 + b_y t + c_y, \tag{7.2-33}$$

which is a polynomial of the *same* complexity (structure) as the forcing function in (7.2-30). The coefficient parameters are selected such that $y_p(t)$ satisfies both sides of (7.2-30). That means, we must have

$$2D^2(a_y t^2 + b_y t + c_y) + 7D(a_y t^2 + b_y t + c_y)$$
$$+ 3(a_y t^2 + b_y t + c_y) \equiv t^2 - 1. \tag{7.2-34}$$

Applying the D operator on the left-hand side yields

$$2D(2a_y t + b_y) + 7(2a_y t + b_y) + 3(a_y t^2 + b_y t + c_y) \equiv t^2 - 1. \tag{7.2-35}$$

Applying one more time D, and rearranging terms yields

$$3a_y t^2 + (14a_y + 3b_y)t + (4a_y + 7b_y + 3c_y) \equiv t^2 - 1. \tag{7.2-36}$$

Thus, equating term by term on both sides of the polynomial equation yields

$$3a_y \equiv 1 \quad \Rightarrow \quad a_y = 1/3, \tag{7.2-37a}$$

$$14a_y + 3b_y \equiv 0 \quad \Rightarrow \quad b_y = -14/9, \tag{7.2-37b}$$

$$4a_y + 7b_y + 3c_y \equiv -1 \quad \Rightarrow \quad c_y = \frac{77}{27}. \tag{7.2-37c}$$

Substituting these values in (7.2-33), then that in (7.2-31) with (7.2-32) yields

$$y(t) = A_y e^{-(1/2)t} + B_y e^{-3t} + \frac{1}{3}t^2 - \frac{14}{9}t + \frac{77}{27}. \tag{7.2-38}$$

Applying the initial conditions to (7.2-38) yields

$$\begin{cases} A_y + B_y + \dfrac{77}{27} = 0, \\ \dfrac{A_y}{2} + 3B_y + \dfrac{14}{9} = 0, \end{cases} \quad \Rightarrow \quad \begin{cases} A_y = -\dfrac{14}{5}, \\ B_y = -\dfrac{7}{135}. \end{cases} \tag{7.2-39}$$

Thus, the final solution for y(t) is

$$y(t) = -\frac{14}{5}e^{-(1/2)t} - \frac{7}{135}e^{-3t} + \frac{1}{3}t^2 - \frac{14}{9}t + \frac{77}{27}. \tag{7.2-40}$$

STEP Two: We proceed with the application of the preceding step on the next variable which in this example is the last variable. Going back to the original set of (7.2-26), we now remove y(t) in favor of x(t). We need to emphasize that it is imperative to start from the original set in order to account for the system coupling or the entire *structural modes* of the system. That means, we begin with a similar equation as that of (7.2-27). Namely, we now multiply the first equation in (7.2-26) by (3D + 9) and the next equation by (D + 3) to get the following set.

$$\begin{cases} (3D+9)(D+1)x(t) + (3D+9)(D+3)y(t) = (3D+9)t, \\ (D+3)(D+2)x(t) + (D+3)(3D+9)y(t) = (D+3)t^2. \end{cases} \tag{7.2-41}$$

Considering similar arguments as in Step One, subtracting the above two equations yields

$$(2D+1)(D+3)x(t) = -3t^2 + 7t + 3, \tag{7.2-42a}$$

whose solution consists of

$$x(t) = x_h(t) + x_p(t). \tag{7.2-42b}$$

Here, $x_h(t)$ is also a solution of $(2D+1)(D+3)x_h(t) \equiv 0$, i.e.,

$$x_h(t) = A_x e^{-(½)t} + B_x e^{-3t}. \tag{7.2-43}$$

If instead we had substituted y(t) in the first equation in (7.2-26), then x(t) would not have included both exponentials which means one less mode and that is not a complete solution. This is why we must go back to the original set to capture the entire structural (natural) modes of the system which in this case are two exponentials $e^{-(½)t}$ and e^{-3t}. Similarly, we choose

$$x_p(t) = a_x t^2 + b_x t + c_x. \tag{7.2-44}$$

From (7.2-42), with $x_p(t)$ as in (7.2-44), we have

$$2D^2(a_x t^2 + b_x t + c_x) + 7D(a_x t^2 + b_x t + c_x)$$
$$+ 3(a_x t^2 + b_x t + c_x) \equiv -3t^2 + 7t + 3, \tag{7.2-45}$$

where (7.2-45) results in

$$a_x = -1, \quad b_x = 7, \quad c_x = -14. \tag{7.2-46}$$

Thus,

$$x(t) = A_x e^{-(½)t} + B_x e^{-3t} - t^2 + 7t - 14. \tag{7.2-47}$$

Applying the initial conditions to (7.2-47) yields

$$\begin{cases} A_x + B_x - 14 = 0, \\ -\dfrac{A_x}{2} - 3B_x + 7 = 0, \end{cases} \Rightarrow \begin{cases} A_x = 14, \\ B_x = 0, \end{cases} \tag{7.2-48}$$

where $B_x = 0$ means x(t) is not affected by one mode as the result of our selected set of initial conditions. Thus, the final numerical solution for x(t) is

$$x(t) = 14e^{-(½)t} - t^2 + 7t - 14. \tag{7.2-49}$$

STEP Final: We continue this process until we recover all variables, which in this problem there is none left.

Comment 7.2-3: The previous example with a rational polynomial forcing function leads us to the case where the forcing function is an arbitrary exponential. For instance, $f(t) = e^{\alpha t}$, provided that α is not a root of the corresponding characteristic equation. Otherwise, refer to Comment 7.2-5. Or, when f(t) is equal to any linear combinations of these types of exponentials, the procedure to compute $x_p(t)$ as outlined below is straightforward and is based on the consequence of the above choice for the particular solution corresponding to a forcing function which is a rational polynomial of t. Again to verify this claim is a purely algebraic exercise.

Example 7.2-3: Solve the following differential equation with $x(0) = 0$ and $\dot{x}(0) = 0$.

$$(D^2 + 3D + 2)x(t) = e^{-3t}. \tag{7.2-50}$$

Solution: Here, we have an exponential forcing function, and although we have not yet solved its characteristic equation, our basic assumption is that -3 is not a root of that equation. In other words, we are not exciting the system with its own **natural mode.** If so, then the analysis becomes more subtle than what we have done so far, and is described in Comment 7.2-5 with an application in Example 7.2-4.

STEP Zero: All pertinent parameters or initial conditions are specified. Also, note that $x(t)$ and $\dot{x}(t)$ are two separate physical quantities. Thus setting $x(0) = \dot{x}(0) = 0$ is not technically correct, and we should show these in two separate equalities, i.e., $x(0) = 0$ and $\dot{x}(0) = 0$.

STEP One: The characteristic equation for this differential equation has two roots as follows.

$$D^2 + 3D + 2 = 0, \quad \Rightarrow \quad D_1 = -1, \quad D_2 = -2. \tag{7.2-51}$$

Thus,

$$x_h(t) = A_1 e^{-t} + A_2 e^{-2t}. \tag{7.2-52}$$

STEP Two: The particular solution is proposed as follows.

$$x_p(t) = ke^{-3t}, \quad \text{for some constant and yet to be specified k.} \tag{7.2-53}$$

The thought behind this choice is that in the context of the previous example, we may use a Taylor series expansion of the exponential to get a set of rational polynomials. Then one must go through the same process as using polynomials to complete the solution. However, that approach soon becomes cumbersome. So, we begin with an exponential at the outset which is the closed form for a series of polynomials and proceed accordingly.

Substituting (7.2-53) in (7.2-50) yields

$$9ke^{-3t} + 3(-3ke^{-3t}) + 2ke^{-3t} = e^{-3t}, \tag{7.2-54}$$

resulting in, k = 1/2.

STEP Final: The complete or total solution is

$$x(t) = x_h(t) + x_p(t) = A_1e^{-t} + A_2e^{-2t} + (½)e^{-3t}. \tag{7.2-55}$$

At t = 0, we have x(0) = 0 and $\dot{x}(0) = 0$. Thus,

$$A_1 + A_2 + \frac{1}{2} = 0, \tag{7.2-56}$$

$$-A_1 - 2A_2 - \frac{3}{2} = 0. \tag{7.2-57}$$

Resulting in, $A_2 = -1$, $A_1 = 1/2$, and

$$x(t) = (½)e^{-t} - e^{-2t} + (½)e^{-3t}. \tag{7.2-58}$$

Comment 7.2-4: Finally, for cases where f(t) = sin ωt or f(t) = cos ωt, or any combinations of these terms, we can use the Euler formula and write

$$\sin \omega t = \frac{e^{j\omega t} - e^{-j\omega t}}{2j}, \tag{7.2-59}$$

$$\cos \omega t = \frac{e^{j\omega t} + e^{-j\omega t}}{2}, \tag{7.2-60}$$

to transform sinusoidals into exponentials. Thus generally we can solve any linear differential equation with constant coefficients and sinusoidal forcing functions. However, one approach when working with sinusoidal forcing inputs is to let x_p be a linear combination of sinusoidals of similar complexity as in (7.7-44) (for rational functions) or (7.7-53) (for exponential functions) and proceed forward.

Comment 7.2-5: If α in $f(t) = e^{\alpha t}$ of Comment 7.2-3 were a root of the corresponding characteristic equation of the system, then we proceed as follows. Instead of choosing a constant K in a typical particular solution like (7.2-53), we assume that

$$x_p(t) = K(t)e^{\alpha t}, \tag{7.2-61}$$

where K is now a function of time. We then proceed to generate an auxiliary differential equation for K(t) with all its relevant initial conditions are set at zero. The simplest particular solution of this differential equation becomes the K(t).

Example 7.2-4: Solve the following differential equation for its complete solution with parametric initial conditions.

$$\ddot{x}(t) + 5\dot{x}(t) + 6x(t) = e^{-2t} + e^{-3t}, \quad t \geq 0. \tag{7.2-62}$$

Solution: We proceed to find a complete solution by generating the corresponding $x_h(t)$ and $x_p(t)$.

STEP Zero: No specific initial condition is given for this second-order differential equation, therefore we assume that $x(0)$ and $\dot{x}(0)$ are two parameters.

STEP One: The characteristic equation and its roots for this system are as follows.

$$D^2 + 5D + 6 = 0, \Rightarrow D_1 = -2, \; D_2 = -3. \tag{7.2-63}$$

Thus,

$$x_h(t) = A_1 e^{-2t} + A_2 e^{-3t}. \tag{7.2-64}$$

STEP Two: Because the forcing function or input consists of two decaying exponentials with the same parameters as the roots of the characteristic equation, we propose the following particular solution.

$$x_p(t) = K_1(t)e^{-2t} + K_2(t)e^{-3t}. \tag{7.2-65}$$

Straightforward differentiation of (7.2-65) yields

$$\dot{x}_p(t) = [\dot{K}_1(t) - 2K_1(t)]e^{-2t} + [\dot{K}_2(t) - 3K_2(t)]e^{-3t}. \tag{7.2-66}$$

$$\ddot{x}_p(t) = [\ddot{K}_1(t) - 4\dot{K}_1(t) + 4K_1(t)]e^{-2t}$$

$$= [\ddot{K}_2(t) - 6\dot{K}_2(t) + 9K_2(t)]e^{-3t}. \tag{7.2-67}$$

Now, constructing (7.2-62) with $x_p(t)$ as in (7.2-65) and its derivatives as in (7.2-66) and (7.2-67) yields

$$\ddot{x}_p(t) + 5\dot{x}_p(t) + 6x_p(t) = [\ddot{K}_1(t) + \dot{K}_1(t)]e^{-2t} + [\ddot{K}_2(t) - \dot{K}_2(t)]e^{-3t}$$

$$\equiv e^{-2t} + e^{-3t}. \tag{7.2-68}$$

Equating both sides of (7.2-68) term by term results in the following differential equations:

$$\ddot{K}_1(t) + \dot{K}_1(t) \equiv 1, \tag{7.2-69}$$

$$\ddot{K}_2(t) - \dot{K}_2(t) \equiv 1. \tag{7.2-70}$$

The particular solutions of the above equations with zero initial conditions are

$$K_1(t) \equiv t, \text{ and } K_2(t) \equiv -t. \tag{7.2-71}$$

Thus, the particular solution for (7.2-62) becomes as follows.

$$x_p(t) = te^{-2t} - te^{-3t}. \tag{7.2-72}$$

STEP Final: The complete or total solution with initial conditions as parameters is

$$x(t) = x_h(t) + x_p(t) = A_1e^{-2t} + A_2e^{-3t} + te^{-2t} - te^{-3t}$$

$$= (t + A_1)e^{-2t} - (t - A_2)e^{-3t}. \tag{7.2-73}$$

7.3. NETWORK ANALYSIS CONTINUED

We now apply the above methodologies to a few cases of interest. What we have covered in differential equations is at minimal, and additional issues are discussed in the following sections. Many engineering students study differential equations thoroughly in their mathematics courses.

Example 7.3-1: Analyze the circuit in Fig. 7.3-1 by setting up a proper group of ordinary differential equations and attempt to solve them. In this process one may need to make certain assumptions. Figuring out what is needed to specify in order to make the problem completely well posed is the main part of this task.

Solution: This circuit can be analyzed by mesh equations, because it has only one voltage source, and to that end we set up the appropriate group of equations.

STEP Zero: We label pertinent variables for a mesh equations approach as shown in Fig. 7.3-1, which are $i_1(t)$, $i_2(t)$ and $i_3(t)$. These currents are obviously functions of time. We also have two capacitors and two inductors which are complex elements and therefore each has its own history. As we gain more experience, we learn that for each complex element which is connected in manners that does not create redundancy, we get what is known in the literature as "one-degree-of-freedom." Because each element creates its own independent dynamic or mode. But, if we have two capacitors in parallel or two inductors in series, *per se,* then these are not independent. Thus, the ultimate differential equation that represents the circuit in Fig. 7.3-1 must be a 4th–order differential equation, because we have no redundancy in this circuit. To solve these equations, we need initial conditions up to the third derivatives for each complex element. These requirements can add up quickly. However, the way that we have set up this problem, i.e., we have not incorporated any switch in the circuit, suggests that all complex elements are "off-the-shelf," with zero initial conditions. We study circuits with switches in order to elaborate on the creation of initial conditions and their effects in the dynamic of these circuits in Section 7.4.

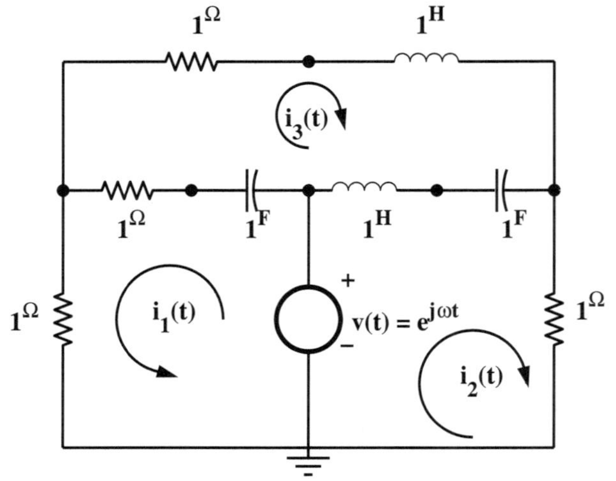

Fig. 7.3-1.

STEP One: We set up "coupled" integro-differential equations to represent this circuit according to the mesh equations methodology. We drop the time dependency for convenience.

For the boundary super loop we have:

$$i_3 + \frac{di_3}{dt} + i_2 - i_1 = 0. \tag{7.3-1}$$

For mesh #1 we have:

$$i_1 + (i_1 + i_3) + \int_{-\infty}^{t} [i_1(\tau) + i_3(\tau)]d\tau = e^{j\omega t}. \tag{7.3-2}$$

For mesh #2 we have:

$$i_2 + \int_{-\infty}^{t} [i_2(\tau) - i_3(\tau)]d\tau + \frac{d}{dt}(i_2 - i_3) = e^{j\omega t}. \tag{7.3-3}$$

STEP Two: Using $D \equiv d/dt$, we differentiate the above set of equations to get the following set of coupled differential equations for currents.

$$-i_1 + i_2 + (D + 1)i_3 = 0, \tag{7.3-4}$$

$$(2D + 1)i_1 + (D + 1)i_3 = De^{j\omega t}, \tag{7.3-5}$$

$$(D^2 + D + 1)i_2 - (D^2 + 1)i_3 = De^{j\omega t}. \tag{7.3-6}$$

From (7.3-4) we have

$$i_1 = i_2 + (D + 1)i_3. \tag{7.3-7}$$

Substituting this value into (7.3-5) results in

$$(2D + 1)[i_2 + (D + 1)i_3] + (D + 1)i_3 = De^{j\omega t},$$

$$(2D + 1)i_2 + 2(D + 1)^2 i_3 = De^{j\omega t}, \tag{7.3-8}$$

Similarly, substituting (7.3-7) into (7.3-6) yields

$$(D^2 + D + 1)i_2 - (D^2 + 1)i_3 = De^{j\omega t}. \tag{7.3-9}$$

The above set of coupled differential equations (7.3-8 & 9) is similar to the case presented in Example 7.4-2. Following that procedure we have:

$$\begin{cases} (D^2 + 1)(2D + 1)i_2 + 2(D^2 + 1)(D + 1)^2 i_3 = (D^2 + 1)De^{j\omega t}, \\ 2(D + 1)^2(D^2 + D + 1)i_2 - 2(D + 1)^2(D^2 + 1)i_3 = 2(D + 1)^2 De^{j\omega t}. \end{cases} \tag{7.3-10}$$

Adding the above two equations yield.

$$[(D^2 + 1)(2D + 1) + 2(D + 1)^2(D^2 + D + 1)]i_2 = [(D^2 + 1) + 2(D + 1)^2]De^{j\omega t}. \tag{7.3-11}$$

STEP Three: Solving (7.3-11), the last decoupled differential equation, for its complete solution incorporating all the necessary initial conditions is the beginning of the process for analyzing the circuit completely. As expected, we now face a 4th–order differential equation whose characteristic equation is as follows.

$$(D^2 + 1)(2D + 1) + 2(D + 1)^2(D^2 + D + 1) \equiv 0. \tag{7.3-12}$$

To solve (7.3-12) we need to use a calculator or some sort of numerical tool in general. But the situation is not so bad as it seems herein. For instance, we note that for $D = -1$, the left-hand side of (7.3-12) is negative, while for $D = -½$, the left-hand side of (7.3-12) is positive. Therefore, there is one root between $-1 < D < -½$. We can narrow this interval with some effort, but we stop here and leave the rest of computation as an exercise. (Problem 7.5.9).

STOP

Comment 7.3-1: Welcome to the real circuit analysis! Analyzing even a simple circuit with only a few complex elements can soon become complicated and requiring computational tools to carry on the task. Thus, we should expect that this trend continues and we face problems which are not so easy to complete by hand. But our strategy is to set aside for a while numerical computation versus the conceptual understanding of the underlying issues as pertain to circuit analysis. To that end we leave heavy computations for future circuit courses.

7.4. CIRCUITS with SEQUENTIAL SWITCHES

This topic is an important and perhaps a complicated part of our elementary circuit course and is reviewed here for two reasons. The first reason is the fact that such circuits are common in practice and often are programmed to engage various segments of a circuit for a particular application. Second, this type of circuits is an excellent academic tool to describe many interesting phenomena regarding the creation and propagation of initial conditions into the evolving dynamics of the circuit. Again, we stand by our earlier claim that we have already covered all essentials of circuit analysis including these circuits, and therefore we begin our presentation by working on several examples.

Example 7.4-1: For the circuit in Fig. 7.4-1 where we close the switch at $t = 0$, find the set of initial conditions up to the second-order derivatives.

Fig. 7.4-1.

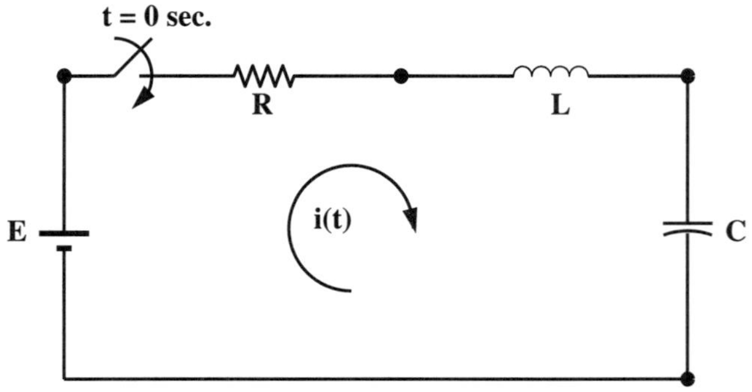

Solution: This circuit has only one switch and is in a form which we can make the following argument. However, we realize that the key to a sound discussion is that for each set of initial conditions the entire set must satisfy the corresponding KVL and KCL. In fact, to develop any higher-order initial condition, we need to solve first the initial values for currents and voltages before proceeding to solve for higher-order terms from the circuit. We must pay special attention to the specific problem under study, since each case has its own situation. For instance, consider a circuit similar to that in Fig. 7.4-2, where we have a loop with capacitors (C_1, E, C_2, and C_3), or a node with inductors (L_1, L_2, L_3, and I). Then we realize that all levels of initial conditions must be developed consistent with the corresponding KVL and KCL associated with this circuit. Otherwise, our analysis is not correct.

Fig. 7.4-2.

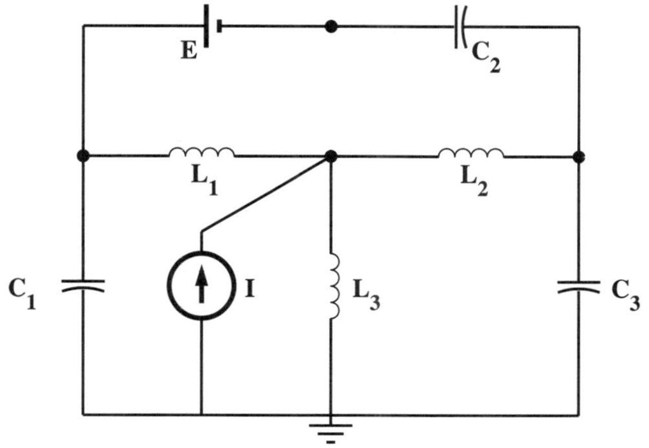

STEP Zero: We evaluate the circuit configuration in Fig. 7.4-1, and note that both energy-stored elements are off-the-shelf and without any initial condition. The main equation governing this circuit when we close the switch is

$$E = Ri(t) + L\frac{di(t)}{dt} + \frac{1}{C}\int i(t)\,dt. \tag{7.4-1}$$

STEP One: Because we close the switch at $t=0$ and $i(t=0^-)=0$. At $t=0^+$, we then have $i(t=0^+)=0$ due to the finite charge and the continuity of the current function. In other words, the physics of the system does not allow a jump-change in current from $t=0^-$ to $t=0^+$. Also, the reason for choosing $t=0^+$ is to avoid differentiation at $t=0$.

STEP Two: To calculate the initial condition of the first derivative of current, we note that $v_C(0)=0$ by assumption. Thus, from (7.4-1) we have:

$$E = Ri(0^+) + L\frac{di(0^+)}{dt} \Rightarrow \frac{di(0^+)}{dt} = \frac{E}{L}. \tag{7.4-2}$$

STEP Final: To calculate higher-order derivatives, say the second-order, we use the essential equation (7.4-1) and its derivative as follows.

$$L\frac{d^2i(t)}{dt^2} + R\frac{di(t)}{dt} + \frac{1}{C}i(t) = 0, \quad (7.4\text{-}3a)$$

$$\frac{d^2i(0^+)}{dt^2} = -\frac{R}{L}\frac{di(0^+)}{dt}. \quad (7.4\text{-}3b)$$

Example 7.4-2: Consider the circuit in Fig. 7.4-3 where the switch is applied at $t=0$. Find the loop currents and their initial values up to the first-order at $t=0^+$. We assume no initial conditions for L and C.

Fig. 7.4-3.

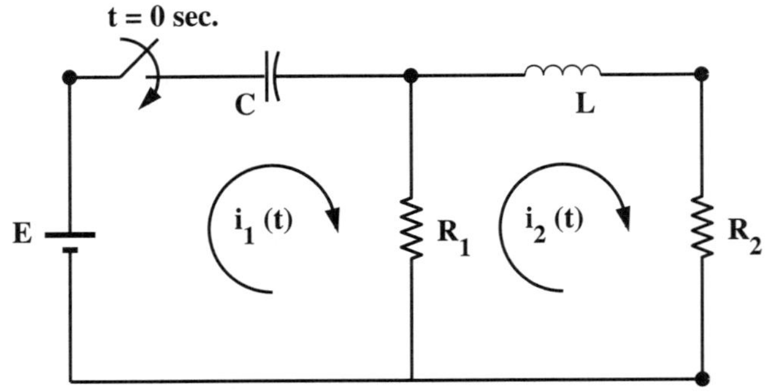

Solution: At $t=0^+$, the capacitor is short and the inductor is open and we are left with the circuit in Fig. 7.4-4.

Fig. 7.4-4.

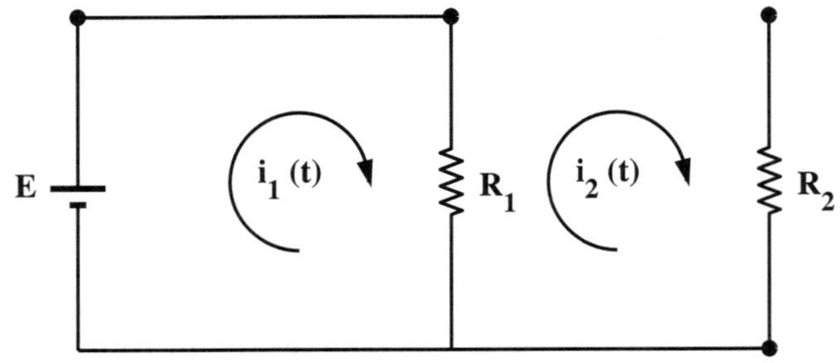

Following similar steps as the previous example, we have:

$$i_1(0^+) = \frac{E}{R_1}, \quad i_2(0^+) = 0. \quad (7.4\text{-}4)$$

The essential mesh equations for the circuit in Fig. 7.4-3 are

$$E = \frac{1}{C}\int i_1(t)dt + R_1[i_1(t) - i_2(t)]. \quad (7.4\text{-}5)$$

$$0 = R_1[i_2(t) - i_1(t)] + R_2 i_2(t) + L\frac{di_2(t)}{dt}. \quad (7.4\text{-}6)$$

Thus, from (7.4-6) the first derivative is

$$\frac{di_2(0^+)}{dt} = \frac{1}{L}[R_1 i_1(0^+) - (R_1 + R_2)i_2(0^+)]$$

$$= \frac{E}{L}. \qquad (7.4\text{-}7)$$

From (7.4-5) we have

$$\frac{i_1}{C} + R_1 \frac{di_1(t)}{dt} - R_1 \frac{di_2(t)}{dt} = 0, \qquad (7.4\text{-}8)$$

$$\frac{di_1(0^+)}{dt} = \frac{E}{L} - \frac{E}{R_1^2 C}. \qquad (7.4\text{-}9)$$

For the second-order term, we start from (7.4-6).

$$\frac{d^2 i_2(t)}{dt^2} = \frac{1}{L}[R_1 \frac{di_1(t)}{dt} - (R_1 + R_2)\frac{di_2(t)}{dt}] \triangleq \text{known values at all time.} \qquad (7.4\text{-}10)$$

Similarly other terms are calculated.

STOP

Remark 7.4-1: The message that we want to convey is that there is no unique way of determining initial conditions and each problem must be studied individually, but always consistent with the governing laws of circuit analysis. The task becomes more involved when we have more than one switches as is described next in Example 7.4-3.

Example 7.4-3: Consider the circuit in Fig. 7.4-5 with two switches. One closes at t = 0 second, and the other closes at t = 10 seconds. Find $v_o(t)$ as indicated for all t.

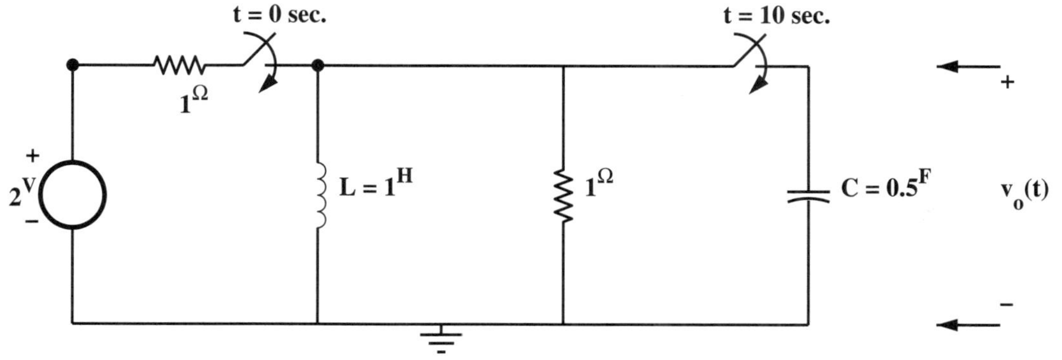

Fig. 7.4-5. A circuit with two sequential switches.

Solution: To analyze a circuit with sequential switches we need to specify the time line that various switches are engaged. What is important in a circuit with sequential switches is *the number of instances* that switches are being engaged, and not the actual number of switches. For instance, in the above example, there are two distinct instances of $t = 0$ and $t = 10$ seconds which change the structure of the circuit. Whether the circuit has two or more critical instances, the analysis is very similar. Therefore, we proceed to draw the time line in each problem, and study the corresponding circuit in each stage of that time line, Fig. 7.4-6.

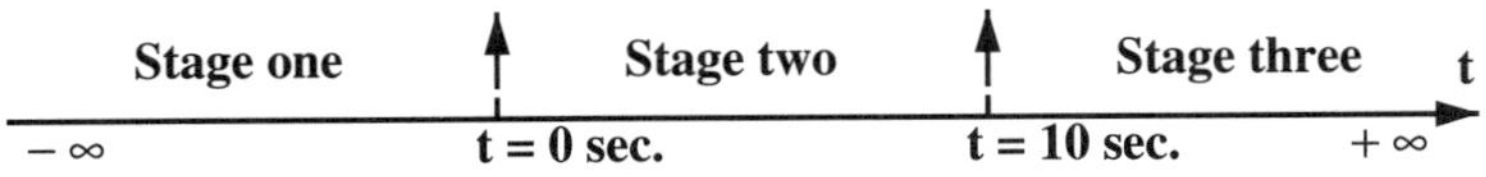

Fig. 7.4-6. Time line for the circuit with sequential switches.

The original problem must now be broken into several sub-problems, each describing the original circuit in one corresponding time domain, which we call stage. Then for each stage, we must analyze the corresponding circuit in the same manner as we do in any such circuit analysis. *The critical issue is to match the circuit in one stage to its adjacent stage through a very careful and deliberate study of the initial and final conditions of these two stages.* Here, the final condition of stage i becomes the initial condition for stage $(i+1)$. The number of stages depend on the number of critical instances when various switches are engaged or disengaged. In the above example we have three stages.

Going back to the circuit in Fig. 7.4-5, with its corresponding time line, we have three stages to contend with. In each stage, we have a circuit which we must analyze in order to monitor its evolution from the beginning to the final time of that stage.

STAGE One: Consider the time from $-\infty$ to the occurrence of the first switch, which is at $t = 0$ second and its corresponding circuit.

For this stage, the circuit in Fig. 7.4-5 becomes that in Fig. 7.4-7.

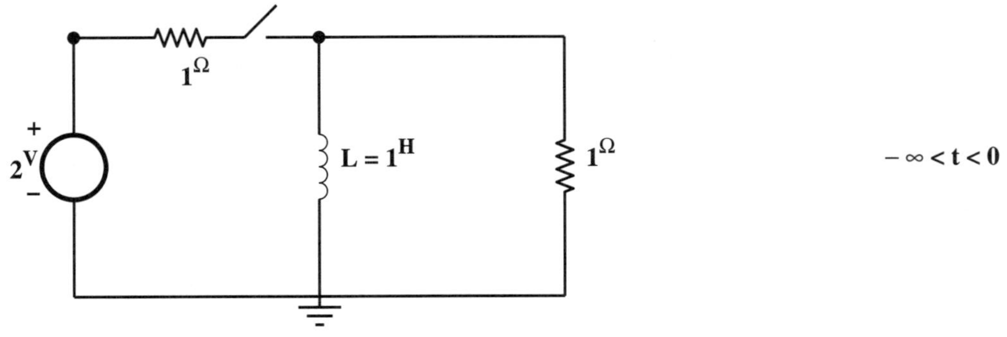

Fig. 7.4-7. The original circuit during its first stage.

Nothing is happening in this stage, because the source is not connected and no initial condition for the inductor is given. Since we have one inductor in the only loop that has no current flowing, and no initial condition is built in this inductor. Note that, if we had an initial condition for the only inductor in this circuit, then there would have been some activity in the loop consisting of the inductor and the 1^{Ω} resistor. As is, this circuit is not functional and no history is being made in the inductor, i.e., the final condition of this stage (referring to the final conditions of all complex elements, which becomes the initial conditions of the next stage) remains zero.

STAGE Two: Consider the time from the occurrence of the first switch until the occurrence of the next switch with its corresponding circuit.

For this stage, the circuit in Fig. 7.4-7 becomes that in Fig. 7.4-8.

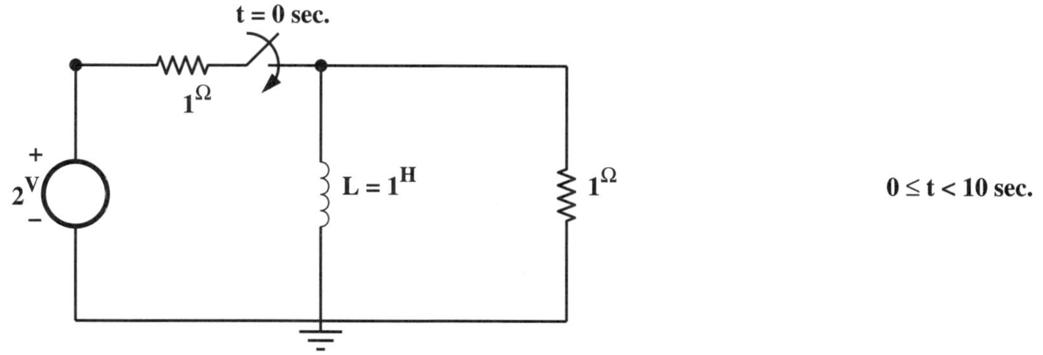

Fig. 7.4-8. The original circuit during its second stage.

We note that before the occurrence of the second switch, a major part of the original circuit is not functional and we do not account for it. We also note well that because of the existence of the switch that is applied at $t = 0$, we can solve this circuit. In other words, because of this switch the source in effect is a step function, and a step function is very different from a constant source. Otherwise, an inductor becomes short circuited when faces a "constant" source, since the derivative of a constant is zero. Many textbooks argue that the circuit has been left "on" for a "long time" (i.e., from $-\infty$ to the present time) to convey the building up of the initial condition in the complex elements. We believe such an argument is unnatural, because what is happening in reality is that we have applied a step function (because of the switch) and as a result the circuit is now excited and we can set a proper set of differential equations to describe its operation. The solution to that set of differential equations, which at the moment has zero initial condition, is then used to monitor and trace the time evolution of the corresponding circuit. Thus, developing its final condition relative to the occurrence of the next switch, if any, gives us a way of building initial conditions and studying its propagation in the circuit which is a natural way of analyzing the underlying circuit. That in turn completes the circuit analysis for this stage of the operation.

For instance, in this stage and when the switch is applied, we have the circuit in Fig. 7.4-9 to analyze.

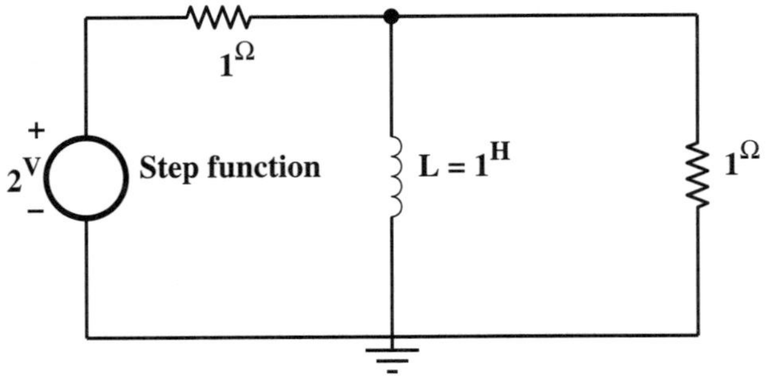

Fig. 7.4-9. The actual circuit during its second stage.

To analyze the circuit in Fig. 7.4-9 we go through the same procedures as in any circuit, namely, we invoke the following steps.

STEP Zero: A simple source transformation brings the circuit in Fig. 7.4-9 to that in Fig. 7.4-10, followed by the circuit in Fig. 7.4-11.

STEP One: The circuit in Fig. 7.4-11 can be easily analyzed by a mesh equation approach as follows.

$$L\frac{di_L}{dt} + (½)i_L = 1^V \text{ (step function).} \qquad (7.4\text{-}11)$$

The solution to this first-order differential equation is as follows.

$$i_L(t) = Ae^{-(½)t} + 2. \qquad (7.4\text{-}12)$$

Note that the first term is the homogeneous solution, while the second term is the particular solution, both generated in a straightforward manner. Invoking the initial condition of i_L, we recall that

the final condition at stage one = the initial condition at stage two.

(7.4-13)

Thus, using $i_L(t = 0) = 0$, we have

$$0 = A + 2 \;\Rightarrow\; A = -2. \qquad (7.4\text{-}14)$$

Therefore, the total solution of the inductor current becomes

$$i_L(t) = -2e^{-(1/2)t} + 2, \quad -\infty < t. \tag{7.4-15}$$

Fig. 7.4-10.

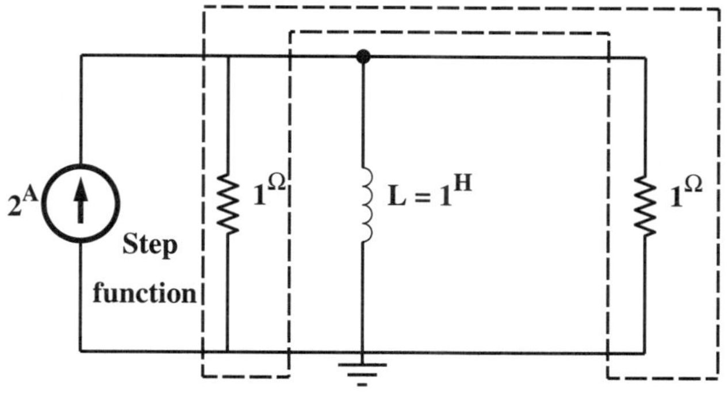

Fig. 7.4-11.

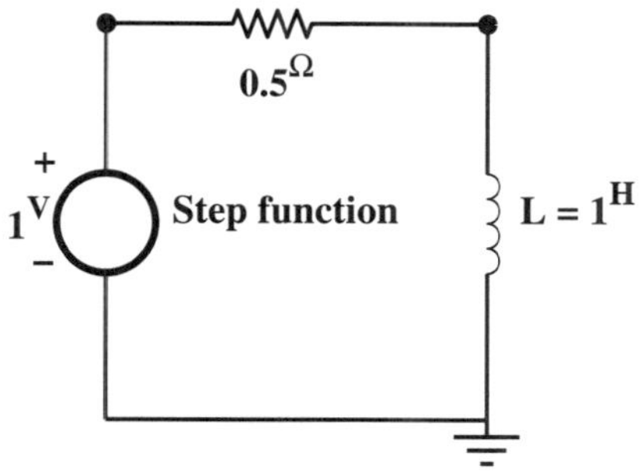

STEP Two: At the end of this stage, when the next switch occurs, i.e., at t = 10 seconds, we have

$$i_L(10) = -2e^{-5} + 2 = 1.986^A. \tag{7.4-16}$$

In some textbooks this sort of "final" value is chosen when $t = \infty$, which is 2^A. We have had difficulties to convey that to the students, and that is why we adhere to the "exact" analytical value. Although some may argue that $e^{-5} \approx 0$.

STEP Three: The above value serves as the current initial value for the next stage. If we had more than one complex element, we would have computed similar terms for other elements as well. Here, we also need to compute the voltage across the inductor at t = 10 seconds, which becomes the initial value for the voltage of that element at the beginning of the next stage.

$$L \frac{di_L}{dt}\bigg|_{t=10} = (1^H)e^{-5} \approx 0. \tag{7.4-17}$$

The above is a reasonable approximation. These values serve as the initial conditions for the next stage, and we have exhausted this stage. It is always easier to shift the time origin and start each stage at t = 0 to simplify our analysis as is described below.

STAGE Three: Consider the time from the occurrence of the second switch until the next switch, which in this case there is none, therefore until $t = \infty$ with its corresponding circuit.

For this stage, the circuit in Fig. 7.4-5 becomes that in Fig. 7.4-12.

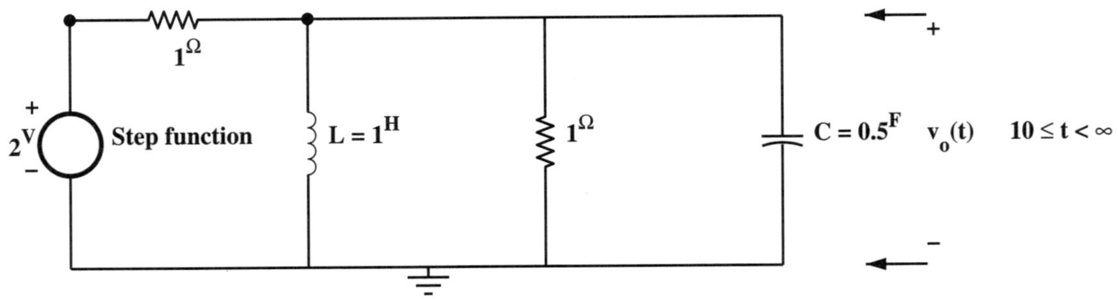

Fig. 7.4-12. The original circuit during its third stage.

Now, we have to analyze the circuit in Fig. 7.4-12 for $t > 10$ seconds. Here, we shift the time origin to $t = 10$ seconds to simplify analysis. Also, we note well that we still consider the original source as a step function, with the inductor having 1.986^A initial current, and the capacitor has no initial condition of its own, however, because this capacitor is now in parallel with the inductor, we need to consider di_L/dt, at $t = 0$ (or $t = 10$ seconds with respect to the old time scale), which has turned out to be very small or zero as indicated in (7.4-17).

In this stage, the circuit in Fig. 7.4-12 becomes that in Fig. 7.4-13 upon one source transformation.

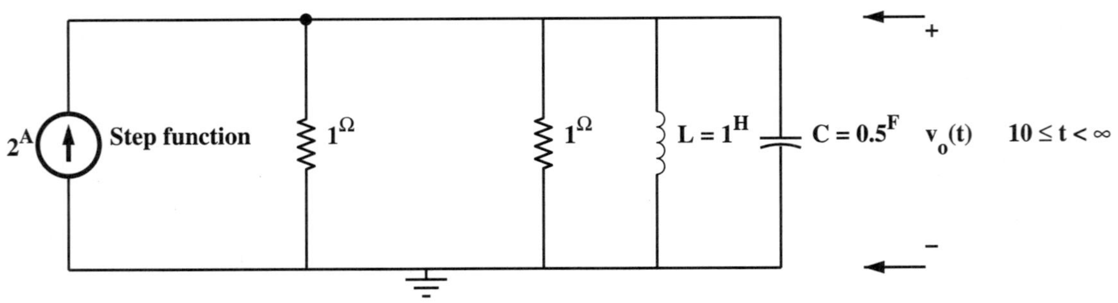

Fig. 7.4-13.

STEP Zero: Another source transformation in Fig. 7.4-13 simplifies this circuit into the circuit in Fig. 7.4-14.

STEP One: The circuit in Fig. 7.4-14 can be analyzed by a node equations approach as follows.

$$2^A = \frac{v_o}{(1/2)} + i_L + i_C$$
$$= 2v_o + i_L + C\frac{dv_C(t)}{dt}$$
$$= 2\left[\frac{di_L(t)}{dt}\right] + i_L + C\frac{d}{dt}\frac{di_L(t)}{dt}$$
$$= \frac{1}{2}\frac{d^2 i_L(t)}{dt^2} + 2\left[\frac{di_L(t)}{dt}\right] + i_L. \tag{7.4-18}$$

Here, we use $v_o = di_L(t)/dt$ (recall that $L = 1^H$). Because we have all pertinent initial conditions for the inductor's current, we set up the above equation in terms of $i_L(t)$.

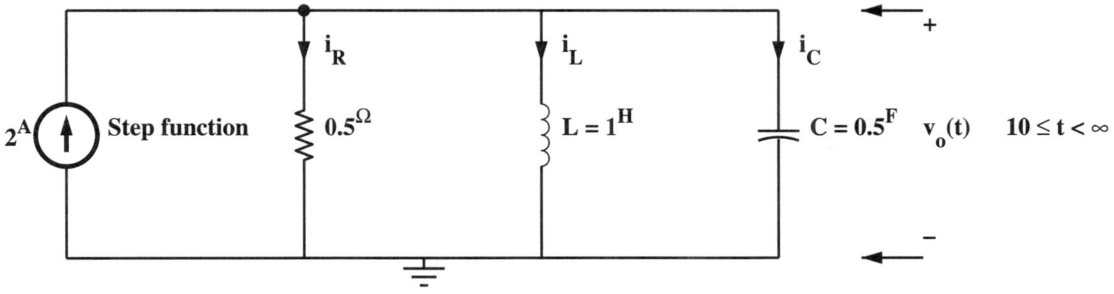

Fig. 7.4-14.

STEP Two: Thus, the current through inductor for which we have all pertinent initial conditions satisfies the following differential equation.

$$[(1/2)D^2 + 2D + 1]i_L(t) = 2^A \text{ (step function)}. \tag{7.4-19}$$

The initial conditions are $i_L(0) = 1.986^A$ and $di_L(0)/dt \approx 0$.

STEP Three: Solving (7.4-19) completes the process of analyzing this circuit. The characteristic equation yields

$$(1/2)D^2 + 2D + 1 = 0 \Rightarrow D_1 = -0.586, \quad D_2 = -3.414. \tag{7.4-20}$$

Thus the complete solution of (7.4-20) becomes

$$i_L(t) = A_1 e^{-0.586t} + A_2 e^{-3.414t} + 2. \tag{7.4-21}$$

Invoking the initial conditions yields

$$1.986 = A_1 + A_2 + 2$$

$$0 = -0.586A_1 - 3.414A_2. \quad (7.4\text{-}22)$$

Thus, upon solving for A_1 and A_2, we have

$$i_L(t) = -0.016e^{-0.586t} + 0.0028e^{-3.414t} + 2. \quad (7.4\text{-}23)$$

From the above equation and the circuit in Fig. 7.2-10, we have

$$v_o(t) = L\frac{di_L(t)}{dt}$$

$$= 0.0094e^{-0.586t} - 0.0096e^{-3.414t}. \quad (7.4\text{-}24)$$

STOP

7.5. PROBLEM SET

7.5.1. Solve the following differential equation with $x(0) = 0$ and $\dot{x}(0) = 0$.

$$\ddot{x}(t) + 3\dot{x}(t) + 2x(t) = e^{-3t}. \quad (7.5\text{-}1)$$

7.5.2. Solve the following differential equation with $x(0)=0$, $\dot{x}(0)=0$ and $\ddot{x}(0)=0$.

$$\frac{d^3x(t)}{dt^3} + 9\frac{d^2x(t)}{dt^2} + 26\frac{dx(t)}{dt} + 24x(t) = t^3 - t^2 + t - 1, \quad (7.5\text{-}2)$$

7.5.3. Solve the following differential equation with $x(0) = 0$ and $\dot{x}(0) = 0$.

$$\ddot{x}(t) + 6\dot{x}(t) + 5x(t) = \cos t - \sin t. \quad (7.5\text{-}3)$$

7.5.4. Solve the following differential equation for x(t) and y(t) with all pertinent initial conditions set at zero values.

$$\begin{cases} (D+2)x(t) + (D+2)y(t) = t, \\ (D+1)x(t) + y(t) = t^2. \end{cases} \quad (7.5\text{-}4)$$

7.5.5. Solve the following set of differential equations for x(t) and y(t) with zero initial conditions.

$$\begin{cases} (D+1)x(t) + (D+4)y(t) = t, \\ (D+2)x(t) + (D+3)y(t) = t^2. \end{cases} \quad (7.5\text{-}5)$$

7.5.6. Solve the following set of differential equations for x(t) and y(t) with zero initial conditions.

$$\begin{cases} (D+1)x(t) + (D+2)y(t) = \sin t, \\ (D+2)x(t) - (D+1)y(t) = \cos t. \end{cases} \tag{7.5-6}$$

7.5.7. Solve the following differential equation with parametric initial conditions.

$$\ddot{x}(t) + 3\dot{x}(t) + 2x(t) = e^{-t} + e^{-2t}. \tag{7.5-7}$$

7.5.8. Solve the following differential equation with parametric initial conditions.

$$\ddot{x}(t) + 7\dot{x}(t) + 6x(t) = e^{-t} + e^{-3t} + e^{-6t}. \tag{7.5-8}$$

7.5.9. Complete the numerical solution in Example 7.3-1.

7.5.10. Given n inductors L_1, L_2, ..., L_n, find an equivalent inductor when these are connected in either parallel or series. Assume all relevant initial conditions are set at zero.

7.5.11. Given n capacitors C_1, C_2, ..., C_n, find an equivalent capacitor when these are connected in either parallel or series. Assume all relevant initial conditions are set at zero.

7.5.12. What is the particular component of the response for the circuit in Fig. 7.5-1 with $v_{in}(t) = 10 \sin t$?

Fig. 7.5-1.

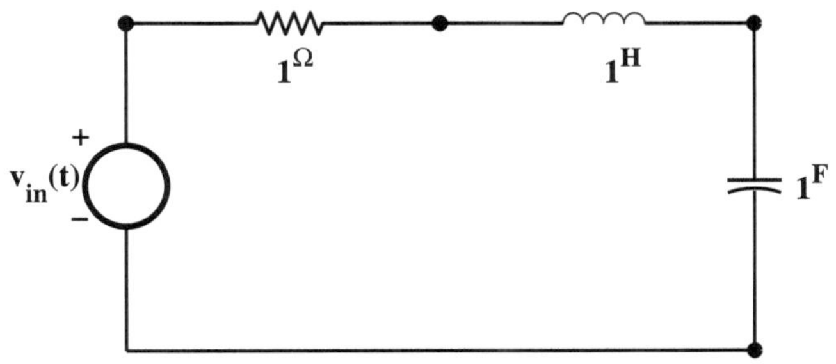

7.5.13. For the circuit in Fig. 7.5-2 and for $v_{in}(t) = 10 \sin t$, find $v_o(t)$.

Fig. 7.5-2.

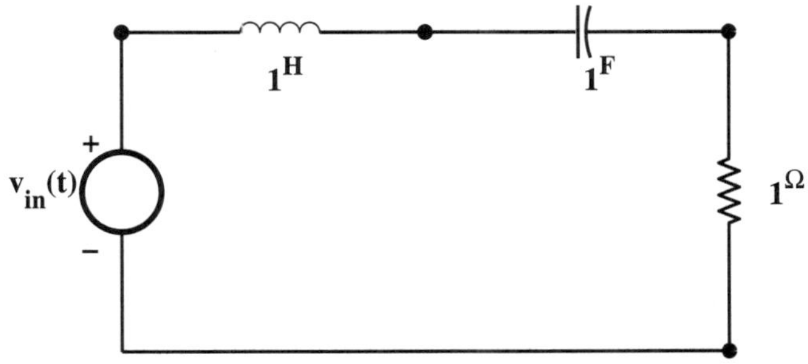

7.5.14. For the circuit in Fig. 7.5-3, find all initial conditions up to the second-order at $t = 0^+$.

Fig. 7.5-3.

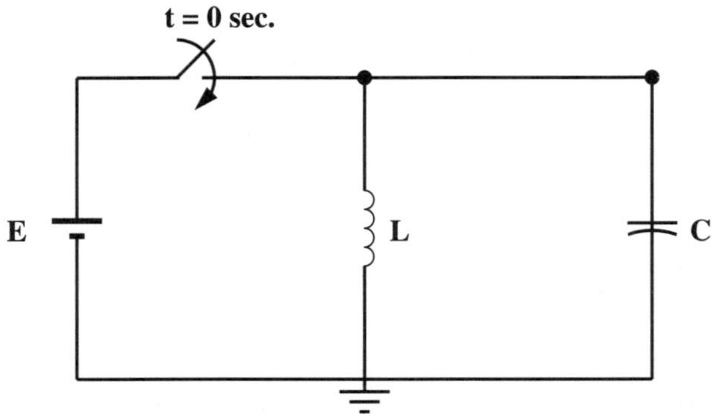

7.5.15. For the circuit in Fig. 7.5-4 we open the switch at $t = 0$. Find the voltage across the switch and its first derivative as $t \to \infty$.

Fig. 7.5-4.

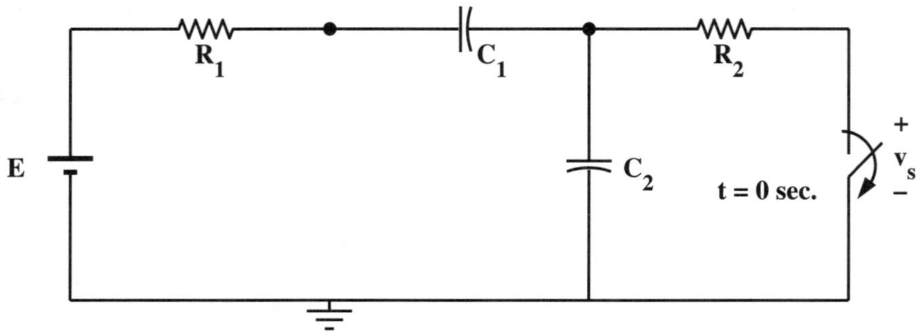

7.5.16. For the circuit in Fig. 7.5-5, find i(t).

Fig. 7.5-5.

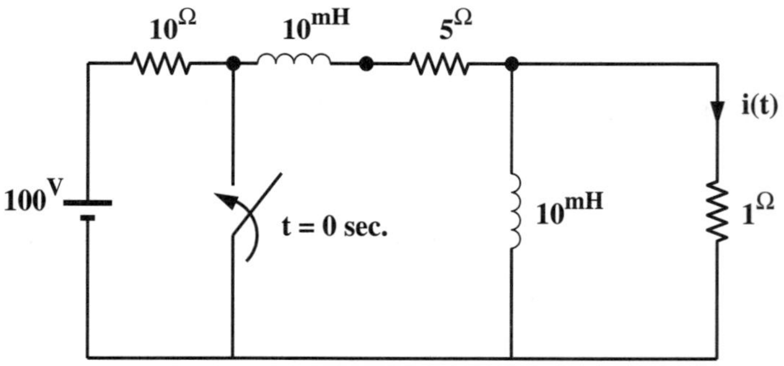

7.5.17. For the circuit in Fig. 7.5-6, and for $\omega = 1000$, find i(t). Assume all relevant initial conditions are set at zero.

Fig. 7.5-6.

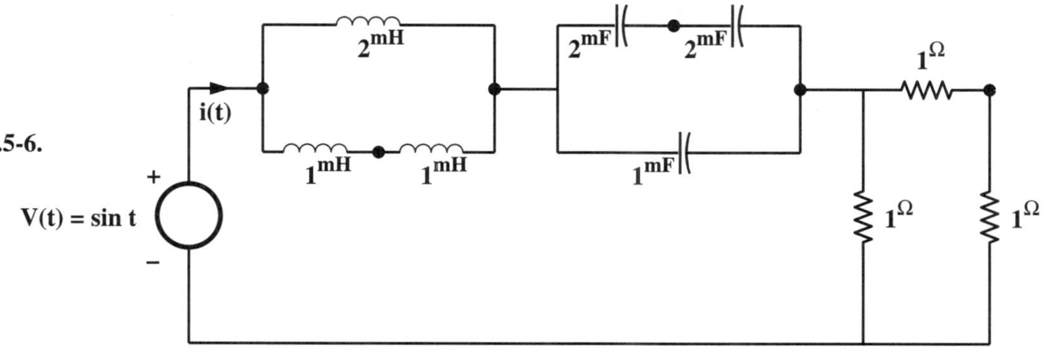

7.5.18. For the circuit in Fig. 7.5-7, and for $\omega = 1000$, find $v_o(t)$. Assume all relevant initial conditions are set at zero.

Fig. 7.5-7.

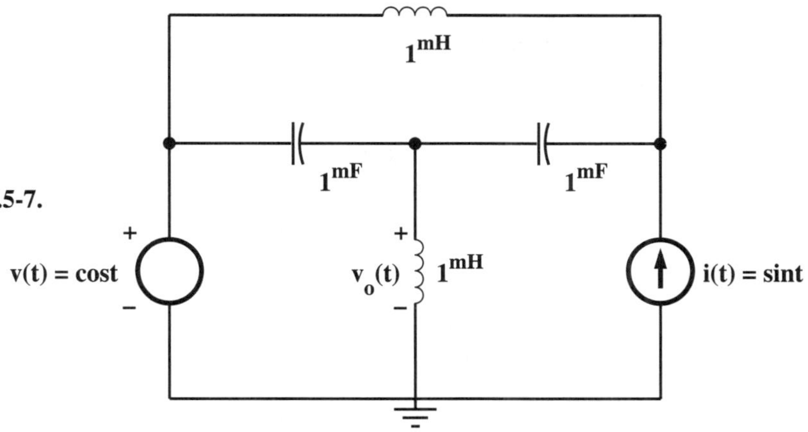

7.5.19. In the circuit of Fig. 7.5-8, the switch is initially at point (a) during $-\infty < t < 0$. At $t \geq 0$ we switch from point (a) to point (b). Find i(t) for $t \geq 0$.

Fig. 7.5-8.

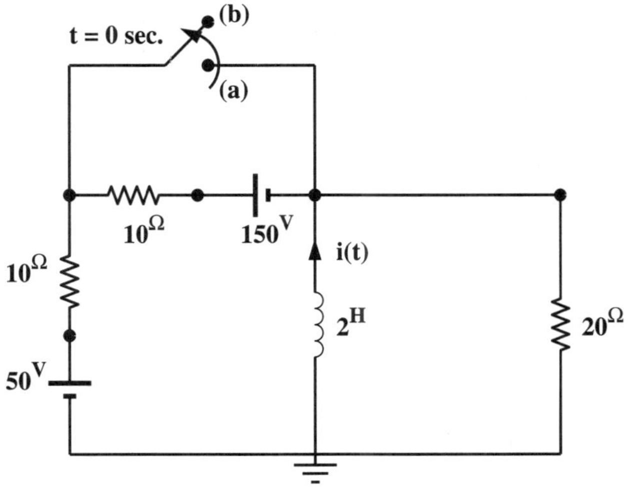

7.5.20. In the circuit of Fig. 7.5-9, the switch is initially at point (a) during $-\infty < t < 0$. At $t \geq 0$ we switch from point (a) to point (b). Find $v_o(t)$ for $t \geq 0$.

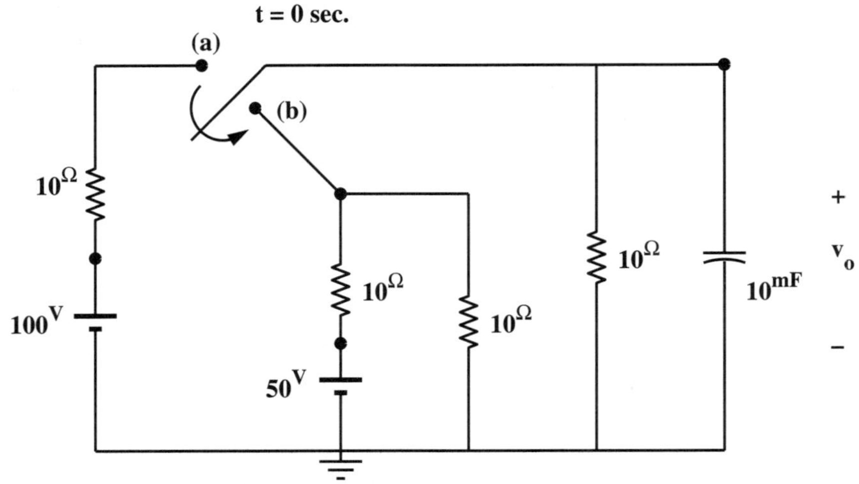

Fig. 7.5-9.

7.5.21. The two switches in the circuit of Fig. 7.5-10 are applied at t = 0 and t = 10 seconds. Find i(t) as indicated.

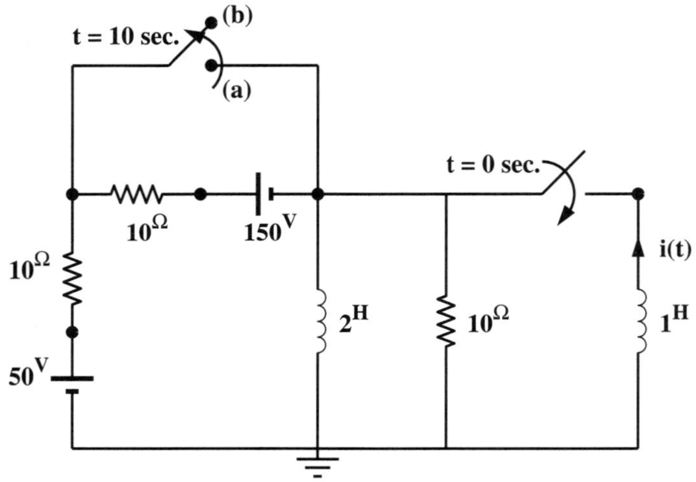

Fig. 7.5-10.

7.5.22. Find v_o and an equivalent circuit with one variable resistor for circuit below.

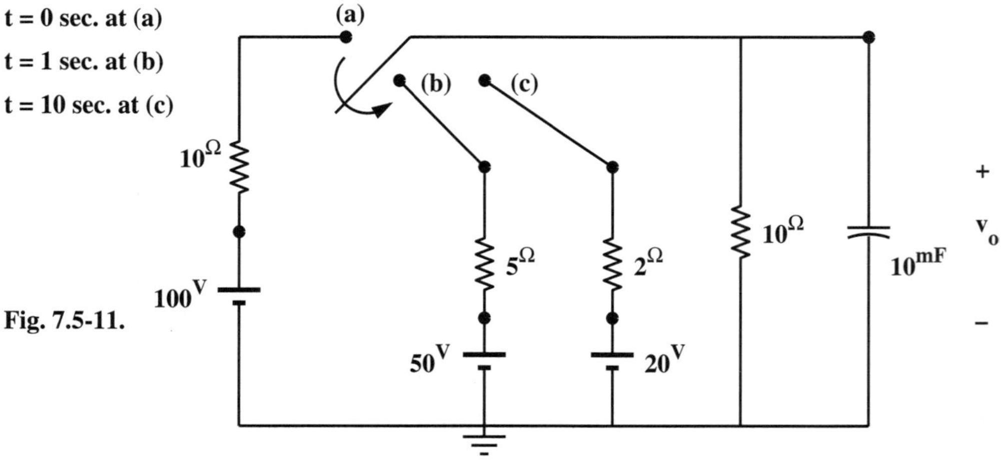

Fig. 7.5-11.

8 MUTUAL INDUCTANCE AND IDEAL TRANSFORMERS

8.1. INTRODUCTION

Whenever we have two (or more) inductors or coils in a circuit, some interactions will take place among them. This interaction is due to the fact that the electromagnetic fields generated in one coil travels through the other ones and affects their performance. The same can be said about the effect of the other coils on the performance of the first one. To account for such interactions we need a special treatment. To that end, suppose we have only two coils called primary and secondary. Then this phenomenon can be shown pictorially by closed paths of magnetic fields as displayed in Fig. 8.1-1.

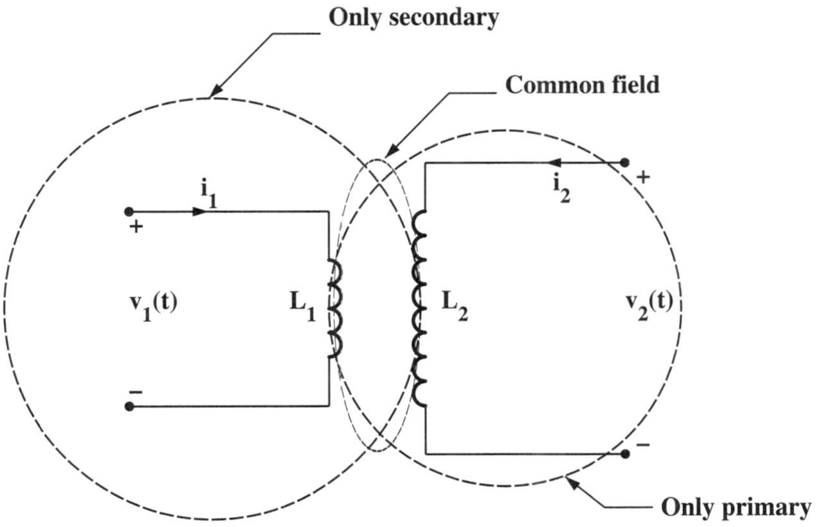

Fig. 8.1-1.

Mathematically, when we have two interacting coils, *per se,* the existence of an interaction is shown by **"dots"** as indicated in Fig. 8.1-2, which are also known as polarity marks. If we assume this interaction does not exist, as that has been the case in examples of our preceding chapters, then we do not use "dots". That means the coils are sealed electromagnetically, and do not interact with each others.

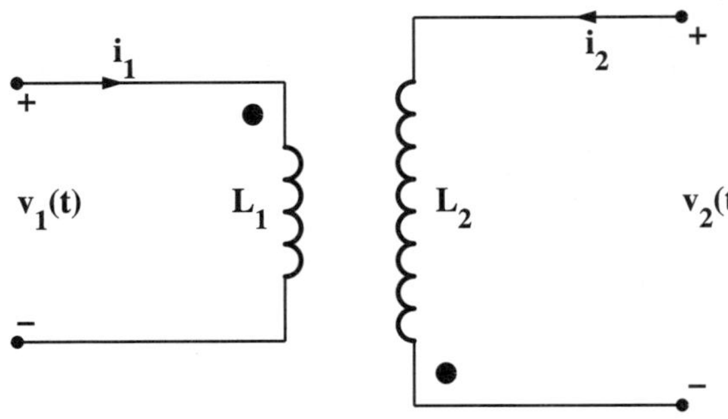

Fig. 8.1-2. Standard symbol for mutual inductances.

In circuit of Fig. 8.1-2, $v_1(t) \neq L_1 di_1(t)/dt$, or similarly, $v_2(t) \neq L_2 di_2(t)/dt$, but rather these are as follows.

$$v_1(t) = L_1 \frac{di_1(t)}{dt} + f(i_2), \tag{8.1-1}$$

$$v_2(t) = L_2 \frac{di_2(t)}{dt} + g(i_1). \tag{8.1-2}$$

Here, $f(i_2)$ represents the indirect contributions of the current in the second coil on the voltage of the first coil. Similarly, $g(i_1)$ shows the contributions of the first current on the voltage across the second coil. The reason for the choice of a current function is that each proposed interacting field is proportional to the current which generates that field. These functions are often approximated by linear terms as is described subsequently. In passing, circuits of this type are also classified under two-port networks which are described in Chapter Eleven. The next topic included in this chapter is an introduction to the ideal transformer which this device complements our overall list of passive elements R, L and C.

8.2. MUTUAL INDUCTANCE

In general we have the following expressions for "f" and "g" in (8.1-1 & 2), which are the first-order approximation to a general function describing this phenomenon.

$$f(i_2) = \pm M_{12} \frac{di_2}{dt}, \tag{8.2-1}$$

$$g(i_1) = \pm M_{21} \frac{di_1}{dt} . \tag{8.2-2}$$

Here, M_{12} and M_{21} are called mutual inductances and are measured in Henry (H – the unit of inductance). These coefficients are mostly identical, in other words, the effect of the first coil on the second one is the same as the second coil on the first. Otherwise, each coefficient is specified accordingly in H. The plus and minus signs reflect the following convention. If both currents enter or leave their corresponding "dot", then M's are positive, otherwise, when one current enters and the other one leaves, the M's are negative.

For the circuit in Fig. 8.1-2, we have

$$v_1(t) = L_1 \frac{di_1(t)}{dt} - M_{12} \frac{di_2}{dt} , \tag{8.2-3}$$

$$v_2(t) = L_2 \frac{di_2(t)}{dt} - M_{21} \frac{di_1}{dt} . \tag{8.2-4}$$

In (8.2-3 & 4), the subscript in M_{12} reads *"in 1, because of 2,"* and similarly, M_{21} reads *"in 2, because of 1"*. Likewise in equations with more than two crossing terms, we develop the underlying coefficients in the same manner. (Problems 8.4.2 to 5).

These equations can be interpreted in terms of a two-port network as follows.

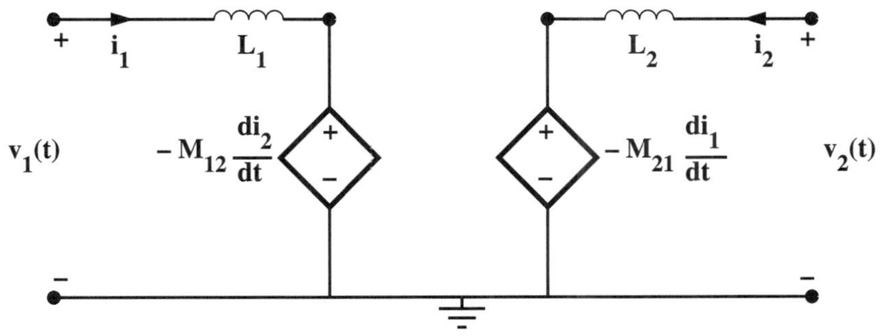

Fig. 8.2-1. A two-port network model for a mutual inductance.

Here, the dependent sources are the actual contributions of the mutual inductances, and just as any such dependent source, their values depend on some other network quantities. Also, note the similarity of circuit in Fig. 8.2-1 to that in Fig. 5.4-9, where the two parts of the circuit are connected through one node which in this case is ground. However, because of dependent sources the corresponding mesh equations in this case are coupled.

Another representation of the circuit in Fig. 8.2-1 which is commonly used is that shown in Fig. 8.2-2. Here, and only for the sake of simplicity and without any loss of generality, we assume that $M_{12} = M_{21} = M$.

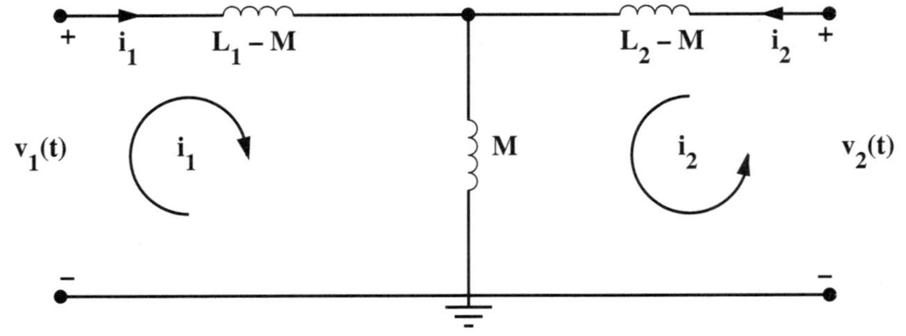

Fig. 8.2-2.

The circuits in Figs. 8.2-1 & 2 are equivalent, because the mesh equations for circuit in Fig. 8.2-2, with three coils $(L_1 - M)$, $(L_2 - M)$ and M which no longer interact, are exactly the same as those for the circuit in Fig. 8.2-1. This representation is correct mathematically, even when $L_1 - M$ or $L_2 - M$ becomes negative.

8.3. IDEAL TRANSFORMERS

This device should be considered as just one new passive element, since it is used in that way for many analytical discussions. This device is constructed from two adjacent coils as shown in Fig. 8.3-1.

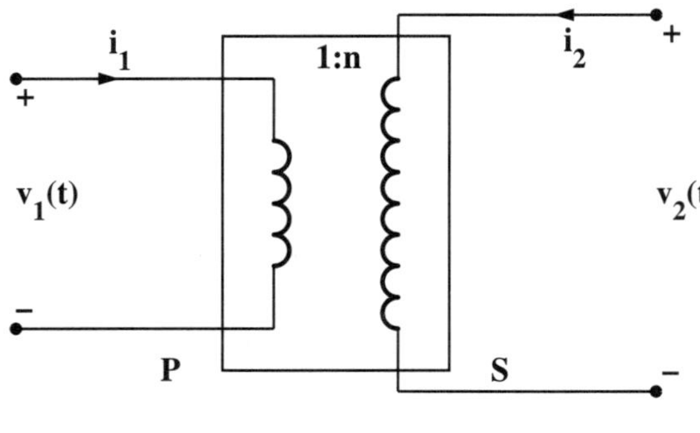

Fig. 8.3-1. An ideal transformer.

Here by convention, we may assume that the number of rings in the primary (P) is "1", and that of the secondary (S) is "n", or we may choose N_1 and N_2 respectively. Here, $N_2 / N_1 = n$ is called the *turn ratio*. Then by the *definition of an ideal transformer,* this device acts as follows.

$$\frac{v_2}{v_1} = \frac{N_2}{N_1} = n \quad \Rightarrow v_2 = nv_1. \qquad (8.3\text{-}1)$$

Furthermore, we may assume that the power transfer from the primary to the secondary is lossless. That means $P_1 = |V_1||I_1|$ and $P_2 = |V_2||I_2|$ are identical, yielding

$$\frac{|I_2|}{|I_1|} = \frac{v_1}{v_2} = \frac{1}{n}. \qquad (8.3\text{-}2)$$

From $R_1 = V_1/|I_1|$, $R_2 = V_2/|I_2|$ and (8.3-2), we have

$$\frac{R_2}{R_1} = (\frac{V_2}{|I_2|})/(\frac{V_1}{|I_1|}) = n^2. \qquad (8.3\text{-}3)$$

Note that in (8.3-1) we have selected lower case letters (time-domain), while in (8.3-2) we have chosen the upper case letters. This is a technical issue based on our chosen direction for the current in the secondary loop. *Three equations (8.3-1) to (8.3-3) are the fundamental equations representing an ideal transformer as depicted in Fig. 8.3-1.* These formulae are also true for complex elements and sources. (Problem 8.4.8).

Comment 8.3-1: When we use an ideal transformer, we identify nodes immediately outside the main box and before any external element is connected to this device as shown for instance in Fig. 8.3-2, next. But we do not put "dots" adjacent to the coils inside the box in Fig. 8.3-1, as we do in typical circuits with inductors, say the circuit in Fig. 8.1-2. The reason is that the symbol in Fig. 8.3-1 in its entirety represents an ideal transformer. Here, the directions of both currents and voltages in the primary and secondary are selected such that both currents enter, or both leave, the designated, though not exposed "dots". The positive polarities for voltages are also selected as shown in Fig. 8.3-1. Some textbooks put "dots" inside this box and adjacent to the coils. For instance, in Fig. 8.3-1, if we were to put "dots", then both "dots" are either on top (both currents are entering) or in the bottom (both currents are leaving). If it turns out that the "dots" and the direction for currents (and correspondingly for voltages) do not meet this standard, then we must replace "n" accordingly in the preceding equations with "$-n$". We never put any "dot" inside the box of an ideal transformer, assuming that the corresponding currents and voltages are exactly as shown in Fig. 8.3-1.

Example 8.3-1: Analyze the circuit in Fig. 8.3-2 containing one ideal transformer.

Solution: This problem consists of two sub-problems. In each side (primary or secondary) of the ideal transformer we need to ask the following question. What is the contribution of the other side to this side? This question is asked in the context of the equivalent circuit analysis. For instance, for someone in the primary who is looking into the secondary sees R_2 and v_2 as the Thevenin's equivalent of the right-hand side of the secondary. Then the question is how much of that equivalent circuit is *loading* the primary circuit? The answer is R_2/n^2 for the load, and v_2/n for the source.

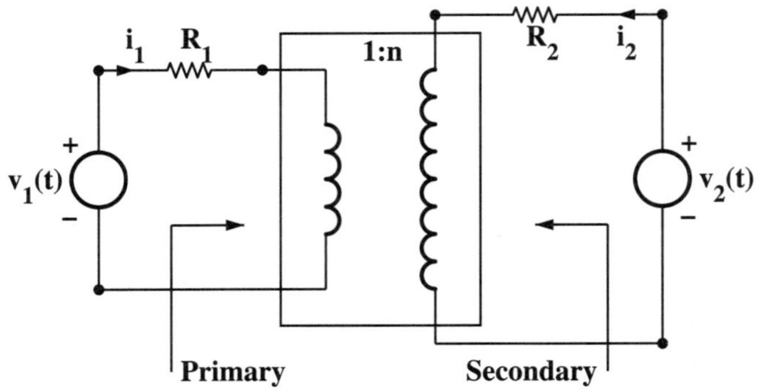

Fig. 8.3-2.

Similarly, for someone in the secondary who is looking into the primary sees R_1 and v_1. These values load the secondary according to n^2R_1 and nv_1. By incorporating these values in the circuit of Fig. 8.3-2, we can decouple the previous problem into two sub-problems as shown in Fig. 8.3-3.

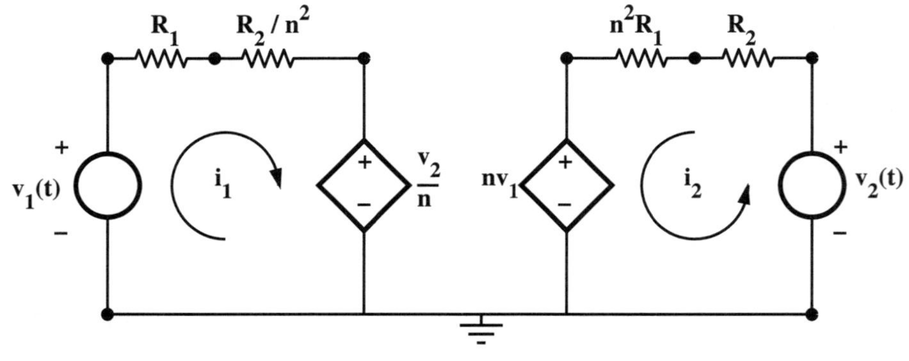

Fig. 8.3-3. An equivalent circuit for an ideal transformer.

Each of the above two circuits, though coupled, can be analyzed using the mesh equations approach to compute the corresponding currents. The results are

$$i_1(t) = \frac{[v_1(t) - v_2(t)/n]}{R_1 + R_2/n^2}, \tag{8.3-4}$$

$$i_2(t) = \frac{[v_2(t) - nv_1(t)]}{R_2 + n^2R_1}. \tag{8.3-5}$$

8.4. PROBLEM SET

Mesh Equations

8.4.1. Write the mesh equations for the circuit in Fig. 8.4-1, with $M = 1^H$.

Fig. 8.4-1.

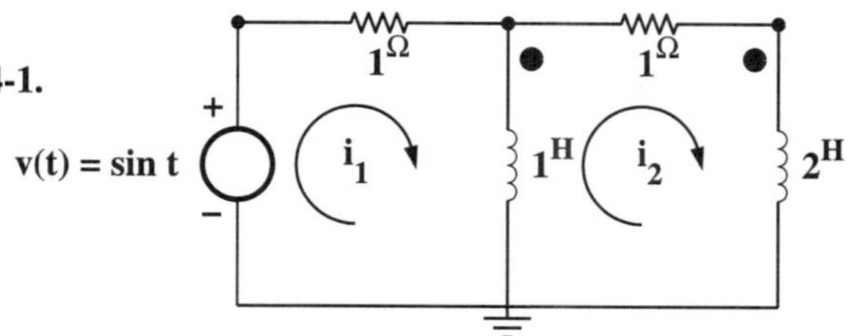

8.4.2. Write the mesh equations for the circuit in Fig. 8.4-2, where $\omega = 1000$ rad/sec and $M = 10^{mH}$ for all three coils.

Fig. 8.4-2.

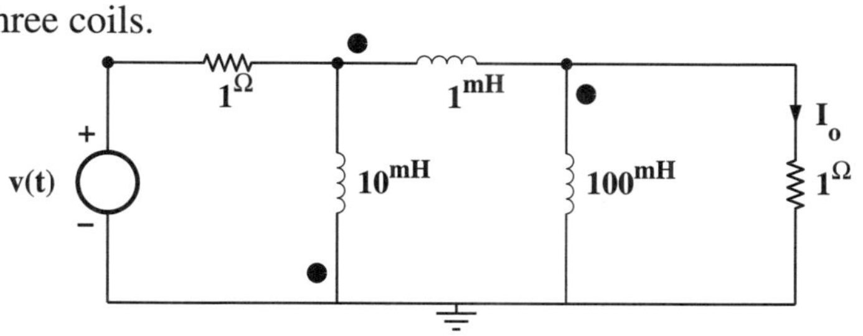

Remark 8.4-1: In most realistic applications the way that coils interact with each other is more complicated than what we have suggested in the above circuit. To consider a correct model for interactions among coils, we propose the next problem where each corresponding pair of polarity marks shows a specific interaction.

8.4.3. Write the mesh equations for the circuit in Fig. 8.4-3, where $\omega = 1000$ rad/sec and $M_{ij} = 10^{mH}$ for all $i, j = 1, 2, 3$. While the polarity marks among coils must be observed as depicted in the circuit.

Fig. 8.4-3.

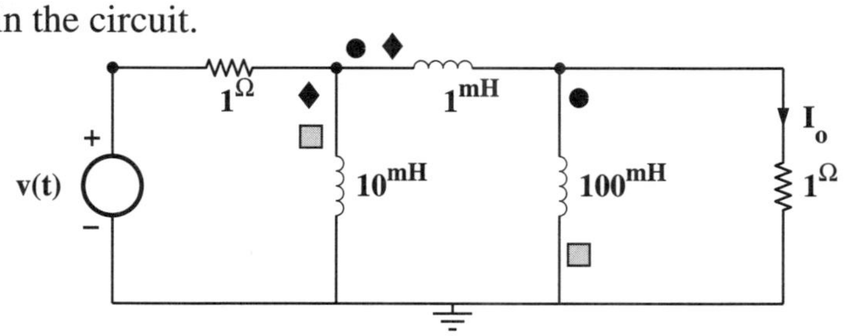

8.4.4. Formulate the mesh equations for the circuit in Fig. 8.4-4, observing the polarity marks and using M_{ij}, $i, j = 1, 2, 3$.

Fig. 8.4-4.

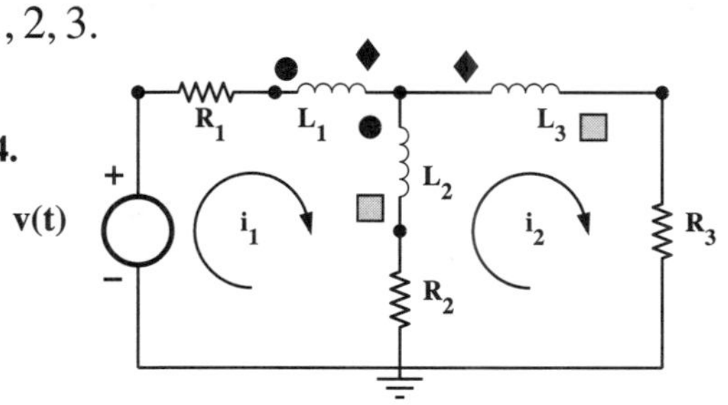

8.4.5. For the circuit in Fig. 8.4-5, write both mesh and loop equations as indicated with two sets of currents.

Fig. 8.4-5.

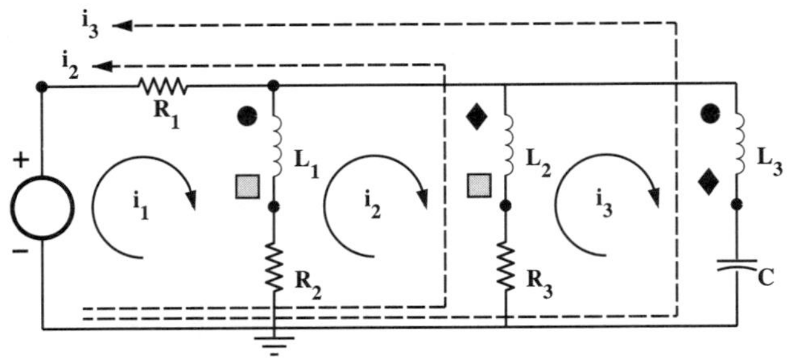

Analysis

8.4.6. Analyze the circuit in Fig. 8.4-6, with $\omega = 1$ rad/sec and $M = 1^H$. Set all pertinent initial conditions at zero.

Fig. 8.4-6.

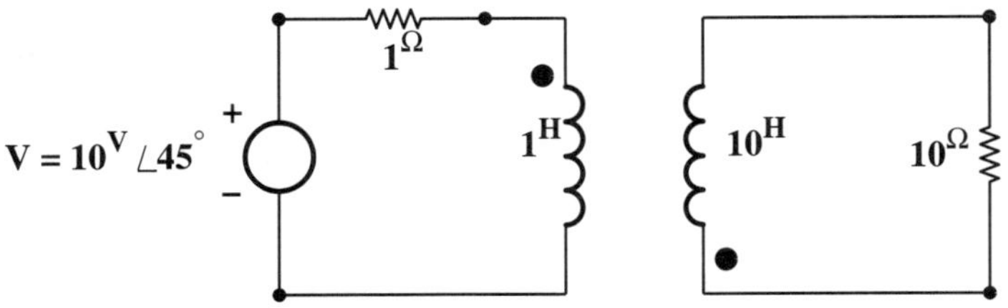

8.4.7. Find $v_o(t)$ for the circuit in Fig. 8.4-7, with $\omega = 1$ rad/sec and $M = 1^H$. Set all pertinent initial conditions at zero.

Fig. 8.4-7.

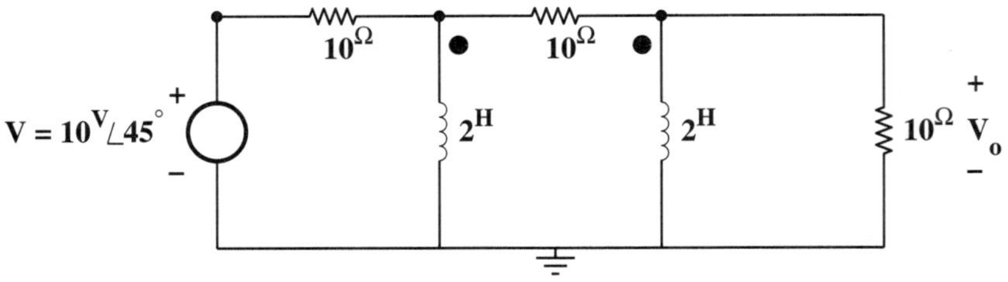

8.4.8. Find I_o in Problem 8.4.2, using $v(t) = 100^V \angle (45^\circ + j\omega t)$, $\omega = 1000$ rad/sec, with $M = 10^{mH}$ for all three coils. Set all pertinent initial conditions at zero.

8.4.9. Find I_o in the circuit of Fig. 8.4-8, with $\omega = 1000$ rad/sec and $M = 1^{mH}$.

Fig. 8.4-8.

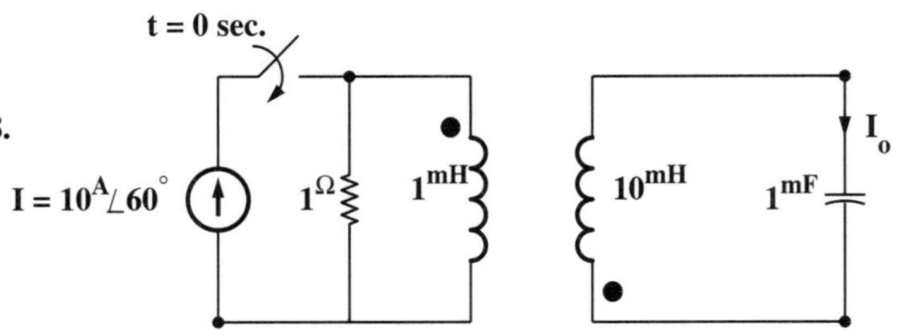

Ideal Transformer

8.4.10. Find $v_o(t)$ for the circuit in Fig. 8.4-9, with an ideal transformer.

Fig. 8.4-9.

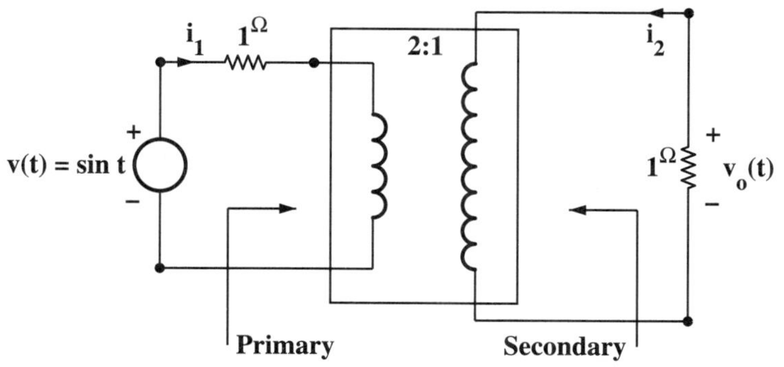

8.4.11. Using source transformation, draw an equivalent circuit for the circuit in Fig. 8.4-10, with only current sources and isolated or sealed inductors. Choose an appropriate direction for currents. This is an opportunity to define a new model for this circuit based on another point of view.

Fig. 8.4-10.

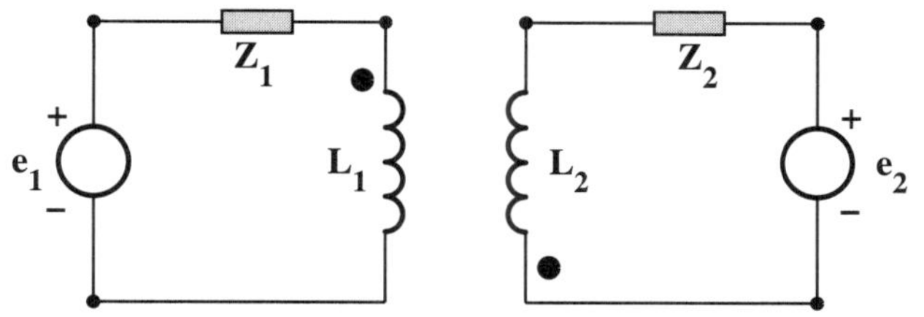

8.4.12. Consider the circuit in Fig. 8.4-11, where the ratio of $L_2/L_1 = ½$, with $M = 4^H$. Compute both currents as shown for $\omega = 100$ rad/sec. Now, let $L_1 \to \infty$ and $L_2 \to \infty$, while the ratio of L_2/L_1 remains the same. Find both currents for these limiting values and compare that with the results had you assumed an ideal transformer with a proper turn ratio n. What is the n? Note that the sign of I_2 is different from our choice in Fig. 8.3-1.

Fig. 8.4-11.

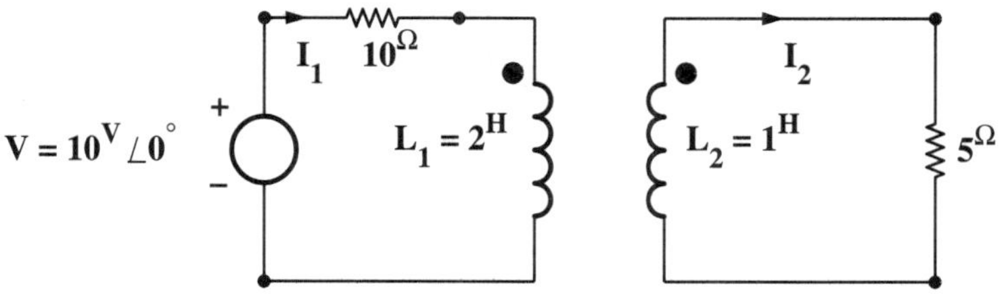

8.4.13. For the circuit in Fig. 8.4-12 with two ideal transformers, find V_o in terms of n_1 and n_2.

Fig. 8.4-12.

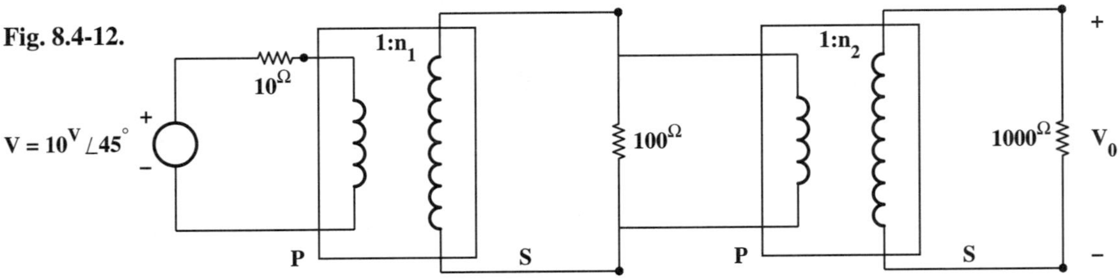

8.4.14. Compute V_o in the circuit of Fig. 8.4-13.

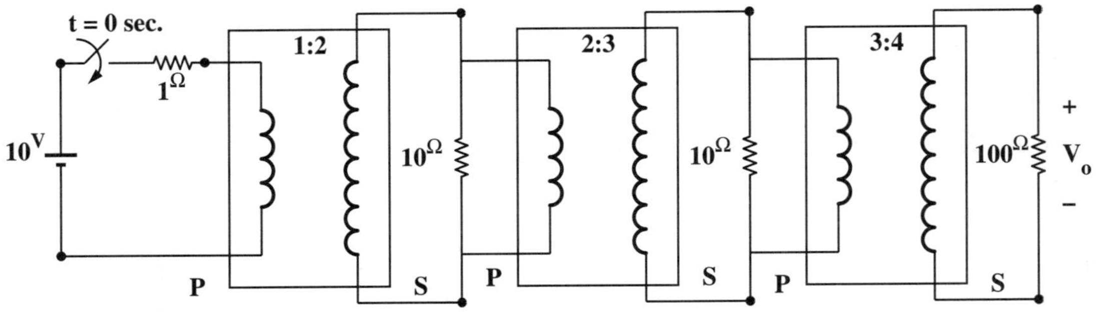

Fig. 8.4-13.

Power

8.4.15. Find Z_l in the circuit of Fig. 8.4-14, with an ideal transformer and at $\omega = 1000$ rad/sec, for the maximum power transfer.

Fig. 8.4-14.

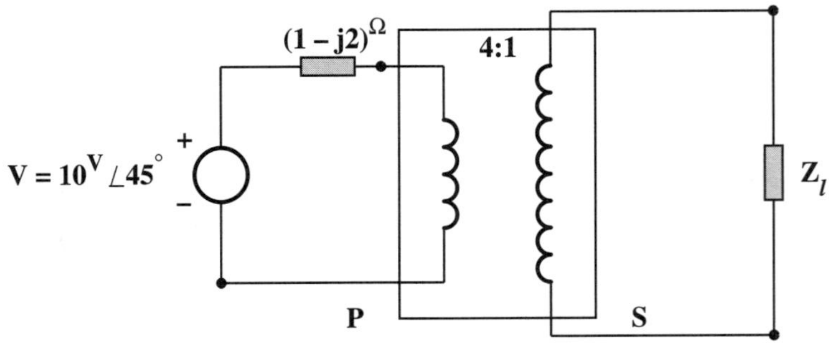

8.4.16. Find dissipated real power in Z_l of the circuit in Fig. 8.4-15 as a function of n. Let $\omega = 1000$ rad/sec. Is there an optimal n for a maximum dissipated power?

Fig. 8.4-15.

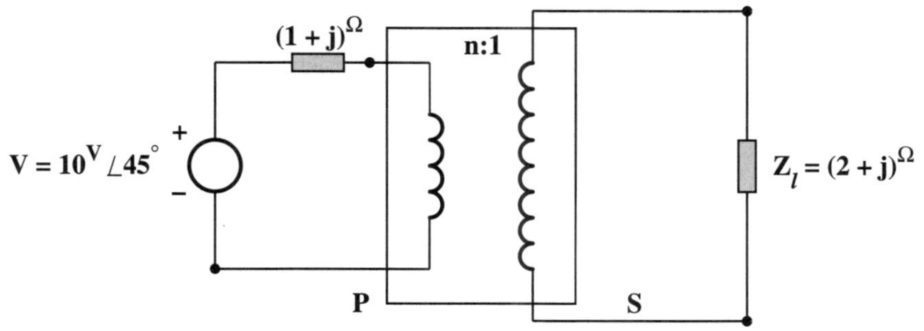

8.4.17. In circuit of Fig. 8.4-16, choose n_1 and n_2 such that the dissipated power in 10^{Ω} and 20^{Ω} be the same.

Fig. 8.4-16.

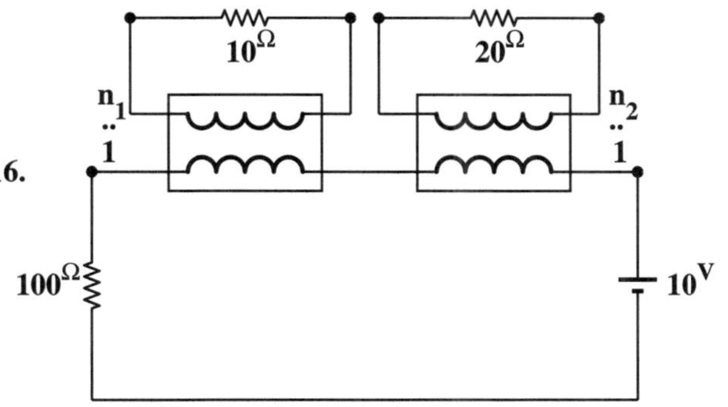

8.4.18. For the circuit in Fig. 8.4-17 and the input function $v_{in}(t) = 10[1(t)]$, find n such that $v_o(t)$ has the maximum amplitude. Repeat this problem for $v_{in}(t) = 10 \sin 100t$.

Fig. 8.4-17.

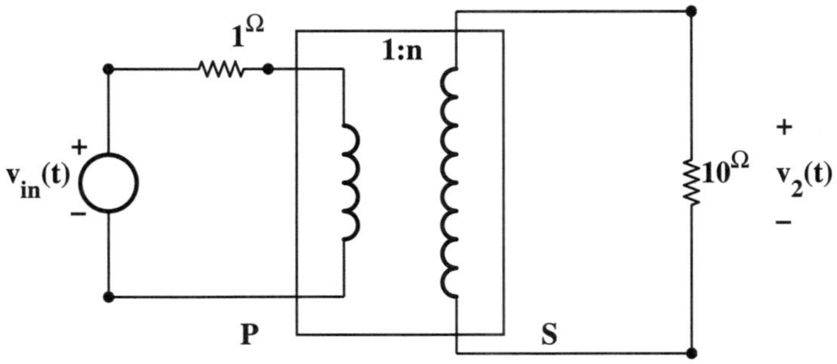

Analysis Continued

8.4.19. Find V_o/V_{in} in the circuit of Fig. 8.4-18. Let $M = 1^H$.

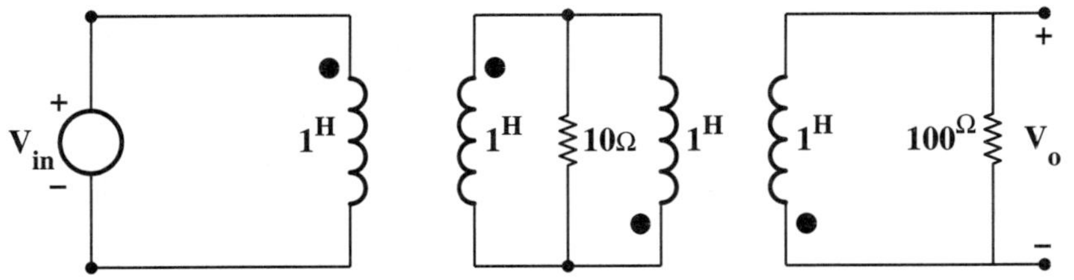

Fig. 8.4-18.

8.4.20. For the circuit in Fig. 8.4-19, find $\frac{di_1(0^+)}{dt}$ and $\frac{di_2(0^+)}{dt}$.

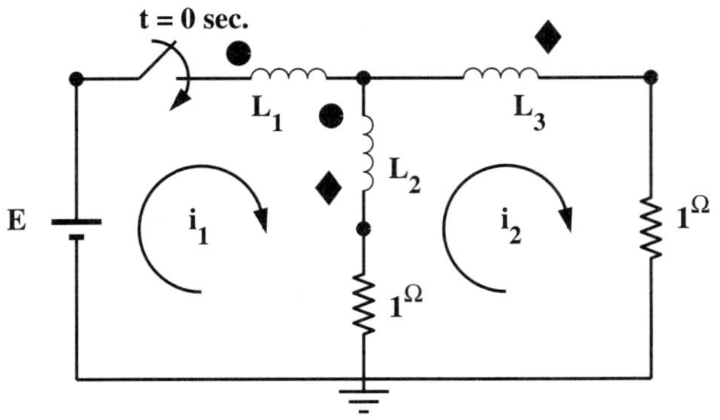

Fig. 8.4-19.

8.4.21. Consider the circuit in Fig. 8.4-20, and let $\omega = 1000$ rad/sec. The switch is initially at (a) for a long time. At $t = 0$, we switch from (a) to (b). Write only all the necessary equations to describe the dynamic of this circuit for $-\infty < t < \infty$. Let $M = 1^{mH}$.

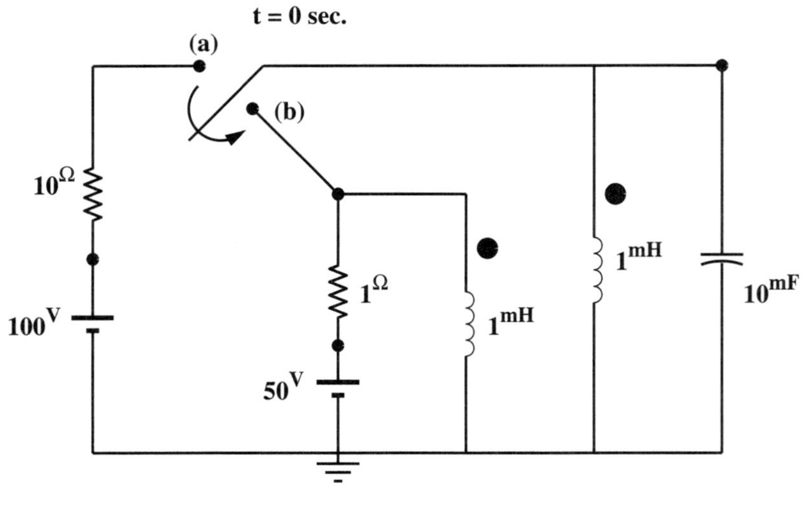

Fig. 8.4-20.

9 LAPLACE TRANSFORMATION

9.1. INTRODUCTION

Solving differential equations which arise naturally in circuit analysis, particularly coupled differential equations, can soon become problematic even for relatively simple circuit cases. Laplace transformations, on the other hand, change these complicated problems into simpler algebraic ones which can be resolved in easier ways than differential equation settings allow. In the following we review the essentials of Laplace transformation techniques as pertain to our basic need in circuit analysis.

Given a piecewise continuous function f(t), which is of exponential order, i.e., $|f(t)| < me^{\alpha t}$, as $t \to \infty$ and for some real numbers $m > 0$ and $\alpha > 0$, the Laplace transformation of this function is defined as follows.

$$\mathcal{L}\{f(t); t \to s\} = \int_0^\infty f(t)e^{-st}dt. \tag{9.1-1}$$

The variable $s = \sigma + j\omega$ is called the complex-frequency or Laplace variable. Its real part ($\mathrm{Re}(s) \triangleq \sigma$) (also known as abscissa of convergence) is chosen such that the above integral to exist. Often the mapping of $t \to s$ is not shown for convenience. If f(t) is an exponentially growing function such as $e^{\alpha t^2}$, then this integral fails to exist. However, in most of our applications in circuits this situation does not arise. Laplace transformation is one of the several integral transformation techniques which we study in an electrical engineering program.

Generally, the above mapping from the time domain to the complex-frequency domain (or frequency domain for short) is shown by upper case letters, while all time-domain variables as stated before are shown by lower case letters. Thus, for all applications we write

$$F(s) \triangleq \int_0^\infty f(t)e^{-st}dt. \qquad (9.1\text{-}2)$$

For instance, the voltage and current in frequency domain are shown as follows.

$$V(s) \triangleq \int_0^\infty v(t)e^{-st}dt, \qquad (9.1\text{-}3)$$

$$I(s) \triangleq \int_0^\infty i(t)e^{-st}dt. \qquad (9.1\text{-}4)$$

9.2. CATALOG of SPECIAL FUNCTIONS

We now begin to compute the Laplace transform of a few important time-domain functions, which are encountered often in circuit analysis.

A. Step Function: As before, this function is mathematically defined and graphically displayed as shown in Fig. 9.2-1.

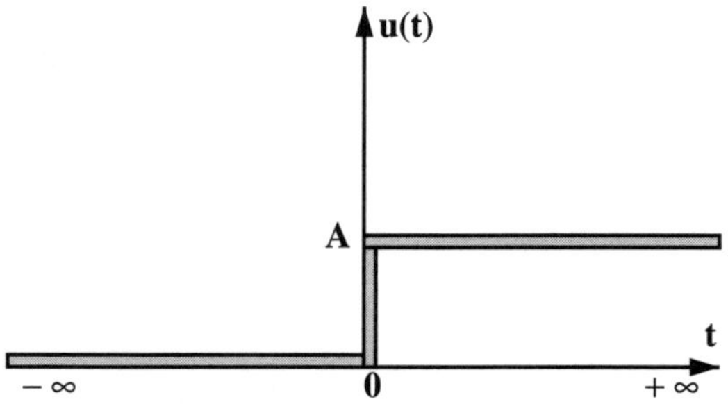

$$u(t) = \begin{cases} A & t > 0 \\ 0 & t \leq 0 \end{cases}$$

Fig. 9.2-1. Step function.

Whose Laplace transform is

$$\mathcal{L}[u(t)] \triangleq \int_0^\infty Ae^{-st}dt$$

$$= A[-\frac{e^{-st}}{s}]_{t=0^+}^{t=\infty} = \frac{A}{s}. \tag{9.2-1}$$

Here, we assume that $\text{Re}(s) \triangleq \sigma > 0$ in order that the upper limit goes to zero. Furthermore, the step function is undefined at $t = 0$, but this fact is not an issue because $\int_{0^-}^{0^+} Ae^{-st}dt = 0$.

B. Unit-Step Function: When $A \equiv 1$, the above is called a unit-step function and is shown by $1(t)$. Thus,

$$\mathcal{L}[1(t)] = \frac{1}{s}. \tag{9.2-2}$$

C. Ramp Function: Also as before, this function is defined as follows.

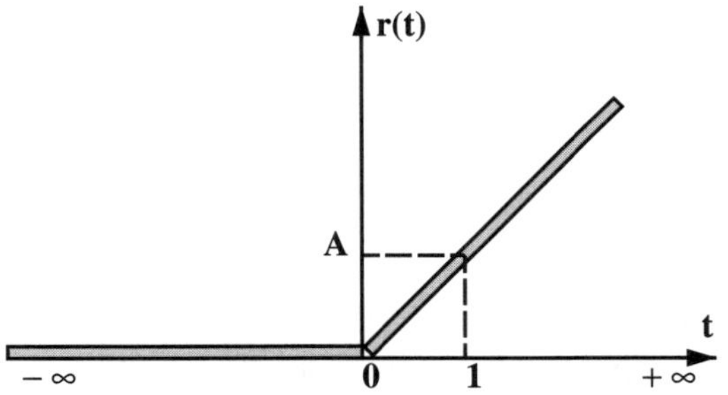

$$r(t) = \begin{cases} At & t > 0 \\ 0 & t \leq 0 \end{cases}$$

Fig. 9.2-2. Ramp function.

Whose Laplace transform upon using the integration by part rule ($\int UdV = UV - \int VdU$), is

$$\mathcal{L}[r(t)] \triangleq \int_0^{\infty} At\, e^{-st}dt$$

$$= A[t\frac{e^{-st}}{-s}]_{t=0}^{t=\infty} - \int_0^{\infty} \frac{Ae^{-st}}{-s}dt$$

$$= \frac{A}{s}\int_0^{\infty} e^{-st}dt = \frac{A}{s^2}. \tag{9.2-3}$$

In general, using similar integration by parts, we can show that

$$\mathcal{L}[t^n] = \frac{n!}{s^{n+1}}. \tag{9.2-4}$$

D. Exponential: This function is defined as follows.

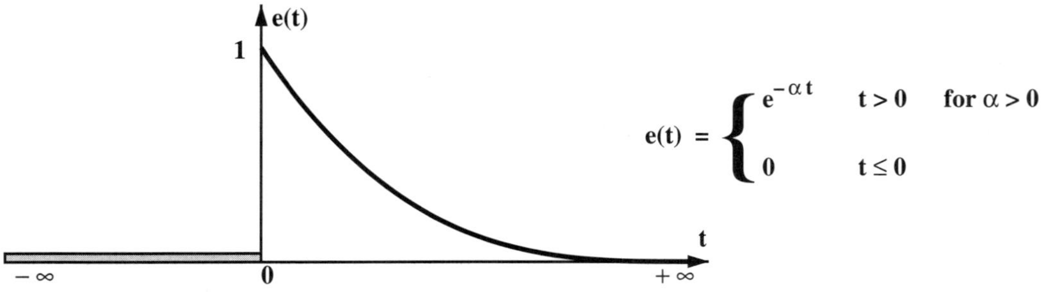

Fig. 9.2-3. Exponential function.

Whose Laplace transform is

$$\mathcal{L}\,[e(t)] = \int_0^{\infty} e^{-\alpha t}\, e^{-st} dt$$

$$= \int_0^{\infty} e^{-(\alpha + s)t} dt = \frac{1}{(s+\alpha)}\,, \qquad (9.2\text{-}5)$$

with the assumption that $\mathrm{Re}(s) \triangleq \sigma > -\alpha$.

E. Sinusoidals: These functions are also defined for $t \geq 0$, Fig. 9.2-4, where T is the period, $\omega = 2\pi/T$ is the angular frequency. Using the formulae in Fig. 9.2-4 and the previous results we can generate the Laplace transformations for $\sin \omega t$ and $\cos \omega t$ as follows.

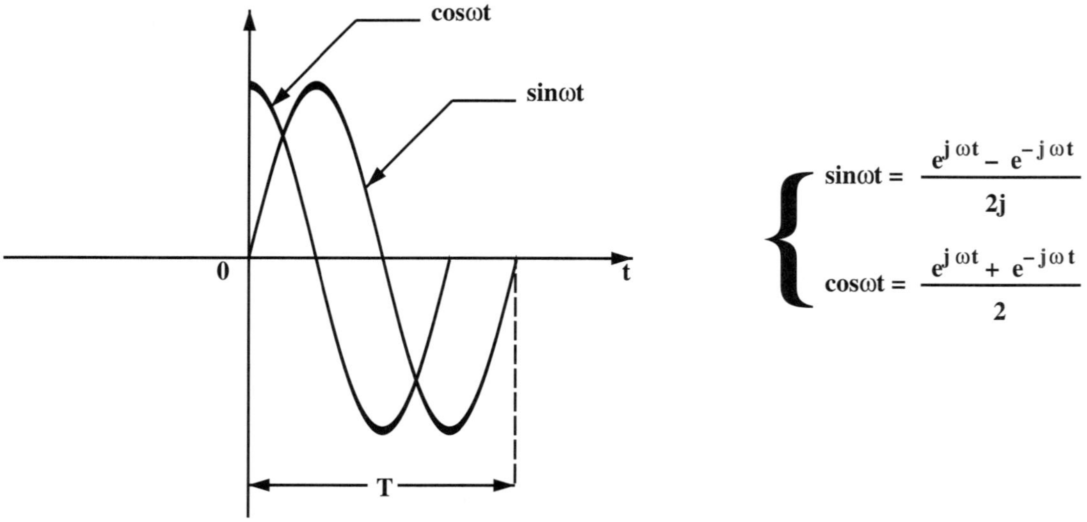

Fig. 9.2-4. Sinusoidals.

$$\mathcal{L}[\sin(\omega t)] = \frac{1}{2j}[\frac{1}{s - j\omega} - \frac{1}{s + j\omega}] = \frac{\omega}{(s^2 + \omega^2)}, \tag{9.2-6}$$

$$\mathcal{L}[\cos(\omega t)] = \frac{1}{2}[\frac{1}{s - j\omega} + \frac{1}{s + j\omega}] = \frac{s}{s^2 + \omega^2}. \tag{9.2-7}$$

The above five functions form the basic catalog of important functions which we must memorize. Together with the forthcoming rules and theorems, we can expand this catalog to cover many other cases of possible applications and interest in circuit analysis.

9.3. LAPLACE TRANSFORM RULES and THEOREMS

Laplace transformation is a methodology which transform time-domain functions into the complex-frequency domain functions. The following operational rules and theorems apply in this transformation. Note that $\mathcal{L}\{Af(t)\} \triangleq AF(s)$, for a constant A.

(1) Summation – Given $f_1(t)$ and $f_2(t)$ where both functions are Laplace transformable, then for two constant numbers A and B, we have

$$\mathcal{L}\{Af_1(t) + Bf_2(t)\} = AF_1(s) + BF_2(s). \tag{9.3-1}$$

This is easy to show by just straightforward application of the definition for Laplace transformation. This rule extends to cases with many, though finite, terms.

(2) Time-Domain Shift – Given $F(s) \triangleq \mathcal{L}[f(t)]$, then $F_T(s) \triangleq \mathcal{L}[f(t - T)]$ is calculated as follows. From the definition we have

$$F_T(s) \triangleq \int_0^\infty f(t - T)e^{-st}dt = \int_T^\infty f(t - T)e^{-st}dt. \tag{9.3-2}$$

The last step follows from the fact that $f(t - T) \equiv 0$ for $t < T$. Now, to evaluate (9.3-2), let $(t - T) = \tau$, then $dt = d\tau$ and (9.3-2) becomes

$$F_T(s) = \int_0^\infty f(\tau)e^{-s(T+\tau)}d\tau$$

$$= e^{-sT}\int_0^\infty f(\tau)e^{-s\tau}d\tau = e^{-sT}F(s). \tag{9.3-3}$$

Thus, a shift of T seconds in time, will multiply the corresponding Laplace transform by e^{-sT}.

An immediate application of this rule is the computation of the Laplace transform of a pulse function.

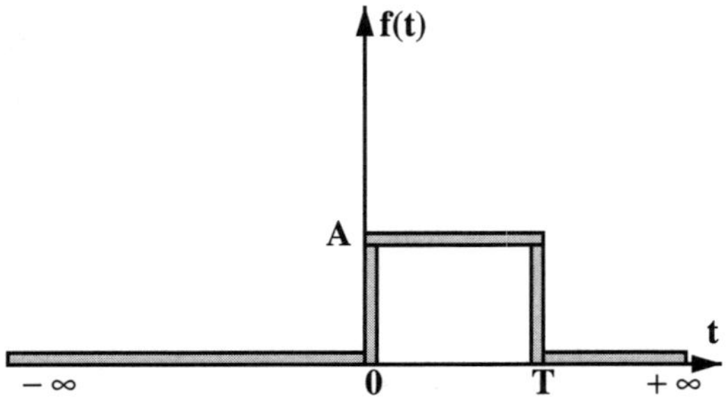

$$\mathbf{f(t)} = \begin{cases} \mathbf{A} & \mathbf{0 < t < T} \\ \mathbf{0} & \textbf{otherwise} \end{cases}$$

Fig. 9.3-1. A pulse function.

This pulse function is the superposition of two step functions as follows.

$$f(t) = Au(t) - Au(t - T). \tag{9.3-4}$$

Thus, its Laplace transform is

$$F(s) = \frac{A}{s} - \frac{Ae^{-sT}}{s} = \frac{A}{s}(1 - e^{-st}). \tag{9.3-5}$$

In a limiting case when $T \to 0$ the above pulse function becomes an **impulse function,** while the area of the rectangular (called impulse power) remains constant at A. This is shown mathematically as follows.

$$f(t) = \begin{cases} \lim\limits_{T \to 0} \dfrac{A}{T} & 0 < t < T, \\ 0 & \text{otherwise.} \end{cases} \tag{9.3-6}$$

Thus, its Laplace transform is

$$\mathcal{L}\,[f(t)] \triangleq \lim_{T \to 0} \Big[\frac{A}{T}\,\frac{1 - e^{-sT}}{s}\Big]$$

$$= \frac{A}{s} \lim_{T \to 0} \frac{\frac{d}{dT}(1 - e^{-sT})}{\frac{dT}{dT}}, \quad \text{(l'Hopital's rule)}$$

$$= \frac{A}{s}\frac{s}{1} = A. \tag{9.3-7}$$

When $A \equiv 1$, this function is called the unit-impulse function and is shown as $\delta(t)$, the Dirac function. Among many properties of a Dirac function, we cite the following:

$$\delta(t - T) = 0, \qquad \text{for } t \neq T, \tag{9.3-8}$$

$$\delta(t - T) = \infty, \qquad \text{for } t = T, \tag{9.3-9}$$

$$\int_{-\infty}^{\infty} \delta(t - T)dt = 1, \tag{9.3-10}$$

$$\int_{-\infty}^{\infty} \delta(\tau - t)g(\tau)d\tau = g(t). \tag{9.3-11}$$

(3) Time-Derivative – Laplace transformation of time-derivative for f(t) is as follows.

$$\mathcal{L}[\frac{df(t)}{dt}] = sF(s) - f(0^+). \tag{9.3-12}$$

Using the integration by part rule ($\int UdV = UV - \int VdU$), the left-hand side of (9.3-12) can be written as follows.

$$\mathcal{L}[\frac{df(t)}{dt}] = \int_0^{\infty} \frac{df(t)}{dt} e^{-st}dt$$

$$= \left[f(t)e^{-st}\right]_0^{\infty} - \int_0^{\infty} f(t)[-se^{-st}]dt$$

$$= s\int_0^{\infty} f(t)e^{-st}dt - f(0^+)$$

$$= sF(s) - f(0^+). \tag{9.3-13}$$

When $f(0^+) \equiv 0$, the above becomes $\mathcal{L}[\frac{df(t)}{dt}] = sF(s)$. That is why "s" is also called a **differentiator** (or a differential operator).

Repeated application of the previous theorem yields the following:

$$\mathcal{L}\left[\frac{d^2f(t)}{dt^2}\right] \triangleq \mathcal{L}\left[\frac{d}{dt}\left[\frac{df}{dt}\right]\right]$$

$$= s\mathcal{L}\left[\frac{df(t)}{dt}\right] - \left[\frac{df(t)}{dt}\right]_{t=0^+}$$

$$= s[sF(s) - f(0^+)] - \left[\frac{df(t)}{dt}\right]_{t=0^+}$$

$$= s^2F(s) - sf(0^+) - \left[\frac{df(t)}{dt}\right]_{t=0^+}. \quad (9.3\text{-}14)$$

This computation can be easily expanded to higher-order derivatives as follows.

$$\mathcal{L}\left[\frac{d^nf(t)}{dt^n}\right] = s^nF(s) - s^{n-1}f(0^+) - s^{n-2}\left[\frac{df(t)}{dt}\right]_{t=0^+} - \cdots - s^0\left[\frac{d^{n-1}f(t)}{dt^{n-1}}\right]_{t=0^+}. \quad (9.3\text{-}15)$$

(4) Integration Theorem – Given $F(s) \triangleq \mathcal{L}[f(t)]$, we can show that

$$\mathcal{L}\left[\int_{-\infty}^{t} f(\tau)d\tau\right] = \frac{F(s)}{s} + \frac{1}{s}\left[\int_{-\infty}^{t} f(\tau)d\tau\right]_{t=0}. \quad (9.3\text{-}16)$$

To show (9.3-16) we use the definition of Laplace transformation and the rule of integration by part.

$$\mathcal{L}\left[\int_{-\infty}^{t} f(\tau)d\tau\right] \triangleq \int_0^{\infty}\int_{-\infty}^{t} f(\tau)d\tau e^{-st}dt$$

$$= \left[\int_{-\infty}^{t} f(\tau)d\tau \frac{e^{-st}}{-s}\right]_{t=0}^{t=\infty} - \int_0^{\infty} f(t)\frac{e^{-st}}{-s}dt$$

$$= \frac{F(s)}{s} + \frac{1}{s}\left[\int_{-\infty}^{t} f(\tau)d\tau\right]_{t=0}. \quad (9.3\text{-}17)$$

If f(t) involves an impulse function at $t = 0$, then we have $\mathcal{L}_+$ and $\mathcal{L}_-$ referring to $t = 0^+$ and $t = 0^-$. Otherwise, we use the above formula.

Furthermore, in many applications the $\int_{-\infty}^{0} f(\tau)d\tau$ is zero and we are left with $F(s)/s$. Similarly, the "$\frac{1}{s}$" is called an **integral** operator.

(5) Multiplication of f(t) by $e^{-\alpha t}$ – Given $F(s) \triangleq \mathcal{L}[f(t)]$, we can show that

$$\mathcal{L}[f(t)\, e^{-\alpha t}] \triangleq \int_0^\infty f(t)\, e^{-\alpha t} e^{-st} dt$$

$$= \int_0^\infty f(t) e^{-(s+\alpha)t} dt$$

$$= F(s+\alpha). \qquad (9.3\text{-}18)$$

For instance,

$$\mathcal{L}[e^{-\alpha t} \sin \omega t] = \frac{\omega}{(s+\alpha)^2 + \omega^2}, \qquad (9.3\text{-}19)$$

$$\mathcal{L}[e^{-\alpha t} \cos \omega t] = \frac{s+\alpha}{(s+\alpha)^2 + \omega^2}. \qquad (9.3\text{-}20)$$

(6) Time-scaling – Given $F(s) \triangleq \mathcal{L}[f(t)]$ and using $\frac{t}{\alpha} = \tau$, we can show that

$$\mathcal{L}[f(\frac{t}{\alpha})] \triangleq \int_0^\infty f(\frac{t}{\alpha}) e^{-st} dt$$

$$= \int_0^\infty f(\tau) e^{-\alpha s \tau} \alpha d\tau$$

$$= \alpha F(\alpha s). \qquad (9.3\text{-}21)$$

(7) Multiplication by t (then followed by t^n) – Given $F(s) \triangleq \mathcal{L}[f(t)]$, we can show that

$$\mathcal{L}[t\, f(t)] \triangleq \int_0^\infty t\, f(t) e^{-st} dt = \int_0^\infty -f(t) \frac{de^{-st}}{ds} dt$$

$$= -\frac{d}{ds} \int_0^\infty f(t) e^{-st} dt$$

$$= -\frac{d}{ds} F(s). \qquad (9.3\text{-}22)$$

For instance,

$$\mathcal{L}[t \cos \omega t] = -\frac{d}{ds} (\frac{s}{s^2 + \omega^2}) = \frac{s^2 - \omega^2}{(s^2 + \omega^2)^2}. \qquad (9.3\text{-}23)$$

Similarly,

$$\mathcal{L}[t^n f(t)] = (-1)^n \frac{d^n}{ds^n} F(s). \tag{9.3-24}$$

(8) Final-Value Theorem – Given f(t) and its derivative which are both Laplace transformable, and if $\lim_{t \to \infty} f(t)$ exists, then

$$\lim_{t \to \infty} f(t) = \lim_{s \to 0} sF(s). \tag{9.3-25}$$

Proof: To prove the above, we note that from the Laplace transformation point of view

$$\lim_{s \to 0} \int_0^\infty [\frac{df(t)}{dt}] e^{-st} dt = \lim_{s \to 0} [sF(s) - f(0)] = \lim_{s \to 0} sF(s) - f(0). \tag{9.3-26}$$

In the left-hand side of (9.3-26) the integration is in terms of "t", while the limit is taken in terms of "s". Thus, we may change these operations to have

$$\lim_{s \to 0} \int_0^\infty [\frac{df(t)}{dt}] e^{-st} dt = \int_0^\infty [\frac{df(t)}{dt}] \lim_{s \to 0} e^{-st} dt$$

$$= \int_0^\infty [\frac{df(t)}{dt}] dt = \lim_{t \to \infty} f(t) - f(0) = \lim_{s \to 0} sF(s) - f(0), \tag{9.3-27}$$

and that completes the proof.

For instance, for $f(t) = e^{-t}$, we have $\lim_{t \to \infty} e^{-t} = 0$, and similarly, $\lim_{s \to 0} sF(s) = \lim_{s \to 0} s/(s+1) = 0$. However, for $f(t) = \sin \omega t$, we have

$$\lim_{s \to 0} sF(s) = \lim_{s \to 0} (s \frac{\omega}{s^2 + \omega^2}) = 0, \tag{9.3-28}$$

while, $\lim_{t \to \infty} (\sin \omega t)$ is not defined, and therefore (9.3-28) is not correct.

(9) Initial-Value Theorem – Similarly, under the above condition for f(t), we have the initial-value theorem as follows.

$$\lim_{t \to 0^+} f(t) = \lim_{s \to \infty} sF(s). \tag{9.3-29}$$

Proof: Here, we note that from the Laplace transformation point of view

$$\mathcal{L}_+[\frac{df(t)}{dt}] = sF(s) - f(0^+). \tag{9.3-30}$$

On the other hand,

$$\lim_{s\to\infty} \int_{0^+}^{\infty} [\frac{df(t)}{dt}]e^{-st}dt \equiv 0. \tag{9.3-31}$$

Thus, equating the right-hand side of (9.3-30) as $s \to \infty$ to zero completes the proof.

The above two theorems help us to compute both initial and final conditions for various complex elements in a circuit.

(10) Convolution Integral – Consider the following integral also known as a convolution integral, which has extensive application in circuit theory and linear systems.

$$y(t) = \int_0^t f_1(t-\tau)f_2(\tau)d\tau. \tag{9.3-32}$$

This integral is a special case of $\int_0^t f_1(t, \tau)f_2(\tau)d\tau$. To compute the Laplace transformation of y(t) we offer the following.

$$\mathcal{L}[y(t)] = \int_0^{\infty} \int_0^t f_1(t-\tau)f_2(\tau)d\tau e^{-st}dt. \tag{9.3-33}$$

Here, we assume that both arguments are positive, i.e., $t - \tau > 0$ and $\tau > 0$. To complete the computation, we use the change of variables as shown in Fig. 9.3-2.

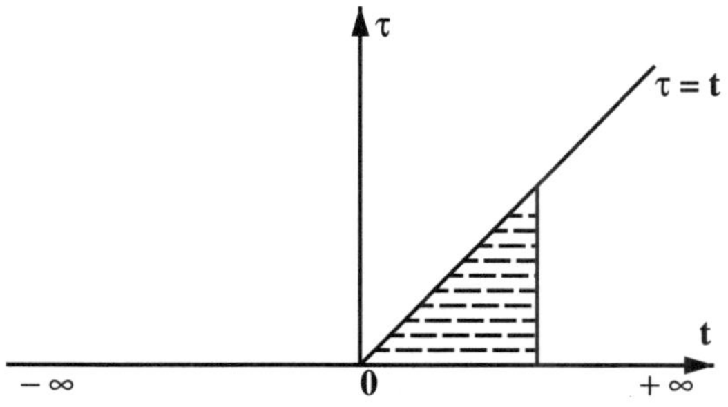

Fig. 9.3-2.

$$\mathcal{L}[y(t)] = \int_0^{\infty} \int_{\tau}^{\infty} f_1(t-\tau)f_2(\tau)d\tau e^{-st}dt. \tag{9.3-34}$$

Let $(t - \tau) = x$ or $t = \tau + x$, and $\tau = y$. Then, the above becomes

$$\mathcal{L}[y(t)] = \int_0^{\infty} \int_0^{\infty} f_1(x)f_2(y)\, e^{-s(x+y)}\, dxdy. \tag{9.3-35}$$

Also, note that $\int_{\tau}^{\infty} = \int_{\tau}^{0} + \int_{0}^{\infty} = \int_{0}^{\infty}$, because f(t) = 0, for t < 0, therefore the corresponding $\int_{\tau}^{0}$ vanishes. Then (9.3-35) becomes

$$\begin{aligned} \mathcal{L}\,[y(t)] &= \int_{0}^{\infty} f_1(x)e^{-sx}dx \int_{0}^{\infty} f_2(y)e^{-sy}dy \\ &= F_1(s)F_2(s). \end{aligned} \tag{9.3-36}$$

It should be noted that $F_1(s)F_2(s) \neq \mathcal{L}\,[f_1(t)f_2(t)]$.

(11) Linear Systems – In general, a linear time-invariant system is described as that depicted in Fig. 9.3-3. The pair of input and output for this system in a Laplace transformation form is related through

$$Y(s) = G(s)\,X(s). \tag{9.3-37}$$

Here, G(s) represents the **transfer function** of the above linear system. This G(s) is represented as the ratio of $\frac{N(s)}{D(s)}$, where N(s) and D(s), representing respectively the numerator and denominator of G(s), are two rational polynomials of s. For a physically realizable system, the kind of system we study in this textbook, N(s) and D(s) are both not only rational polynomials of s, but also have constant real coefficients. The time-domain counterpart of g(t), based on the convolution integral we have just studied in (9.3-32), is

$$y(t) = \mathcal{L}^{-1}[G(s)\,X(s)] = \int_{0}^{t} g(t-\tau)x(\tau)d\tau. \tag{9.3-38}$$

We explore the many applications of this type inverse relationship next.

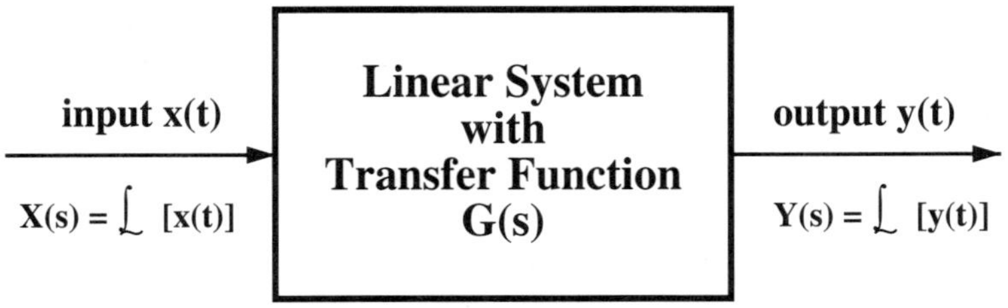

Fig. 9.3-3. A linear system.

9.4. INVERSE TRANSFORMATION

The cardinal rule of writing this textbook has been to build new knowledge from the knowledge gained. However, we have just utilized a definition in (9.3-38) that has not been formally defined yet. Furthermore, we have not explicitly declared our intention or reasons as why do we use the Laplace transformation, since as is this complex function cannot be of much use. We prefer to work with numbers in a real-time setting. So going from f(t) → F(s) is only a step toward devising this methodology, but what we really want to know is how do we get back from F(s) → f(t). In other words, given some F(s), how do we find the corresponding f(t)? To compute f(t) we go through a process that is called the inverse Laplace transformation and is shown formally as follows.

$$f(t) = \mathcal{L}^{-1}[F(s);\ s \to t]. \tag{9.4-1}$$

Mathematically, when all the necessary conditions are met, the above f(t) can be computed as follows.

$$f(t) = \frac{1}{2\pi j}\int_{c-j\infty}^{c+j\infty} F(s)e^{st}ds, \quad t > 0. \tag{9.4-2}$$

Here, c (also called the abscissa of convergence) is a real number and is chosen such that it is larger than the real parts of all **singular points (poles)** of F(s) as is described subsequently. This construction can be shown pictorially as that in Fig. 9.4-1.

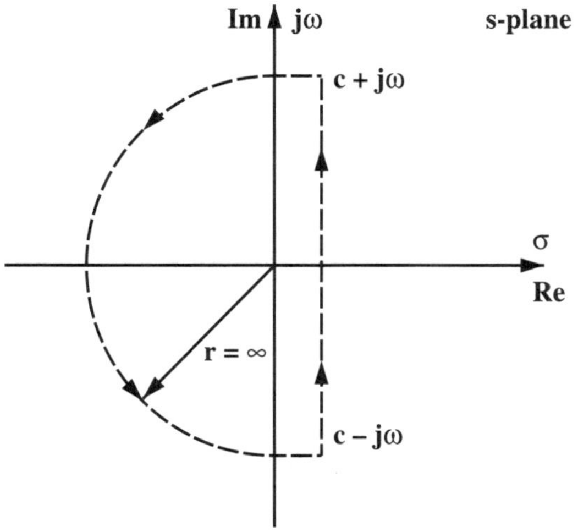

Fig. 9.4-1. The complex s-plane.

Here, all singular points of F(s) are supposedly in the left-hand side of the imaginary axis, i.e., inside the left-half circle with infinite radius. The integral over the closed path by Cauchy's theorem is equal to the sum of all residues of F(s). In other words,

$$\frac{1}{2\pi j}\int_{\text{closed path}} F(s)e^{st}ds = \sum_{\text{all singular points}} \text{Residues of } F(s)e^{st}. \tag{9.4-3}$$

Applying (9.4-3) to (9.4-2), we add the integration over the left-half circle in Fig. 9.4-1 to make a closed path. Since the integral over that half circle with infinite radius goes to zero, we are then left with our original integral over the right-hand line (path) from $c - j\infty$ to $c + j\infty$. Thus,

$$f(t) = \sum_{\text{all singular points}} \text{Residues of } F(s)e^{st}. \tag{9.4-4}$$

Comment 9.4-1: The above formula supported by the Cauchy's Theorem is not so straightforward as the method which is described next. As an engineer we search for simple ways of calculating the inverse Laplace transformations, which is based on (9.4-4), but uses a different orientation. One such method is devised by Oliver Heaviside, an accomplished mathematician and electrical engineer of the last century.

9.5. PARTIAL FRACTION EXPANSIONS

The outcome of the preceding methodology, that works perfectly in all cases, can be generated using the following approach known as partial fraction expansion technique, which is substantially more straightforward than that stemming from (9.4-2).

In a large class of problems (including those in circuit analysis), the type of Laplace transformed functions which we face are of the following nature.

$$F(s) = \frac{N(s)}{D(s)}, \quad \text{where } \deg[N(s)] \le \deg[D(s)] = n. \tag{9.5-1}$$

Here, "deg $(\cdot)$" refers to the degree of $(\cdot)$, the corresponding rational polynomial, which as stated before both N(s) and D(s) are rational polynomials of s. When we set $N(s) \equiv 0$, the roots of this equation are called the **zeros** of F(s) (or the zeros of the system). Similarly, when we set $D(s) \equiv 0$, the resulting values for s are called the **poles** of F(s) (or the poles of the system).

Given $F(s) = \dfrac{N(s)}{D(s)}$, with the condition that $\deg[N(s)] \le \deg[D(s)] = n$, we set $D(s) \equiv 0$ and study its roots before proceeding to develop its corresponding partial fraction technique.

9.5.1. Simple Poles

First we consider the case where every pole of F(s) is simple, i.e., we have n distinct poles shown by p_1 to p_n. Then the denominator of F(s) can be factored as follows.

$$F(s) = \frac{N(s)}{(s+p_1)\cdots(s+p_n)}. \tag{9.5-2}$$

In this case, we can easily show that the above F(s) can be written as

$$F(s) = \frac{k_1}{s+p_1} + \frac{k_2}{s+p_2} + \cdots + \frac{k_n}{s+p_n}, \tag{9.5-3}$$

where,

$$k_1 = [(s+p_1)F(s)]\Big|_{s=-p_1},$$

$$\vdots$$

$$k_n = [(s+p_n)F(s)]\Big|_{s=-p_n}. \tag{9.5-4}$$

The proof is simple. For instance, to calculate k_i, on one hand we have

$$[(s+p_i)F(s)]\Big|_{s=-p_i} = \frac{N(s)(s+p_i)}{(s+p_1)\cdots(s+p_i)\cdots(s+p_n)}\Big|_{s=-p_i}, \tag{9.5-5}$$

where the right-hand side of (9.5-5) has a factor of $(s+p_i)$ in its numerator and denominator, which the net effect is a cancellation of this factor and we are left with a number upon substituting for $s = -p_i$ in the remaining part of this side. On the other hand, the right-hand side of (9.5-3), upon multiplying both side of that equation by $(s+p_i)$ becomes

$$[(s+p_i)F(s)]\Big|_{s=-p_i} = \left[\frac{k_1(s+p_i)}{s+p_1} + \cdots + k_i + \cdots + \frac{k_n(s+p_i)}{s+p_n}\right]\Big|_{s=-p_i}, \tag{9.5-6}$$

which is equal to k_i, and in turn this k_i is equal to the number resulted from the right-hand side of (9.5-5), and that completes the assertion.

Comment 9.5-1: The above procedure to compute k_i works whether the poles are simple real or simple complex numbers as shown subsequently.

Going back to (9.5-3), now that we have all k_i's, we can apply the inverse Laplace operator to both sides of that equation to get

$$\mathcal{L}^{-1}[F(s)] = \mathcal{L}^{-1}\Big[\frac{k_1}{s+p_1} + \frac{k_2}{s+p_2} + \cdots + \frac{k_n}{s+p_n}\Big]$$

$$= \mathcal{L}^{-1}[\frac{k_1}{s+p_1}] + \mathcal{L}^{-1}[\frac{k_2}{s+p_2}] + \cdots + \mathcal{L}^{-1}[\frac{k_n}{s+p_n}]. \tag{9.5-7}$$

The key in structuring (9.5-7) is that to fracture a typical F(s) into pieces that are easily detectable members of our catalog for Laplace transform functions. In (9.5-7), we recognize each term as the Laplace transform of an exponential function. Thus,

$$\mathcal{L}^{-1}[F(s)] = k_1 e^{-p_1 t} + k_2 e^{-p_2} + \cdots + k_n e^{-p_n t}. \tag{9.5-8}$$

Example 9.5-1 (Simple Real Poles): Given $F(s) = \frac{4}{(s+1)(s+2)(s+3)}$, compute f(t) in both ways of partial fraction expansion and the method stemming from the Cauchy's theorem.

Solution: First, we study the poles of F(s). Here, the denominator is given in a factor form and as such we note that the poles are distinct and real. Thus,

$$F(s) = \frac{k_1}{s+1} + \frac{k_2}{s+2} + \frac{k_3}{s+3}, \tag{9.5-9}$$

where

$$k_1 = (s+1)F(s)\Big|_{s=-1}$$

$$= (s+1)\frac{4}{(s+1)(s+2)(s+3)}\Big|_{s=-1}$$

$$= \frac{4}{(s+2)(s+3)}\Big|_{s=-1} = 2. \tag{9.5-10}$$

$$k_2 = (s+2)F(s)\Big|_{s=-2} = \frac{4}{(s+1)(s+3)}\Big|_{s=-2} = -4. \tag{9.5-11}$$

$$k_3 = (s+3)F(s)\Big|_{s=-3} = \frac{4}{(s+1)(s+2)}\Big|_{s=-3} = 2. \tag{9.5-12}$$

Thus,

$$F(s) = \frac{2}{s+1} - \frac{4}{s+2} + \frac{2}{s+3}, \tag{9.5-13}$$

resulting in

$$\mathcal{L}^{-1}[F(s)] = \mathcal{L}^{-1}\left[\frac{2}{s+1} - \frac{4}{s+2} + \frac{2}{s+3}\right]$$

$$= 2e^{-t} - 4e^{-2t} + 2e^{-3t}. \tag{9.5-14}$$

Alternative Solution to Example 9.5-1: From inverse formula we have

$$f(t) = \frac{1}{2\pi j}\int_{c-j\infty}^{c+j\infty} F(s)e^{st}ds$$

$$= \frac{1}{2\pi j}\int_{c-j\infty}^{c+j\infty} \frac{4e^{st}}{(s+1)(s+2)(s+3)}ds. \tag{9.5-15}$$

Here, c is real and is chosen such that it is larger than the real parts of all **singular points (poles)** of the integrand which are located at $s = -1$, $s = -2$ and $s = -3$ as shown in Fig. 9.5-1.

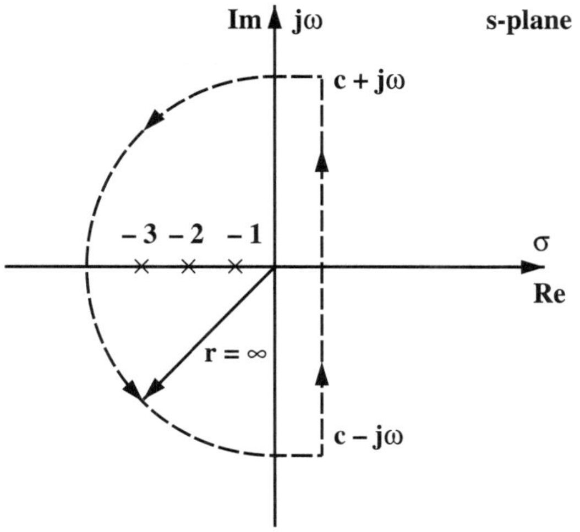

Fig. 9.5-1.

Therefore from (9.4-4) and for (9.5-15) we have

$$f(t) = \sum_{\text{all singular points}} \text{Residues of } \frac{4e^{st}}{(s+1)(s+2)(s+3)}$$

$$= 2e^{-t} - 4e^{-2t} + 2e^{-3t}. \qquad (9.5\text{-}16)$$

Comment 9.5-2: When some of the poles in a physically realizable system are complex, they will appear as pairs of complex conjugate numbers. For instance, if we have $p_i = \sigma_i + j\omega_i$, then there is another pole which we call $\bar{p}_i$ that is the complex conjugate of p_i, i.e., $\bar{p}_i = \sigma_i - j\omega_i$. This comes from the physics of the system and the beauty of it is that if k_i corresponds to p_i, then for $\bar{p}_i$ we have $\bar{k}_i$, i.e., the complex conjugate of k_i, which simplifies our computation. This result is due to the fact that all physically realizable systems (those we can build in the laboratory) have transfer functions which are the ratio of two rational polynomials of s, and as such there is no complex coefficient present, resulting in either real or complex conjugate pairs of poles. We show this concept by an example.

Example 9.5-2 (Simple Complex Poles): Consider $F(s) = \dfrac{1}{(s+1)(s^2+s+1)}$, find f(t) by a partial fraction expansion.

Solution: First, we let $(s+1)(s^2+s+1) = 0$, resulting in $s = -1$, $s = -\frac{1}{2} + \frac{j\sqrt{3}}{2}$ and $s = -\frac{1}{2} - \frac{j\sqrt{3}}{2}$. Thus,

$$F(s) = \frac{A}{s+1} + \frac{B}{s + \frac{1}{2} + \frac{j\sqrt{3}}{2}} + \frac{C}{s + \frac{1}{2} - \frac{j\sqrt{3}}{2}}. \qquad (9.5\text{-}17)$$

Straightforward computation yields

$$A = \Big[(s+1)F(s)\Big]\Big|_{s=-1} = 1 \qquad (9.5\text{-}18)$$

$$B = \left[(s + \frac{1}{2} + \frac{j\sqrt{3}}{2})F(s)\right]\Bigg|_{s=-\frac{1}{2}-\frac{j\sqrt{3}}{2}} = \left[\frac{1}{(s+1)(s + \frac{1}{2} - \frac{j\sqrt{3}}{2})}\right]\Bigg|_{s=-\frac{1}{2}-\frac{j\sqrt{3}}{2}}$$

$$= \frac{1}{(\frac{1}{2} - \frac{j\sqrt{3}}{2})(-j\sqrt{3})} = \frac{1}{-\frac{3}{2} - \frac{j\sqrt{3}}{2}} = \frac{-2}{3 + j\sqrt{3}}$$

$$= \frac{-2(3 - j\sqrt{3})}{(3 + j\sqrt{3})(3 - j\sqrt{3})} = \frac{-3 + j\sqrt{3}}{6}. \tag{9.5-19}$$

Similarly,

$$C = \left[(s + \frac{1}{2} - \frac{j\sqrt{3}}{2})F(s)\right]\Bigg|_{s=-\frac{1}{2}+\frac{j\sqrt{3}}{2}} = \left[\frac{1}{(s+1)(s + \frac{1}{2} + \frac{j\sqrt{3}}{2})}\right]\Bigg|_{s=-\frac{1}{2}+\frac{j\sqrt{3}}{2}}$$

$$= \frac{1}{(\frac{1}{2} + \frac{j\sqrt{3}}{2})(+j\sqrt{3})} = \frac{1}{-\frac{3}{2} + \frac{j\sqrt{3}}{2}} = \frac{-2}{3 - j\sqrt{3}}$$

$$= \frac{-2(3 + j\sqrt{3})}{(3 - j\sqrt{3})(3 + j\sqrt{3})} = \frac{-3 - j\sqrt{3}}{6} = \bar{B} \text{ (as expected)}. \tag{9.5-20}$$

Thus,

$$F(s) = \frac{1}{s+1} + \frac{\frac{-3 + j\sqrt{3}}{6}}{s + \frac{1}{2} + \frac{j\sqrt{3}}{2}} + \frac{\frac{-3 - j\sqrt{3}}{6}}{s + \frac{1}{2} - \frac{j\sqrt{3}}{2}}, \tag{9.5-21}$$

resulting in

$$f(t) = e^{-t} - \left(\frac{3 - j\sqrt{3}}{6}\right)e^{-(\frac{1}{2} + \frac{j\sqrt{3}}{2})t} - \left(\frac{3 + j\sqrt{3}}{6}\right)e^{-(\frac{1}{2} - \frac{j\sqrt{3}}{2})t}. \tag{9.5-22}$$

Comment 9.5-3: Certainly, we must regroup (9.5-22) into a more desirable form. For instance, to have the complex number j in a real-time expression is not desirable. However, in each case when we face a pair of complex poles the best way to handle this situation is to keep the pairs in a group as we show in the following derivation.

Alternative Solution to Example 9.5-2: Whenever, we have a pair of complex poles, we use the following approach whose origin is in the passive network synthesis to extract various poles. For instance, in the above case we write

$$F(s) = \frac{A}{s+1} + \frac{Bs + C}{s^2 + s + 1}, \tag{9.5-23}$$

i.e., we group each pair of complex conjugate poles together. Or simply we write

$$F(s) \equiv \frac{A}{s+1} + R(s), \tag{9.5-24}$$

where R(s) is referred to as the *remaining* terms. Since we have already shown that $A \equiv 1$, therefore

$$R(s) = F(s) - \frac{1}{s+1} = \frac{1 - s^2 - s - 1}{(s+1)(s^2 + s + 1)}. \tag{9.5-25}$$

What we have done in (9.5-25) is that we have extracted one pole from the system, here this pole is at $s = -1$. As a result of this extraction we have generated a common factor between the numerator and denominator in (9.5-25). In other words, if we have not made any mistake in our algebra, then the numerator in (9.5-25) is divisible by $(s + 1)$. If that is not the case, then we must have made a mistake. Using this concept, we can advance our computation substantially in order to expedite the process of constructing a partial fraction expansion through extraction of poles which are already recognized. For instance, we can show that the numerator in (9.5-25) is $-s(s + 1)$, thus,

$$\begin{aligned} R(s) &= \frac{-s(s+1)}{(s+1)(s^2 + s + 1)} \\ &= \frac{-s}{s^2 + s + 1} \equiv \frac{Bs + C}{s^2 + s + 1}. \end{aligned} \tag{9.5-26}$$

In (9.5-26), $B \equiv -1$ and $C \equiv 0$. Clearly, its denominator is similar to that in (9.3-19) or (9.3-20) of our catalog. Indeed, when we recognize the denominator, we rearrange the numerator such that it fits correspondingly with its denominator. For instance, we add and subtract ½ in the numerator of (9.5-26), followed by some other adjustments to make it look like one of the member of our catalog as follows.

$$R(s) = \frac{-s}{s^2 + s + 1} = \frac{\frac{1}{2} - (s + \frac{1}{2})}{(s + \frac{1}{2})^2 + \frac{3}{4}}$$

$$= \frac{\sqrt{3}}{3} \times \frac{\frac{\sqrt{3}}{2}}{(s + \frac{1}{2})^2 + (\frac{\sqrt{3}}{2})^2} - \frac{s + \frac{1}{2}}{(s + \frac{1}{2})^2 + (\frac{\sqrt{3}}{2})^2}. \qquad (9.5\text{-}27)$$

We recognize each term in (9.5-27) as Laplace transform of terms in (9.3-19 & 20) of our catalog. Thus, as expected we have

$$f(t) = e^{-t} + \frac{\sqrt{3}}{3} e^{-\frac{t}{2}} \sin(\frac{\sqrt{3}}{2} t) - e^{-\frac{t}{2}} \cos(\frac{\sqrt{3}}{2} t), \qquad (9.5\text{-}28)$$

which is in a more desirable form than (9.5-22).

9.5.2. Multiple Poles

Here, we have one or more poles with multiplicities of, say r. For instance,

$$F(s) = \frac{N(s)}{\cdots (s + p_i)^r \cdots}, \qquad (9.5\text{-}29)$$

where p_i is a multiple pole of F(s) whose multiplicity is r. In (9.5-29), we can expand F(s) with respect to terms corresponding to this multiple poles as follows.

$$F(s) = \cdots + \frac{k_{i1}}{(s + p_i)^r} + \frac{k_{i2}}{(s + p_i)^{r-1}} + \cdots + \frac{k_{ir}}{s + p_i} + \cdots. \qquad (9.5\text{-}30)$$

In (9.5-30), the $\cdots$ refers to all the other poles which may have their own complexities similar to that of p_i. Furthermore, we label k_{ij}'s in the order in which they must be computed. For instance,

$$k_{i1} = (s + p_i)^r F(s)\Big|_{s=-p_i}. \qquad (9.5\text{-}31)$$

This coefficient is exactly the same as that for a simple pole. The following k_{i2} to k_{ir}, however, must be calculated according to the following formulae.

$$k_{i2} = \left\{ \frac{d}{ds} [(s + p_i)^r F(s)] \right\}_{s=-p_i},$$

$$k_{i3} = \frac{1}{2!} \left\{ \frac{d^2}{ds^2} [(s + p_i)^r F(s)] \right\}_{s=-p_i},$$

.
.
.

$$k_{ir} = \frac{1}{(r-1)!} \left\{ \frac{d^{r-1}}{ds^{r-1}} [(s+p_i)^r F(s)] \right\}_{s=-p_i}. \tag{9.5-32}$$

We show through our examples that using the previously defined extraction method we can simplify this process, and possibly we may not need to memorize the above set of formulae.

Example 9.5-3 (Multiple Real Poles): Given $F(s) = \dfrac{1}{s^2(s+1)(s+2)}$, find f(t) by a partial fraction expansion.

Solution: The poles of the system are at the origin with multiplicities of two, plus two other poles at $s = -1$ and $s = -2$. Using partial fraction expansion yields

$$F(s) = \frac{A}{s+1} + \frac{B}{s+2} + \frac{C_{11}}{s^2} + \frac{C_{12}}{s}. \tag{9.5-33}$$

The above A, B and C_{11} can be computed by inspection.

$$A = \frac{1}{s^2(s+2)}\bigg|_{s=-1} = 1, \tag{9.5-34}$$

$$B = \frac{1}{s^2(s+1)}\bigg|_{s=-2} = -\tfrac{1}{4}, \tag{9.5-35}$$

$$C_{11} = \frac{1}{(s+1)(s+2)}\bigg|_{s=0} = \tfrac{1}{2}. \tag{9.5-36}$$

However, the computation of C_{12} is a challenge and for that we propose to use the formula in (9.5-32).

$$C_{12} = \frac{d}{ds}\left[s^2F(s)\right] = \frac{d}{ds}\left[\frac{1}{(s+1)(s+2)}\right]\bigg|_{s=0}$$

$$= -\left[\frac{2s+3}{(s+1)^2(s+2)^2}\right]\bigg|_{s=0} = -\frac{3}{4}. \tag{9.5-37}$$

Substituting these values in (9.5-33) and taking the inverse transform results in

$$f(t) = e^{-t} - \frac{1}{4}e^{-2t} + \frac{1}{2}t - \frac{3}{4}1(t). \tag{9.5-38}$$

Alternative Solution to Example 9.5-3: We may use the pole extraction approach in this example as follows.

$$F(s) = \frac{1}{s+1} - \frac{1}{4} \times \frac{1}{s+2} + R(s). \tag{9.5-39}$$

$$R(s) = \frac{1}{s^2(s+1)(s+2)} - \frac{1}{s+1} - \frac{1}{4} \times \frac{1}{s+2}$$

$$= \frac{1 - s^2(s+2) + \frac{1}{4}s^2(s+1)}{s^2(s+1)(s+2)}$$

$$= \frac{-\frac{3}{4}s^3 - \frac{7}{4}s^2 + 1}{s^2(s+1)(s+2)} \equiv \frac{A(s+1)(s+2)(s+B)}{s^2(s+1)(s+2)}. \tag{9.5-40}$$

In other words, looking at the numerator of (9.5-40) we note that we have a third-order polynomial. Since this numerator must have two zeros at $s = -1$ and $s = -2$, we impose upon R(s) to be factored as such in (9.5-40), by having two parameters A and B, in order that the extraction of two poles at $s = -1$ and $s = -2$ are completed. Thus, we must have

$$As^3 + A(B+3)s^2 + A(3B+2)s + 2AB \equiv -\frac{3}{4}s^3 - \frac{7}{4}s^2 + 1. \tag{9.5-41}$$

This equation holds if and only if the coefficients on both sides are identical. That means, starting from the obvious ones, we must have

$$A = -\frac{3}{4}, \text{ and } 2AB = 1 \Rightarrow B = -\frac{2}{3}. \tag{9.5-42}$$

There are two other constraint equations, namely,

$$A(B+3) = -\frac{7}{4}, \tag{9.5-43}$$

$$A(3B+2) = 0, \tag{9.5-44}$$

which both are also satisfied by the previous values for A and B. Thus,

$$R(s) = \frac{-\frac{3}{4}(s+1)(s+2)(s-\frac{2}{3})}{s^2(s+1)(s+2)} = -\frac{3}{4}\frac{s-\frac{2}{3}}{s^2}$$

$$= -\frac{3}{4} \times \frac{1}{s} + \frac{1}{2} \times \frac{1}{s^2}, \tag{9.5-45}$$

which indeed both terms match the preceding results of (9.5-38).

Example 9.5-4 (Mixed Poles): Given $F(s) = \dfrac{s^2 - 3s + 2}{(s+2)^2(s^2+2s+2)}$, find f(t) by a partial fraction expansion.

Solution: Clearly, the poles of the system can be identified by inspection and we have the following mix:

$$F(s) = \frac{k_{11}}{(s+2)^2} + \frac{k_{12}}{s+2} + \frac{k_3}{s+1+j} + \frac{k_4}{s+1-j}. \tag{9.5-46}$$

Straightforward computation yields

$$k_{11} = \frac{s^2 - 3s + 2}{s^2 + 2s + 2}\bigg|_{s=-2} = 6, \tag{9.5-47}$$

$$k_3 = \frac{s^2 - 3s + 2}{(s+2)^2(s+1-j)}\bigg|_{s=-1-j} = \frac{(1+j)^2 + 3(1+j) + 2}{(-1-j+2)^2(-1-j+1-j)}$$

$$= \frac{5j+5}{(1-j)^2(-2j)} = -\frac{5}{4}(1+j). \tag{9.5-48}$$

Similarly,

$$k_4 = \bar{k}_3 = -\frac{5}{4}(1-j). \tag{9.5-49}$$

For k_{12} we use (9.5-32).

$$k_{12} = \frac{d}{ds}\left[\frac{s^2 - 3s + 2}{s^2 + 2s + 2}\right]\bigg|_{s=-2}$$

$$= \left[\frac{(2s-3)(s^2+2s+2) - (2s+2)(s^2-3s+2)}{(s^2+2s+2)^2}\right]\bigg|_{s=-2}$$

$$= \frac{-7(2) - (-2)(12)}{(2)^2} = \frac{5}{2}. \tag{9.5-50}$$

Thus,

$$f(t) = 6te^{-2t} + \frac{5}{2}e^{-2t} - \frac{5}{4}(1+j)e^{-(1+j)t} - \frac{5}{4}(1-j)e^{-(1-j)t}. \tag{9.5-51}$$

Regrouping these terms by the order of exponentials yields

$$f(t) = (6t + \frac{5}{2})e^{-2t} - [\frac{5}{4}(e^{-jt} + e^{jt}) + \frac{5j}{4}(e^{-jt} - e^{jt})]e^{-t}$$

$$= (6t + \frac{5}{2})e^{-2t} - \frac{5}{2}(\cos t + \sin t)e^{-t}. \qquad (9.5\text{-}52)$$

Example 9.5-5 (Mixed Poles): Consider $F(s) = \dfrac{4}{(s+1)^2(s+2)(s^2+s+1)}$, find f(t) by a partial fraction expansion.

Solution: Here, by inspection we have

$$F(s) = \frac{4}{(s+1)^2} + \frac{A_{12}}{s+1} + \frac{\frac{4}{3}}{s+2} + \frac{Cs + D}{s^2+s+1}. \qquad (9.5\text{-}53)$$

Then, we use extraction approach as follows.

$$R(s) = F(s) - \frac{4}{(s+1)^2} - \frac{\frac{4}{3}}{s+2}, \qquad (9.5\text{-}54)$$

which is slightly more complicated than (9.5-25), because in contrast with (9.5-25), we are extracting two poles at a time. The result is

$$R(s) = \frac{4 - 4(s+2)(s^2+s+1) - \frac{4}{3}(s+1)^2(s^2+s+1)}{(s+1)^2(s+2)(s^2+s+1)}. \qquad (9.5\text{-}55)$$

The numerator must have a factor of **at least** $(s + 1)(s + 2)$ (or possibly multiples of $(s + 1)$), if we had not made a mistake. Indeed, this is easy to verify and we may divide the numerator by multiplication of these two factors to get

$$R(s) = \frac{(s+1)(s+2)(-\frac{4}{3}s^2 - 4s - \frac{8}{3})}{(s+1)^2(s+2)(s^2+s+1)} = \frac{-\frac{4}{3}s^2 - 4s - \frac{8}{3}}{(s+1)(s^2+s+1)}. \qquad (9.5\text{-}56)$$

The next natural step is to fracture this R(s) as follows.

$$R(s) \equiv \frac{A_{12}}{s+1} + \frac{Cs + D}{s^2 + s + 1}. \tag{9.5-57}$$

Here, with some straw of good luck the numerator in (9.5-56) has another factor of (s + 1), i.e., the numerator is $(s + 1)(-\frac{4}{3}s - \frac{8}{3})$, thus, because of the cancellation of this factor $A_{12} \equiv 0$, and we are left with

$$R(s) = \frac{-\frac{4}{3}s - \frac{8}{3}}{s^2 + s + 1}. \tag{9.5-58}$$

Using (9.5-58) together with $A_{12} \equiv 0$ in (9.5-53) yields

$$F(s) = \frac{4}{(s+1)^2} + \frac{4}{3}\times\frac{1}{s+2} - \frac{4}{3}\times\frac{s+2}{s^2+s+1}.$$

$$= \frac{4}{(s+1)^2} + \frac{4}{3}\times\frac{1}{s+2} - \frac{4}{3}\times\frac{s + \frac{1}{2} + \frac{3}{2}}{(s+\frac{1}{2})^2 + (\frac{\sqrt{3}}{2})^2}$$

$$= \frac{4}{(s+1)^2} + \frac{4}{3}\times\frac{1}{s+2} - \frac{4}{3}\times\frac{s + \frac{1}{2}}{(s+\frac{1}{2})^2 + (\frac{\sqrt{3}}{2})^2} - \frac{4\sqrt{3}}{3}\times\frac{\frac{\sqrt{3}}{2}}{(s+\frac{1}{2})^2 + (\frac{\sqrt{3}}{2})^2}. \tag{9.5-59}$$

Resulting in,

$$f(t) = 4te^{-t} + \frac{4}{3}e^{-2t} - \frac{4}{3}e^{-\frac{t}{2}}\left[\cos(\frac{\sqrt{3}}{2}t) + \sqrt{3}\sin(\frac{\sqrt{3}}{2}t)\right]. \tag{9.5-60}$$

Example 9.5-6 (Mixed Poles): Now, consider $F(s) = \frac{4}{(s+1)^3(s+2)(s^2+s+1)}$, and find f(t) by a partial fraction expansion.

Solution: Here, we have

$$F(s) = \frac{A_{11}}{(s+1)^3} + \frac{A_{12}}{(s+1)^2} + \frac{A_{13}}{s+1} + \frac{B}{s+2} + \frac{Cs + D}{s^2 + s + 1}. \tag{9.5-61}$$

Using (9.5-32) we have

$$A_{11} = \frac{d^0}{ds^0}\left[\frac{4}{(s+2)(s^2+s+1)}\right]\Bigg|_{s=-1} = 4, \tag{9.5-62}$$

$$A_{12} = \frac{d}{ds}\left[\frac{4}{(s+2)(s^2+s+1)}\right]\Bigg|_{s=-1}$$

$$= -4\left[\frac{s^2+s+1+(s+2)(2s+1)}{(s+2)^2(s^2+s+1)^2}\right]\Bigg|_{s=-1} = 0, \tag{9.5-63}$$

$$A_{13} = \frac{1}{2!}\frac{d^2}{ds^2}\left[\frac{4}{(s+2)(s^2+s+1)}\right]\Bigg|_{s=-1}$$

$$= 2\frac{d}{ds}\left[\frac{-3(s^2+2s+1)}{(s+2)^2(s^2+s+1)^2}\right]\Bigg|_{s=-1} = 0. \tag{9.5-64}$$

The last equality is because the numerator is of the order of $(s+1)^2$. Similarly, we get

$$B = \left[\frac{4}{(s+1)^3(s^2+s+1)}\right]\Bigg|_{s=-2} = -\frac{4}{3}. \tag{9.5-65}$$

Substituting these values in (9.5-61) yields

$$F(s) = \frac{4}{(s+1)^3} - \frac{\frac{4}{3}}{s+2} + R(s), \tag{9.5-66}$$

where,

$$R(s) = \frac{4}{(s+1)^3(s+2)(s^2+s+1)} - \frac{4}{(s+1)^3} + \frac{\frac{4}{3}}{s+2}$$

$$= 4\times\frac{1-(s+2)(s^2+s+1)+\frac{1}{3}(s+1)^3(s^2+s+1)}{(s+1)^3(s+2)(s^2+s+1)}$$

$$= 4\times\frac{1-(s+1+1)(s^2+s+1)+\frac{1}{3}(s+1)^3(s^2+s+1)}{(s+1)^3(s+2)(s^2+s+1)}$$

$$= 4 \times \frac{-(s+1)(s^2+s+1) - s(s+1) + \frac{1}{3}(s+1)^3(s^2+s+1)}{(s+1)^3(s+2)(s^2+s+1)}$$

$$= -4 \times \frac{(s+1)(s^2+s+1+s) - \frac{1}{3}(s+1)^3(s^2+s+1)}{(s+1)^3(s+2)(s^2+s+1)}$$

$$= -4 \times \frac{(s+1)^3 - \frac{1}{3}(s+1)^3(s^2+s+1)}{(s+1)^3(s+2)(s^2+s+1)}$$

$$= -4 \times \frac{1 - \frac{1}{3}(s^2+s+1)}{(s+2)(s^2+s+1)} = \frac{4}{3} \times \frac{s^2+s+1-3}{(s+2)(s^2+s+1)}$$

$$= \frac{4}{3} \times \frac{(s+2)(s-1)}{(s+2)(s^2+s+1)} = \frac{4}{3} \times \frac{s-1}{s^2+s+1}. \tag{9.5-67}$$

Thus, as expected, we have extracted multiple poles at $s = -1$ and a simple pole at $s = -2$, and are left with the following.

$$F(s) = \frac{4}{(s+1)^3} - \frac{\frac{4}{3}}{s+2} + \frac{\frac{4}{3}(s + \frac{1}{2} - \frac{3}{2})}{s^2+s+1}. \tag{9.5-68}$$

Resulting in,

$$f(t) = 2t^2e^{-t} - \frac{4}{3}e^{-2t} + \frac{4}{3}e^{-\frac{t}{2}}[\cos(\frac{\sqrt{3}}{2}t) - \sqrt{3}\sin(\frac{\sqrt{3}}{2}t)]. \tag{9.5-69}$$

Example 9.5-7 (Mixed Poles): Given $X(s) = \frac{4}{(s+1)^3(s^2+s+1)}$, find x(t) and $\lim_{t\to\infty} x(t)$.

Solution: Here, we certainly need x(t) first, and to compute that, we use a partial fraction expansion. By inspection we have

$$X(s) = \frac{4}{(s+1)^3} + R_1(s), \tag{9.5-70}$$

where

$$R_1(s) = \frac{4}{(s+1)^3(s^2+s+1)} - \frac{4}{(s+1)^3}$$

$$= \frac{4\,[1 - s^2 - s - 1]}{(s+1)^3(s^2+s+1)} = \frac{-4s}{(s+1)^2(s^2+s+1)}$$

$$= \frac{4}{(s+1)^2} + R_2(s). \tag{9.5-71}$$

Similarly,

$$R_2(s) = \frac{-4s}{(s+1)^2(s^2+s+1)} - \frac{4}{(s+1)^2} = \frac{-4s - 4(s^2+s+1)}{(s+1)^2(s^2+s+1)}$$

$$= \frac{-4(s+1)^2}{(s+1)^2(s^2+s+1)} = -\frac{4}{s^2+s+1}. \tag{9.5-72}$$

Thus,

$$F(s) = \frac{4}{(s+1)^3} + \frac{4}{(s+1)^2} - \frac{4}{s^2+s+1}. \tag{9.5-73}$$

Resulting in,

$$x(t) = (2t^2 + 4t)e^{-t} - \frac{8\sqrt{3}}{3}\, e^{-\frac{t}{2}} \sin(\frac{\sqrt{3}}{2}\, t). \tag{9.5-74}$$

Therefore,

$$\lim_{t \to \infty} x(t) = 0, \tag{9.5-75}$$

which can also be confirmed using the final value theorem, i.e.,

$$\lim_{t \to \infty} x(t) = \lim_{s \to 0} sX(s) = 0. \tag{9.5-76}$$

Example 9.5-8 (Pairs of Complex Conjugate Poles): Given $F(s) = \dfrac{1}{(s^2+s+1)(s^2+1)}$, find f(t) by a partial fraction expansion.

Solution: Clearly, the poles of the system are two pairs of complex conjugate numbers. First, we use the following approach, letting four simple poles be counted individually.

$$F(s) = \frac{A}{s + \frac{1}{2} + j\frac{\sqrt{3}}{2}} + \frac{B}{s + \frac{1}{2} - j\frac{\sqrt{3}}{2}} + \frac{C}{s+j} + \frac{D}{s-j}. \tag{9.5-77}$$

Here,

$$A = \frac{1}{(s + \frac{1}{2} - j\frac{\sqrt{3}}{2})(s^2 + 1)}\Bigg|_{s=-\frac{1}{2}-j\frac{\sqrt{3}}{2}} = \frac{1}{\frac{3}{2} - j\frac{\sqrt{3}}{2}}. \quad (9.5\text{-}78)$$

$$B = \frac{1}{(s + \frac{1}{2} + j\frac{\sqrt{3}}{2})(s^2 + 1)}\Bigg|_{s=-\frac{1}{2}+j\frac{\sqrt{3}}{2}} = \frac{1}{\frac{3}{2} + j\frac{\sqrt{3}}{2}} = \bar{A} \text{ (as expected).} \quad (9.5\text{-}79)$$

$$C = \frac{1}{(s^2 + s + 1)(s - j)}\Bigg|_{s=-j} = \frac{1}{(-j)(-2j)} = -\frac{1}{2}. \quad (9.5\text{-}80)$$

$$D = \frac{1}{(s^2 + s + 1)(s + j)}\Bigg|_{s=j} = \frac{1}{(j)(2j)} = -\frac{1}{2} = \bar{C} \text{ (as expected).} \quad (9.5\text{-}81)$$

The rest of the analysis is straightforward.

Alternative Solution to Example 9.5-8: Or, we may choose the following approach to fracture F(s) by grouping pairs of complex poles together.

$$F(s) = \frac{As + B}{s^2 + s + 1} + \frac{Cs + D}{s^2 + 1}$$

$$= \frac{(As + B)(s^2 + 1) + (s^2 + s + 1)(Cs + D)}{(s^2 + s + 1)(s^2 + 1)}$$

$$= \frac{(A + C)s^3 + (B + C + D)s^2 + (A + C + D)s + (B + D)}{(s^2 + s + 1)(s^2 + 1)}. \quad (9.5\text{-}82)$$

Equating (9.5-82) term by term with the original F(s) results in

$$0 \equiv A + C, \quad (9.5\text{-}83)$$

$$0 \equiv B + C + D, \quad (9.5\text{-}84)$$

$$0 \equiv A + C + D, \quad (9.5\text{-}85)$$

$$1 \equiv B + D. \quad (9.5\text{-}86)$$

Solving the above equations yields, $D = 0$, $B = 1$, $C = -1$ and $A = 1$. Substituting these values in (9.5-82) yields

$$F(s) = \frac{s+1}{s^2+s+1} - \frac{s}{s^2+1}$$

$$= \frac{s+\frac{1}{2}+\frac{1}{2}}{s^2+s+1} - \frac{s}{s^2+1}, \qquad (9.5\text{-}87)$$

resulting in

$$f(t) = e^{-\frac{t}{2}}[\cos(\frac{\sqrt{3}}{2}t) + \frac{\sqrt{3}}{3}\sin(\frac{\sqrt{3}}{2}t)] - \cos t. \qquad (9.5\text{-}88)$$

9.6. SOLUTION of DIFFERENTIAL EQUATIONS

Clearly, our goal in developing the Laplace transformation techniques has been to use this methodology as an alternative approach to solve differential equations. To that end, we consider the following n^{th}–order *ODE* with constant coefficients arising from circuit problems.

$$\frac{d^n x(t)}{dt^n} + a_{n-1}\frac{d^{n-1}x(t)}{dt^{n-1}} + \cdots + a_1\frac{dx(t)}{dt} + a_0 x(t) = f(t). \qquad (9.6\text{-}1)$$

We assume that all pertinent initial conditions are specified, say at t = 0. To solve this equation, we take the Laplace transformation of its both sides to get

$$\mathcal{L}[\frac{d^n x(t)}{dt^n}] + a_{n-1}\mathcal{L}[\frac{d^{n-1}x(t)}{dt^{n-1}}] + \cdots + a_1\mathcal{L}[\frac{dx(t)}{dt}] + a_0\mathcal{L}[x(t)] = \mathcal{L}[f(t)].$$

(9.6-2)

Let $\mathcal{L}[x(t)] \triangleq X(s)$ and $\mathcal{L}[f(t)] \triangleq F(s)$. Then, using the Laplace transformation for the derivatives of x(t), we have:

$\mathcal{L}[\frac{dx(t)}{dt}] = sX(s) - x(0^+)$, this is the only formula we must remember,

(9.6-3)

because all other higher-order terms can be generated from (9.6-3). For instance

$$\mathcal{L}[\frac{d^2x(t)}{dt^2}] = \mathcal{L}\left\{\frac{d}{dt}[\frac{dx(t)}{dt}]\right\}$$

$$= s\mathcal{L}[\frac{dx(t)}{dt}] - [\frac{dx(t)}{dt}]_{t=0^+} = s^2X(s) - sx(0^+) - \dot{x}(0^+). \qquad (9.6\text{-}4)$$

Similarly, the Laplace transformations of all other higher-order derivatives are generated recursively in the above manner.

Putting all these terms together, we have

$$X(s) = \frac{F(s) + \{\text{terms coming from initial conditions}\}}{D(s)}, \qquad (9.6\text{-}5)$$

where D(s) is the same as the characteristic polynomial in (9.6-1), when we replace $D \equiv \frac{d}{dt}$ by s. The remaining task of generating x(t) from X(s) is already described in the preceding examples of this chapter.

Example 9.6-1: Solve the following differential equation

$$\ddot{x}(t) + \dot{x}(t) + x(t) = \sin t, \qquad (9.6\text{-}6)$$

with $x(0) = 0$ and $\dot{x}(0) = 0$.

Solution: Taking the Laplace transformation of both sides in (9.6-6), and incorporating the initial conditions yields

$$(s^2 + s + 1)X(s) = \frac{1}{s^2 + 1}. \qquad (9.6\text{-}7)$$

Thus,

$$X(s) = \frac{1}{(s^2 + s + 1)(s^2 + 1)}. \qquad (9.6\text{-}8)$$

We have already developed the inverse Laplace transform of this transfer function in Example 9.5-8. The result for x(t) is given in (9.5-88).

Example 9.6-2 (Example 7.4-2 Revisited): Consider the following set of coupled ordinary differential equations with zero initial conditions.

$$\begin{cases} (D + 1)x(t) + (D + 3)y(t) = t, \\ (D + 2)x(t) + (3D + 9)y(t) = t^2. \end{cases} \qquad (9.6\text{-}9)$$

Solve for x(t) and y(t) using Laplace transformation techniques.

Solution: Using the formula for the Laplace transformation of a derivative in (9.6-3), we transform (9.6-9) into the following set

$$\begin{pmatrix} s+1 & s+3 \\ s+2 & 3s+9 \end{pmatrix} \times \begin{Bmatrix} X(s) \\ Y(s) \end{Bmatrix} = \begin{Bmatrix} 1/s^2 \\ 2/s^3 \end{Bmatrix} = \frac{1}{s^3} \times \begin{Bmatrix} s \\ 2 \end{Bmatrix}. \tag{9.6-10}$$

To find the solution for X(s) and Y(s) we need to find the inverse of the above coefficient matrix, which is straightforward (cf., Section 5.7). The result is

$$\begin{Bmatrix} X(s) \\ Y(s) \end{Bmatrix} = \frac{1}{s^3(2s^2+7s+3)} \times \begin{pmatrix} 3s+9 & -(s+3) \\ -(s+2) & s+1 \end{pmatrix} \times \begin{Bmatrix} s \\ 2 \end{Bmatrix}$$

$$= \frac{1}{s^3(2s+1)(s+3)} \times \begin{Bmatrix} s(3s+9) - 2(s+3) \\ -s(s+2) + 2(s+1) \end{Bmatrix}$$

$$= \frac{1}{s^3(2s+1)(s+3)} \times \begin{Bmatrix} 3s^2+7s-6 \\ -s^2+2 \end{Bmatrix}. \tag{9.6-11}$$

Starting with X(s), we have

$$X(s) = \frac{3s^2+7s-6}{s^3(2s+1)(s+3)}$$

$$= \frac{A_x}{2s+1} + \frac{B_x}{s+3} + \frac{C_x}{s^3} + \frac{D_x}{s^2} + \frac{E_x}{s}. \tag{9.6-12}$$

By inspection, we have

$$A_x = 28, \quad B_x = 0, \quad C_x = -2. \tag{9.6-13}$$

We expect $B_x = 0$, from (7.4-41), nevertheless the result develops as we proceed. However, we use this information to cancel the common factor (s + 3) between the numerator and denominator of X(s) as we proceed in the following. Now, we define,

$$R_x(s) = X(s) - \frac{A_x}{2s+1} - \frac{C_x}{s^3}$$

$$= \frac{3s-2}{s^3(2s+1)} - \frac{28}{2s+1} + \frac{2}{s^3}$$

$$= \frac{3s - 2 - 28s^3 + 4s + 2}{s^3(2s+1)} = -\frac{7s(2s+1)(2s-1)}{s^3(2s+1)}$$

$$= -\frac{7(2s-1)}{s^2} = \frac{7}{s^2} - \frac{14}{s}, \qquad (9.6\text{-}14)$$

resulting in $D_x = 7$ and $E_x = -14$. Thus,

$$X(s) = \frac{14}{s + ½} - \frac{2}{s^3} + \frac{7}{s^2} - \frac{14}{s}. \qquad (9.6\text{-}15)$$

The corresponding x(t) which matches exactly (7.4-41) is as follows.

$$x(t) = 14e^{-\frac{t}{2}} - t^2 + 7t - 14\,[1(t)]. \qquad (9.6\text{-}16)$$

Here, we use 1(t), instead of just one, to follow the Laplace transform rules requiring to have $t > 0$.

Similarly, we have

$$Y(s) = \frac{-s^2 + 2}{s^3(2s+1)(s+3)}$$

$$= \frac{A_y}{2s+1} + \frac{B_y}{s+3} + \frac{C_y}{s^3} + \frac{D_y}{s^2} + \frac{E_y}{s}. \qquad (9.6\text{-}17)$$

Here also by inspection, we have

$$A_y = -\frac{28}{5}, \quad B_y = -\frac{7}{135}, \quad C_y = \frac{2}{3}. \qquad (9.6\text{-}18)$$

We define,

$$R_y(s) = Y(s) - \frac{A_y}{2s+1} - \frac{B_y}{s+3} - \frac{C_y}{s^3}$$

$$= \frac{-s^2+2}{s^3(2s+1)(s+3)} + \frac{28}{5(2s+1)} + \frac{7}{135(s+3)} - \frac{2}{3s^3}$$

$$= \frac{-135s^2 + 270 + 28\times 27s^3(s+3) + 7s^3(2s+1) - 90(2s+1)(s+3)}{135s^3(2s+1)(s+3)}$$

$$= \frac{-135s^2 + 7s^3[108(s+3) + (2s+1)] - 90(2s^2+7s)}{135s^3(2s+1)(s+3)}$$

$$= \frac{-135s + 7s^2(110s+325) - 180s - 630}{135s^2(2s+1)(s+3)}$$

$$= \frac{7(110s^3 + 325s^2 - 45s - 90)}{135s^2(2s+1)(s+3)}$$

$$= \frac{7(2s+1)(s+3)(55s-30)}{135s^2(2s+1)(s+3)}$$

$$= \frac{7(55s-30)}{135s^2} = -\frac{14}{9s^2} + \frac{77}{27s}, \tag{9.6-19}$$

resulting in $D_y = -\frac{14}{9}$ and $E_y = \frac{77}{27}$. Thus,

$$Y(s) = \frac{-14}{5(s+½)} - \frac{7}{135(s+3)} + \frac{2}{3s^3} - \frac{14}{9s^2} + \frac{77}{27s}. \tag{9.6-20}$$

As expected we have the following y(t), which is the same as that in (7.4-32).

$$y(t) = -\frac{14}{5}e^{-\frac{t}{2}} - \frac{7}{135}e^{-3t} + \frac{1}{3}t^2 - \frac{14}{9}t + \frac{77}{27}[1(t)]. \tag{9.6-21}$$

Comment 9.6-1: Although two examples are hardly enough to support making an assertion, we claim that we can solve all linear *ODE*'s which are arising from circuit analysis in a rather straightforward manner using Laplace transformation techniques. This is a very powerful methodology that will empower us to solve very complicated circuit problems which otherwise will not have been possible so easily.

9.7. APPLICATIONS of LAPLACE TRANSFORMATION in CIRCUITS

The Laplace transformation settings for both KCL and KVL go through in exactly the same manner as in their counterparts for the time-domain settings. The results are summarized in Table 9.7-1.

Table 9.7-1. Time-domain vs. frequency-domain theorems.

KCL	$\sum i(t) = 0$	$\sum I(s) = 0$
KVL	$\sum v(t) = 0$	$\sum V(s) = 0$
Ohm's Law	$v(t) = Ri(t)$	$V(s) = RI(s)$

For an inductor, we have $v_L(t) = L\dfrac{di_L(t)}{dt}$. Taking Laplace transformation of both sides of this equation yields

$$V_L(s) = L\,\mathcal{L}[\frac{di_L}{dt}] = L[sI_L(s) - i_L(0)]. \tag{9.7-1}$$

Letting $i_L(0) = 0$, we have

$$V_L(s) = (sL)I_L(s) \triangleq Z_L(s)I_L(s), \tag{9.7-2}$$

with the following relationship for the **impedance**

$$Z_L(s) \triangleq sL. \tag{9.7-3}$$

Similarly, for a capacitor we have $i_C(t) = C\dfrac{dv_C(t)}{dt}$, resulting in

$$I_C(s) = C\,\mathcal{L}[\frac{dv_C}{dt}] = C[sV_C(s) - v_C(0)]. \tag{9.7-4}$$

For $v_C(0) = 0$, we have

$$I_C(s) = (sC)V_C(s) \triangleq Y(s)V_C(s). \tag{9.7-5}$$

Here Y(s), the **admittance** which is the inverse of impedance, is as follows.

$$Y_C(s) \triangleq sC = \frac{1}{Z_C(s)}. \tag{9.7-6}$$

If we take into account the initial conditions, then the above two cases change and we have respectively the following relationships. For inductor, we have

$$V_L(s) = sLI_L(s) - Li_L(0), \tag{9.7-7}$$

or

$$I_L(s) = \frac{i_L(0)}{s} + \frac{V_L(s)}{sL}. \tag{9.7-8}$$

The above relationship is depicted in Fig. 9.7-1.

Similarly, for a capacitor we have

$$I_C(s) = sC\,V_C(s) - Cv_C(0), \tag{9.7-9}$$

or

$$V_C(s) = \frac{v_C(0)}{s} + \frac{I_C(s)}{sC}. \tag{9.7-10}$$

Here, this relationship is shown in Fig. 9.7-2.

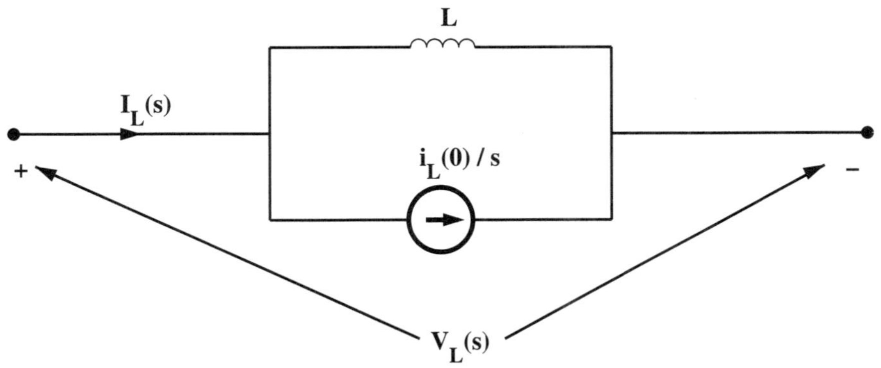

Fig. 9.7-1. An inductor with the initial condition.

Comment 9.7-1: Putting all these together, we can analyze a linear network with complex elements. We may do so either using Laplace transformation of various variables and proceed accordingly. Or we may use time-domain variables and set up the problem initially in a time-domain setting, then use Laplace transformation to get the corresponding desired solution in the frequency domain. In the final analysis, however, when all is said and done, we must get the final desired result in the time-domain setting. That means we use inverse Laplace transformation to get back in the time domain. What we want to emphasize is that when we use impedances and admittances, the circuit analysis is in exactly the same manner as that for the resistive networks, which we have seen exhaustively, and that empowerment is the beauty of the Laplace transformation methodology.

Comment 9.7-2: In the following, we show how circuits involving R, L and C's are formulated in the Laplace transformation settings and how we may solve the corresponding circuits. But first, let us review the time behavior of these elements as summarized in Table 9.7-2.

Comment 9.7-3: In Table 9.7-2, the initial conditions, when exist, can also be shown by δ functions, since these functions are valid only at time $t = 0$. For instance, we may consider the initial condition of an inductor as a delta-current source that is applied at $t = 0^+$, which is effective only at that moment. This representation has its advantages when is used correctly. Similarly, we can state identical comment about the initial voltage of a capacitor as a δ-voltage source applied at $t = 0^+$. The appropriate box in Table 9.7-2 illustrates how the above δ sources are to be implemented in each case.

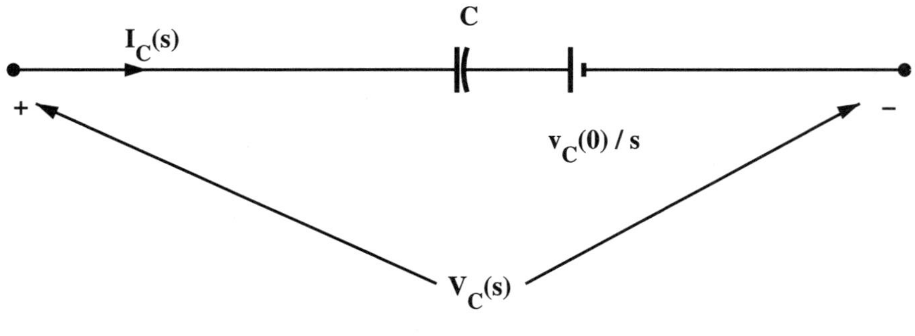

Fig. 9.7-2. A capacitor with the initial condition.

Table 9.7-2. Time behavior of circuit elements.

Elements		$t = 0^+$	$t = \infty$
Resistor	R	R	R
Inductor	L	OC	SC
Inductor with initial condition	L, $i(0^+)$	$I_0 = i(0^+)$	SC, $I_0 = i(0^+)$
Capacitor	C	SC	OC
Capacitor with initial condition	C, + −, $v(0^+)$	+ −, $V_0 = v(0^+)$	OC, + −, $V_0 = v(0^+)$

Some interpretations of these quantities are now in order. For instance, from $v_L(t) = L\dfrac{di_L}{dt}$ we have $V_L(s) = sLI_L(s) - Li_L(0^+)$. If $i_L(0^+) = 0$, then $Z_L(s) = sL$. At $t = 0^+$, $Z_L \to \infty$ (open circuit – OC) and at $t = \infty$, $Z_L \to 0$ (short circuit – SC).

Similarly, from $i_C(t) = C\dfrac{dv_C}{dt}$ we have $I_C(s) = sC\,V_C(s) - Cv_C(0^+)$. If $v_C(0^+) = 0$, then $Y_C(s) = sC$ and at $t = 0$, $Y_C \to \infty$ (short circuit), while as $t \to \infty$, $Y_C \to 0$, i.e., open circuit.

Example 9.7-1: Determine the transfer function of a linear time-invariant system whose input and output are respectively $x(t) = 1(t)$ (a unit-step function) and $y(t) = t^2 + t$.

Solution: From $y(t) = \int_0^t g(t - \tau)x(\tau)d\tau$, we have

$$Y(s) = G(s)\,X(s) \;\Rightarrow\; G(s) = Y(s)/X(s). \tag{9.7-11}$$

Here, $X(s) = \frac{1}{s}$ and $Y(s) = \frac{2}{s^3} + \frac{1}{s^2} = \frac{s+2}{s^3}$. Thus,

$$G(s) = \frac{s+2}{s^2}. \tag{9.7-12}$$

There are a host of ways to come up with a class of linear systems whose transfer function is (9.7-12). For instance, if we choose the linear system as a circuit, then that circuit can be constructed with L's and C's such that its transfer function becomes (9.7-12), we leave the details for the reader.

Example 9.7-2: Suppose $x(t) = \sin t$ is the input voltage, and $y(t) = t \sin t$ is the output current, then find $G(s) = \mathcal{L}[y(t)]/\mathcal{L}[x(t)] \triangleq$ admittance.

Solution: Taking the Laplace transformations of various parts in this case result in

$$G(s) = Y(s)/X(s) = \frac{2s}{(s^2+1)^2} \Big/ \frac{1}{s^2+1}$$

$$= \frac{2s}{s^2+1} = \frac{1}{\frac{s}{2} + \frac{1}{2s}}. \tag{9.7-13}$$

The $\frac{s}{2}$ corresponds to an inductor with $L = (½)^H$, and $\frac{1}{2s}$ is a series capacitor with $C = 2^F$. The corresponding circuit is shown in Fig. 9.7-3.

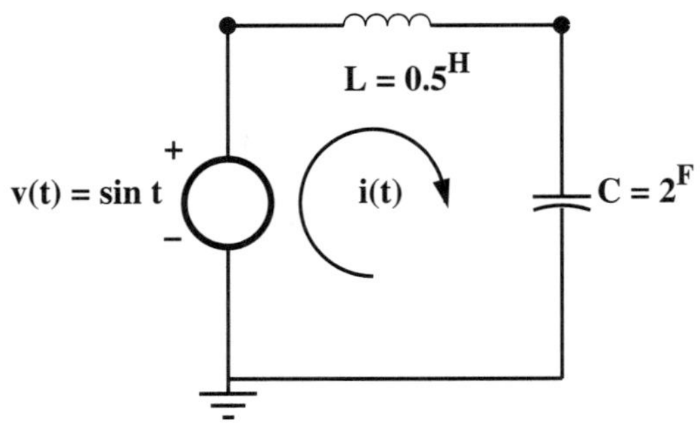

Fig. 9.7-3.

Example 9.7-3: Consider the circuit of Fig. 9.7-4. The switch becomes open at $t = 0$. What is the circuit equivalent for this operation?

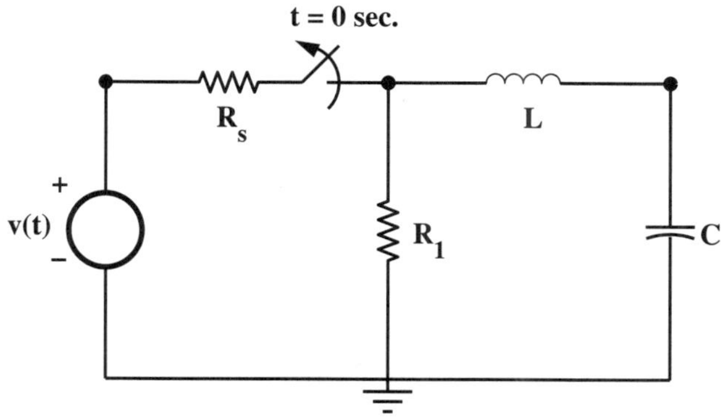

Fig. 9.7-4.

Solution: The above circuit is equivalent of the following two circuits.

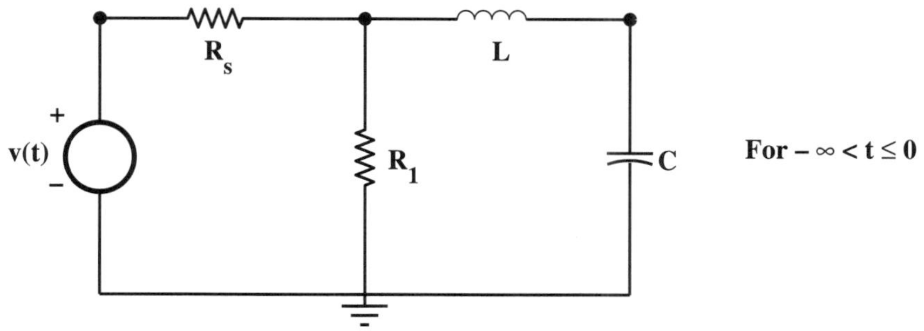

Fig. 9.7-5.

The task of circuit in Fig. 9.7-5 is to generate initial conditions for the next stage of its operation, which those values are the final conditions for this circuit. Because, going from the initial time at $t = -\infty$ to $t = 0$ is the same as going from an initial time of $t = 0$ to the final time of $t = \infty$.

In the next stage, and when the switch opens, we no longer have the independent voltage source v(t) and the operating circuit is that shown in Fig. 9.7-6. This circuit works because of the initial conditions of its elements, which are computed from the final conditions of the previous circuit just before the switching.

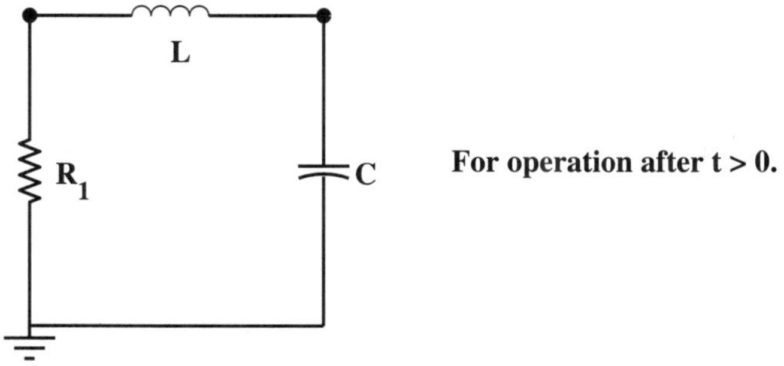

Fig. 9.7-6.

Example 9.7-4: Discuss the solution of circuit in Fig. 9.7-7, with appropriate set of initial conditions.

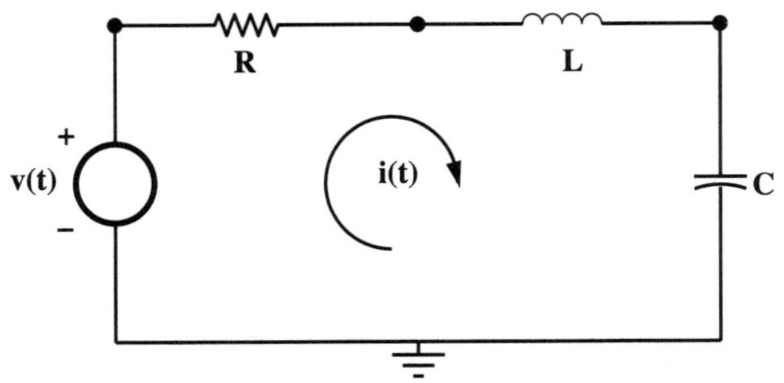

Fig. 9.7-7.

Solution: Writing a mesh equation yields

$$v(t) = Ri(t) + L\frac{di}{dt} + \frac{1}{C}\int_{-\infty}^{t} i(\tau)d\tau. \quad (9.7\text{-}14)$$

$$V(s) = RI(s) + sLI(s) - Li(0^+) + \frac{1}{C}[\frac{I(s)}{s} + v_C(0^+)]. \quad (9.7\text{-}15)$$

Here, we may substitute for V(s) as given specifically, and study I(s), provided that all parameters are specified. Then we may compute I(s) followed by i(t).

Example 9.7-5: For v(t) = sin t and t > 0, with zero initial conditions, find i(t) in the circuit of Fig. 9.7-8.

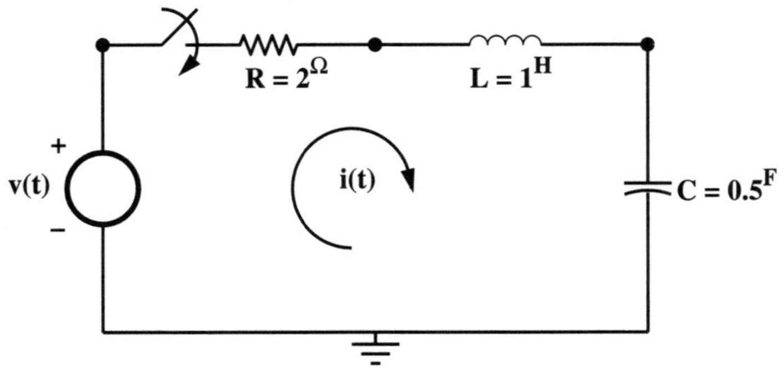

Fig. 9.7-8.

Solution: Because of the switch at t = 0, we have

$$V(s) = (R + sL + \frac{1}{sC})I(s) = (2 + s + \frac{2}{s})I(s). \tag{9.7-16}$$

On the other hand,

$$V(s) = \frac{1}{s^2 + 1}. \tag{9.7-17}$$

Thus,

$$I(s) = \frac{s}{(s^2 + 2s + 2)(s^2 + 1)} \equiv \frac{as + b}{s^2 + 2s + 2} + \frac{cs + d}{s^2 + 1}$$

$$= \frac{as^3 + as + bs^2 + b + cs^3 + 2cs^2 + 2cs + ds^2 + 2ds + 2d}{(s^2 + 2s + 2)(s^2 + 1)}. \tag{9.7-18}$$

Equating term by term on both sides of (9.7-18) yields

$$\begin{cases} a + c = 0, \\ b + 2c + d = 0, \\ a + 2c + 2d = 1, \\ b + 2d = 0, \end{cases} \Rightarrow \begin{cases} c = -a, \\ -2a + b + d = 0, \\ -a + 2d = 1, \\ b + 2d = 0, \end{cases} \Rightarrow \begin{cases} d = -\frac{b}{2}, \\ -2a + b - \frac{b}{2} = 0, \\ -a - b = 1, \end{cases} \Rightarrow$$

$$\begin{cases} -4a + b = 0, \\ -a - b = 1, \end{cases} \Rightarrow$$

$$a = -\frac{1}{5}, \quad b = -\frac{4}{5}, \quad c = \frac{1}{5}, \quad d = \frac{2}{5}. \tag{9.7-19}$$

Substituting the above values in (9.7-18) yields

$$I(s) = \frac{-\frac{1}{5}s - \frac{4}{5}}{s^2 + 2s + 2} + \frac{\frac{1}{5}s + \frac{2}{5}}{s^2 + 1}$$

$$= \frac{-\frac{1}{5}(s + 1) - \frac{3}{5}}{(s + 1)^2 + 1} + \frac{\frac{1}{5}s + \frac{2}{5}}{s^2 + 1}. \tag{9.7-20}$$

$$i(t) = -\frac{1}{5}e^{-t}(\cos t + 3\sin t) + \frac{1}{5}(\cos t + 2\sin t). \tag{9.7-21}$$

Example 9.7-6: Consider the circuit in Fig. 9.7-9 for $-\infty < t < 0$. At $t = 0^+$, we open the switch. Find $I_2(s)$ as shown.

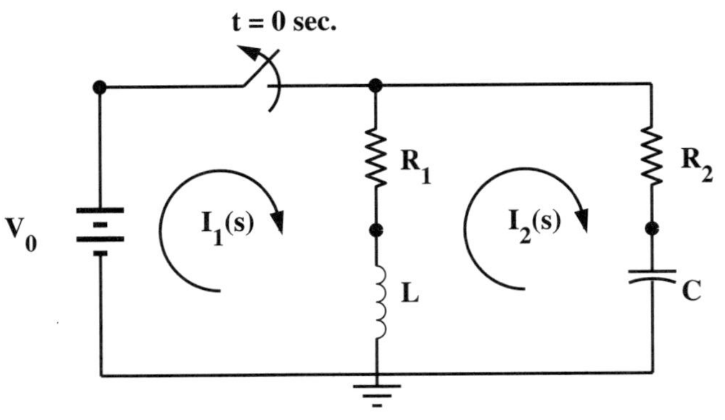

Fig. 9.7-9.

Solution: To complete this problem, we need to determine the initial conditions $i_L(0^-)$ and $v_C(0^-)$, which these values are the final conditions for the first stage and just before t = 0. At $t = 0^-$, the inductor is short and the capacitor is open, because the circuit is in the final condition of its previous stage. Thus, we have the circuit in Fig. 9.7-10.

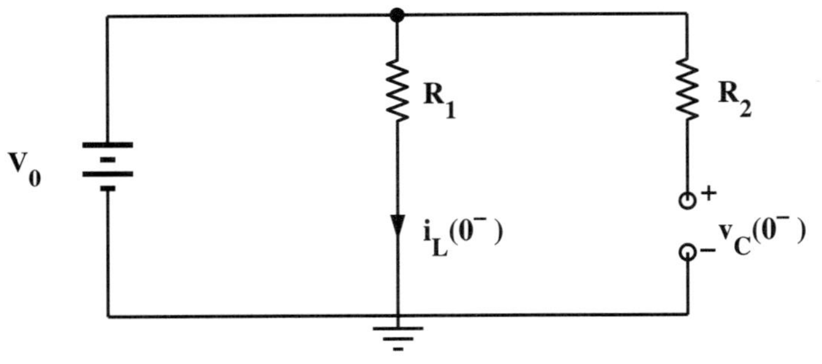

$$\begin{cases} i_L(0^-) = \dfrac{V_0}{R_1} \\ v_C(0^-) = V_0 \end{cases}$$

Fig. 9.7-10.

At $t \geq 0$, we open the switch which is equivalent to applying a step function in the reverse time and therefore the circuit will be activated as shown in Fig. 9.7-11.

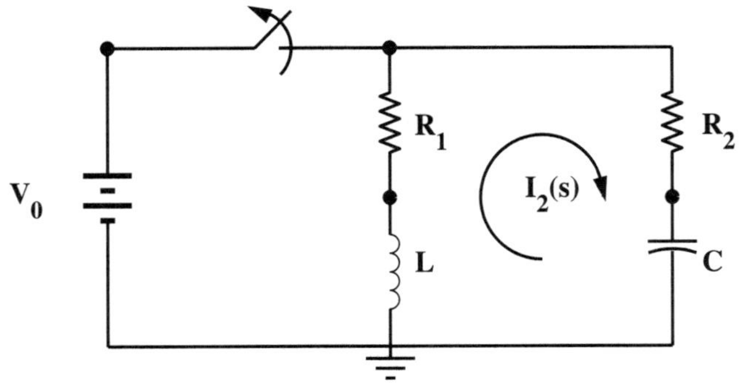

Fig. 9.7-11.

The only operating part of this circuit is the right-hand loop, whose set of initial conditions is the set of final conditions for the preceding circuit. Here, using the mesh equation yields

$$(R_1 + R_2)i_2(t) + L\frac{di_2}{dt} + \frac{1}{C}\int_{-\infty}^{t} i_2(\tau)d\tau = 0. \tag{9.7-22}$$

Taking the Laplace transformation from the both sides yields

$$(R_1 + R_2)I_2(s) + sLI_2(s) - Li_L(0^-) + \frac{1}{C}\frac{I_2(s)}{s} + \frac{v_C(0^-)}{s} = 0. \tag{9.7-23}$$

In (9.7-23), we also note that the $i_L(0^-)$ is in the opposite direction of $i_2(t)$, and that is confirmed by the Laplace transformation.

Substituting the corresponding numerical values in (9.7-23) results in the final expression for $I_2(s)$ as follows.

$$I_2(s) = \frac{L\frac{V_0}{R_1} - \frac{V_0}{s}}{(R_1 + R_2) + sL + \frac{1}{sC}}$$

$$= \frac{V_0(sLC - R_1C)}{LCR_1s^2 + (R_1 + R_2)R_1Cs + R_1}. \tag{9.7-24}$$

Example 9.7-7: Analyze the circuit in Fig. 9.7-12 with zero initial conditions.

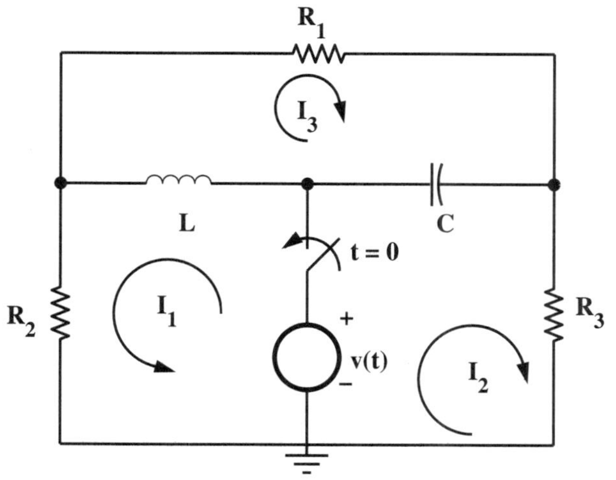

Fig. 9.7-12.

Solution: The mesh equations approach is the most suitable method in this case, resulting in:

$$V(s) = (sL + R_2)I_1(s) + sLI_3(s). \tag{9.7-25}$$

$$V(s) = -\frac{1}{sC} I_3(s) + (R_3 + \frac{1}{sC})I_2(s). \tag{9.7-26}$$

Also, taking the super loop around the circuit yields

$$R_1I_3(s) + R_3I_2(s) - R_2I_1(s) = 0. \tag{9.7-27}$$

These three currents are calculated from the following matrix equation.

$$\begin{Bmatrix} I_1(s) \\ I_2(s) \\ I_3(s) \end{Bmatrix} = \begin{pmatrix} sL + R_2 & 0 & sL \\ 0 & R_3 + \dfrac{1}{sC} & -\dfrac{1}{sC} \\ -R_2 & R_3 & R_1 \end{pmatrix}^{-1} \times \begin{Bmatrix} V(s) \\ V(s) \\ 0 \end{Bmatrix}. \tag{9.7-28}$$

9.8. CONCLUDING REMARKS and IMPORTANT THEOREMS

Going back to circuit analysis using Laplace transformation methods, consider the circuit in Fig. 9.8-1, where each element is shown in its Laplace transformation form.

The analysis of this circuit is straightforward, but the point which we want to make is that when each branch is shown by either its corresponding impedance or admittance, the rest of the analysis becomes no different from what we have seen for their counterpart in resistive circuits. In fact, all of our analysis go through in the same manner as resistive networks when we utilize either Z's or Y's in each branch as the net load in that branch.

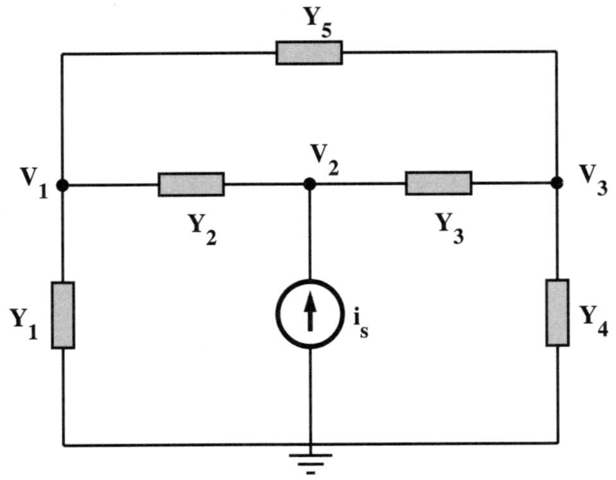

Fig. 9.8-1.

What about the phasor analysis? Can we benefit from Laplace transformation to simplify a phasor analysis? The answer is yes. We may use the Laplace transformation settings to compute a particular quantity first, then replace $s = \sigma + j\omega$ by only $s = j\omega$, and express that quantity in a phasor form by simply calculating its magnitude and phase or angle evaluated at the operating angular frequency ω.

We close this section by the following important theorems which are valid in both time- and frequency-domain measurement settings.

Theorem 9.8-1 (Reciprocity): A linear circuit with one voltage source is **reciprocal** if the following property known as reciprocity holds. Let $v_l(t)$ (or $V_l(s)$) be the source in link *l*, which causes a current $i_k(t)$ (or $I_k(s)$) to flow in link *k*, Fig. 9.8-2.

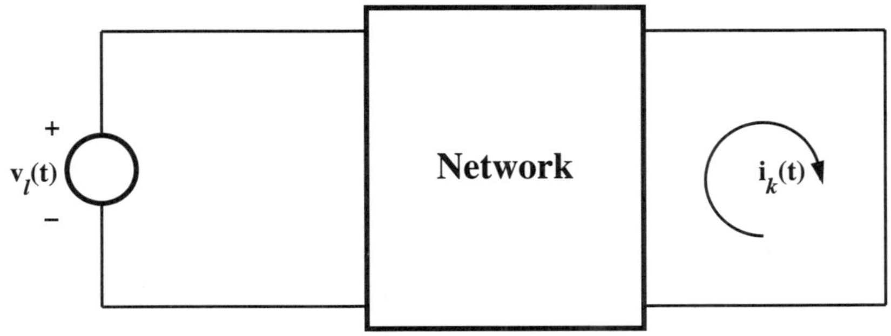

Fig. 9.8-2.

Now, we put this voltage source in link *k* and measure the resulting current in link *l*, which is shown as $i_l(t)$ in Fig. 9.8-3.

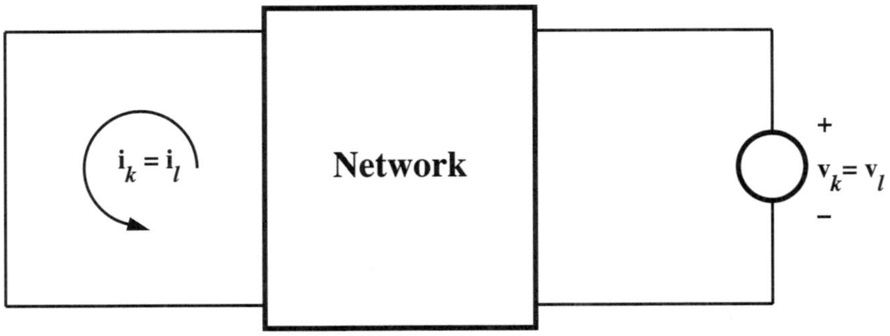

Fig. 9.8-3.

If $i_l(t) = i_k(t)$ for all *l* and *k*, then this circuit is called reciprocal.

Remark 9.8-1: In both Figs 9.8-2 & 3, we use the same direction for the designated currents, namely they both enter the positive polarity of the designated sources. Also, by link we mean the designated branch.

Remark 9.8-2: At the core of this theorem is the requirement for network symmetry. For instance, if we write the mesh equations, then for any I_k we have

$$I_k = \frac{\Delta_{1k}}{\Delta} V_1 + \frac{\Delta_{2k}}{\Delta} V_2 + \cdots + \frac{\Delta_{lk}}{\Delta} V_l + \cdots, \tag{9.8-1}$$

where Δ is the determinant of system and each Δ_{lk} is the corresponding cofactor. If all sources except V_l are zero, then $I_l = \frac{\Delta_{jk}}{\Delta} V_l$. Similarly, if all sources except V_j are zero, then $I_j = \frac{\Delta_{kj}}{\Delta} V_j$. When the mesh equations are symmetric, we have $\Delta_{jk} = \Delta_{kj}$, resulting in

$$\frac{I_k}{V_j} = \frac{I_j}{V_k}. \tag{9.8-2}$$

All passive network are symmetric, and we invite the reader to study Problem 9.9.35.

Next, we offer Tellegen's theorem as one of the most celebrated theorem in circuit theory.

Theorem 9.8-2 (Tellegen [4]): Consider a network with *k* branches. Choose arbitrary directions and values for currents and voltages of these elements provided that both the corresponding KCL and KVL satisfy these choices. Then

$$\sum_{j=1}^{k} v_j(t)\, i_j(t) = 0. \tag{9.8-3}$$

This theorem has been generalized in several directions and has been the subject of many research projects. One interesting extension is as follows. The above result also holds for voltages measured at time t_1, and currents measured at t_2 in a manner that the sum of their products also vanish. That is,

$$\sum_{j=1}^{k} v_j(t_1)\, i_j(t_2) = 0. \tag{9.8-4}$$

Another extension of this result says that if we have two networks N_1 and N_2 with the same topology, i.e., the same structure but not necessarily the same element types or values. Then we can choose a set of currents and voltages for their branches with the main conditions described earlier in the theorem as far as satisfying both KVL and KCL, then

$$\sum_{j=1}^{k} v_j^{N_1}(t_1)\, i_j^{N_2}(t_2) = \sum_{j=1}^{k} v_j^{N_2}(t_1)\, i_j^{N_1}(t_2) = 0. \tag{9.8-5}$$

The above two powerful theorems from circuit theory are very important results which we should remember for many future applications.

In the same manner as above, all source transformations in Laplace transformation as well as in complex-domain settings are exactly the same as their counterparts in the real-domain setting. So are the Thevenin's and Norton's theorems, which hold in any domain setting with its application for circuits in that domain.

The Thevenin's and Norton's theorem are two important and powerful theorems in circuit theory, which these two seemingly different theorems are indeed complementary of each other, as stated in Section 5.3 for resistive networks and in the following for the most general complex networks.

Theorem 9.8-3 (Thevenin): Given any two-terminal (port) of a network composed of linear elements as well as independent voltage and current sources, we can find an equivalent voltage source and an equivalent series impedance which behave exactly the same as the original network at this given port.

We emphasize that depending on where we pull these two terminals in a given network, we will have different equivalent circuits. Thus Thevenin's equivalent network is a function of the selected port in the circuit, and it only represents that port's property.

How do we measure this voltage source and series impedance?

As stated in Section 5.3 for resistive network, the voltage source is equal to the potential difference between the above two terminals caused by the network with no external elements connected to these terminals. The series impedance is equal to the impedance looking into the above two terminal points with all independent power sources within the terminal pair set equal to zero (i.e., the voltage sources are short and the current sources are opened).

Theorem 9.8.-4 (Norton): Given any two-terminal (port) in a network composed of linear elements as well as independent voltage and current sources, we can find an equivalent current source with an equivalent parallel impedance which behave exactly the same as the original network at this given port.

To emphasize the port dependency of these theorems consider Fig. 9.8-4. In this figure, for someone who is looking into networks (a), (b) and (c) at port aa', he/she sees exactly the same input voltage and input current in all these three networks. In other words, the equivalence between these networks holds or is defined only at one specifically selected port.

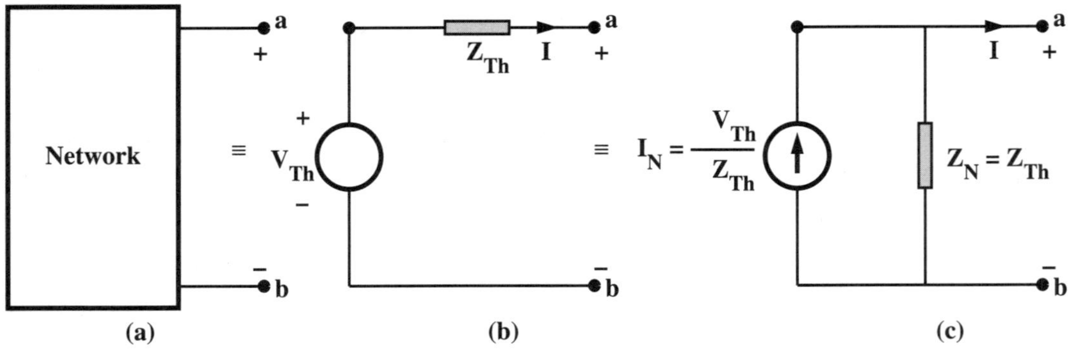

Fig. 9.8-4.

Finally, most of the materials presented in Laplace transformations thus far are generated for one-port networks. We are about to branch out to some elementary analysis of the so-called two-port networks in Chapter Eleven. Most of the preceding results which are valid in any time- or frequency-domain measurement setting are also true for multi-port networks.

9.9. PROBLEM SET

Our fundamental objective in this course or for that matter in this textbook is to solve numerically a practical circuit. We want to do so in order to evaluate how a particular circuit works, and ultimately how to design new circuits with new functions. To reach that level of competence requires some preparations. Again, in the spirit of openness and sharing our thoughts with the readers, we feel a need to give some insights as why are we about to do things below differently from traditional settings. One of the truths about the contemporary ways of living is that for most people there is seldom a moment to reflect. Everyone is busy to press ahead and only the most fortunate ones have the luxury and pleasure of pausing from time to time to reflect. We are among those rare groups, fortunate enough to pause, reflect and ask questions. We strongly believe that always there must be a sense of purpose in one's conduct and to that end we raise the subsequent questions. Starting with,

How do we know that a new method or methods which we are studying are any more powerful or versatile than the one we have just labored to learn? For instance, methodologies stemming from techniques in solving differential equations are more than sufficient for solving any conceivable circuit problem and yet we have ventured to master the current methodology in Laplace-transformation techniques. What is the thought behind such a desire? Now that we have labored to learn this new method of Laplace transformation, which is one of the several integral transformation methodologies taught in an electrical engineering program, is that improving our skills as far as reaching the goal of generating numerical solutions more systematic and faster than before? In other words, are we learning a new powerful and general methodology which is applicable to all sorts of circuit problems or not? If not, we are not interested to continue. Fortunately, the answer is definitely positive, and one proven way to confirm such a claim and emphasize the utility of this or any other methodology which we learn is to compare its outcomes with a previous method, when applied to the same set of problems. It would be worthwhile to establish the utility of each method in any given situation by comparing earlier solutions of some of the problems below which come from preceding chapters, with solutions stemming from the current Laplace-transformational methods. To that end, we give some insights for each group of problems selected below as we proceed.

Group A – Fundamentals: Here, we practice the actual generation of Laplace transformed functions with pencil and paper. We need to practice a few such functions to appreciate its course of development.

9.9.1. With the least computations, find the Laplace transformations for the following functions starting from the definitions and theorems for this transformational methodology, and not from the catalog of the functions which are presented in this chapter. Clearly, we assume that these functions are Laplace transformable and are defined for $t \geq 0$.

$$f(t) = t^3 - \frac{1}{t}. \tag{9.9-1}$$

$$f(t) = e^{-t} + 1(t). \tag{9.9-2}$$

$$f(t) = 2t - te^{-\alpha t}, \quad \alpha > 0. \tag{9.9-3}$$

$$f(t) = e^{-jt} + je^{-jt}. \tag{9.9-4}$$

$$f(t) = t^3 e^{-2t}. \tag{9.9-5}$$

$$f(t) = t^4 + 4t^3 + 12t^2 + 24t. \tag{9.9-6}$$

$$f(t) = \int_0^t t \sin t \, dt. \tag{9.9-7}$$

$$f(t) = e^{-3t}(\sin 2t + \cos 2t). \tag{9.9-8}$$

$$f(t) = \sin^2 2t - \cos^2 2t. \tag{9.9-9}$$

$$f(t) = 2te^{-t} \cos 2t + e^{-t} \sin 2t. \tag{9.9-10}$$

$$f(t) = t \sin t \cos 2t. \tag{9.9-11}$$

$$f(t) = 1(t) - 1(t-1) + 1(t-2) - 1(1-3) + \cdots. \tag{9.9-12}$$

Remark 9.9-1: In many practical situations we face graphical or experimental values of the function and based on that information we need to construct F(s). For instance, (9.9-12) is a train of pulses which is described analytically. As another example consider the periodic waveform displayed in Fig. 9.9-1.

Fig. 9.9-1.

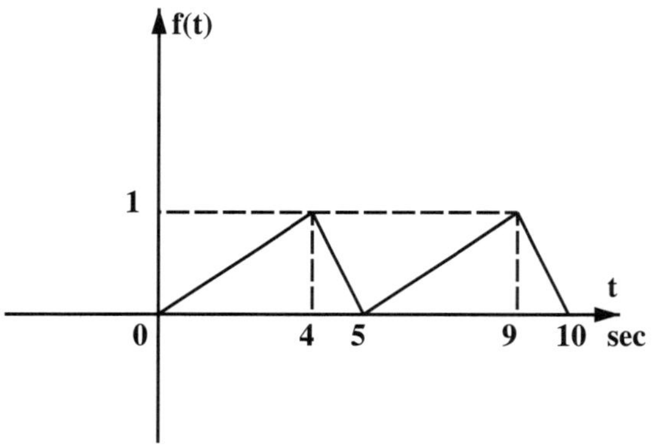

9.9.2. Find the Laplace transformation of the waveform shown in Fig. 9.9-1.

Group B – Inverse Transformations: We have stated many times that ultimately our goal is to find numerical responses in the time-domain setting. Thus, we need to practice again with pencil and paper, how the actual inverse transformations are performed using our main tool of partial fraction expansion.

9.9.3. Given the following set of F(s)'s, where each $F(s) = \mathcal{L}[f(t); t \to s]$, find the corresponding inverse Laplace transformed f(t)'s using partial fraction expansion methods and the catalog of known functions.

$$F(s) = \frac{10(s+1)}{(s+2)(s+3)(s+4)}. \tag{9.9-13}$$

$$F(s) = \frac{(s-1)}{(s-2)(s-3)(s+4)}. \tag{9.9-14}$$

$$F(s) = \frac{2(s-1)(s+1)}{s(s+2)(s+3)(s+4)}. \tag{9.9-15}$$

$$F(s) = \frac{8}{(s+1)(s+2)(s+3)^2}. \tag{9.9-16}$$

$$F(s) = \frac{5(s+1)}{(s+2)(s^2+s+4)}. \tag{9.9-17}$$

$$F(s) = \frac{6(s+1)}{s^2(s^2+s+4)}. \tag{9.9-18}$$

$$F(s) = \frac{1}{(s+2)^2(s+3)^3(s+4)^2}. \tag{9.9-19}$$

$$F(s) = \frac{\frac{1}{2}(s^2+3s+1)}{s(s^2+4s+3)}. \tag{9.9-20}$$

$$F(s) = \frac{(s^2-s+1)}{s^2(s+2)^2(s^2+s+4)}. \tag{9.9-21}$$

$$F(s) = \frac{100(s+1)}{(s+2)^2(s+3)^2(s^2+s+1)}. \tag{9.9-22}$$

$$F(s) = \frac{100(s+1)}{(s^2+s+1)(s^2+3s+1)}. \tag{9.9-23}$$

$$F(s) = \frac{10(s+1)}{(s^2+2s+4)^2(s^2+s+10)^2}. \tag{9.9-24}$$

Remark 9.9-2: Using Laplace transformation rules and theorems we can simplify many numerical problems. To that end consider the next two problems.

9.9.4. Given $Y(s) = \frac{G(s)R(s)}{1+G(s)H(s)}$, where $R(s) = \frac{10}{s}$, $G(s) = \frac{s+20}{s^2+4s+16}$, $H(s) = \frac{K}{s(s+5)}$, find $z = \int_0^\infty y(t)dt$ in terms of K.

9.9.5. Suppose $I(s) = \frac{1}{s^2+2s+5}$, find $i(t=0)$ and $i(t=\infty)$.

9.9.6. Given $T(s) = \dfrac{3s + 2}{s + 1}$, and $e(t) = t$, find $r(t)$, where $R(s) = T(s)E(s)$.

Group C – Differential Equations: Ultimately, when everything is said and done, we need to solve differential equations to complete any circuit analysis. Laplace transformation is a powerful method to assist us with this goal and we practice that first, before the associated differential equations arising in circuit analysis are considered. This practice involves the combination of the preceding two groups by first taking the Laplace transformation of forcing functions and differentials incorporating all pertinent initial conditions, thus transforming the differential equation into an algebraic one in terms of the Laplace transformation of the differentiated variable. Followed by generating the inverse transformation of the differentiated variable as follows.

9.9.7. Using Laplace transformation methods and with a full knowledge of the catalog for special functions and theorems, solve the following differential equations with their corresponding set of initial conditions, revisited from Section 7.5. Here, $D \equiv d/dt$. It would be worthwhile to compare the computational efforts of using operational methods (a quasi-traditional approach) in Chapter Seven, versus that of the current integral transformational method.

$$\ddot{x}(t) + 3\dot{x}(t) + 2x(t) = e^{-3t}, \quad x(0) = 0, \quad \dot{x}(0) = 0. \tag{9.9-25}$$

$$\begin{cases} D^3x(t) + 9D^2x(t) + 26Dx(t) + 24x(t) = t^3 - t^2 + t - 1, \\ x(0) = 0, \quad \dot{x}(0) = 0 \text{ and } \ddot{x}(0) = 0. \end{cases} \tag{9.9-26}$$

$$\begin{cases} \ddot{x}(t) + 6\dot{x}(t) + 5x(t) = \cos t - \sin t, \\ x(0) = 0 \text{ and } \dot{x}(0) = 0. \end{cases} \tag{9.9-27}$$

$$\begin{cases} (D + 2)x(t) + (D + 2)y(t) = t, \\ (D + 1)x(t) + y(t) = t^2, \\ x(0) = 0, \quad \dot{x}(0) = 0, \quad y(0) = 0, \quad \dot{y}(0) = 0. \end{cases} \tag{9.9-28}$$

$$\begin{cases} (D + 1)x(t) + (D + 4)y(t) = t, \\ (D + 2)x(t) + (D + 3)y(t) = t^2, \\ x(0) = 0, \quad \dot{x}(0) = 0, \quad y(0) = 0, \quad \dot{y}(0) = 0. \end{cases} \tag{9.9-29}$$

$$\begin{cases} (D+1)x(t) + (D+2)y(t) = \sin t, \\ (D+2)x(t) - (D+1)y(t) = \cos t, \\ x(0) = 0,\ \dot{x}(0) = 0,\ y(0) = 0,\ \dot{y}(0) = 0. \end{cases} \tag{9.9-30}$$

$$\ddot{x}(t) + 3\dot{x}(t) + 2x(t) = e^{-t} + e^{-2t}, \text{ with } x(0), \dot{x}(0) \text{ as parameters.} \tag{9.9-31}$$

$$\ddot{x}(t) + 7\dot{x}(t) + 6x(t) = e^{-t} + e^{-3t} + e^{-6t}, \text{ with } x(0), \dot{x}(0) \text{ as parameters.} \tag{9.9-32}$$

Group D – Applications in Circuits: Now that we have mastered both direct and inverse Laplace transformations with applications in solving differential equations, the moment of truth, i.e., to demonstrate the utility of these transformational techniques in circuit analysis, is upon us. Thus, we study again in the context of current methodology several examples and problems from the preceding chapters, but with a new order of presentation as it becomes evident below. To avoid repeating the first line of Problem 9.9.7, we declare that from this point forward and in all subsequent problems our opening line of using Laplace-transformation method is:

Using Laplace transformation methods and with a full knowledge of the catalog for special functions and theorems, numerically solve for the time-domain setting of the corresponding differential equations arising from the proposed method of analysis.

9.9.8. Redraw the circuit in Fig. 9.9-2 in the Laplace domain setting with all pertinent initial conditions displayed accordingly.

Fig. 9.9-2.

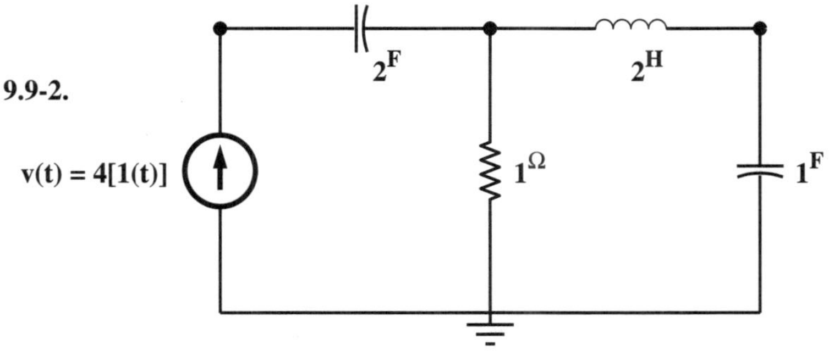

9.9.9. Given the circuit in Fig. 9.9-3, find $Z_{eq}(s)$ incorporating all initial conditions.

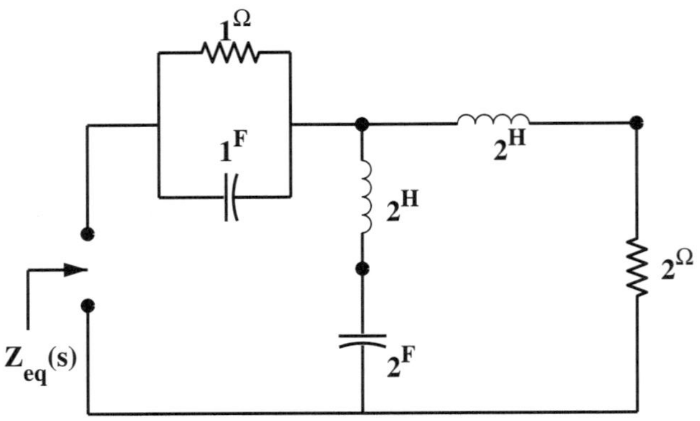

Fig. 9.9-3.

9.9.10. For the circuit in Fig. 9.9-4 and with a unit-impulse current source, draw $v_o(t)$ versus time, upon first calculating $V_o(s)$.

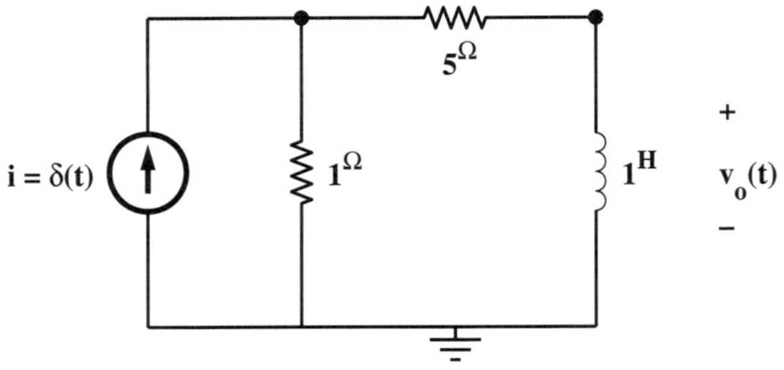

Fig. 9.9-4.

9.9.11. Given $v_{in}(t)$ as a train of periodic pulses shown in Fig. 9.9-5, find $v_o(t)$ in the circuit of Fig. 9.9-6. Let the initial condition of $v_C(0) = 0$. Use Laplace transformation method.

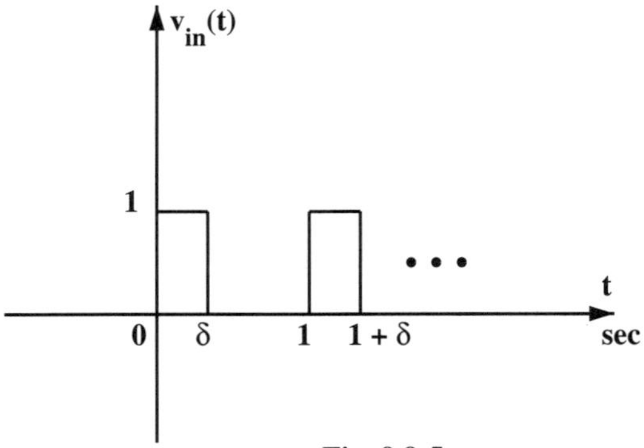

Fig. 9.9-5.

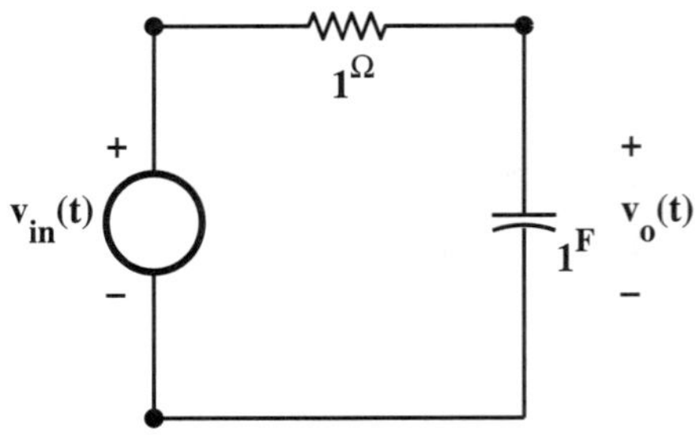

Fig. 9.9-6.

9.9.12. Numerically solve three differential equations arising from the mesh equations analysis of circuits in Figs. 9.9-7 to 9.9-9. Let all relevant initial conditions be set at zero. Clearly, we must establish the corresponding mesh equations, using Laplace transformations techniques first, before attempting to solve them.

Fig. 9.9-7. The Fig. 6.4-7 revisited.

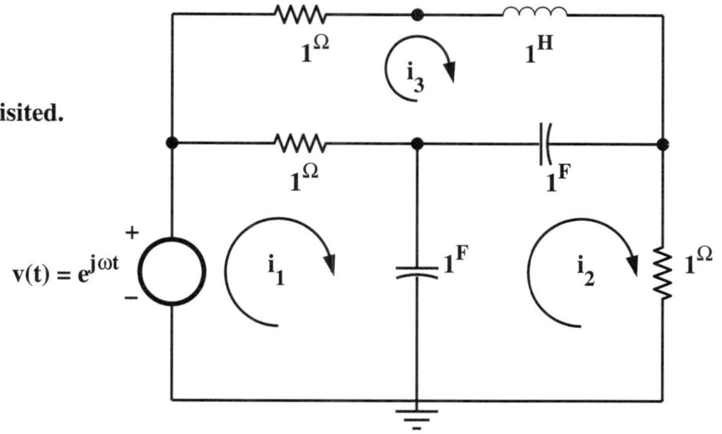

Fig. 9.9-8. The Fig. 6.6-8 revisited.

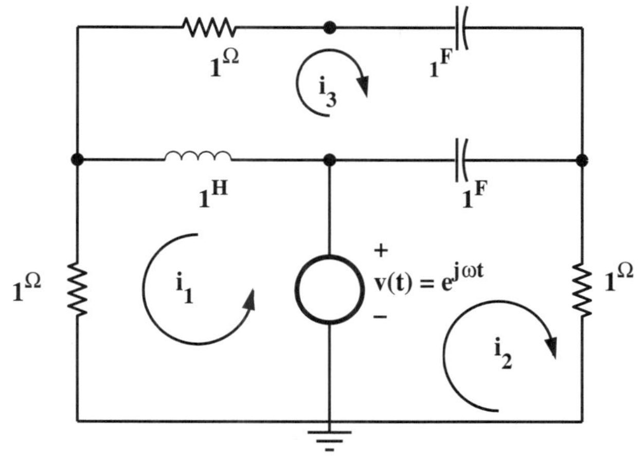

Fig. 9.9-9. The Fig. 6.6-9 revisited.

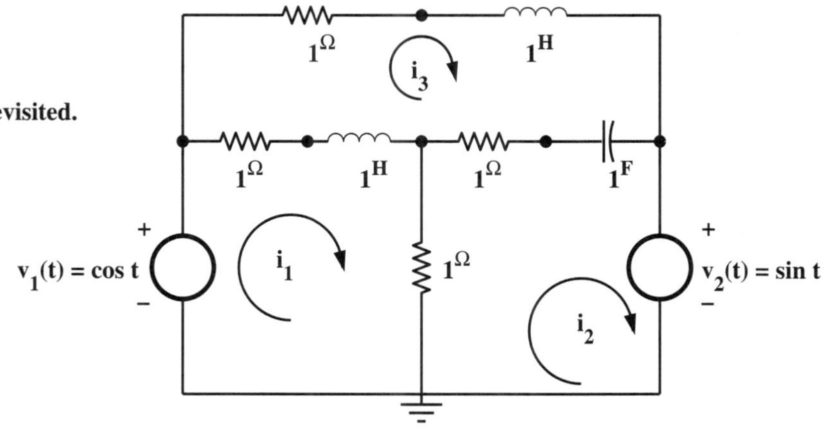

9.9.13. Write the set of mesh equations for the circuit in Fig. 9.9-10, with $M = 1^H$ in a Laplace-transformation setting. Let all relevant initial conditions be set at zero.

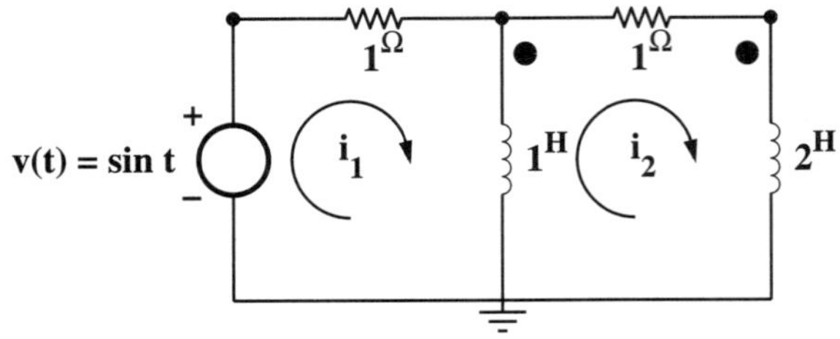

Fig. 9.9-10. The Fig. 8.4-1 revisited.

9.9.14. Numerically solve three differential equations arising from the node equations analysis of circuits in Fig. 9.9-11, where all L's are 10^{mH} and all C's are $10^{\mu F}$, while all R's are 2^{Ω}. Set all pertinent initial conditions at zero.

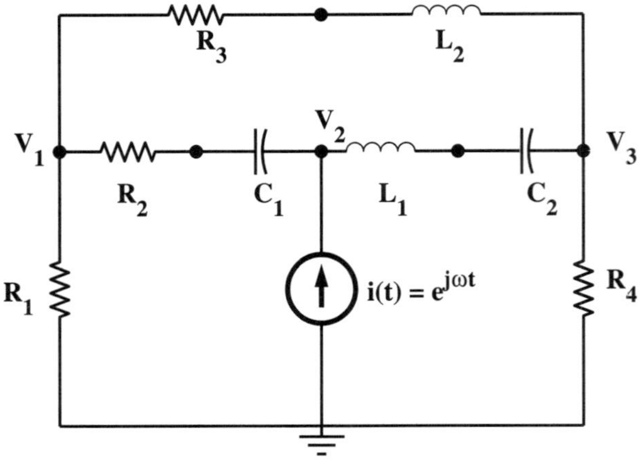

Fig. 9.9-11. The Fig. 6.4-5 revisited.

9.9.15. Completely analyze the circuit in Fig. 9.9-12, where all L's are 10^{mH} and all C's are $10^{\mu F}$, while all R's are 2^{Ω}. Set all pertinent initial conditions at zero. *Comment:* Our main objective is to develop a complete numerical solution and to that end we propose the mesh equations approach first. Then using the time-domain responses of three mesh currents finish the analysis completely.

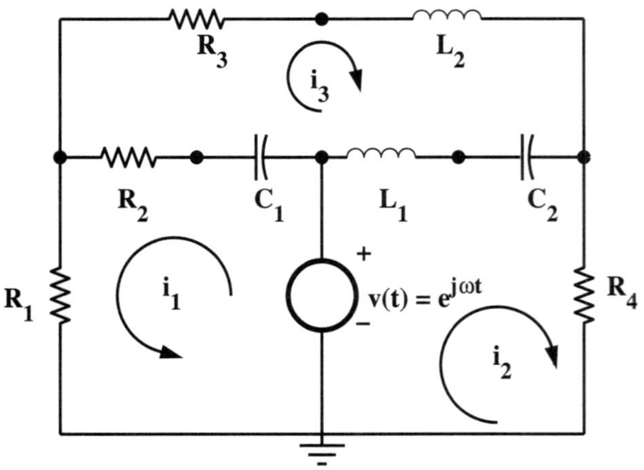

Fig. 9.9-12. The Fig. 6.4-4 revisited.

9.9.16. Completely analyze the circuit in Fig. 9.9-13 in as many different ways as you can. Set all initial conditions at zero. *Comment:* Clearly, this is a major task, and we suggest that in each method only those equations which represents the essential variables in Laplace-transformation forms be generated. For numerical purposes, only the one solution resulted from superposition rule should be carried out in its complete numerical details, leading to a numerical solution for the circuit in a time-domain setting.

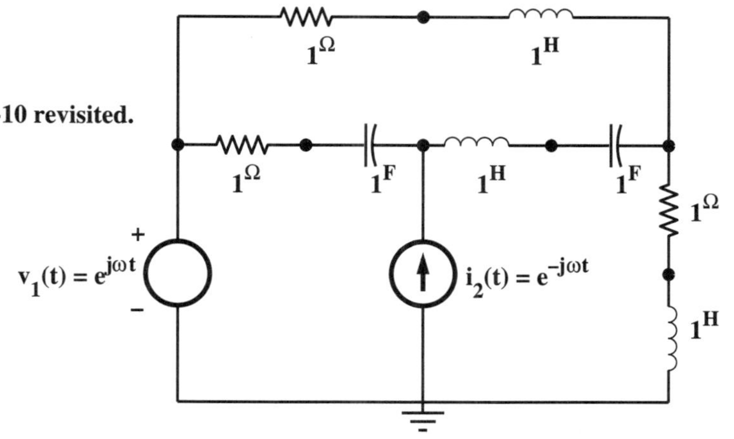

Fig. 9.9-13. The Fig. 6.6-10 revisited.

Remark 9.9-3: As Problem 9.9.16 shows in a tangible way, we now know how various methods which are used to analyze a circuit with resistive elements are easily amenable for circuits with complex elements when these circuits are set in a Laplace-transformation setting. In other words, the most complex circuits when are set in a Laplace-domain setting can be analyzed with the same simplicity and straightforwardness as those with resistive circuits, which this observation conveys the generality and versatility of this approach.

Remark 9.9-4: What about the circuits with complex sources and the like? For instance in Chapter Six, we have analyzed complex circuits using phasor forms to describe complex quantities. Can we use Laplace variable $s=\sigma + j\omega$ instead to set a phasor-form problem in the Laplace-transformation setting, then solve the resulting problem, and when we are done, put $\sigma=0$ or $s=j\omega$? The answer is definitely positive. Furthermore, in most cases this symbolic analysis with "s" is more systematic than working with complex numbers. To that end, we now know how to replace impedances and admittances of complex elements with their Laplace variable counterparts, but what about complex sources? Here, each source, say $V=|V|\angle V$, is replaced with $Ve^{j\omega t}=|V|\angle(\angle V+j\omega t)$, which only affects the angle part of the source and not its amplitude. When the analysis is completed in the Laplace-transformation setting, we replace "s" with $j\omega$, and substitute for ω, the operating frequency which is assigned to the problem. To demonstrate this point consider the following additional circuits repeated from Chapter Six.

9.9.17. When the complete solution in Problem 9.9.16 is developed in terms of a Laplace-transformation setting, let $s=j\omega$ with $\omega=1$ rad/sec and find the corresponding numerical solutions. Compare likewise results with those generated in Problem 6.6.11, where the same circuit is solved similarly in many different numerical ways, but each with the complex numbers in the phasor-form representation.

9.9.18. Analyze the circuit in Fig. 9.9-14 in a Laplace-transform setting. Then evaluate the final result at $s=j1000$ numerically. Here, both sources are functions of time through $e^{j\omega t}$ and all relevant initial conditions are set at zero.

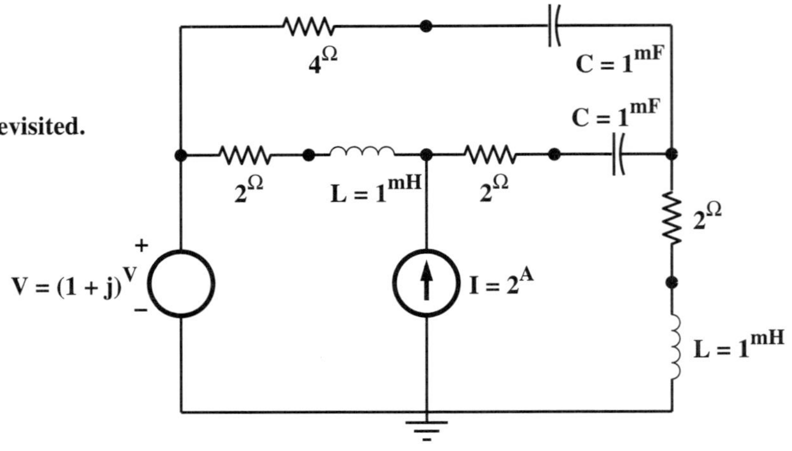

Fig. 9.9-14. The Fig. 6.3-4 revisited.

9.9.19. For the circuit in Fig. 9.9-15, and for $\omega=1000$, find $v_o(t)$ using a Laplace-transformation method. Assume all relevant initial conditions are set at zero.

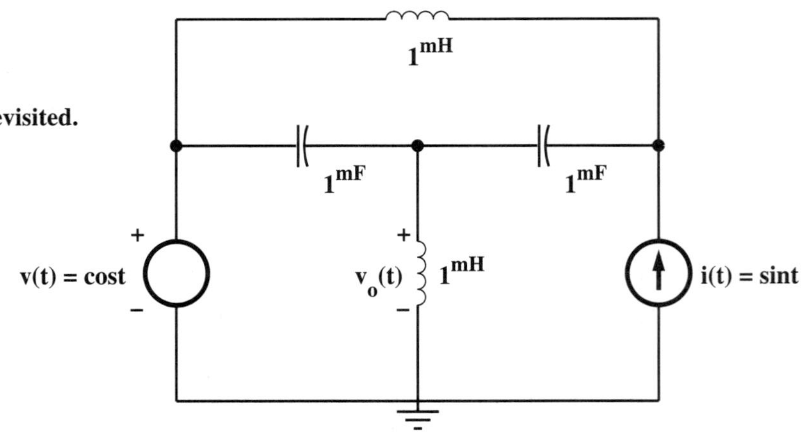

Fig. 9.9-15. The Fig. 7.5-7 revisited.

9.9.20. For the circuit in Fig. 9.9-16, find V_o/V_1 at $\omega=1$, using Laplace-transformation technique. Here, both sources are functions of time through $e^{j\omega t}$ and we set the initial conditions at zero.

Fig. 9.9-16. The Fig. 6.6-13 revisited.

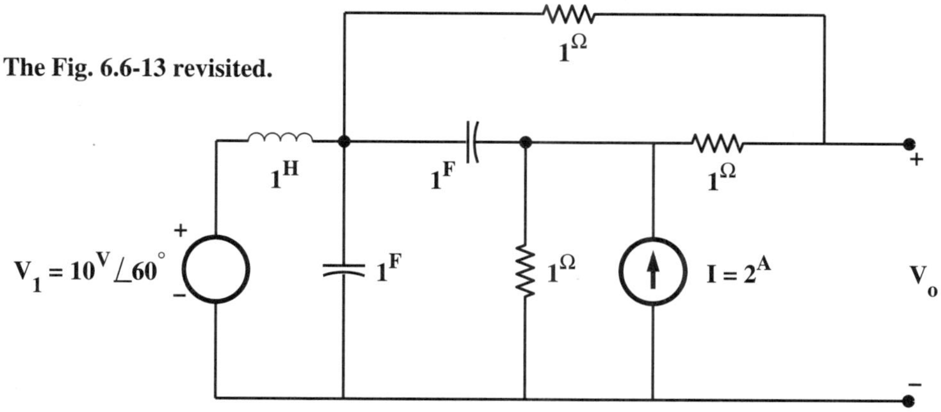

9.9.21. Find $v_o(t)$ for the circuit in Fig. 9.9-11 with $M = 1^H$, using Laplace-transformation method. Let $\omega = 1$ rad/sec and compare your result with that in Problem 8.4.7.

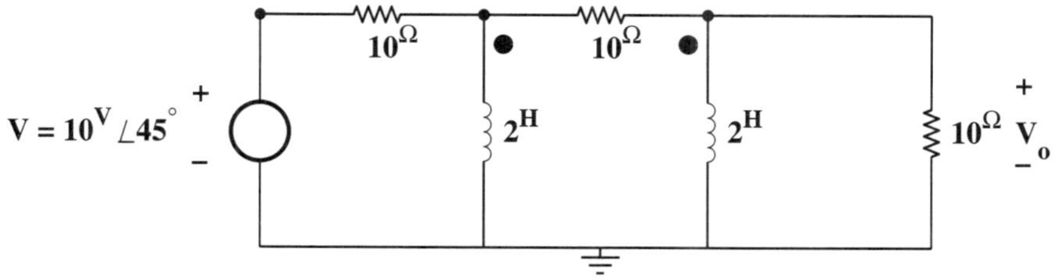

Fig. 9.9-17. The Fig. 8.4-8 revisited.

9.9.22. Write the mesh equations for the circuit in Fig. 9.9-18, with $M = 10^{mH}$ for all three coils, in a Laplace-transformation setting. Let $\omega = 1000$ rad/sec, find I_o when $v(t) = \sin 1000t$.

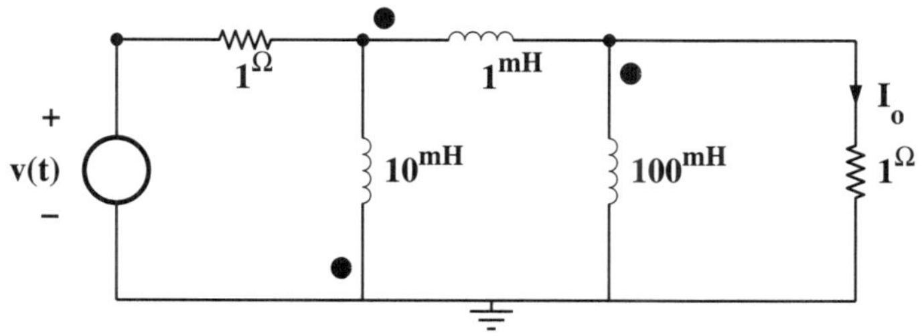

Fig. 9.9-18. The Fig. 8.4-2 revisited.

Group E – Switching Circuits: Laplace-transformation methodology is the most effective tool to use in circuits with sequential switches where both initial- and final-condition theorems can provide us with numerical answers to transitional issues in each operating stage of the circuit. Here, we may use with a good approximation the final value theorem, if the switching duration is longer than a few, say five, circuit's time constant. Of course, in many instances we do not know the circuit's time constant (or natural mode) prior to some analysis, but as we gain experience we can make the judgment and use the final value theorem as though $t \to \infty$ which is intended in that theorem.

9.9.23. Given the circuit in Fig. 9.9-19, with $v_{in}(t) = \sin t$, find $v_o(t)$ using Laplace transformation method.

Fig. 9.9-19.

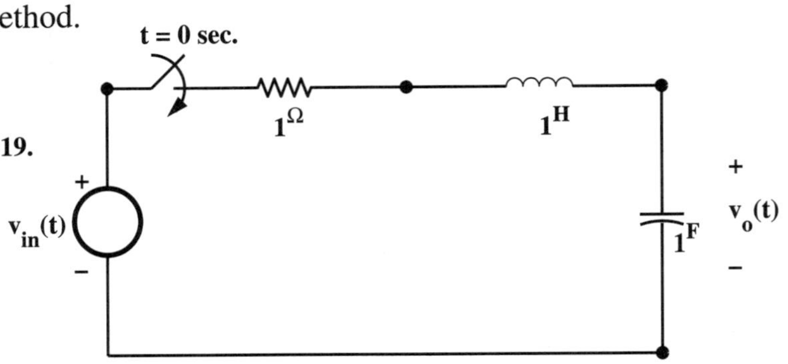

9.9.24. What is the response $v_o(t)$ in the circuit of Fig. 9.9-20 if the switch is opened at $t=0$. Use a Laplace-transformation setting.

Fig. 9.9-20.

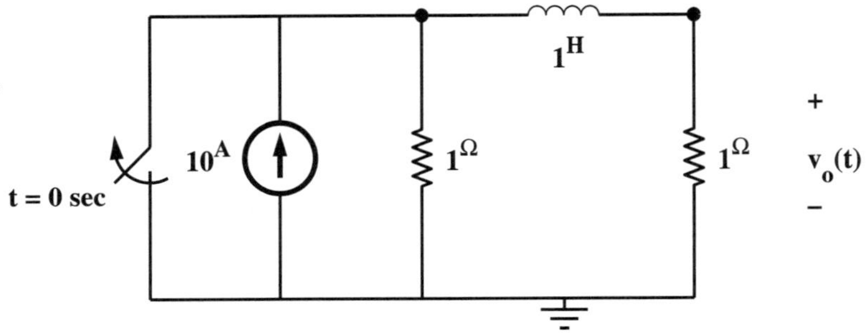

9.9.25. Using Laplace transformation method, and for the circuit in Fig. 9.9-21, find the initial and final conditions for $v_1(t)$ and $v_2(t)$. Also compute $\dfrac{d^2v_1(t=0^+)}{dt^2}$.

Fig. 9.9-21.

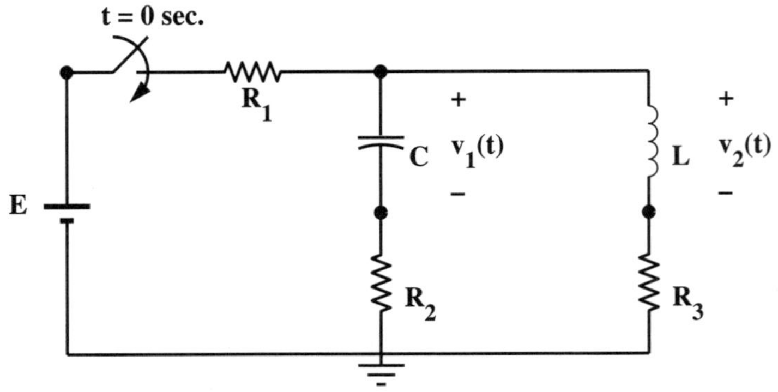

9.9.26. Consider the circuit in Fig. 9.9-22 with two switches. One closes at t = 0 second, and the other closes at t = 10 seconds. Find $v_o(t)$ as indicated for all t using a Laplace-transformation method.

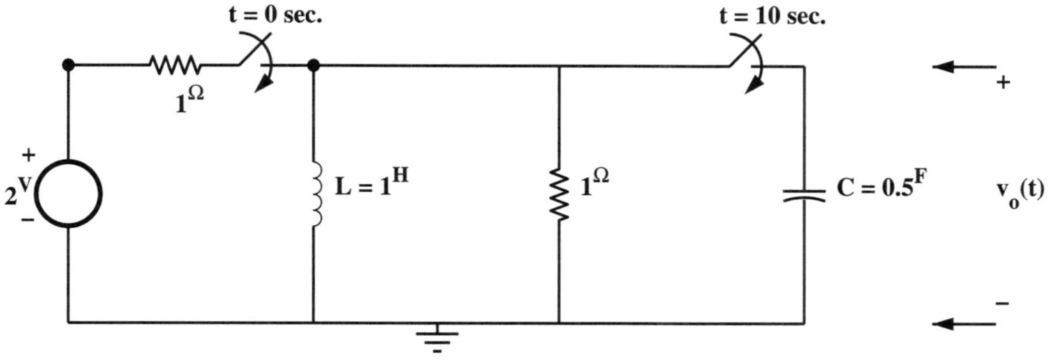

Fig. 9.9-22. The Fig. 7.4-5 revisited.

9.9.27. For the circuit in Fig. 9.9-23, find i(t) using a Laplace-transformation method.

Fig. 9.9-23. The Fig. 7.5-5 revisited.

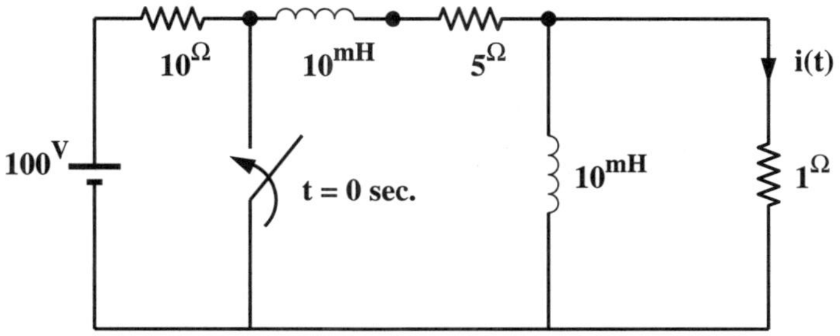

9.9.28. In the circuit of Fig. 9.9-24, the switch is initially at point (a) during $-\infty < t < 0$. At $t \geq 0$ we switch from point (a) to point (b). Find i(t) for $t \geq 0$ using a Laplace-transformation method.

Fig. 9.9-24. The Fig. 7.5-8 revisited.

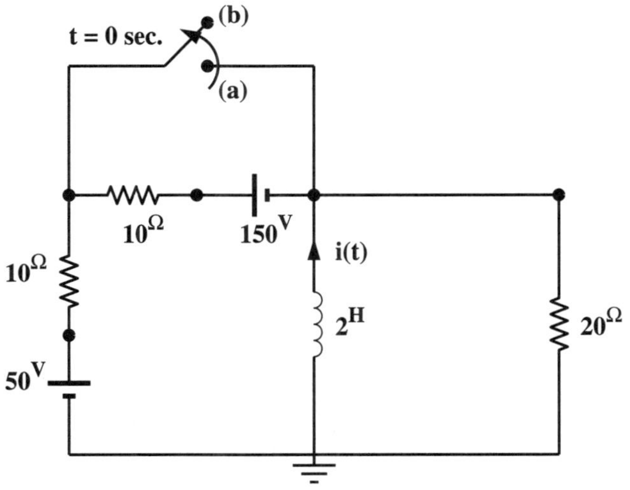

9.9.29. In the circuit of Fig. 9.9-25, the switch is initially at point (a) during $-\infty < t < 0$. At $t \geq 0$ we switch from point (a) to point (b). Find $v_o(t)$ for $t \geq 0$ using a Laplace-transformation method.

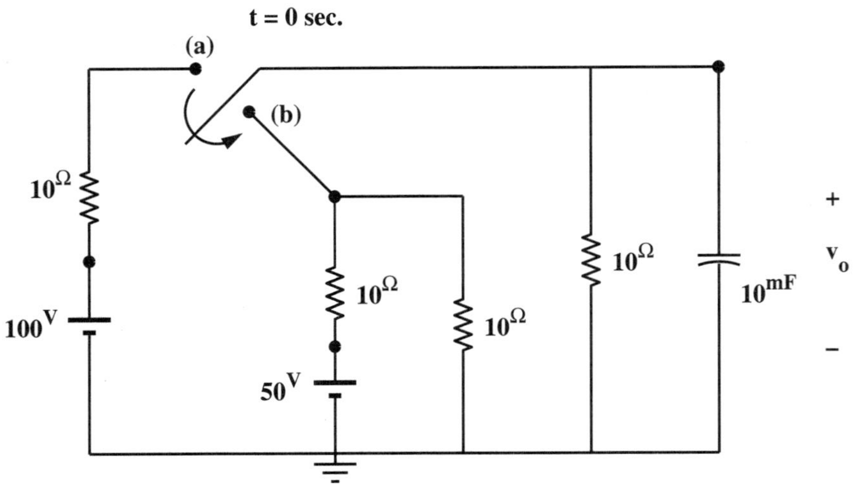

Fig. 9.9-25. The Fig. 7.5-9 revisited.

9.9.30. The two switches in the circuit of Fig. 9.9-26 are applied at t = 0 and t = 10 seconds. Find i(t) as indicated using a Laplace-transformation method.

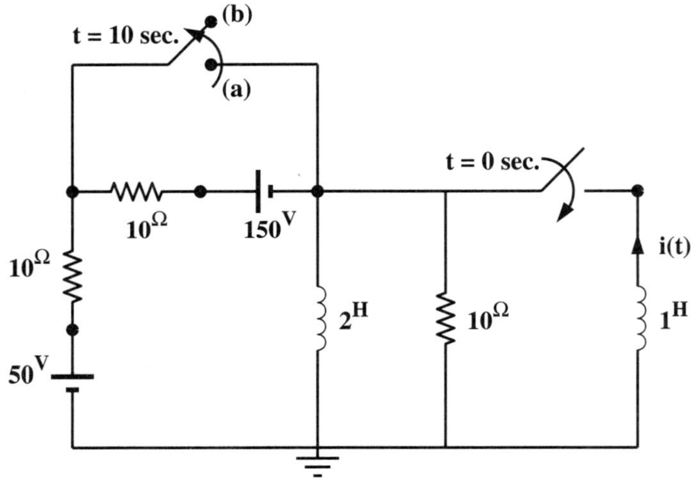

Fig. 9.9-26. The Fig. 7.5-10 revisited.

9.9.31. Consider the circuit in Fig. 9.9-27, and let $\omega = 1000$ rad/sec, $M = 1^{mH}$. The switch is initially at (a) for a long time. At t = 0, we switch from (a) to (b). Write only all the necessary set of equations to describe the dynamic of this circuit for $-\infty < t < \infty$, using a Laplace-transformation method.

Group F – General Theorems: Laplace-transformation techniques work with most theorem we have studied so far. The following problems are a few samples of cases which are presented to complement the topics described previously.

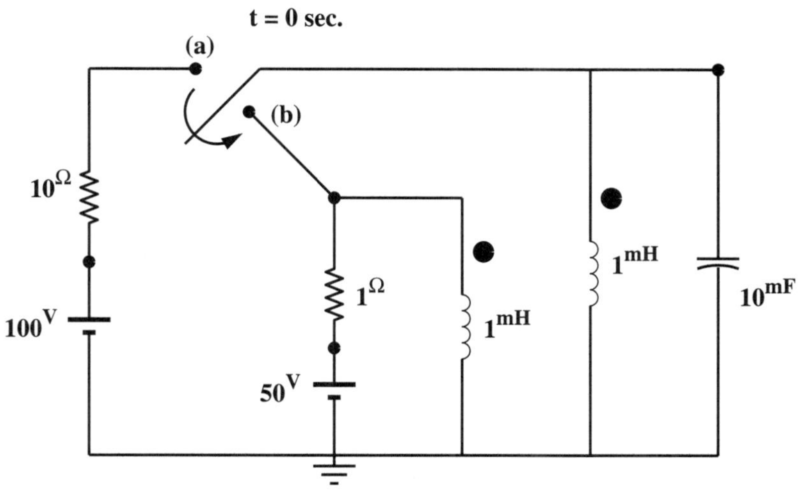

Fig. 9.9-27. The Fig. 8.4-21 revisited.

9.9.32. Consider the circuit in Fig. 9.9-28 with an input pulse function as its current source. The initial condition for the $1^{\mu F}$ capacitor is 10^{V}. Analyze the delivery of power and energy from the current source to the capacitor for one complete period. Use the Laplace domain setting.

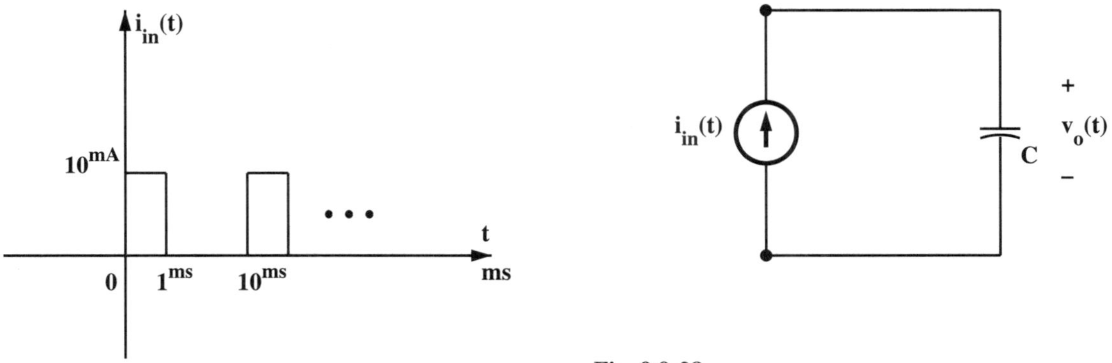

Fig. 9.9-28.

9.9.33. Consider the circuit in Fig. 9.9-29 with a step function as its input voltage source. Suppose the initial condition of i(t) is zero. Analyze the delivery of power and energy from the source to the inductor for $t \in [0, 2^{ms}]$.

Fig. 9.9-29.

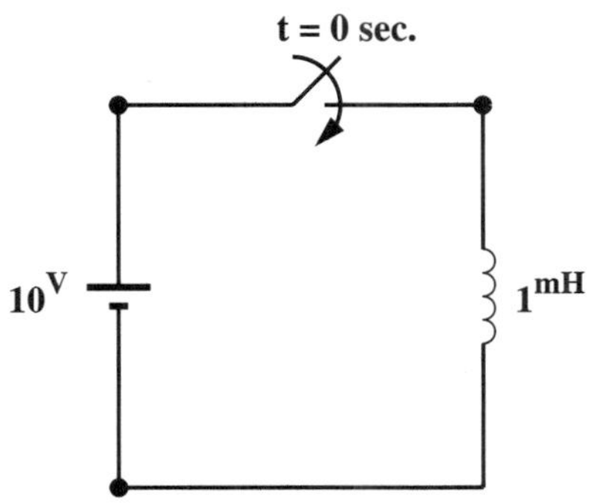

9.9.34. For the circuit in Fig. 9.9-30, find Z_l at $\omega = 1$ rad/sec for the maximum power transfer, using Laplace-transformation techniques.

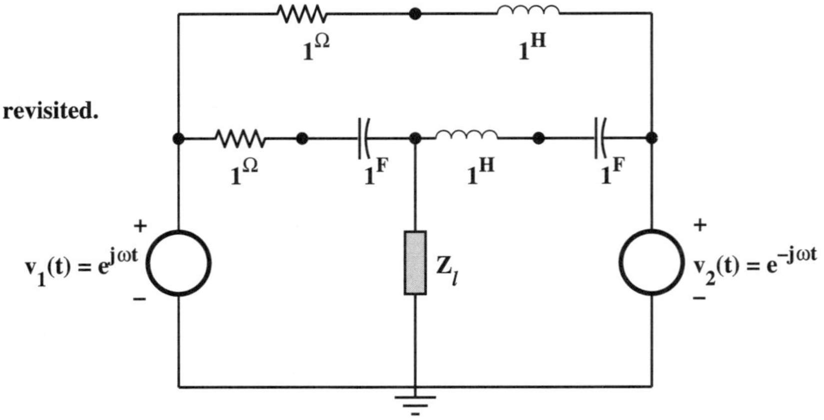

Fig. 9.9-30. The Fig. 6.6-14 revisited.

9.9.35. Consider the circuit in Fig. 9.9-31. Set up the node equations approach and verify the reciprocity theorem by first applying only I_1 and studying V_3, and second by applying I_2 and studying V_1. Use the Laplace transformation technique for this verification. Will the result on reciprocity be different if we use other analytical techniques?

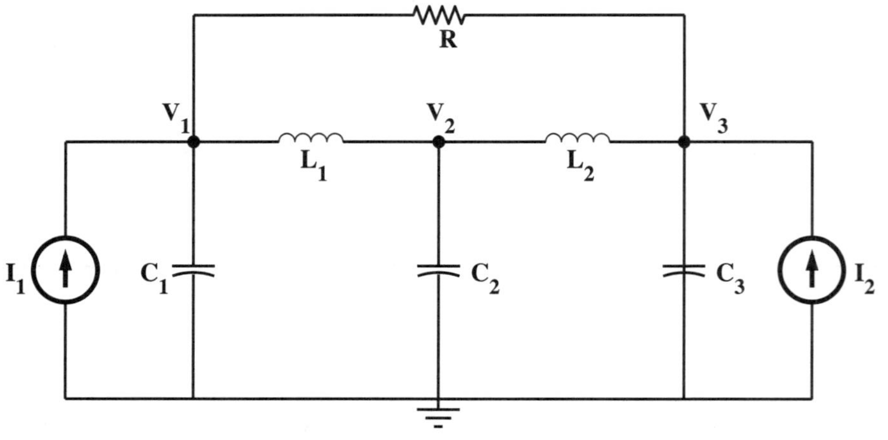

Fig. 9.9-31.

9.9.36. Verify the three forms of Tellegen's theorem for the circuits in Fig. 9.9-32.

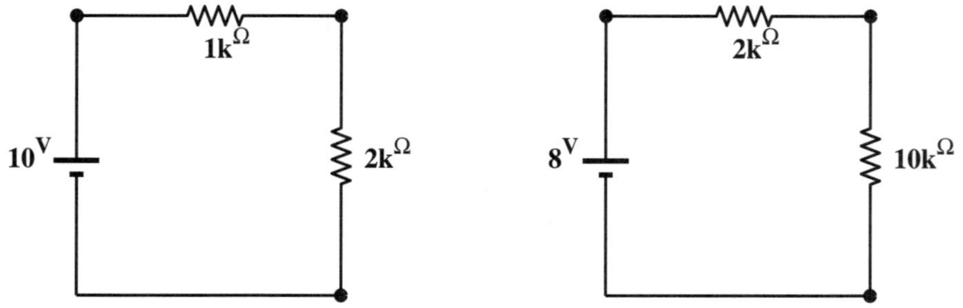

Fig. 9.9-32.

10 FREQUENCY RESPONSE

10.1. INTRODUCTION

An integral part of any elementary circuit analysis is a basic understanding of issues in frequency response and the associated topics in a network. Here, we use a sinusoidal waveform $s(t) = A \sin \omega t$ as depicted in Fig. 10.1-1 to excite the network and study its corresponding steady-state output. In this figure, the period T is measured in second (sec) and is equal to 1/f, where f is the frequency measured in hertz (Hz), while $\omega = 2\pi f$ is the angular frequency measured in radian per second (rad/sec). The Laplace transformation of s(t) is $\mathcal{L}[s(t)] = A\omega/(s^2 + \omega^2)$. The following materials should be studied in conjunction with that presented in Chapter Six, in particular, discussions regarding the magnitude and angle of a complex number.

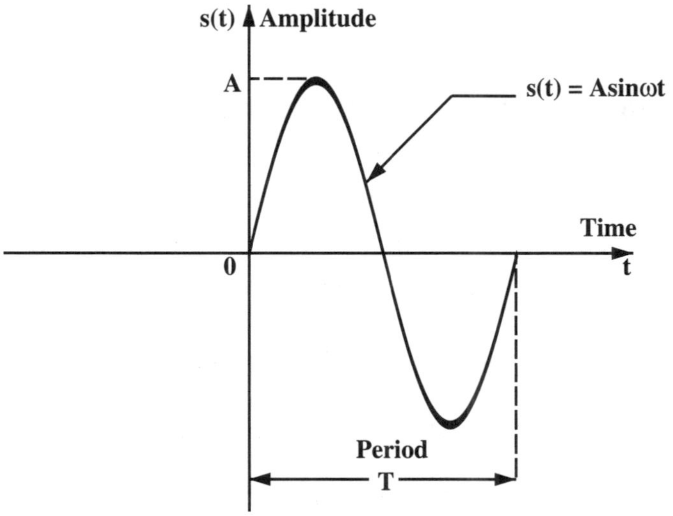

Fig. 10.1-1. A sinusoidal waveform.

Consider a linear system with the transfer function Y(s) (admittance), where I(s) = Y(s)V(s), with V(s) as its input and I(s) as its output, Fig. 10.1-2.

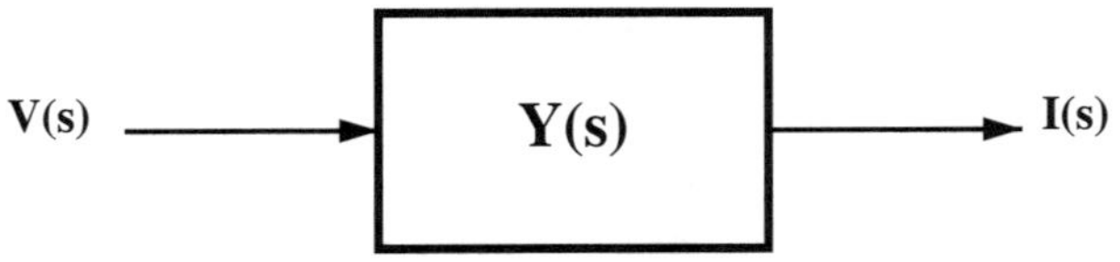

Fig. 10.1-2.

In Fig. 10.1-2, if $V(s) = \frac{A\omega}{s^2 + \omega^2}$, then

$$I(s) = Y(s)\,\frac{A\omega}{s^2 + \omega^2}\,. \qquad (10.1\text{-}1)$$

Taking the inverse Laplace transformation (an analysis similar to that given in Example 9.5-4, *per se*) yields

$$i(t) = \mathcal{L}^{-1}[Y(s)\,\frac{A\omega}{s^2 + \omega^2}]$$

$$= \text{transient response} + \text{steady} - \text{state response}. \qquad (10.1\text{-}2)$$

The transient response dies out after a few system's time constant (structural mode), but the steady-state response remains and for the above case that becomes

$$i_{ss}(t) = A|Y(j\omega)|\sin(\omega t + \phi), \qquad \phi = \angle Y(j\omega). \qquad (10.1\text{-}3)$$

In other words, the output is proportionally a replica of the input shifted in phase according to the angle of the transfer function Y(s) – a standard property of the linear system.

10.2. FREQUENCY-DOMAIN REPRESENTATION

Before presenting our main motivation for studying frequency domain, we need to review some general approaches to represent typical transfer functions in this setting. This review complements our presentation in Chapter Six.

In general, when we have $F(s) \triangleq \frac{N(s)}{D(s)}$ for two rational polynomials N(s) and D(s), where $s = \sigma + j\omega$ is the Laplace variable. This representation is the same as $F(s) \triangleq \text{Re}[F(s)] + j\text{Im}[F(s)]$, where "Re" and "Im" stand for the real and imaginary parts respectively. Since we now work on the $j\omega$-axis where $\sigma = 0$, we have

$$F(s)\Big|_{s=j\omega} = \mathrm{Re}[F(j\omega)] + j\,\mathrm{Im}[F(j\omega)] \tag{10.2-1a}$$

$$= |F(j\omega)|\angle F(j\omega) \triangleq \rho\angle\theta. \tag{10.2-1b}$$

In (10.2-1b) just as in any complex number, we note that

$$\rho \triangleq \sqrt{[\mathrm{Re}\,F(j\omega)]^2 + [\mathrm{Im}\,F(j\omega)]^2}, \quad \text{and } \theta \triangleq \angle \mathrm{tg}^{-1}\frac{\mathrm{Im}[F(j\omega)]}{\mathrm{Re}[F(j\omega)]}. \tag{10.2-1c}$$

In general, given $F(s) = \dfrac{N(s)}{D(s)}$, we may assume that both N(s) and D(s) are factored as follows.

$$F(s) \triangleq \frac{N(s)}{D(s)} = \frac{\prod_{k=1}^{m}(s + z_k)}{\prod_{l=1}^{n}(s + p_l)}. \tag{10.2-2}$$

Here, z_k's are the zeros of the system, while p_l's are the poles of the system. To have the above factor form is a reasonable assumption, although in practice we may not be able to factor a parametric rational polynomial into this form quickly. When possible, we may write F(s) in terms of its real and imaginary parts or magnitude and angle. We choose the second approach (10.2-1b), and write F(s) as follows.

$$F(s) = \frac{\prod_{k=1}^{m}|s + z_k|}{\prod_{l=1}^{n}|s + p_l|}\left\{\angle\prod_{k=1}^{m}(s + z_k) - \angle\prod_{l=1}^{n}(s + p_l)\right\}$$

$$= \frac{\prod_{k=1}^{m}|s + z_k|}{\prod_{l=1}^{n}|s + p_l|}\left\{\sum_{k=1}^{m}\angle(s + z_k) - \sum_{l=1}^{n}\angle(s + p_l)\right\}. \tag{10.2-3}$$

The angle on the right-hand side of (10.2-3) follows from the fact that such a quantity for a product of complex numbers is the sum of the angle for each term.

Comment 10.2-1: To interpret quantities such as (10.2-3) in a numerical sense, a new unit and way of measuring is introduced which substantially enhances our grasp of these often large numerical values. To introduce such a unit, let P_r be a reference power level, and P be the power that we want to measure. Then we introduce a new unit which is ten times the logarithm of the ratio between these two power levels, i.e., $10\log P/P_r$. In circuit theory, the electric power is proportional to the voltage times current; and current is the ratio of voltage divided by resistance, thus the ratio of two power levels becomes as follows.

$$10 \log \frac{P}{P_r} = 10 \log \frac{VI}{V_r I_r}$$

$$= 10 \log \frac{R_r V^2}{R V_r^2} = 10 \log (\frac{R_r}{R})(\frac{V}{V_r})^2. \tag{10.2-4}$$

If we assume that $R = R_r$, then we have:

$$10 \log (\frac{P}{P_r}) = 20 \log (\frac{V}{V_r}). \tag{10.2-5}$$

This idea has led to the introduction of a new quantity called decible (dB), which is 20 times the logarithm of the magnitude of a complex number. Here, we note that although dB is introduced to compare two power levels, often we use this unit in an abstract sense for quantities which may not directly be associated with any power level. Furthermore and without any loss of generality, by letting $R = R_r$ we also let $P_r = 1^W$ (one Watt). Thus, we now study directly the following expression, as a new way of evaluating a complex quantity.

$$20 \log |F(s)| = 20 \log \frac{\prod_{k=1}^{m} |s + z_k|}{\prod_{l=1}^{n} |s + p_l|}$$

$$= 20 \sum_{k=1}^{m} \log |s + z_k| - 20 \sum_{l=1}^{n} \log |s + p_l|. \tag{10.2-6}$$

These quantities must now be drawn on logarithmic scales and such plots are generally called Bode's diagrams. Here, the plural refers to both magnitude [such as (10.2-6)] and angle [such as the angle portion in (10.2-3)] versus ω diagrams for each complex number. Referring to Bode's diagrams, say for the magnitude, we use for the horizontal axis the logarithm of ω in the form of $\omega = (10)^r$, where r is the desired reference range. For instance, in Fig. 10.3-1, the range of r is from -1 to 4. On the vertical axis we may use either a linear scale (as in Fig. 10.3-1) or a logarithmic scale. A graph with axes in logarithmic scales is called a *log-log plot.* A graph with linear vertical axis while logarithmic horizontal axis is called a *semilog plot.* On a logarithmic axis, say the horizontal axis in Fig. 10.3-1, if we select two angular frequencies ω_1 and ω_2 such that $\omega_2/\omega_1 = 10$, then the distance between ω_1 and ω_2 is called one *decade.* If the ratio $\omega_2/\omega_1 = 2$, then this distance is called one *octave.* Finally, if instead of base ten logarithm we use base $e = (2.718\ldots)$, then we have a unit called *"neper"* instead of dB. We use dB exclusively.

The beauty of such transformations from ordinary numbers to the logarithmic numbers is that we can draw a wide range of variables, in this case frequency, in just one sheet of paper. We can add and/or subtract a few simple numbers (which are the logarithm of actual variables) versus multiplying huge numerical values for these variables. If we recall that more than six decades ago this concept was introduced, then we realize how ingenious this approach had been to simplify computation in circuit design. In some corners we hear that these concepts are obsolete, and we should not even bother teaching them. We believe these are important concepts which are very instrumental in learning the whole circuits and seeing the entire picture from the eyes of those who had created its theoretical foundation.

10.3. BODE DIAGRAMS

To develop experience in drawing and manipulating Bode diagrams for complicated transfer functions, we need to review such diagrams for a few reference cases first. To do so, in the following we review the magnitude and phase of four important cases, and display their magnitudes in Fig. 10.3-1. These cases form the nucleus for drawing Bode diagrams of complicated transfer functions. There is one other important case besides these four which merits special attention and that case follows subsequently.

(1) $F_1(s) = K > 0$, for example $K = 10$.

For this constant case, and certainly for all $s = j\omega$, we have

$$20 \log |F_1(j\omega)| = 20 \log K = 20 \log 10 = 20 \text{ dB}. \tag{10.3-1a}$$

$$\theta_1(\omega) \triangleq \angle F_1(j\omega) \equiv 0^\circ. \tag{10.3-1b}$$

(2) $F_2(s) = s$.

Here, and for $s = j\omega$, we have

$$20 \log |F_2(j\omega)| = 20 \log \omega, \quad \omega > 0. \tag{10.3-2a}$$

$$\theta_2(\omega) \triangleq \angle F_2(j\omega) \equiv 90^\circ. \tag{10.3-2b}$$

(3) $F_3(s) = \dfrac{1}{Ts + 1}$.

For this case, and for $s = j\omega$, we have

$$20\log|F_3(j\omega)| = 20\log\frac{1}{\sqrt{1+T^2\omega^2}} \tag{10.3-3a}$$

$$= \begin{cases} 20\log\dfrac{1}{\omega T}, & \text{for large } \omega, \\ -3.01, & \omega = \dfrac{1}{T}, \\ 0, & \text{for small } \omega. \end{cases}$$

$$\theta_3(\omega) \triangleq \angle F_3(j\omega) = -\mathrm{tg}^{-1}\omega T. \tag{10.3-3b}$$

(4) $F_4(s) = Ts + 1.$

Here, and for $s = j\omega$, we have

$$20\log|F_4(j\omega)| = 20\log\sqrt{1+T^2\omega^2} \tag{10.3-4a}$$

$$= \begin{cases} 20\log\omega T, & \text{for large } \omega, \\ 3.01, & \omega = \dfrac{1}{T}, \\ 0, & \text{for small } \omega. \end{cases}$$

$$\theta_4(\omega) \triangleq \angle F_4(j\omega) = \mathrm{tg}^{-1}\omega T. \tag{10.3-4b}$$

All the above four cases in their so-called **asymptotic** forms are shown in Fig. 10.3-1. In this figure we have not displayed the above ± 3.01 dB, only for the sake of simplicity. In actual computations we account for these correction values.

Referring to Fig. 10.3-1, we note that a rather wide range of frequencies and magnitude are shown in a manageable space on the logarithmic scales, which could not be shown so easily otherwise. Furthermore, the point $\omega = 1/T$ is called the **corner frequency** that plays the central role in this drawing. The corner frequency refers to the case when the corresponding slope changes. While, the *cut off frequency* generally refers to the corner when the output magnitude reaches 3 dB below its maximum value, i.e., the corner where magnitude changes in order that this reduction to occur.

As stated before, there is one other case which merits special classification, although we may argue that this is a combination of the above four cases.

(5) $F_5(s) = \dfrac{\omega_n^2}{s^2 + 2\zeta\omega_n s + \omega_n^2}.$

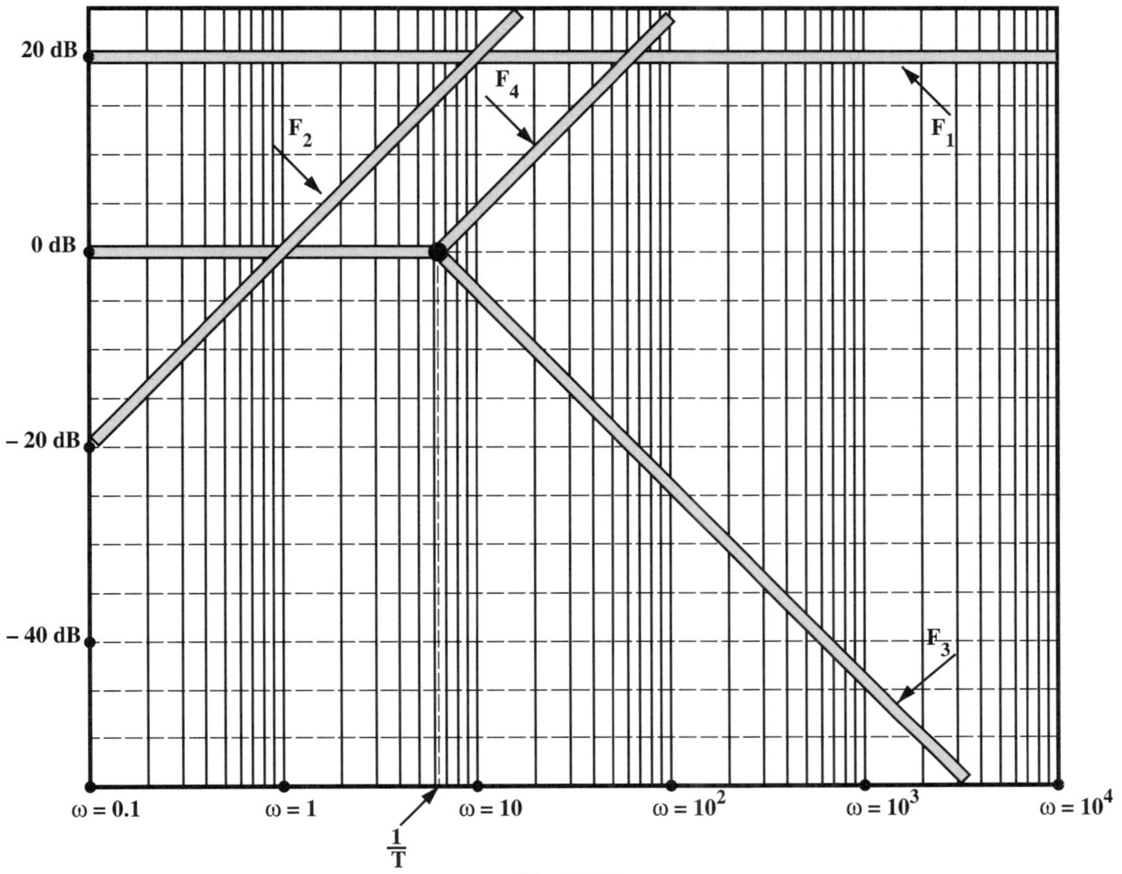

Fig. 10.3-1.

This is a transfer function with a second-order denominator versus that of the first-order in $F_3(s)$. For this second-order case to be studied extensively in future courses, ω_n is called the natural frequency, and ζ is called the damping ratio. We assume that the two poles of this system are a pair of complex numbers p_1 and $\bar{p}_1$. Thus, we have:

$$20\log|F_5(j\omega)| = 20\log\left[\frac{\omega_n^2}{|s+p_1||s+\bar{p}_1|}\right]\Bigg|_{s=j\omega}$$

$$= 40\log\omega_n + 20\log\frac{1}{|j\omega+p_1|} + 20\log\frac{1}{|j\omega+\bar{p}_1|}. \qquad (10.3\text{-}5)$$

From (10.3-5), we note that theoretically we need to combine one function of the type $F_1(s)$ with two functions of the type $F_3(s)$ to construct an asymptotic representation for $F_5(s)$, which is exactly why this logarithmic approach is so appealing.

However, we may take another look at $F_5(s)$ and note that upon normalizing this transfer function with respect to the natural frequency ω_n – a common practice to tackle this type of problem – we have:

$$F_5(s)\Big|_{s=j\omega} = \frac{1}{1-(\frac{\omega}{\omega_n})^2 + j2\zeta\frac{\omega}{\omega_n}}$$

$$= \frac{1}{\sqrt{[1-(\frac{\omega}{\omega_n})^2]^2 + 4\zeta^2(\frac{\omega}{\omega_n})^2}} \angle -\mathrm{tg}^{-1}\frac{2\zeta\frac{\omega}{\omega_n}}{1-(\frac{\omega}{\omega_n})^2}. \qquad (10.3\text{-}6)$$

Clearly, we note that the magnitude in (10.3-6) has *similar* asymptotic characteristic as that in (10.3-3), namely:

$$20\log\left|F_5(j\omega)\right| = 20\log\frac{1}{\sqrt{[1-(\frac{\omega}{\omega_n})^2]^2 + 4\zeta^2(\frac{\omega}{\omega_n})^2}} \qquad (10.3\text{-}7)$$

$$= \begin{cases} 20\log\dfrac{1}{(\frac{\omega}{\omega_n})^2} = 40\log\dfrac{1}{\frac{\omega}{\omega_n}}, & \text{for large } \omega, \\ 20\log\dfrac{1}{2\zeta}, & \omega = \omega_n, \\ 0, & \text{for small } \omega. \end{cases}$$

We also note two distinct differences between the asymptotic values in (10.3-7) versus those in (10.3-3). For large ω we note that in (10.3-7) the slope is *twice* its counterpart in (10.3-3), which means as we expect when frequency increases $F_5(s)$ goes to zero faster than $F_3(s)$.

The second distinction is that at the corner frequency ω_n, we have the peak value $20\log\frac{1}{2\zeta}$, which is a function of ζ. When $\zeta < 0.5$, this peak value is above the 0 dB line. Otherwise, the peak value is below this line.

In summary, for cases similar to $F_5(s)$ *we can draw an asymptotic magnitude plot similar to that for* $F_3(s)$ *except with twice the slope.*

Similarly, we may introduce the following case, which it may seem something new but in reality this is not really any different from what we have seen before.

(6) $F_6(s) = K\dfrac{s(T_1 s+1)\omega_n^2}{(T_2 s+1)(s^2+2\zeta\omega_n s+\omega_n^2)}, \quad K>0,$

whose magnitude can be written as

$$20\log|F_6(j\omega)| = 20\log K + 20\log|j\omega| + 20\log|j\omega T_1 + 1|$$

$$+ 20\log\frac{1}{|j\omega T_2 + 1|} + 20\log\left|\frac{\omega_n^2}{s^2 + 2\zeta\omega_n s + \omega_n^2}\right|_{s=j\omega}. \qquad (10.3\text{-}8)$$

Now, the message is clear that we can draw Bode diagrams of any complicated transfer function by judiciously combining the first four cases shown in Fig. 10.3-1, plus another function which we call $F_5(s)$. Although, $F_5(s)$ is similar to $F_3(s)$

Remark 10.3-1: Before presenting a numerical example to draw their corresponding Bode diagrams (Magnitude and angle), we emphasize that as shown in Fig. 10.3-1, in general the so-called corner frequencies play the central role in this process. These are the first items which must be specified before proceeding further. The locations of corner points in a Bode diagram indicate those *critical frequencies* in which the slope of each segment in the diagram changes. The rest is just adding up algebraically the vertical quantities, whose values are either positive or negative, in order to get their net sum for the overall transfer function.

Example 10.3-1: Draw Bode magnitude and phase diagrams for the following transfer function.

$$F(s) = \frac{16(s+1)(s+2)}{(s+3)(s^2+s+1)}. \qquad (10.3\text{-}9)$$

Solution: As stated before we must specify the corner frequencies before proceeding forward. In this case we have the transfer function in a factored form with two real zeros and three poles, which one is real and two are complex conjugate. Each factor is amenable with the reference cases we have studied in the preceding section. Thus, we begin to identify each component and proceed accordingly as follows.

$$F(s) = \frac{32(s+1)(\frac{1}{2}s+1)}{3(\frac{1}{3}s+1)(s^2+s+1)} \triangleq \frac{K(T_{41}s+1)(T_{42}s+1)\omega_n^2}{(T_3 s+1)(s^2+2\zeta\omega_n s+\omega_n^2)}. \qquad (10.3\text{-}10)$$

Each term in (10.3-10) is in the standard form of our reference cases and as such we have the following transfer functions to contend with:

$$F(s) \triangleq F_1(s)F_3(s)F_{41}(s)F_{42}(s)F_5(s). \qquad (10.3\text{-}11)$$

Each of the above component is processed in Tables 10.3-1 & 2 (for magnitude and angle), and is drawn **asymptotically** and respectively on Figs. 10.3-3 & 4 as follows.

Table 10.3-1. Asymptotic behavior of magnitudes in Example 10.3-1.

Components	Asymptotic behavior of magnitudes in dB versus $\omega > 0$
$F_1(s) = 32/3$	$20 \log \lvert F_1(j\omega)\rvert = 20.56$ dB for all ω
$F_3(s) \triangleq \dfrac{1}{T_3 s + 1} = \dfrac{1}{\frac{1}{3}s + 1}$	$20 \log \lvert F_3\rvert = 20 \log \dfrac{1}{\sqrt{1 + T_3^2\omega^2}} = \begin{cases} 20 \log \dfrac{3}{\omega}, & \text{large } \omega, \\ -3.01, & \omega = \dfrac{1}{T_3} = 3, \\ 0, & \text{for small } \omega. \end{cases}$
$F_{41}(s) \triangleq T_{41}s + 1 = s + 1$	$20 \log \lvert F_{41}\rvert = 20 \log \sqrt{1 + T_{41}^2\omega^2} = \begin{cases} 20 \log \omega, & \text{large } \omega, \\ 3.01, & \omega = \dfrac{1}{T_{41}} = 1, \\ 0, & \text{for small } \omega. \end{cases}$
$F_{42}(s) \triangleq T_{42}s + 1 = \frac{1}{2}s + 1$	$20 \log \lvert F_{42}\rvert = 20 \log \sqrt{1 + T_{42}^2\omega^2} = \begin{cases} 20 \log \dfrac{\omega}{2}, & \text{large } \omega, \\ 3.01, & \omega = \dfrac{1}{T_{42}} = 2, \\ 0, & \text{for small } \omega. \end{cases}$
$F_5(s) \triangleq \dfrac{\omega_n^2}{s^2 + 2\zeta\omega_n s + \omega_n^2} = \dfrac{1}{s^2 + s + 1}$, $\omega_n \triangleq 1,\ \zeta \triangleq ½$	$20 \log \lvert F_5\rvert = 20 \log \dfrac{1}{\sqrt{(1-\omega^2)^2 + \omega^2}} = \begin{cases} 40 \log \dfrac{1}{\omega}, & \text{large } \omega, \\ 0, & \omega = 1, \\ 0, & \text{for small } \omega. \end{cases}$

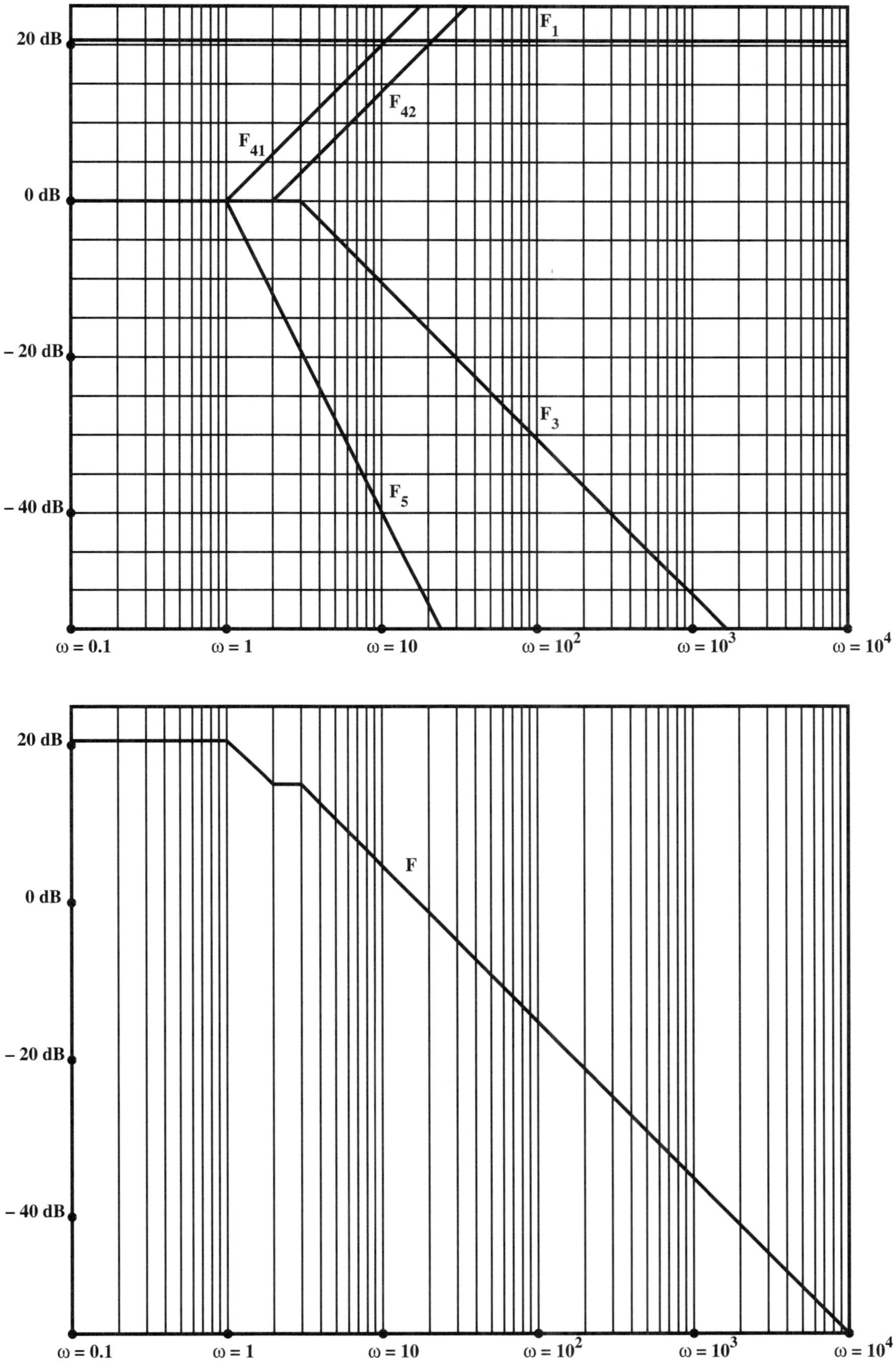

Fig. 10.3-2. Asymptotic magnitudes–top components, above total.

The computations for asymptotic angles are similar to the magnitudes, which the results are summarized in Table 10.3-2 and are drawn *asymptotically* on Fig. 10.3-3. Certainly in future applications these drawings should be performed in consultation with the many programs which are easily available in various computational packages. Likewise, using Matlab™ (Matrix Laboratory) (a trademark of Mathworks, Inc.) which is a widely available computational package, we have simulated all components of F(s) and this function itself as depicted in Figs. 10.3-4 to 9. Perhaps one point that should be made clear is all our results are for asymptotic behaviors, while simulated results are exact values. The accuracy of our figures will improve with more intermediate computations, particularly around corner frequencies.

Table 10.3-2. Asymptotic behavior of angles in Example 10.3-1.

Components	Asymptotic behavior of angles in degrees versus $\omega > 0$
$F_1(s) = 32/3$	$\theta_1(\omega) \equiv 0^\circ$ for all ω
$F_3(s)$	$\theta_3(\omega) \triangleq \angle \mathrm{tg}^{-1} \dfrac{\mathrm{Im}[F_3(j\omega)]}{\mathrm{Re}[F_3(j\omega)]} = -\mathrm{tg}^{-1}\dfrac{\omega}{3} = \begin{cases} -90^\circ, & \text{large } \omega, \\ -45^\circ, & \omega = \dfrac{1}{T_3} = 3, \\ 0^\circ, & \text{for small } \omega. \end{cases}$
$F_{41}(s)$	$\theta_{41}(\omega) \triangleq \angle \mathrm{tg}^{-1} \dfrac{\mathrm{Im}[F_{41}(j\omega)]}{\mathrm{Re}[F_{41}(j\omega)]} = \mathrm{tg}^{-1}\omega = \begin{cases} 90^\circ, & \text{large } \omega, \\ 45^\circ, & \omega = \dfrac{1}{T_{41}} = 1, \\ 0^\circ, & \text{for small } \omega. \end{cases}$
$F_{42}(s)$	$\theta_{42}(\omega) \triangleq \angle \mathrm{tg}^{-1} \dfrac{\mathrm{Im}[F_{42}(j\omega)]}{\mathrm{Re}[F_{42}(j\omega)]} = \mathrm{tg}^{-1}\dfrac{\omega}{2} = \begin{cases} 90^\circ, & \text{large } \omega, \\ 45^\circ, & \omega = \dfrac{1}{T_{42}} = 2, \\ 0^\circ, & \text{for small } \omega. \end{cases}$
$F_5(s)$	$\theta_5(\omega) \triangleq \angle \mathrm{tg}^{-1} \dfrac{\mathrm{Im}[F_5(j\omega)]}{\mathrm{Re}[F_5(j\omega)]} = -\mathrm{tg}^{-1}\dfrac{\omega}{1-\omega^2} = \begin{cases} -180^\circ, & \text{large } \omega, \\ -90^\circ, & \omega = 1, \\ 0^\circ, & \text{for small } \omega. \end{cases}$

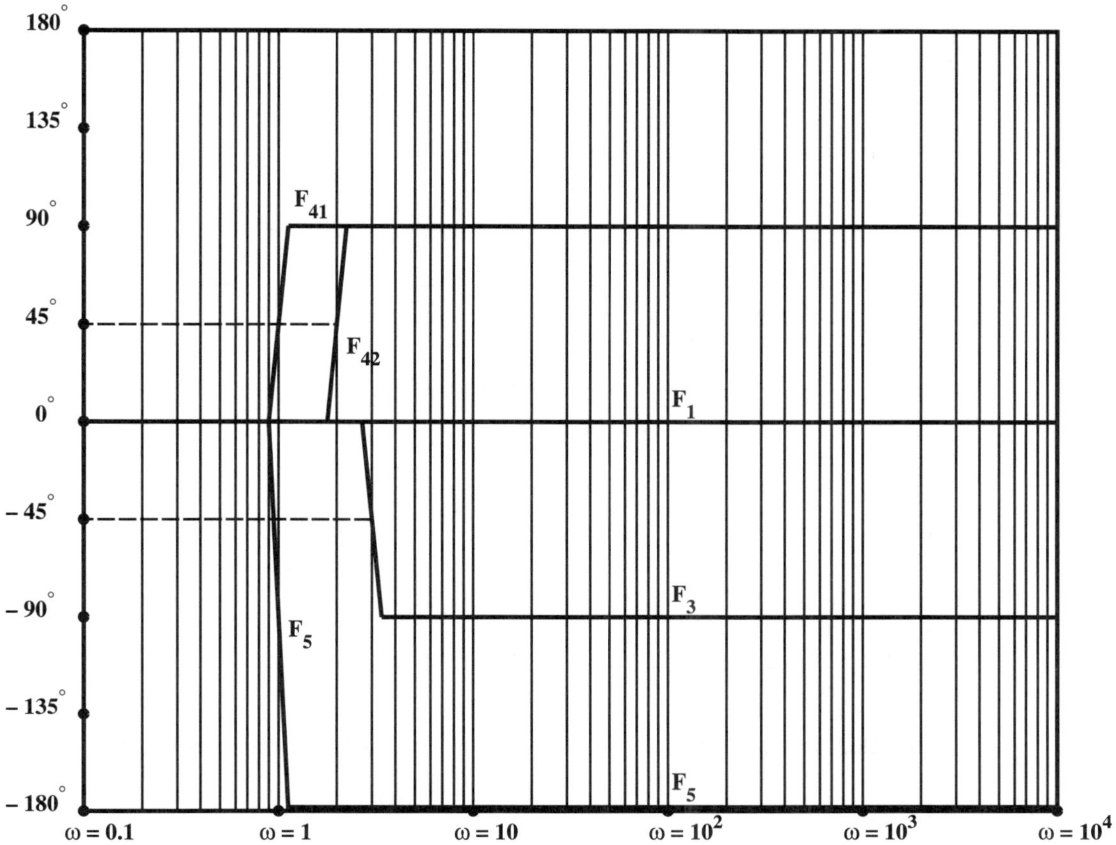

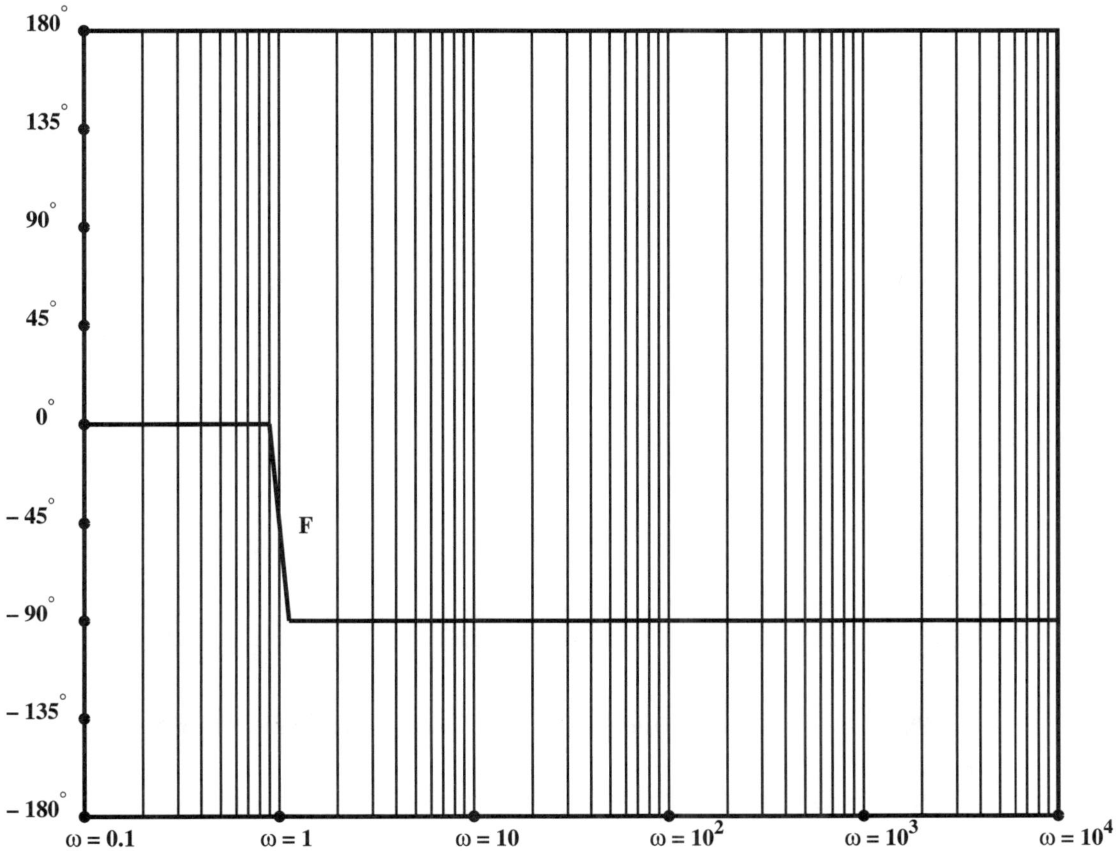

Fig. 10.3-3. Asymptotic angles–top components, above total.

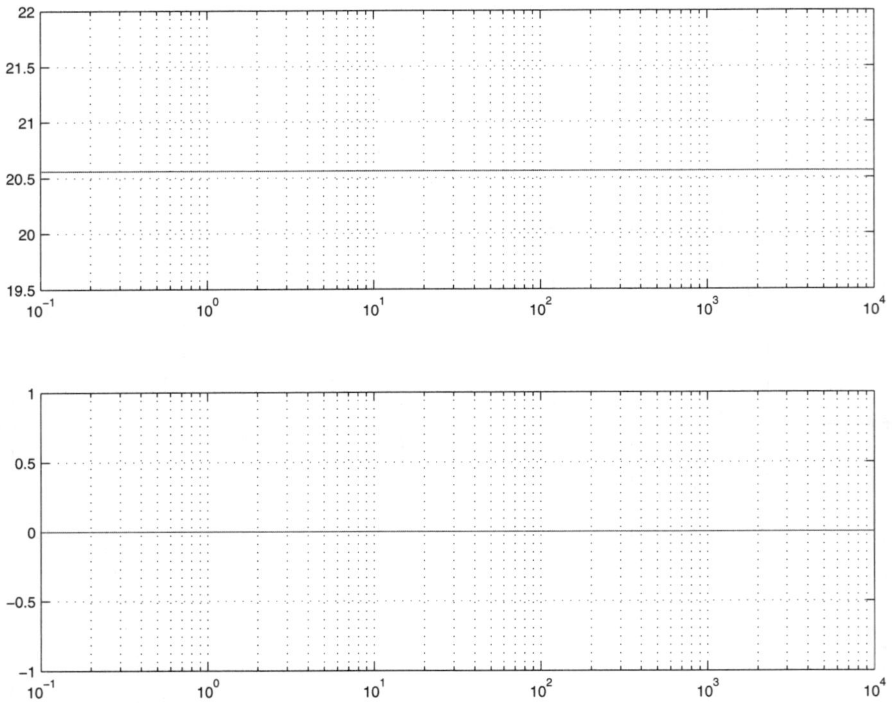

Fig. 10.3-4. Magnitude (top) in dB's and phase (above) in degrees of $F_1(s)$.

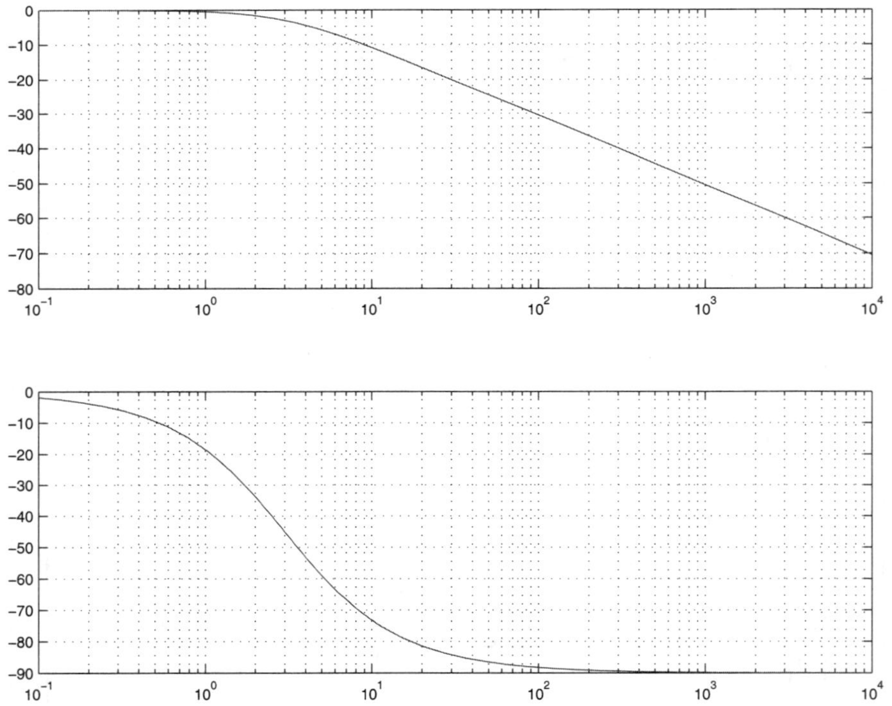

Fig. 10.3-5. Magnitude (top) in dB's and phase (above) in degrees of $F_3(s)$.

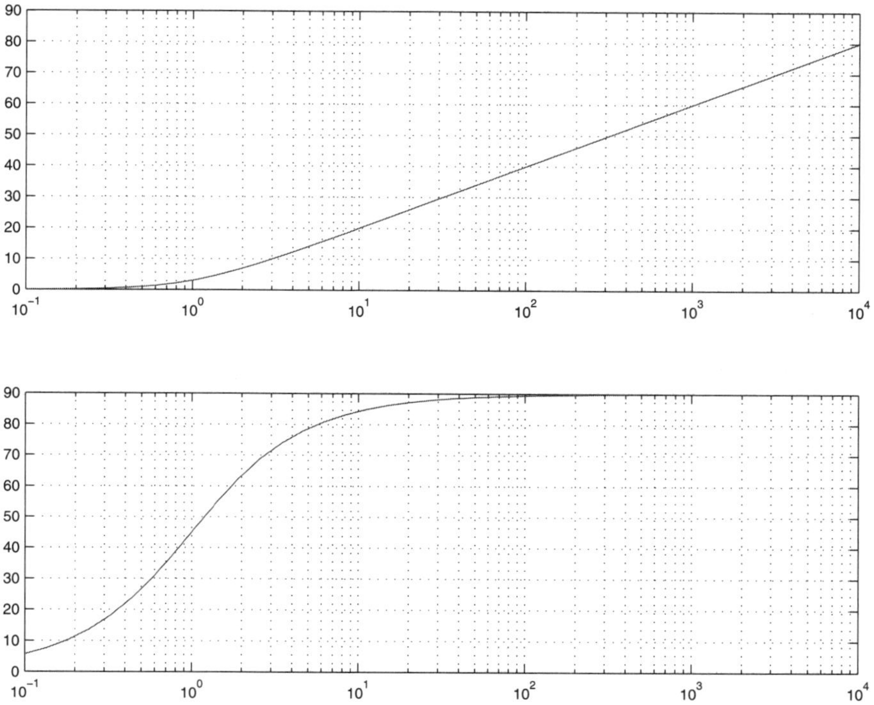

Fig. 10.3-6. Magnitude (top) in dB's and phase (above) in degrees of $F_{41}(s)$.

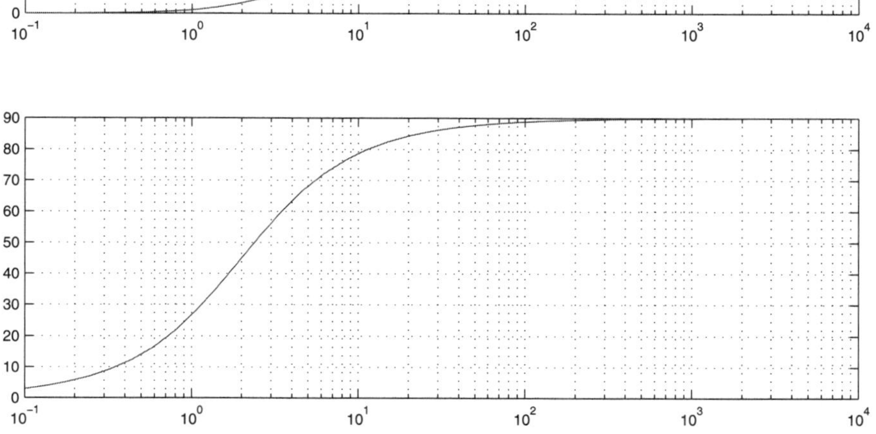

Fig. 10.3-7. Magnitude (top) in dB's and phase (above) in degrees of $F_{42}(s)$.

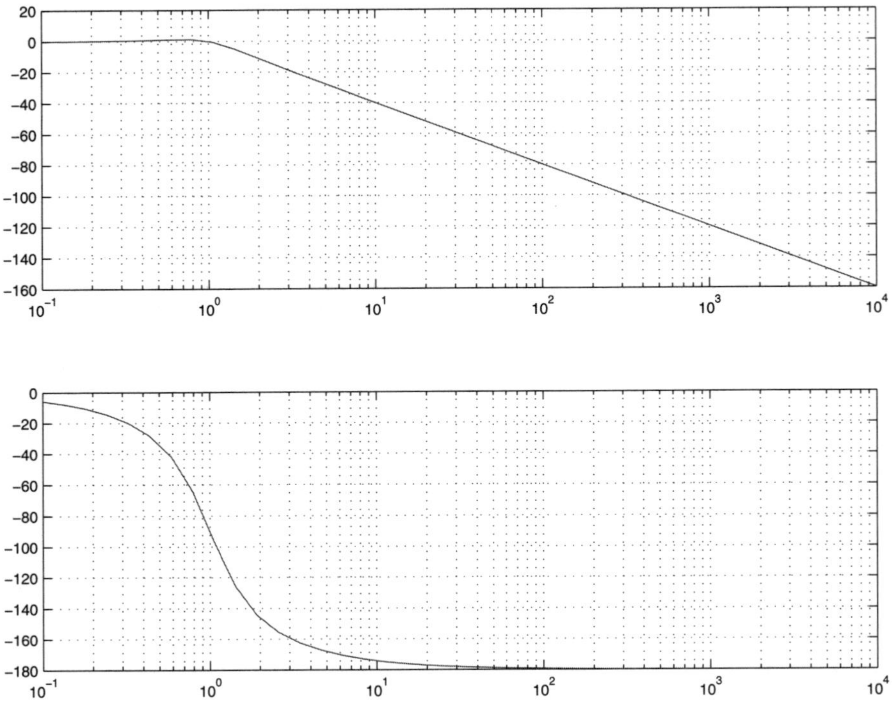

Fig. 10.3-8. Magnitude (top) in dB's and phase (above) in degrees of $F_5(s)$.

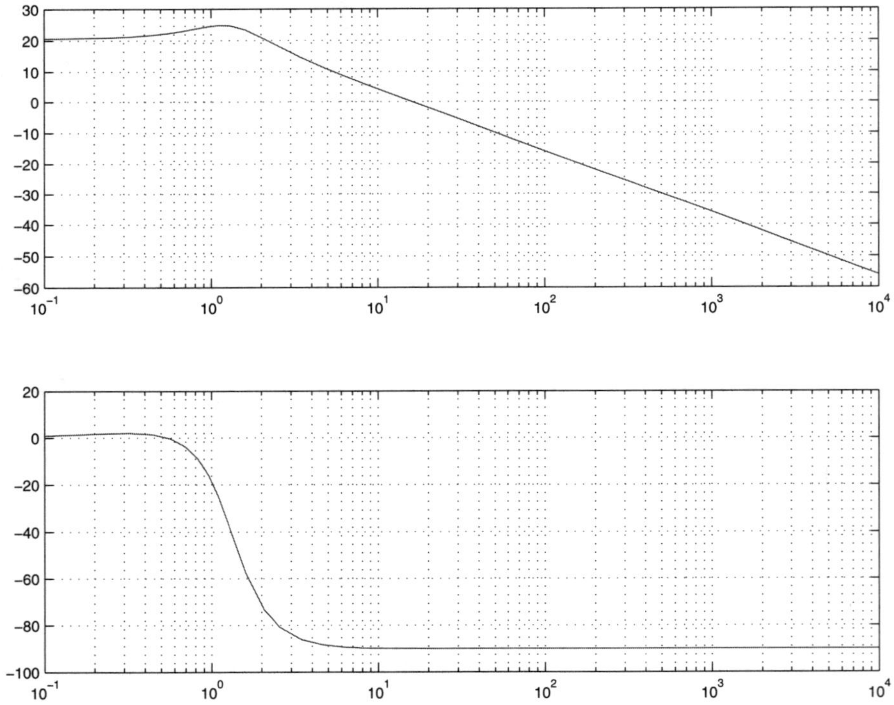

Fig. 10.3-9. Magnitude (top) in dB's and phase (above) in degrees of F(s).

10.4. A FEW REFERENCE TRANSFER FUNCTIONS

Before presenting our main motivation for studying frequency response, we review a few special transfer functions (also called filters, as we elaborate further in Section 10.5), whose input-output relationships are of interest. These transfer functions for a rational polynomial D(s) are shown by their magnitude versus frequency only for the sake of presentation, Figs. 10.4-1 to 5, below. Also, in these figures and for the sake of presentation, we normalize the output magnitude to one.

(1) $T_l(s) = \dfrac{K}{D(s)}$, $K > 0$ is a constant: *Low-Pass Filter*

This is called a low-pass filter, because when we draw $|T_l(j\omega)|$ versus ω, for frequencies below ω_l its frequency response or output is nonzero, while for frequencies above ω_l the response becomes identically zero, Fig. 10.4-1.

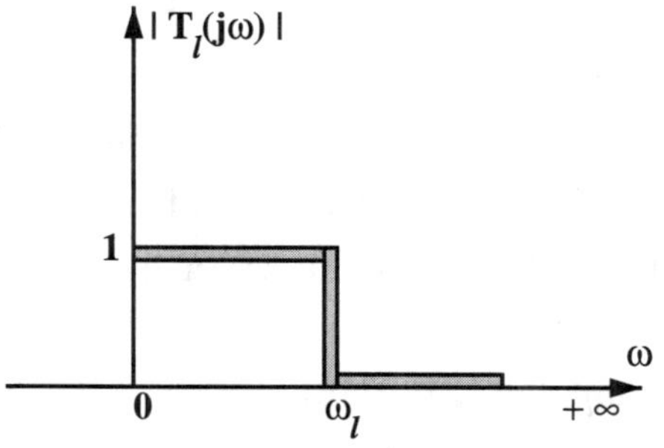

Fig. 10.4-1. An ideal low-pass filter.

(2) $T_h(s) = \dfrac{Ks^n}{D(s)}$, $K > 0$, $n = \deg D(s)$: *High-Pass Filter*

This is called a high-pass filter, because for any frequency above ω_h its frequency response or output is nonzero, while for frequencies below ω_h this response is identically zero, Fig. 10.4-2.

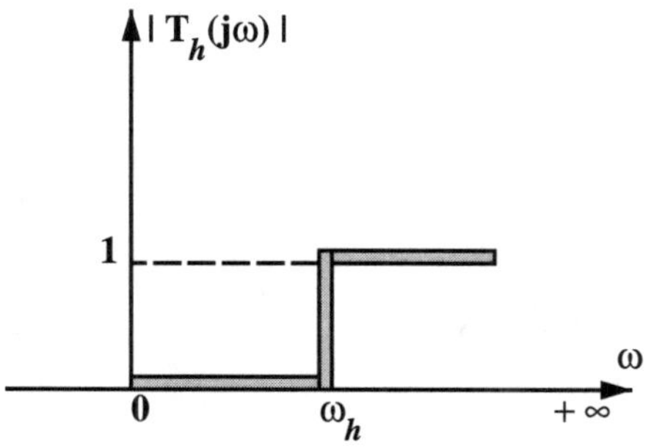

Fig. 10.4-2. An ideal high-pass filter.

(3) $T_{bp}(s) = \dfrac{Ks^{n/2}}{D(s)}$, $K > 0$, $n = \deg D(s)$: *Band-Pass Filter*

This filter is called a band-pass filter, because of its frequency response which only permits signals with frequencies between $(\omega_l - \omega_h)$ go through without attenuation, Fig. 10.4-3. This filter is built from cascading cases (1) and (2) in the above when $\omega_l > \omega_h$.

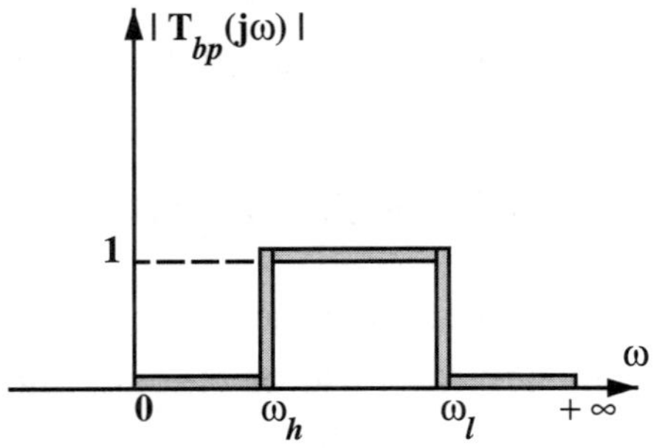

Fig. 10.4-3. An ideal band-pass filter.

In the above, we have already used the magic words of "cascading" and "attenuation". Let us pause for a moment and reflect on these words and their meaning. Referring to Fig. 10.4-4, when we cascade an ideal low-pass filter and an ideal high-pass filter, and depending on the values of ω_l or ω_h relative to each other we may get either Fig. 10.4-3 above, or Fig. 10.4-5 below. In Fig. 10.4-3, we keep signals whose frequencies are between $\omega_l - \omega_h > 0$, while in Fig. 10.4-5 we attenuate or discard signals whose frequencies are in the range of $\omega_h - \omega_l > 0$. But the real message is, we can shape the outcome of a circuit by modifying its input or output signal by passing it through a filter. We select this filter by adjusting its parameters, which in this case is its corner frequency. In this process we get a desired performance from a circuit whose output requires adjustment. This idea is something new to us, because so far we have done only analysis and with no input from us. Clearly, we now see that there are ample opportunities for us to design circuits to perform as we wish.

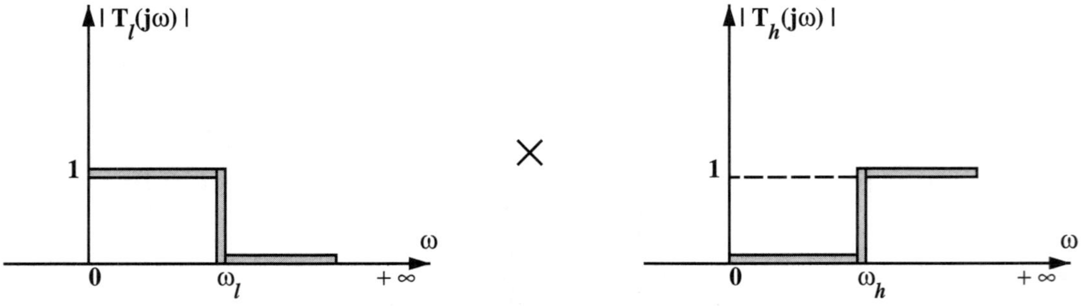

An ideal low-pass filter. **An ideal high-pass filter.**

Fig. 10.4-4. Cascading filter.

(4) *Band-Stop Filter*

It is now evident that a band-pass filter can be built by cascading a low-pass filter with a high-pass filter when $\omega_h > \omega_l$. This filter has, of course, application when we want the output to be zero for $\omega_l < \omega < \omega_h$, Fig. 10.4-5. In other words, we stop signals whose frequencies are between $\omega_h - \omega_l$.

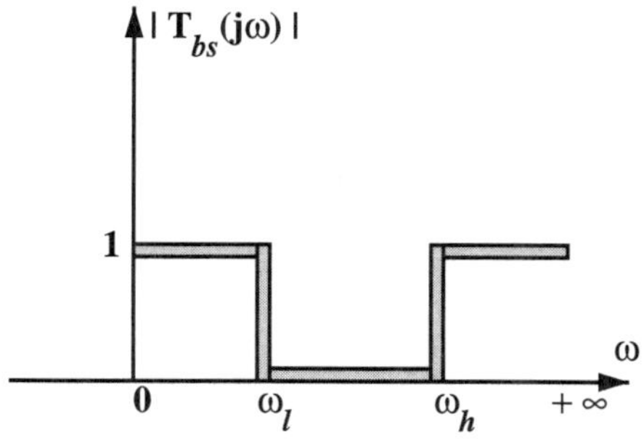

Fig. 10.4-5. An ideal band-stop filter.

Comment 10.4-1: In reality, the response curves of practical filters that we design or buy are not so sharp as those shown in the above four cases. In fact, instead of Fig. 10.4-1, *per se,* we have the following typical practical filter shown in Fig. 10.4-6. In a practical low-pass filter, instead of the maximum normalized output 1.00 at ω_l, we have 0.707 which is 3 dB below the maximum output level, and similarly the output does not vanish so sharp as what is shown in Fig. 10.4-1, but rather it vanishes gradually as indicated in Fig. 10.4-6. We may state similar imperfection for other three cases stated above. The main point is that in reality we do not have filters as shown in Figs. 10.4-1 to 5 with sharp corners. What we have instead are filters whose characteristic curves are gradual and *within* some limiting boundaries as shown in Fig. 10.4-7 for a low-pass filter. The more compressed the gutters become, the more closer we get to an ideal filter.

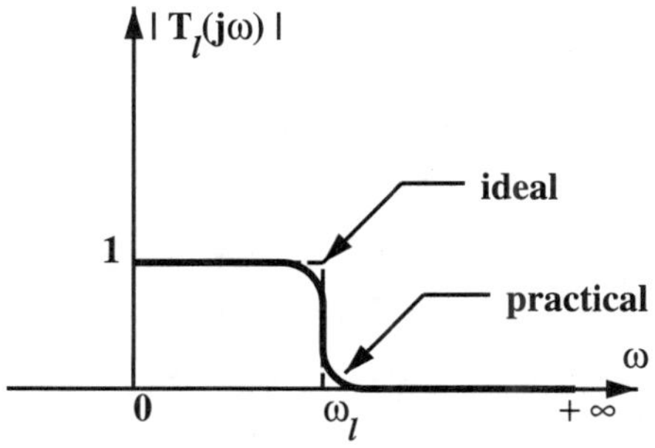

Fig. 10.4-6. A practical low-pass filter.

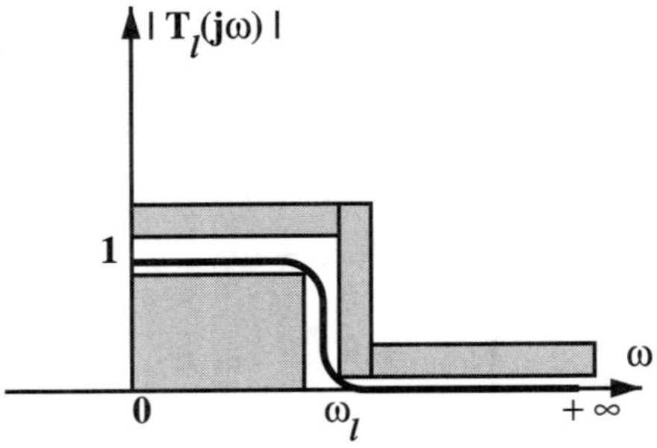

Fig. 10.4-7. Design boundaries for a low-pass filter.

10.5. WAVE SHAPING CIRCUITS

Perhaps the main reason for introducing frequency response is to acquaint ourselves with the very basic notion of wave shaping circuits and filter theory. It is just that natural to study some expressions of fundamentals in circuit or filter design and the thought behind it, that otherwise we may not get the right orientation in understanding the whole picture of a basic circuit course. For instance, although we have been emphasizing only circuit analysis thus far, we have accumulated enough background to prepare us for entertaining the basic features in circuits design, which can best be described in the frequency-domain settings. As such we appreciate the versatility of circuit analysis even more.

Here, there are two issues meriting further discussions. First, what is the meaning of wave shaping? From the above remarks we see that by an appropriate choice or choices of T(s), we can modify or shape the frequency response of a network to meet our desired performance. This opportunity enables us to create any output that we like from almost any signal by passing that signal through an appropriate or what we may call an optimal filter. For instance, when we want the output waveform vanishes between a particular frequency range, we pass that signal through a band-stop filter tuned to that particular frequency range of interest. We may cite similar examples using many other reference cases described in Section 10.4 and in § 11.7.5. Therefore the message that we really want to convey is that it is perfectly alright to dream about what we want, because with some effort we most likely are able to fulfill our dream using a plethora of design methodologies already in the market, or we may devise our own to complete the task. But first, let us dream.

Second, we have not shown any circuit that acts like a low-pass filter, *per se*, or for that matter any other reference cases which we have reviewed in Figs. 10.4-1 to 5.

So, if we have come up with a particular desired performance, then how do we actually build the corresponding circuit? Or what are some of the typical steps we must take to complete a design task? To answer this question requires several courses which their contents are beyond the scope of this textbook. There are so many parameters to evaluate which makes it very difficult to give one concrete methodology for design. If we can give one word to initiate the process however, then that word would be *experience.* Here, we rely on what we have learned in the analysis and begin the process for a new design from our previous design experience. To elucidate this point we give the following case study.

Let us choose a simple RC-circuit as our filter and study its wave shaping properties as follows. For the sake of argument, we assume that the network which generates our input signal for this filter is represented by its Thevenin equivalent circuit. Now, we pass the output of that circuit through our filter and before feeding it into another network for further study. Depending on how this filter (RC-network) has been designed (chosen) we may have different kinds of output at port $a'b'$, which is the input to the remaining network in our application, Fig. 10.5-1.

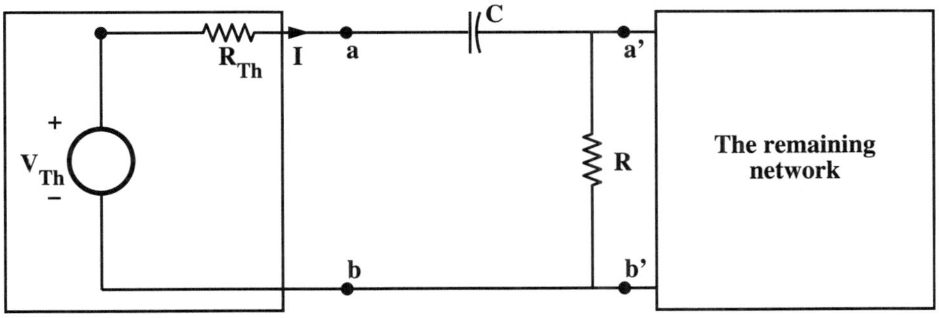

Fig. 10.5-1.

From the above remarks we know that in order to design a filter we must have the corresponding input-output relationship. In general, this relationship is expressed by the ratio of two rational polynomials in s, the Laplace transformation variable. It is easy to show that the input-output relationship for our filter in Fig. 10.5-1 is

$$V_{a'b'}(s) = T(s)V_{ab}(s), \tag{10.5-1}$$

with

$$T(s) \triangleq \frac{RCs}{1 + RCs}. \tag{10.5-2}$$

Clearly, this transfer function represents a high-pass filter. Depending on how do we select values for three parameters R, C and s, however, we may have some rather interesting properties for this circuit. For instance, suppose we have $|RCs| \ll 1$, resulting in $T(s) \approx RCs$, then this filter acts as a differentiator, which may not be desirable depending on the noise level in the system. This phenomenon can actually be created and tested in the laboratory. For example, if we apply a square wave input with the frequency of say 1 KHz with an amplitude of one volt, then we can show that by changing anyone of these three variables, R, C and ω ($s = j\omega$) we can change the shape of this square wave input rather drastically. Since a unit of variable capacitor is rather expensive, while a unit of variable resistor is widely available, we can keep the value of C for the purpose of this experiment constant and change only values of R. Or we may just change the frequency in steps of 100 (rad/sec) (or equivalently in Hz) to observe waveforms on the oscilloscope and record the cases where the RC time constant relative to the period of the input signal is such that the output across the resistor is: a series of spikes; undistorted square wave, and square waves with a slight tilt on the top. We can show that in each case the shape of the output depends on the selected values of these three parameters R, C and ω. Thus, it seems that even a simple RC-circuit can give us a wide spectrum of outputs when tuned properly.

Should we interchange R and C and study the output across the capacitor, then the transfer function becomes as follows.

$$T(s) = \frac{1}{1 + RCs}, \tag{10.5-3}$$

which is a low-pass filter. Here, for $|RCs| >> 1$, this circuit acts like an integrator. Contrary to the differentiator, an integrator is often a stable circuit.

Remark 10.5-1: Consider a transfer function $T(s) = N(s)/D(s)$, such that $N(s) = D(-s)$ with $D(s) = Re[D(s)] + jIm[D(s)]$. Then $|T(j\omega)| \equiv 1$ (all pass) for all ω, and the angle $\theta_T(j\omega) = -2tg^{-1}\{Im[D(j\omega)]/Re[D(j\omega)]\}$. Thus, if we use this filter in another circuit, then the magnitude of the overall circuit remains the same, but its overall phase changes. For example, let

$$T(s) = \frac{-s+1}{s+1} \Rightarrow |T(j\omega)| \equiv 1, \text{ and } \theta_T(j\omega) = -2tg^{-1}\omega. \tag{10.5-4}$$

Now, if we apply a sine-wave input to this transfer function, its magnitude at the output remains the same as the input, but its phase changes.

As another example, suppose we want to change the phase of a low-pass filter, say, slightly shift that to the right. Then we cascade that filter with a typical (10.5-4) shifted to the right. These filters are also called *"phase-correction filters."*

The crux of the matter is that even with our limited design experience we can come up with rather interesting circuits acting as filters, and use them to shape our signals as we please and that is a remarkable experience.

10.6. BASIC PROPERTIES of A SECOND-ORDER SYSTEM

The second-order system presented in the preceding discussions under $F_5(s)$ in Section 10.3, has vast applications in circuits and systems. Here, we review some of its essential properties. As a typical transfer function of the type $F_5(s)$, consider a system whose input R(s) and output C(s) are related as follows.

$$\frac{C(s)}{R(s)} = \frac{\omega_n^2}{s^2 + s\zeta\omega_n s + \omega_n^2} \quad . \tag{10.6-1}$$

Or equivalently in the time-domain representation we have

$$c(t) = \mathcal{L}^{-1}\{\omega_n^2 R(s)/(s^2 + 2\zeta\omega_n s + \omega_n^2)\}. \tag{10.6-2}$$

The above two equivalent expressions are general input and output relationships for this type of system. This particular transfer function represents what is called a low-pass filter and has received much attention in system theory and in network design. The higher-order transfer functions under certain conditions are decomposed into several blocks of second-order (as well as first-order) transfer functions of various types studied in Sections 10.3 & 4. Studying each of these blocks separately will enable us, in general, to gain the necessary information about the overall system transfer function. We thus analyze various possible forms of the above transfer function with respect to its two parameters as follows.

Consider (10.6-1), where ω_n is called the natural frequency and ζ is called the damping ratio. Knowing these two parameters is sufficient to analyze the complete behavior of this transfer function. Depending on the values of these parameters, the entire system behavior changes as is described next.

A. Underdamped Case ($0 < \zeta < 1$)

When $\zeta < 1$, and for $t \geq 0$, we have

$$\frac{C(s)}{R(s)} = \frac{\omega_n^2}{(s + \zeta\omega_n + j\omega_d)(s + \zeta\omega_n - j\omega_d)} \ , \tag{10.6-3}$$

where $\omega_d = \omega_n\sqrt{1-\zeta^2}$ is called the damped natural (damping) frequency. The unit-step response of this system (i.e., when $r(t) = 1(t)$ or $R(s) = 1/s$) becomes $C(s) = \omega_n^2/s(s + \zeta\omega_n + j\omega_d)(s + \zeta\omega_n - j\omega_d)$. From which we have

$$c(t) = 1 - (\cos \omega_d t + \frac{\zeta}{\sqrt{1-\zeta^2}} \sin \omega_d t) \exp(-\zeta \omega_n t)$$

$$= 1 - \frac{\exp(-\zeta \omega_n t)}{\sqrt{1-\zeta^2}} \sin (\omega_d t + \tan^{-1} \frac{\sqrt{1-\zeta^2}}{\zeta}). \tag{10.6-4}$$

It is easy to show that for this system e(t) = [r(t) – c(t)] → 0 as t → ∞. If $\zeta = 0$, then we have a pure sinusoidal output [oscillation] as follows: $c(t) = 1 - \cos \omega_n t$.

B. Critically Damped Case ($\zeta = 1$)

If the two poles of the transfer function are close to each other, which is the same as $\zeta \to 1$, then we have $C(s) = \omega_n^2 R(s)/(s + \omega_n)^2$. The unit-step response of this system becomes $c(t) = 1 - [\exp(-\omega_n t)](1 + \omega_n t)$.

C. Overdamped Case ($\zeta > 1$)

In this case, the unit-step response of the system becomes

$$C(s) = \omega_n/s(s + \zeta\omega_n + \omega_n\sqrt{\zeta^2 - 1})\,(s + \zeta\omega_n - \omega_n\sqrt{\zeta^2 - 1}). \tag{10.6-5}$$

From (10.6-5) and for $t \geq 0$, we have

$$c(t) = 1 + \frac{1}{2\sqrt{\zeta^2 - 1}\,(\zeta + \sqrt{\zeta^2 - 1})} \exp[-(\zeta + \sqrt{\zeta^2 - 1})\,\omega_n t]$$

$$- \frac{1}{2\sqrt{\zeta^2 - 1}\,(\zeta - \sqrt{\zeta^2 - 1})} \exp[-(\zeta - \sqrt{\zeta^2 - 1})\omega_n t]. \tag{10.6-6}$$

Clearly c(t) consists of two decaying exponential terms. When $\zeta >> 1$, one of these two exponential terms goes to zero much faster than the other one. Thus $c(t) = 1 - \alpha \exp(-p_1 t)$, where p_1 is the *dominant* (closer to the imaginary-axis) pole of the system and α is a constant.

10.6.1. Second-Order Passive Networks

As an example to the general results presented in the above part of this section, consider the simple RLC network shown in Fig. 10.6-1.

Fig. 10.6-1.

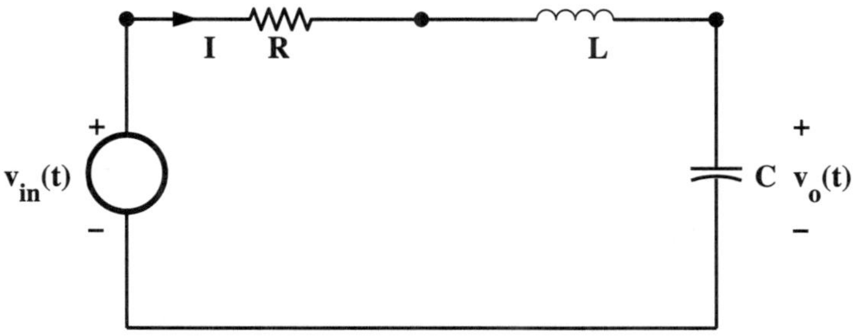

It is easy to show that in the circuit of Fig. 10.6-1, we have

$$\frac{V_o(s)}{V_{in}(s)} = \frac{1/sC}{R + sL + 1/sC}$$

$$= \frac{1/LC}{s^2 + (R/L)s + 1/LC} \,. \tag{10.6-7}$$

Comparing (10.6-7) with (10.6-1) yields the natural frequency and the damping ratio as follows.

$$\omega_n = \frac{1}{\sqrt{LC}} \,, \tag{11.6-8}$$

$$\zeta = \frac{R}{2}\sqrt{\frac{C}{L}} \,. \tag{11.6-9}$$

Thus by changing these two parameters, or equivalently changing the elements, we can study $v_o(t)$ and observe all the cases with regard to the system being underdamped, critically damped, or overdamped.

On the other hand, if we have no idea about the values of at most two parameters – say, R and L, then we can measure and/or record $v_o(t)$ and by studying its transient response behavior we can identify these two parameters, although we have not discussed such issues in this textbook. For example, we may wish to have a particular behavior in the system, and correspondingly we may adjust any one of these parameters R, L or C to get such a desired performance. It is also interesting to note that for the circuit in Fig. 10.6-1, at $s = j\omega_n$, we have a real constant

$$|V_o(j\omega_n)/V_{in}(j\omega_n)| = 1/2\zeta \,. \tag{10.6-10}$$

10.7. PARALLEL RESONANCE

There is an important phenomenon in RLC circuits which we have not formally explored, although it has been reviewed in physics that is the prerequisite for this course. Consider the parallel RLC circuit shown in Fig. 10.7-1.

Fig. 10.7-1.

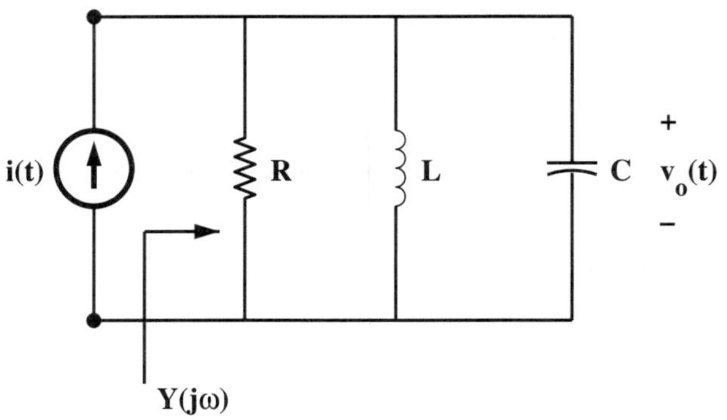

When we study the input admittance of this circuit versus frequency, we realize that by adjusting the operational frequency, we may select a particular frequency under which this circuit acts as a purely resistive one. This phenomenon known as resonance may not always be desirable, but is not avoidable either. In other words, in any RLC circuit and between one L and one C, the possibility of running into a particular frequency which yields this phenomenon exist. Again, this frequency tuning may only occur between one pair of L and C, and at one specific frequency. Also, before proceeding any further we note that the circuit in Fig. 10.7-1 is not the only configuration which yields this phenomenon, but this is the classic model to describe this concept.

To expand on this idea, we reconsider the circuit in Fig. 10.7-1, where we have

$$Y(j\omega) = \frac{1}{R} + j(\omega C - \frac{1}{\omega L}). \tag{10.7-1}$$

This admittance is purely resistive if and only if the imaginary part vanishes, i.e., at $\omega \equiv \omega_o$ such that

$$\omega_o C - \frac{1}{\omega_o L} = 0 \ \Rightarrow \ \omega_o = \frac{1}{\sqrt{LC}}, \tag{10.7-2}$$

we have a purely resistive load. This condition is called resonance. At this *resonant* frequency, the input current and the output voltage are in phase and the amplitude of the admittance (response) is maximum. Another way of looking at this phenomenon is, at resonance we have a rather large response at a very narrow frequency band.

If we look back at (10.7-1) in terms of its Laplace transformation, we have

$$Y(s) = \frac{1}{R} + sC + \frac{1}{sL}$$

$$= \frac{C(s+\alpha-j\omega_d)(s+\alpha+j\omega_d)}{s}, \qquad (10.7\text{-}3)$$

where

$$\alpha = \frac{1}{2RC} \qquad \omega_d = \sqrt{\omega_o^2 - \alpha^2}, \qquad (10.7\text{-}4)$$

with ω_o as in (10.7-2). In (10.7-4), we note that the damping frequency ω_d is always smaller than the resonant frequency ω_o.

Going back to the output voltage $V_o(j\omega)$ (here, we change correctly the notation to an upper case letter) and draw it versus ω (or we may generate the response experimentally as the case may be), we have a schematic diagram curve as shown in Fig. 10.7-2. If this curve were generated analytically, then we can calculate $d|V_o(j\omega)|/d\omega$ and set that identically equal to zero to find the frequency which corresponds to the maximum magnitude which is ω_o. The *width* of the resonant response curve is defined as the two half-power frequency ω_1 and ω_2, which their corresponding magnitudes are $0.707 (= 1/\sqrt{2})$ (or 3 dB) below the maximum value of "1" for $|V_o|$.

If the curve in Fig. 10.7-2 becomes more like a funnel, then the difference of $\omega_2 - \omega_1$ becomes small. This observation leads to the introduction of a new parameter called *quality factor* shown by Q.

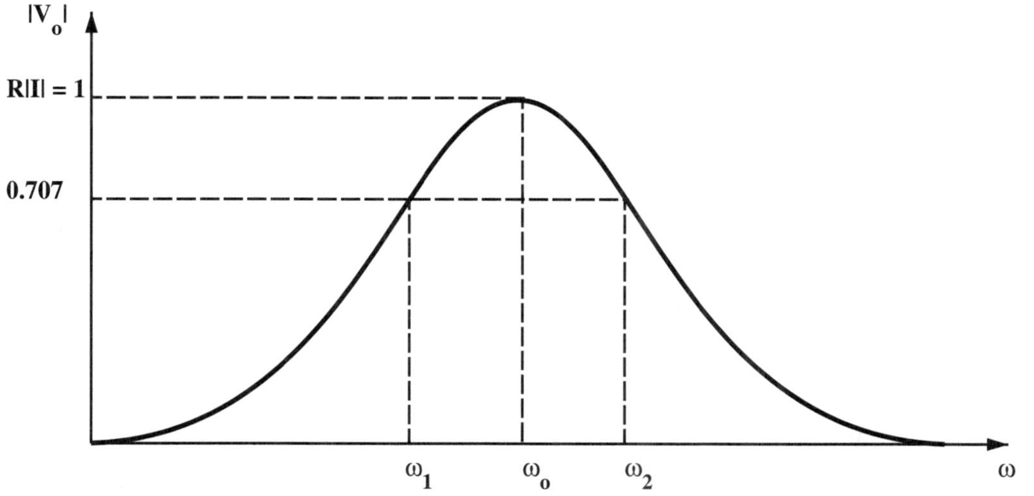

Fig. 10.7-2. Output voltage versus frequency in a parallel RLC circuit.

The sharpness of the response of any resonant circuit is determined by the maximum amount of energy stored per energy that is lost during one complete period of the response. That means, in general Q is defined as follows.

$$Q = 2\pi \frac{\text{maximum energy stored}}{\text{total energy lost per period}}. \quad (10.7\text{-}5)$$

From physics, the instantaneous energy stored in a capacitor is $W_C = (½)Cv_C^2$, and that in an inductor is $W_L = (½)Li_L^2$. Thus to compute Q for a resonance parallel RLC network at the resonant frequency of ω_o given in (10.7-2), we have

$$Q_o = 2\pi \frac{(W_C + W_L)_{max}}{P_R T}. \quad (10.7\text{-}6)$$

In circuit of Fig. 10.7-1 and for $i(t) = I_m \cos \omega_o t$, we have

$$v_o(t) = RI_m \cos \omega_o t \Rightarrow W_C = (½)Cv_o^2 = (½)I_m^2R^2C \cos^2 \omega_o t. \quad (10.7\text{-}7)$$

Similarly,

$$W_L = (½)Li_L^2 = (½)L(\frac{1}{L}\int_0^t v_o(t)dt)^2 = (½)I_m^2R^2C \sin^2 \omega_o t. \quad (10.7\text{-}8)$$

The total instantaneous energy is $W_C + W_L = (½)I_m^2R^2C$. Since,

$$P_R = (½)I_m^2R \Rightarrow P_RT = (½)I_m^2RT. \quad (10.7\text{-}9)$$

Thus,

$$Q_o = 2\pi \frac{I_m^2R^2C/2}{I_m^2RT/2} = \frac{2\pi}{T}RC = \omega_o RC. \quad (10.7\text{-}10)$$

Recalling ω_o from (10.7-2), we have

$$Q_o = \omega_o RC = R\sqrt{\frac{C}{L}}. \quad (10.7\text{-}11)$$

As expected, Q_o is a dimensionless constant that is a function of three elements in a parallel resonant circuit.

A circuit may have many triple $\{R, L, C\}$, but each exhibits its own resonant property, and we may have a set of $\{\omega_o\}$ for different segments of the circuit.

Going back to Fig. 10.7-2, the height of the maximum depends on R, but the width or steepness of the curve shown depends on L and C. We may relate these quantities to the circuit parameters. For instance, we now have

$$\alpha = \frac{1}{2RC} = \frac{\omega_o}{2Q_o}. \tag{10.7-12}$$

Also,

$$\omega_d = \sqrt{\omega_o^2 - \alpha^2} = \omega_o\sqrt{1 - (\frac{1}{2Q_o})^2}. \tag{10.7-13}$$

If we look back at $Y(j\omega)$ in the circuit of Fig. 10.7-1, then we have

$$\begin{aligned} Y(j\omega) &= \frac{1}{R} + j(\omega C - \frac{1}{\omega L}) \\ &= \frac{1}{R}[1 + jQ_o(\frac{\omega}{\omega_o} - \frac{\omega_o}{\omega})]. \end{aligned} \tag{10.7-14}$$

Since at the resonant frequency we have $|Y(j\omega_o)| = \frac{1}{R}$, the only way that $|Y(j\omega)| = \frac{\sqrt{2}}{R}$ is when

$$Q_o(\frac{\omega}{\omega_o} - \frac{\omega_o}{\omega}) = \pm 1. \tag{10.7-15}$$

Note that the Fig. 10.7-2 is drawn in terms of V_o, while we now are looking at the admittance $Y(j\omega)$ whose inverse is proportion to V_o. If we let

$$Q_o(\frac{\omega_2}{\omega_o} - \frac{\omega_o}{\omega_2}) = 1 \Rightarrow \omega_2 = \omega_o[\sqrt{1 + (\frac{1}{2Q_o})^2} + \frac{1}{2Q_o}], \tag{10.7-16}$$

and if we let

$$Q_o(\frac{\omega_1}{\omega_o} - \frac{\omega_o}{\omega_1}) = -1 \Rightarrow \omega_1 = \omega_o[\sqrt{1 + (\frac{1}{2Q_o})^2} - \frac{1}{2Q_o}]. \tag{10.7-17}$$

The last parameter of interest is the bandwidth (BW), which for the resonant circuit in Figs. 10.7-1 & 2 is defined as $\omega_2 - \omega_1$. Using (10.7-16) and (10.7-17), we have

$$BW = \omega_2 - \omega_1 = \frac{\omega_o}{Q_o}. \tag{10.7-18}$$

The BW represents the width of the response curve, although the response is extended over the infinite horizon. Thus, for a resonant circuit with high Q_o (quality factor), we have a narrow BW and vice versa. The narrow bandwidth means more *frequency selectivity,* and more specialized utilization of the filter design than otherwise.

10.8. SCALING

Consider the circuit in Fig. 10.7-1, and its corresponding frequency response depicted in Fig. 10.7-2, where we have selected the maximum value of $|V_o|$ to be "1". Now, suppose we draw another frequency response associated with this circuit, namely its input impedance Z(s), and show that in Fig. 10.8-1. Clearly, this response that is drawn for one set of element values is proportion to the output voltage in Fig. 10.7-2.

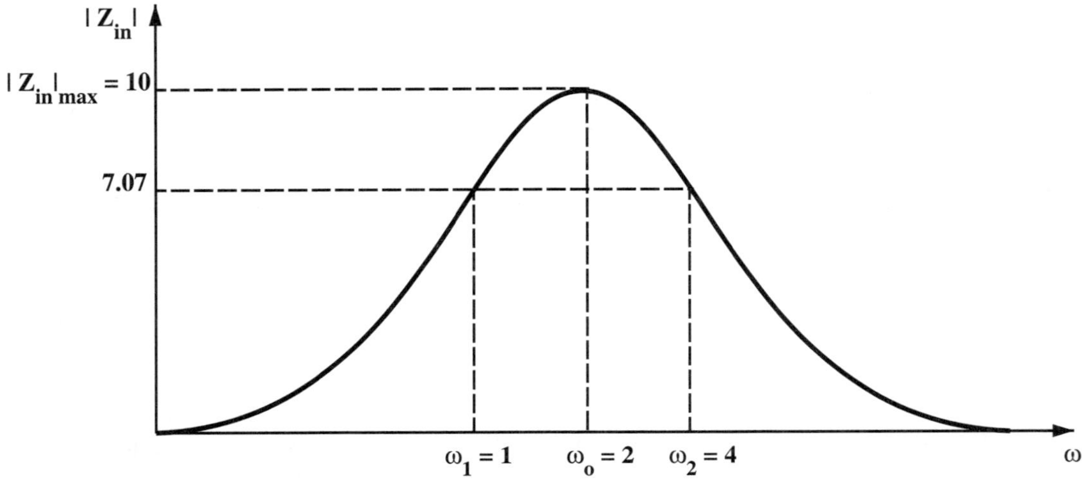

Fig. 10.8-1. Input impedance versus frequency in a parallel RLC circuit.

Referring to Fig. 10.8-2, or any similar frequency response associated with a circuit, we may wish to maintain its basic characteristics, namely its shape and its frequency distribution. But we may not be satisfied with its scale. In other words, we may wish that the actual maximum value were larger or smaller than its current value of 10, which incidentally this value is selected only for the sake of discussion. Similarly, we may wish that its corresponding frequency distribution (ω_1, ω_o and ω_2) were wider or more compressed than their current values. Clearly, as the analysis in the preceding section shows these values are all functions of the actual circuit elements as well as the operational frequency for the circuit. Is there anything we can do to change these characteristic values without changing the shape of the graph? Before, we answer this question, we recall that throughout this textbook we have used circuits whose element values are chosen solely to simplify the numerical analysis, rather than showing concern for the actual manufacturing of the circuit with all its physical constraints. We have acknowledged this concern and our rationale in Comment 5.8-1. Now, we want to address the real underlying issues constructively.

Putting all our concerns regarding the change of magnitude or the range of the operational frequency together, with or without changing the shape of the response graph, yields a multi-faceted design problem. Some of the adjustment to a typical graph as in Fig. 10.7-2, or Fig. 10.8-1, *per se,* whether is expanding its frequency range or is funneling its maximum is addressed in problems of Section 10.9. But an issue that has not been addressed for a general circuit pertaining the above matters is called *scaling,* and that concept is the subject of this section.

Suppose we want that the actual maximum magnitude of 10, as shown in Fig. 10.8-1, were 75. However, we are satisfied with the shape of this impedance as a function of frequency. In other words, we still want the maximum to occur at ω_o. How do we scale up this response function to a new maximum value for its magnitude without changing the shape of its graph and its frequency distribution? In other words, at $\omega_o = 2$ rad/sec, instead of the maximum magnitude of 10, we now have the new maximum 75, which is $K_m = 75/10$ times larger than before, while the entire frequency distribution still remains the same. That means as shown in Fig. 10.8-2, not only ω_o remains the same, but also do ω_1, ω_2, and the BW. This concept is called *magnitude scaling,* which refers to the process of changing the magnitude for an impedance of a one-port network by a factor of K_m, while its frequency distribution remains the same. Again, let us be explicit that this change is done only by scaling element values and not changing the actual circuit configuration. Incidentally, this K_m can be both larger or smaller than one, resulting in scaling *up* or *down* for the underlying circuit.

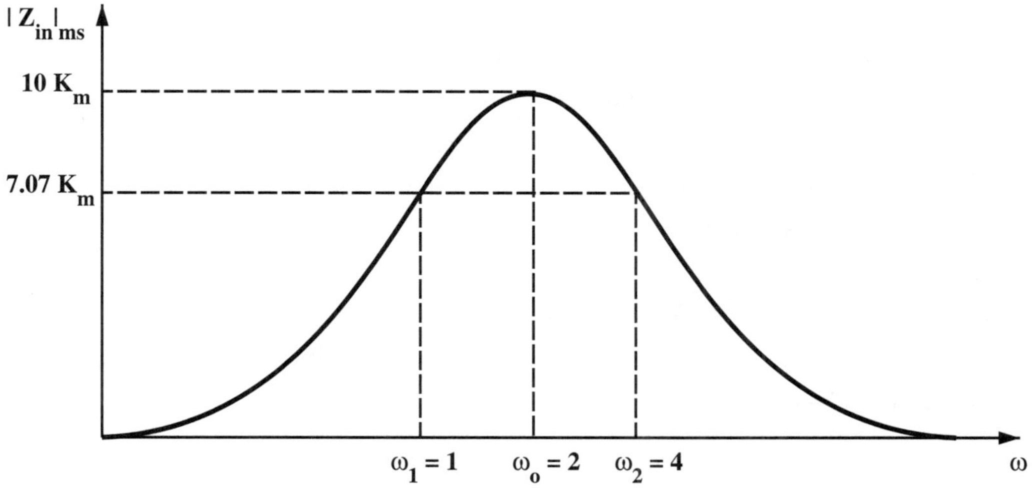

Fig. 10.8-2. Magnitude-scaled input impedance versus frequency in a parallel RLC circuit.

To implement such a magnitude scaling, we start with elements beginning with resistors. To scale a resistor up or down, all we need to do is to let $R \rightarrow K_m R$. Similarly, for an inductor, we let $L \rightarrow K_m L$. But for a capacitor, because Z_C is proportion to the inverse of C, we let $C \rightarrow C/K_m$.

Now, suppose we are not satisfied with the actual frequency distribution, while correspondingly we wish to maintain the shape of the frequency response. For instance, we want the maximum to occur at $K_f \times \omega_o$ instead of ω_o, and similarly both ω_1 and ω_2 must also be scaled by the same K_f to their new values of $K_f\omega_1$ and $K_f\omega_2$, as shown in Fig. 10.8-3. How do we accomplish this task?

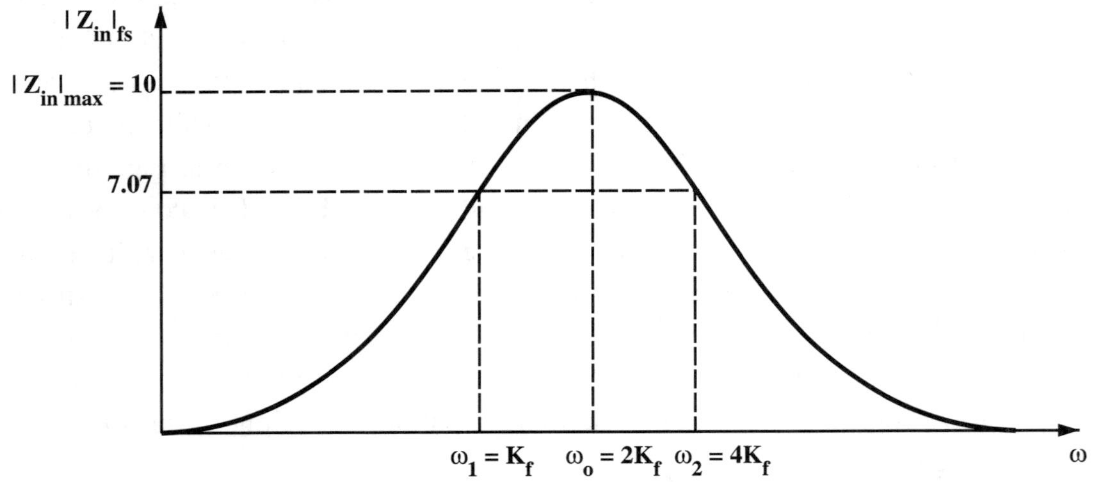

Fig. 10.8-3. Frequency-scaled input impedance versus frequency in a parallel RLC circuit.

Referring to Fig. 10.8-3, it is imperative to remember that we want to make this scaling only by changing the elements' values and not the frequency of the applied signal. How do we reach this goal? The answer is to perform a *frequency scaling,* which is as follows. We begin with each element and develop a procedure to make the overall changes necessary to accomplish our task. Starting with resistors, these are not frequency dependent elements, so they are not going to change in this process. Thus, we let each $R \rightarrow R$. While, impedances of both inductor and capacitor change as a function of frequency. Here, to accommodate this frequency scaling we let $L \rightarrow L/K_f$ and $C \rightarrow C/K_f$.

What if we want to perform both magnitude and frequency scaling simultaneously, then what is the procedure? The answer is we apply successively K_m and K_f as described above to every element accordingly. This procedure is summarized in Table 10.8-1.

Table 10.8-1. Magnitude and frequency scalings.

Element	Magnitude scaling	Frequency scaling	Both scalings
R	$K_m R$	R	$K_m R$
L	$K_m L$	L / K_f	$(K_m / K_f) L$
C	C / K_m	C / K_f	$C / K_m K_f$

Comment 10.8-1: In many cases, we start with an expression for the input impedance Z(s) of a one-port network and without its actual circuit. So, we cannot perform the above scaling directly to element by element. In that case, the procedure to modify or scale an impedance function is as follows. Multiply Z(s) by K_m to get $Z_{ms}(s) = K_m Z(s)$ as the magnitude-scaled transfer function. Then replace each s in $Z_{ms}(s)$ with (s / K_f) to get the new impedance $Z_{ms}^{fs}(s) = K_m Z(s / K_f)$, which corresponds to both magnitude- and frequency-scaled transfer function. We then construct a new circuit based on this new transfer function, which such a discussion is outside the scope of this textbook.

10.9. PROBLEM SET

Bode Diagrams

10.9.1. Draw Bode diagrams for the following transfer functions on Figs. 10.9-1 & 2.

$$F(s) = \frac{1}{s(2s+1)} . \qquad (10.9\text{-}1)$$

$$F(s) = \frac{s+1}{(s+2)(s+3)} . \qquad (10.9\text{-}2)$$

$$F(s) = \frac{1}{s(s^2+s+1)} . \qquad (10.9\text{-}3)$$

$$F(s) = \frac{1}{s^2(10s + 1)}. \tag{10.9-4}$$

$$F(s) = \frac{s^2 + s + 2}{s^2(s + 10)}. \tag{10.9-5}$$

$$F(s) = \frac{18(2s + 1)(3s + 1)(4s + 1)}{s^3(10s + 1)(5s^2 + s + 1)}. \tag{10.9-6}$$

$$F(s) = \frac{(s + 2)(s + 3)(s^2 + s + 1)}{s(s + 4)(s^2 + s + 10)}. \tag{10.9-7}$$

$$F(s) = \frac{(s^2 + 1)(s^3 + 3s^2 + 1)e^{-s}}{s^3(s + 10)(s^3 + s + 1)}. \tag{10.9-8}$$

Comment 10.9-1: In (10.9-8), we have both non-factored new expressions as well as an exponential term which is called a *delay.* In many practical systems we have this sort of delays which make the system "not linear", and "not linear" is different from "nonlinear". The system is definitely not lumped and we cannot classify it under what we have studied in this textbook. But on a positive note, this example shows both the limitation as well as the opportunity in utilization of Bode's diagrams. Here, computing procedures can be used to complete this assignment. However, always a hybrid approach when computing procedures are used only partially is the most desirable way to *learn* the subtle and intricate parts of an engineering topic.

Wave Shaping Circuits

10.9.2. In the circuit of Fig. 10.5-1, let the input signal be $v_{ab}(t) = \sin t$, find the parameters of the RC shaping network such that the output signal $v_{a'b'}$ be the scaled replica of the input function, if possible. Then change the element values such that the output becomes the differential of the input signal, and explain how.

10.9.3. In the circuit of Fig. 10.5-1, change the place of two elements R and C. Let the input signal be $v_{ab}(t) = \sin t$, find the parameters of the RC shaping network such that the output signal $v_{a'b'}$ be the scaled replica of the input function, if possible. Then change the element values of this shaping filter such that the output becomes the integral of the input signal, and explain how.

10.9.4. Let us cascade a low-pass filter $T_1(s) = 1/(s + 1)$ with an all-pass filter $T_2(s) = (-s + 1)/(s + 1)$. Use a sinusoidal waveform like $\sin t$ as an input to $T_1(s)$. Draw the output of $T_1(s)$ and apply that to $T_2(s)$. Draw the output of $T_2(s)$ and show the phase shift with respect to its input.

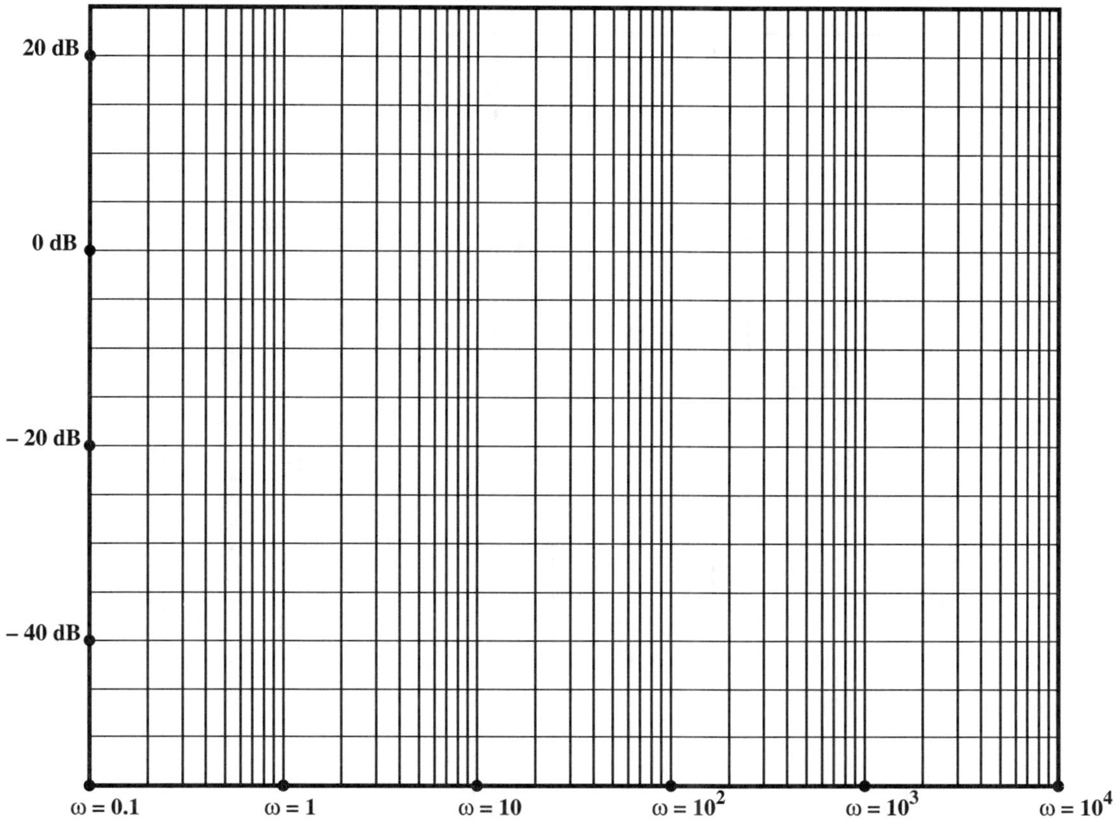

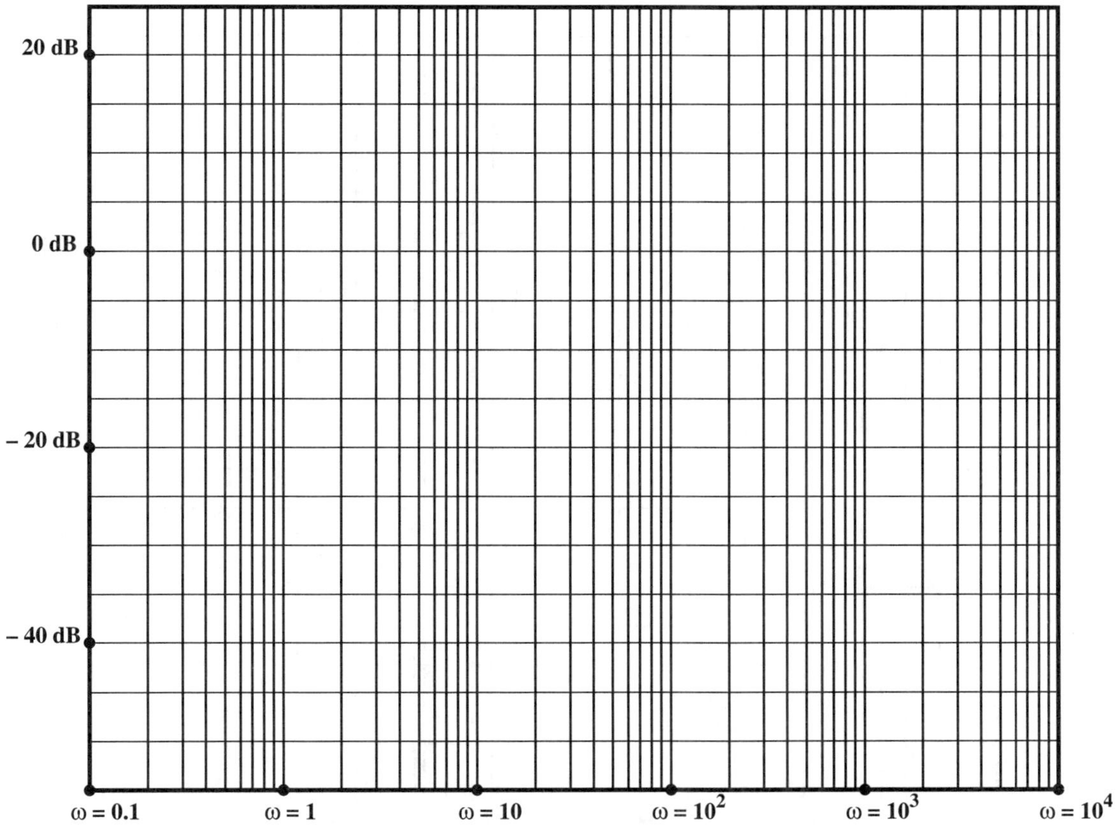

Fig. 10.9-1. Asymptotic magnitudes–top components, above total.

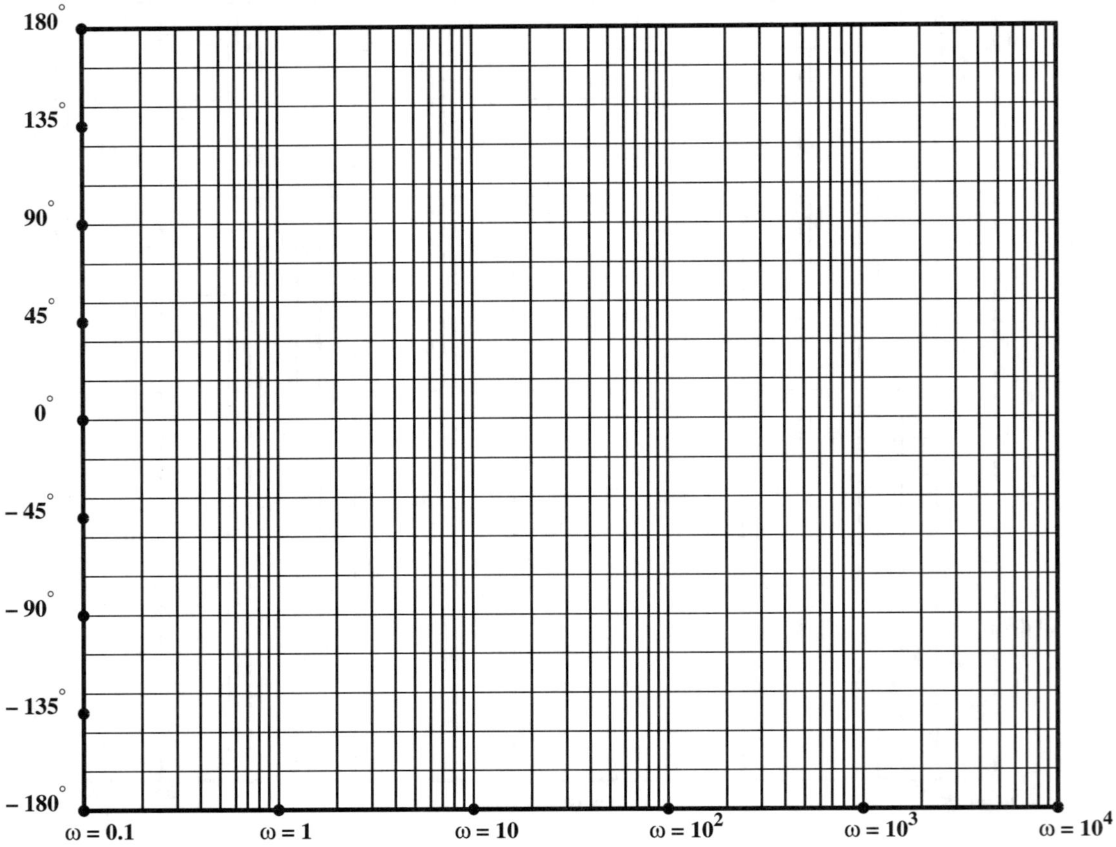

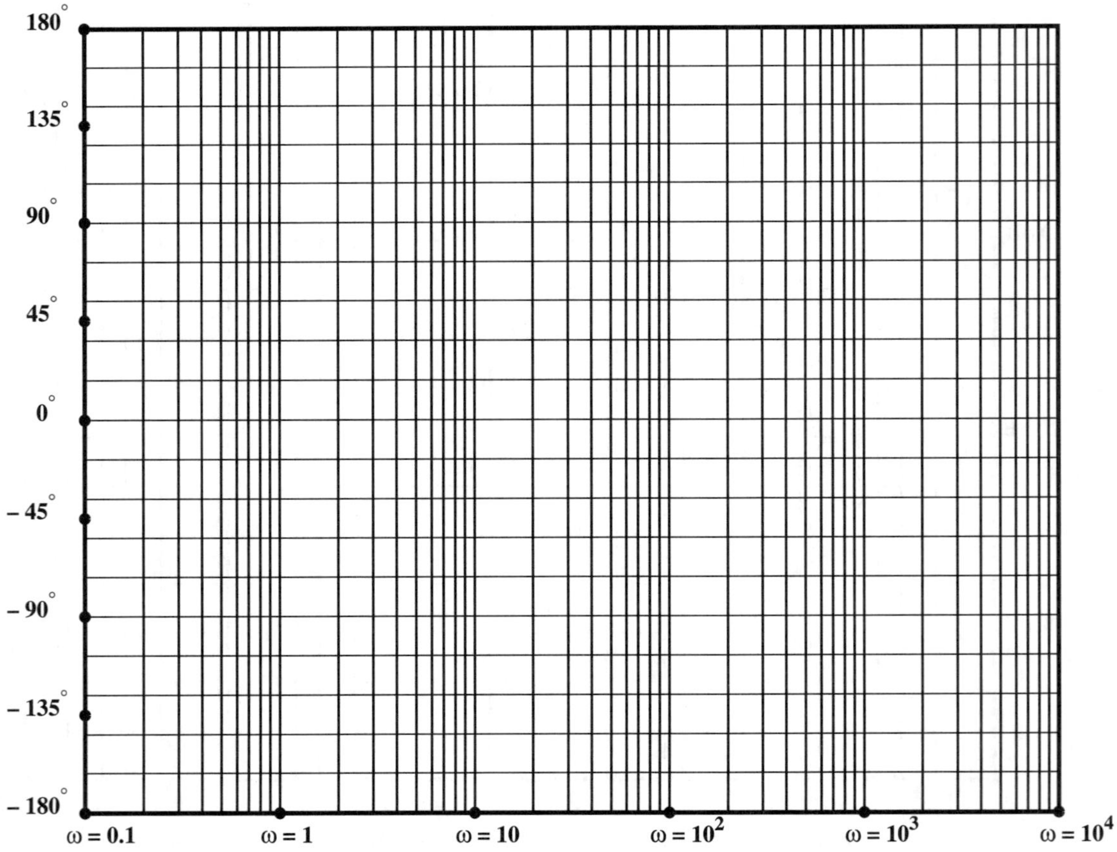

Fig. 10.9-2. Asymptotic angles–top components, above total.

10.9.5. Consider the circuit in Fig. 10.9-3, also known as a Twin-T filter, consisting of a high-pass filter and a low-pass filter. Select a set of values for R's and C's such as $R = 1K^{\Omega}$ and $C = 1^{\mu F}/2\pi$, and draw the magnitude Bode diagram for $|V_o(j\omega)|/|V_{in}(j\omega)|$. Specify the critical frequency at which Bode diagram changes substantially. Discuss your observation for the performance of this circuit and its potential application as a wave shaping filter.

Fig. 10.9-3.

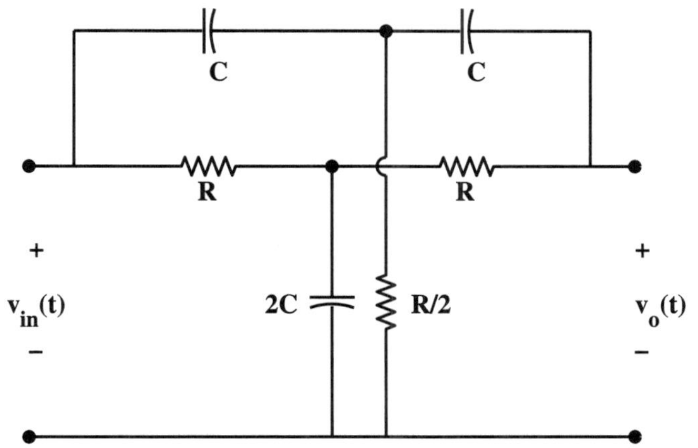

10.9.6. Given $T(s) = \dfrac{s^2 + 1}{s^2 + (½)s + 1}$, draw its magnitude versus frequency diagram and discuss its utility as a wave-shaping circuit.

Input-Output Ratio

10.9.7. Consider two cascaded first-order filters with the same characteristic as shown in Fig. 10.9-4. Let $R = 10K^{\Omega}$ and $C = 1^{\mu F}$. Draw the magnitude Bode diagram for the overall transfer function $|V_2(j\omega)|/|V_1(j\omega)|$.

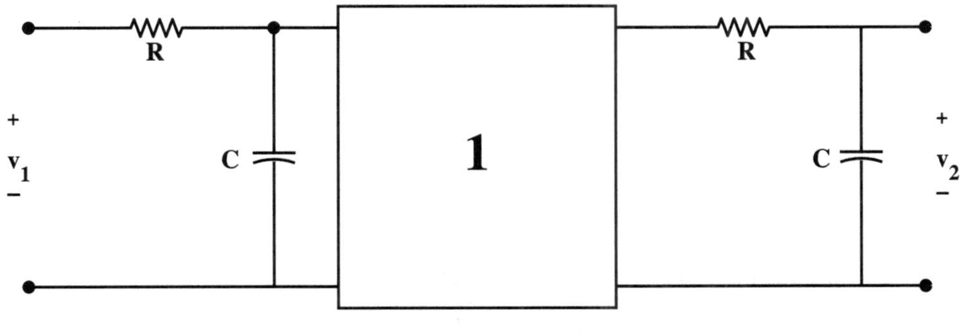

Fig. 10.9-4.

10.9.8. Draw $\dfrac{|V_o(j\omega)|}{V_{in}(j\omega)|}$ for each of the two circuits shown in Figs. 10.9-5 & 6.

Compare the results parameter by parameter, say critical frequency by the critical frequency and the like. Let $R_1 = R_2 = 1^{\Omega}$, $L = 1^{H}$ and $C = 1^{F}$ and $v_{in}(t) = 1(t)$.

Fig. 10.9-5.

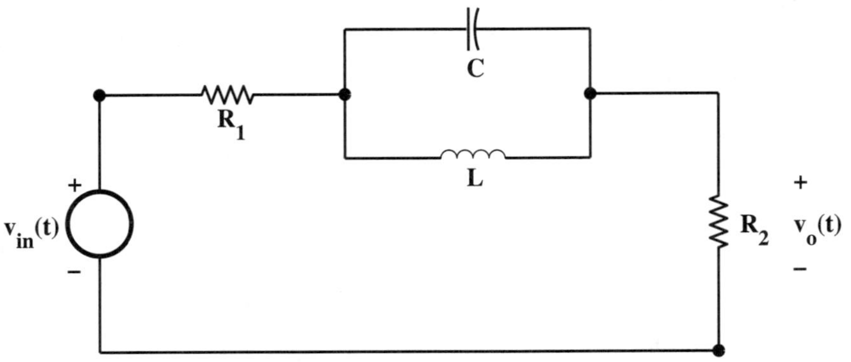

Fig. 10.9-6.

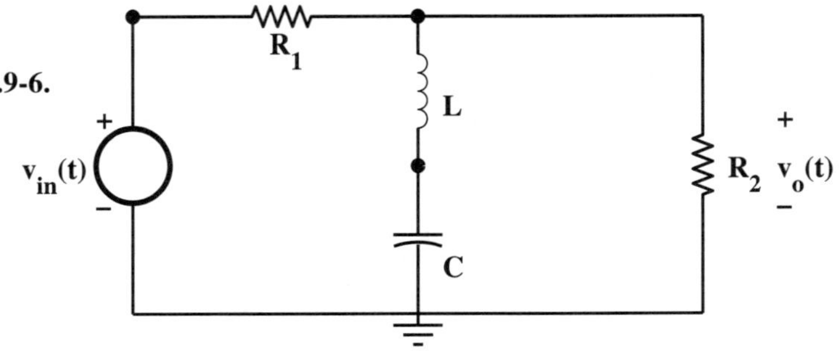

10.9.9. In the circuit of Fig. 10.9-7, the input current source is a train of pulses. Draw $\frac{|V_o(j\omega)|}{|I_{in}(j\omega)|}$ for $\omega = 0.1$ to $\omega = 1000$ rad/sec.

Fig. 10.9-7.

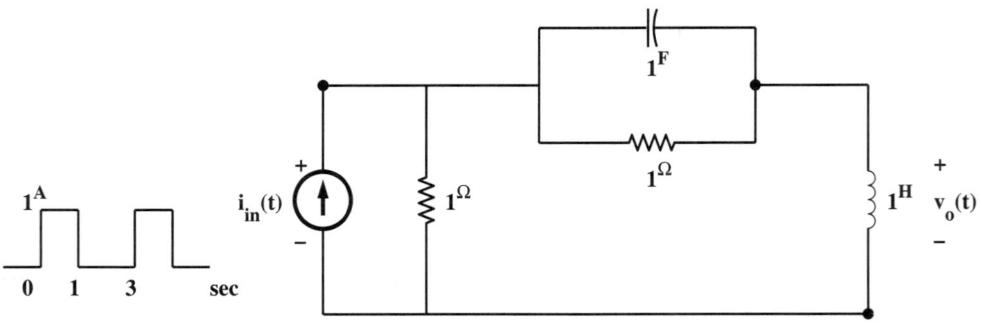

10.9.10. Consider the two circuits in Figs. 10.9-8 & 9. Choose R, L, C in the circuit of Fig. 10.9-9 such that for the same input function both circuits have identical output.

Fig. 10.9-8.

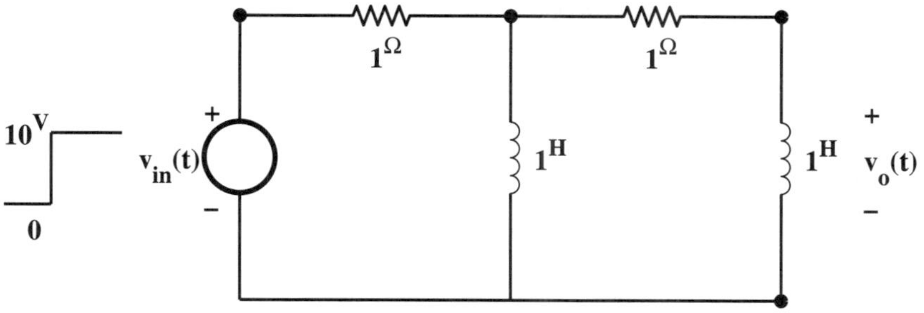

Fig. 10.9-9.

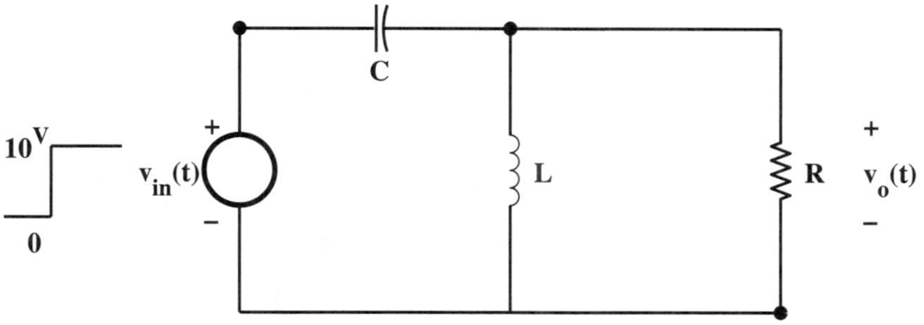

Second-Order Circuits

10.9.11. Consider the RLC circuit in Fig. 10.6-1, and let $R = 10^{\Omega}$, $L = 10^{mH}$ and $C = 10^{\mu F}$. Find $v_o(t)$, when $v_{in}(t) = 1(t)$. Draw $v_o(t)$ as a function of time and identify circuit parameters ω_n and ζ. Then change twice the value of R, in order to have all three cases of underdamped, critically damped and overdamped responses for this circuit.

10.9.12. For the circuit in Fig. 10.9-10, compute $V_o(s)/V_{in}(s)$ in terms of its parameters. Let $R_1 = R_2 = 1^{\Omega}$, and choose two sets of values for L and C. Draw the corresponding $|V_o(j\omega)| / |V_{in}(j\omega)|$ for each of these two sets and discuss their effects on the output.

Fig. 10.9-10.

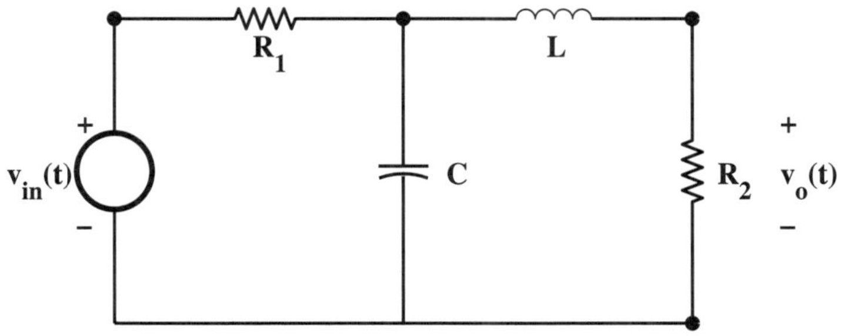

10.9.13. For three different values of L and with $v_{in}(i) = 1(t)$, draw $V_o(j\omega)| / |V_{in}(j\omega)|$ for the circuit in Fig. 10.9-11.

Fig. 10.9-11.

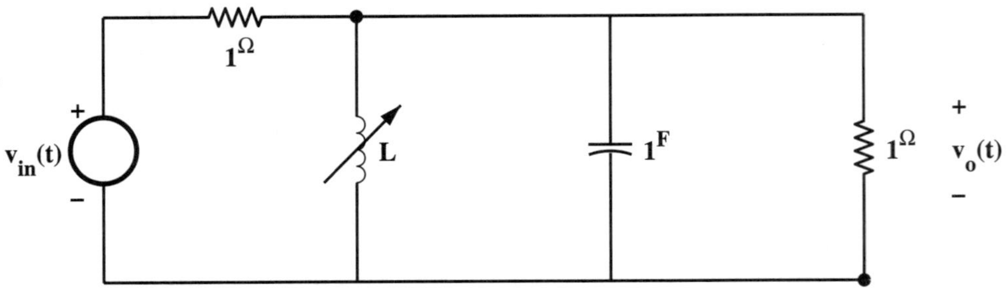

Resonant Circuits and Scaling

10.9.14. Consider the parallel RLC circuit in Fig. 10.7-1, and let $R = 10^{\Omega}$, $L = 1^{H}$ and $C = 1^{F}$.

(a) Draw its output voltage $|V_o(j\omega)|$ versus ω and find all its pertinent parameters described in Section 10.7, such as ω_o, ω_d, Q_o and BW.

(b) Change the element values to have a new $Q = 10Q_o$, and explain your rationale.

(c) Change the element value to have a new BW twice that of the original circuit, and explain your selection rationale.

10.9.15. Consider the RLC circuit in Fig. 10.9-11, and let $L = 10^{mH}$. Draw its transfer function. Apply magnitude scaling first to scale its magnitude up by ten fold. Second, apply a frequency scaling to expand its frequency range by 1000 times. Third, combine these two scalings. In each case find the corresponding circuit elements and draw the transfer function.

11 TWO-PORT NETWORKS

11.1. INTRODUCTION

Two-port networks are important parts of the overall circuit theory, because this class of circuits constitutes the fundamentals of circuits which are commonly used in electronics applications. Thus, the readers must be exposed to these rather advanced topics, in order that their prospective be broaden. We believe they are ready, and we must explore this opportunity and show them a few interesting topics.

In this chapter the fundamentals of two-port networks are reviewed. The common *algebra* for three-terminal active devices is included in this study as well. Although this theory is developed for future utilization in active two-port networks with transistors and similar devices or electronics components, no direct mention of such devices is made. However, the theory of two-port networks is equally applicable to circuits having *operational amplifiers* as is described in Section 11. 7.

11.2. FUNDAMENTALS of TWO-PORT NETWORKS

11.2.1. Description of Two-Port Networks

A general description in both time- and frequency-domain representations of two-port network is shown in Fig. 11.2-1. In this case four parameters, as compared to only two in a typical one-port network, are to be specified. We call ports ab and $a'b'$ the input and output port, respectively. In general, we expect the input and output variables to be related *via* some functional relationships. We may symbolically show this set of variables as follows.

$$V_1 = f_1(I_1, I_2, V_2), \tag{11.2-1}$$

$$I_1 = g_1(V_1, I_2, V_2), \tag{11.2-2}$$

$$V_2 = f_2(I_2, I_1, V_1), \tag{11.2-3}$$

$$I_2 = g_2(V_2, I_1, V_1). \tag{11.2-4}$$

Fig. 11.2-1.

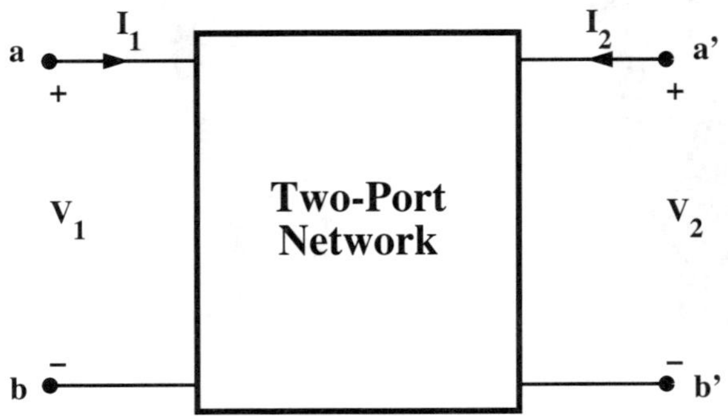

Clearly, there is a redundancy in (11-2-1) to (11.2-4). For example, from $V_2 = f_2(I_2, I_1, V_1)$ and $V_1 = f_1(I_1, I_2, V_2)$, we have $V_1 = f_1[I_1, I_2, f_2(I_2, I_1, V_1)]$. In this equation, and upon solving for V_1 results in

$$V_1 = V_1(I_1, I_2). \tag{11.2-5a}$$

Similarly, from $I_2 = g_2(V_2, I_1, V_1)$ and $V_1 = f_1[I_1, g_2(V_2, I_1, V_1), V_2]$, we have

$$V_1 = V_1'(I_1, V_2). \tag{11.2-5b}$$

In other words, depending on how do we proceed, each variable is at most a function of only two other variables and not three. Similarly, we have the following pairs:

$$I_1 = I_1(V_1, I_2) \text{ or } I_1 = I_1'(V_1, V_2), \tag{11.2-6}$$

$$V_2 = V_2(I_2, I_1) \text{ or } V_2 = V_2'(I_2, V_1), \tag{11.2-7}$$

$$I_2 = I_2(V_2, I_1) \text{ or } I_2 = I_2'(V_2, V_1). \tag{11.2-8}$$

The above functional relationships are the most general representations for any two-port network. These relationships are also valid for the two-port three-terminal networks shown in Fig. 11.2-2, as well. Here, the input and output pairs are connected and that is why this configuration is called a two-port, three-terminal network.

In the case of **linear two-port networks,** we have the **linearized versions** of (11.2-5) to (11.2-8). Depending on the choice of input and output representations for any given two-port network, we may have one of the following parameter sets.

Fig. 11.2-2.

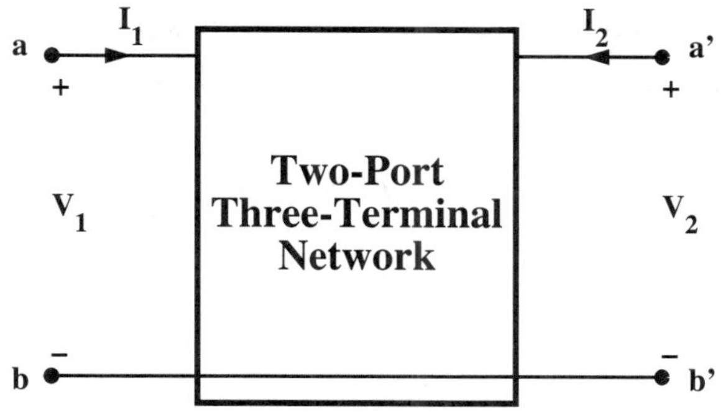

Starting with V_1 and V_2, we get

$$V_1 = V_1(I_1, I_2) \triangleq (V_1 \text{ is a linear function of } I_1, I_2) = \alpha I_1 + \beta I_2, \quad (11.2\text{-}9)$$

$$V_2 = V_2(I_1, I_2) \triangleq (V_2 \text{ is a linear function of } I_1, I_2) = \alpha' I_1 + \beta' I_2. \quad (11.2\text{-}10)$$

Obviously, the physical dimensions of these four parameters $(\alpha, \beta, \alpha', \beta')$ can be associated with impedances. Thus the natural name for these quantities is *Z*-parameters. Similarly, we may choose other combinations of these input and output quantities to get other sets of parameters (assuming the linearity, of course). These parameter sets are summarized in Table 11.2-1. It is to be noted that for the *ABCD*-parameters the sign of I_2 has been changed relative to that in Fig. 11.2-1. The physical dimensions of these sets of parameters can be directly accounted for in any given situation. For instance, the sets *H*- and *G*- as well as the singles C and B are hybrid (impedances and admittances), while A and D are dimensionless. There is another set called *A'B'C'D'*-parameters which is the reverse of the set for *ABCD*-parameters, i.e., going from the output port back to the input port. The *ABCD*- parameters are also called the *transmission parameters*. The matrix *ABCD* is called the *transmission matrix*, while the matrix *A'B'C'D'* is its inverse.

The self-explanatory Table 11.7-2 shows the relationships among these sets of parameters. It is very important to realize that not every two-port network has all of these six sets of parameters, because during the conversion we divide some entries by others, and if those divisors are zeros, we cannot proceed. To elucidate this matter consider the two circuits in Problem 11.8.1.

These linearized two-port models are also the first step, or the first-order approximation, toward analyzing any general nonlinear two-port network such as the circuit corresponding to a transistor, when such a device is linearized.

Example 11.2-1: Consider the circuit in Fig. 11.2-3. Derive both *Z*- and *H*-parameters for this circuit. (Problem 11.8.3).

Table 11.2-1. Parameter sets for a two-port network.

$$\begin{bmatrix} V_1 \\ V_2 \end{bmatrix} = \begin{pmatrix} Z_{11} & Z_{12} \\ Z_{21} & Z_{22} \end{pmatrix} \begin{bmatrix} I_1 \\ I_2 \end{bmatrix}, \ Z\text{ - parameters,}$$

$$\begin{bmatrix} V_1 \\ I_2 \end{bmatrix} = \begin{pmatrix} H_{11} & H_{12} \\ H_{21} & H_{22} \end{pmatrix} \begin{bmatrix} I_1 \\ V_2 \end{bmatrix}, \ H\text{ - parameters,}$$

$$\begin{bmatrix} I_1 \\ V_2 \end{bmatrix} = \begin{pmatrix} G_{11} & G_{12} \\ G_{21} & G_{22} \end{pmatrix} \begin{bmatrix} V_1 \\ I_2 \end{bmatrix}, \ G\text{ - parameters,}$$

$$\begin{bmatrix} I_1 \\ I_2 \end{bmatrix} = \begin{pmatrix} Y_{11} & Y_{12} \\ Y_{21} & Y_{22} \end{pmatrix} \begin{bmatrix} V_1 \\ V_2 \end{bmatrix}, \ Y\text{ - parameters,}$$

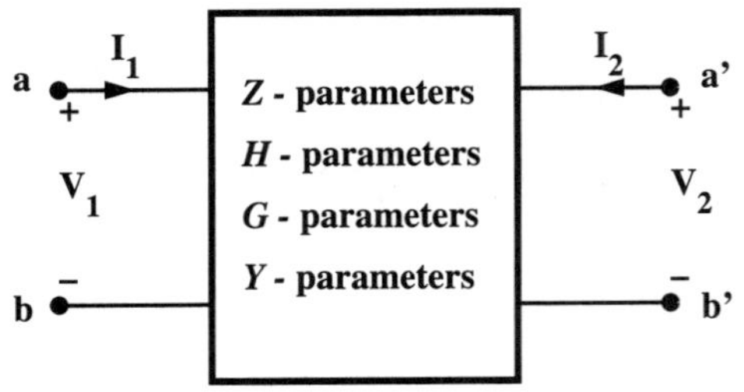

$$\begin{bmatrix} V_1 \\ I_1 \end{bmatrix} = \begin{pmatrix} A & B \\ C & D \end{pmatrix} \begin{bmatrix} V_2 \\ -I_2 \end{bmatrix}, \ ABCD\text{ - parameters.}$$

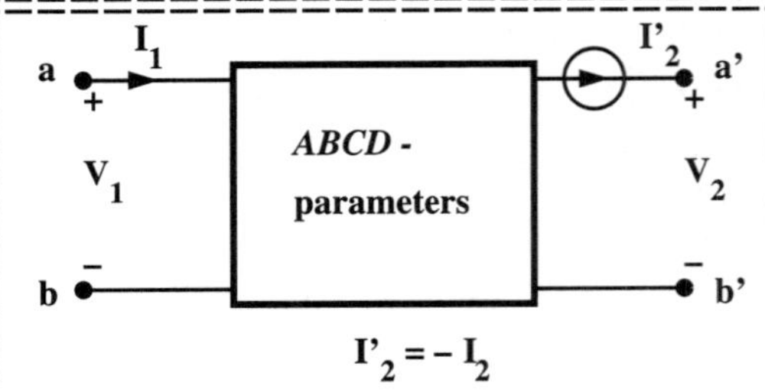

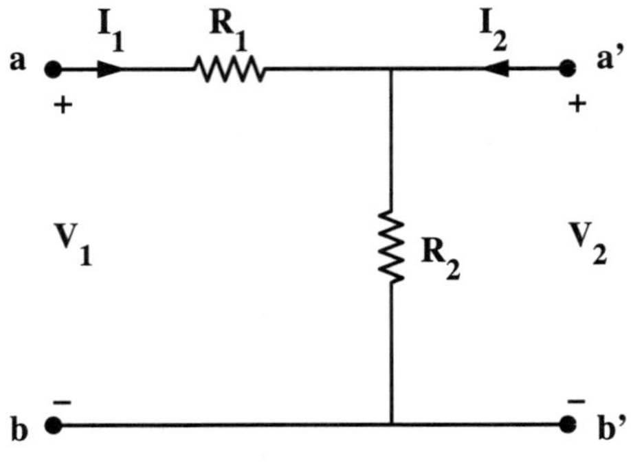

Fig. 11.2-3.

Solution: From KVL we have

$$V_1 = (R_1 + R_2)I_1 + R_2 I_2, \tag{11.2-11}$$

$$V_2 = R_2 I_1 + R_2 I_2. \tag{11.2-12}$$

Thus:

$$Z \triangleq \begin{pmatrix} R_1 + R_2 & R_2 \\ R_2 & R_2 \end{pmatrix}. \tag{11.2-13}$$

Also from (11.2-12) we have $I_2 = -I_1 + V_2/R_2$. Substituting this I_2 into (11.2-11) yields $V_1 = R_1 I_1 + V_2$. Thus:

$$H \triangleq \begin{pmatrix} R_1 & 1 \\ -1 & 1/R_2 \end{pmatrix}. \tag{11.2-14}$$

11.2.2. Interconnection of Two-Port Networks

An interesting feature of this analysis is that we can break a two-port network into several less complicated such networks which are interconnected in certain identifiable manners. By knowing the appropriate parameters for each of these sub-networks and the ways which are connected together, we can express the overall network parameters, though to acquire this level of competency requires some preparations. As is described subsequently, this *algebra* makes the feedback amplifier circuits simpler to analyze than before. In the following we consider the interconnection of several two-port networks and the corresponding overall set of parameters.

Suppose a two-port network N consists of two identifiable two-port sub-networks N_1 and N_2 that are connected series to series as shown in Fig. 11.2-4. Here, we assume that N_1 and N_2 can be easily analyzed and this assumption makes our analytical approach fruitful.

Also, in Fig. 11.2-4, the implication of having two ideal and 1:1 transformers is that these two-port sub-networks are such that the current which gets into N_1 is exactly the same as that getting into N_2. Similarly, I_2 is equal to I_2'. If these conditions are met, then for N_1 we have:

$$\begin{bmatrix} V_1 \\ V_2 \end{bmatrix} = \begin{pmatrix} Z_{11} & Z_{12} \\ Z_{21} & Z_{22} \end{pmatrix} \begin{bmatrix} I_1 \\ I_2 \end{bmatrix}, \tag{11.2-15}$$

and for N_2 we have:

$$\begin{bmatrix} V_1' \\ V_2' \end{bmatrix} = \begin{pmatrix} Z_{11}' & Z_{12}' \\ Z_{21}' & Z_{22}' \end{pmatrix} \begin{bmatrix} I_1' \\ I_2' \end{bmatrix}. \tag{11.2-16}$$

Since,

$$\begin{bmatrix} V_{in} \\ V_o \end{bmatrix} = \begin{bmatrix} V_1 + V_1' \\ V_2 + V_2' \end{bmatrix} \quad \& \quad \begin{bmatrix} I_{in} \\ I_o \end{bmatrix} = \begin{bmatrix} I_1 \\ I_2 \end{bmatrix} = \begin{bmatrix} I_1' \\ I_2' \end{bmatrix}. \tag{11.2-17}$$

Thus the overall network parameters become:

$$\begin{bmatrix} V_{in} \\ V_o \end{bmatrix} = \begin{pmatrix} Z_{11} + Z_{11}' & Z_{12} + Z_{12}' \\ Z_{21} + Z_{21}' & Z_{22} + Z_{22}' \end{pmatrix} \begin{bmatrix} I_{in} \\ I_o \end{bmatrix}. \tag{11.2-18}$$

This is a very strong result in the sense that we have the overall two-port network parameters for the original circuit from that of its sub-networks, and this is done almost by inspection.

We bear in mind that the above algebra is valid in its entirety when we have exactly the indicated interconnection with proper isolations. If anyone of these conditions fails, then this analysis is not valid.

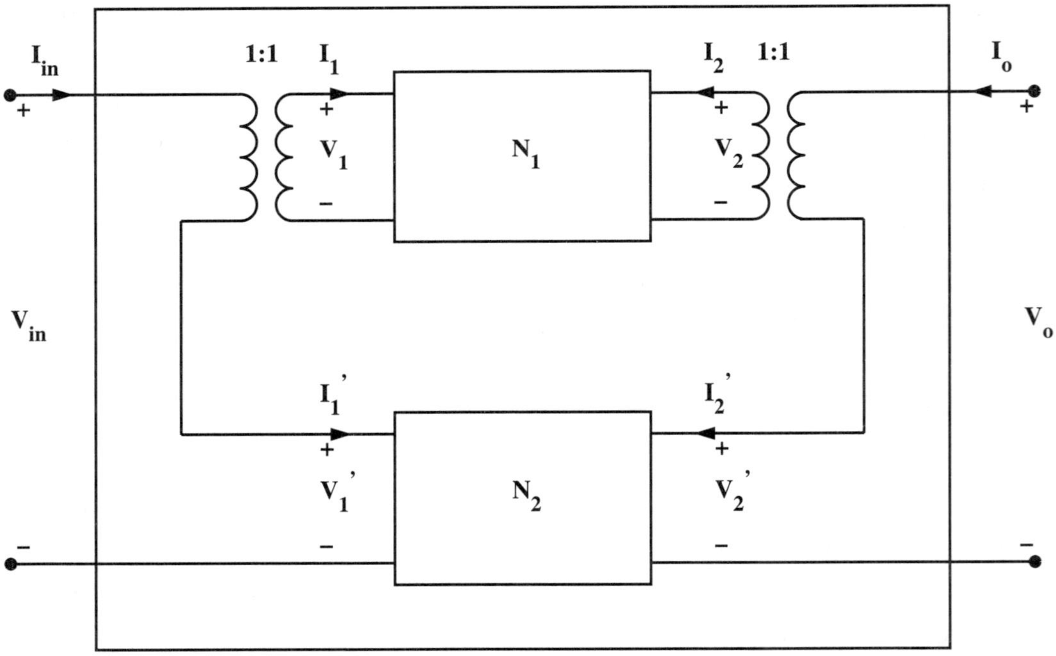

Fig. 11.2-4. Network N with two series to series interconnected sub-networks N_1 and N_2.

On the other hand, if the two sub-networks N_1 and N_2 were interconnected parallel to parallel with valid port currents isolation conditions as depicted in Fig. 11.2-5 (with similar reasoning as given in the preceding case), then we have the following overall *Y*-parameters for the original two-port network N. But first we note that here,

$V_{in} = V_1 = V_1'$ and $V_o = V_2 = V_2'$, while, $I_{in} = I_1 + I_1'$, and $I_o = I_2 + I_2'$. Thus,

$$\begin{bmatrix} I_{in} \\ I_o \end{bmatrix} = \begin{pmatrix} Y_{11} + Y_{11}' & Y_{12} + Y_{12}' \\ Y_{21} + Y_{21}' & Y_{22} + Y_{22}' \end{pmatrix} \begin{bmatrix} V_{in} \\ V_o \end{bmatrix}. \tag{11.2-19}$$

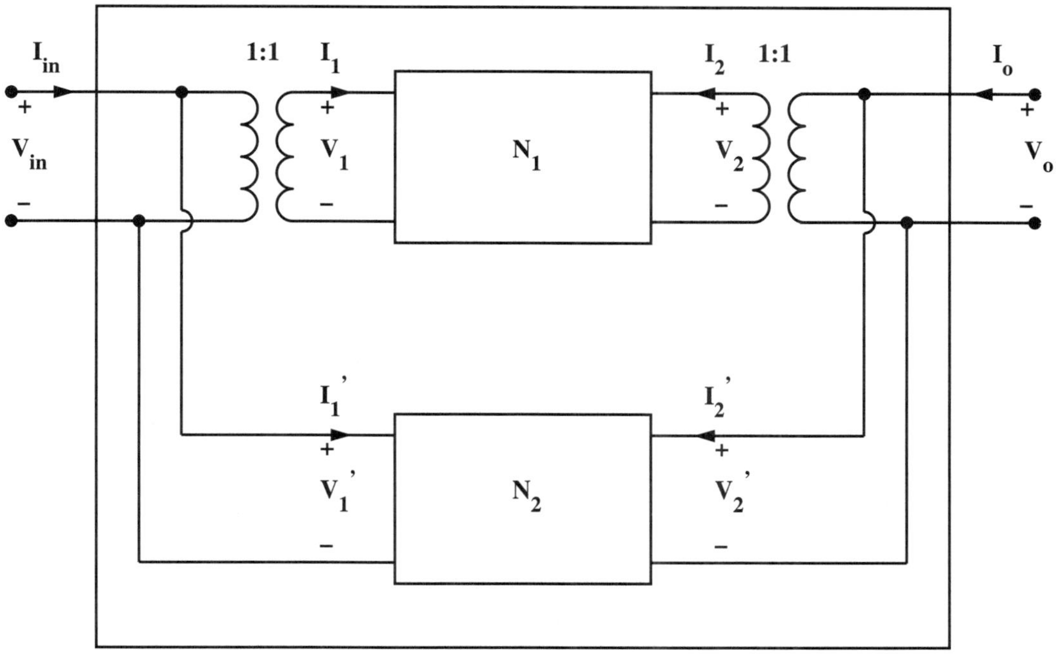

Fig. 11.2-5. Network N with two parallel to parallel interconnected sub-networks N_1 and N_2.

The other three cases of interest occur when we have *hybrid* interconnections (series-parallel and parallel-series) and the cascaded interconnection under the same port currents isolation conditions. For series-parallel interconnections *H*-parameters, for parallel-series interconnection *G*-parameters, and for cascaded interconnection *ABCD*-parameters are used to describe the overall two-port network parameters.

Utilization of these techniques in interconnecting networks have many applications in *feedback* amplifiers. Suppose circuit N_1 plays the role of the *feed-forward* or *open-loop* amplifier and the circuit N_2 plays the role of the feedback amplifier. Then using the preceding described circuit algebra with the corresponding interconnection conditions, we can analyze the overall network parameters more simply than that possible using almost all conventional techniques. To that effect we review some basic aspects of feedback theory in the next section. But first Table 11.2-2 shows how to convert one set of parameters to another.

Table 11.2-2. Conversion chart for two-port networks [8].

All matrices in each row are equal.

Parameter Sets Determinant	Z $\det Z = \Delta_Z$	H $\det H = \Delta_H$	G $\det G = \Delta_G$	Y $\det Y = \Delta_Y$	$ABCD$ $= A\ldots$ $\det A\ldots = \Delta_A$	$A'B'C'D'$ $= A'\ldots$ $\det A'\ldots = \Delta_{A'}$
Z	$\begin{matrix} z_{11} & z_{12} \\ z_{21} & z_{22} \end{matrix}$	$\begin{matrix} \frac{\Delta_H}{h_{22}} & \frac{h_{12}}{h_{22}} \\ -\frac{h_{21}}{h_{22}} & \frac{1}{h_{22}} \end{matrix}$	$\begin{matrix} \frac{1}{g_{11}} & -\frac{g_{12}}{g_{11}} \\ \frac{g_{21}}{g_{11}} & \frac{\Delta_G}{g_{11}} \end{matrix}$	$\begin{matrix} \frac{y_{22}}{\Delta_Y} & -\frac{y_{12}}{\Delta_Y} \\ -\frac{y_{21}}{\Delta_Y} & \frac{y_{11}}{\Delta_Y} \end{matrix}$	$\begin{matrix} \frac{A}{C} & \frac{\Delta_A}{C} \\ \frac{1}{C} & \frac{D}{C} \end{matrix}$	$\begin{matrix} \frac{D'}{C'} & \frac{1}{C'} \\ \frac{\Delta_{A'}}{C'} & \frac{A'}{C'} \end{matrix}$
H	$\begin{matrix} \frac{\Delta_Z}{z_{22}} & \frac{z_{12}}{z_{22}} \\ -\frac{z_{21}}{z_{22}} & \frac{1}{z_{22}} \end{matrix}$	$\begin{matrix} h_{11} & h_{12} \\ h_{21} & h_{22} \end{matrix}$	$\begin{matrix} \frac{g_{22}}{\Delta_G} & -\frac{g_{12}}{\Delta_G} \\ -\frac{g_{21}}{\Delta_G} & \frac{g_{11}}{\Delta_G} \end{matrix}$	$\begin{matrix} \frac{1}{y_{11}} & -\frac{y_{12}}{y_{11}} \\ \frac{y_{21}}{y_{11}} & \frac{\Delta_Y}{y_{11}} \end{matrix}$	$\begin{matrix} \frac{B}{D} & \frac{\Delta_A}{D} \\ -\frac{1}{D} & \frac{C}{D} \end{matrix}$	$\begin{matrix} \frac{B'}{A'} & \frac{1}{A'} \\ -\frac{\Delta_{A'}}{A'} & \frac{C'}{A'} \end{matrix}$
G	$\begin{matrix} \frac{1}{z_{11}} & -\frac{z_{12}}{z_{11}} \\ \frac{z_{21}}{z_{11}} & \frac{\Delta_Z}{z_{11}} \end{matrix}$	$\begin{matrix} \frac{h_{22}}{\Delta_H} & -\frac{h_{12}}{\Delta_H} \\ -\frac{h_{21}}{\Delta_H} & \frac{h_{11}}{\Delta_H} \end{matrix}$	$\begin{matrix} g_{11} & g_{12} \\ g_{21} & g_{22} \end{matrix}$	$\begin{matrix} \frac{\Delta_Y}{y_{22}} & \frac{y_{12}}{y_{22}} \\ -\frac{y_{21}}{y_{22}} & \frac{1}{y_{22}} \end{matrix}$	$\begin{matrix} \frac{C}{A} & -\frac{\Delta_A}{A} \\ \frac{1}{A} & \frac{B}{A} \end{matrix}$	$\begin{matrix} \frac{C'}{D'} & -\frac{1}{D'} \\ \frac{\Delta_{A'}}{D'} & \frac{B'}{D'} \end{matrix}$
Y	$\begin{matrix} \frac{z_{22}}{\Delta_Z} & -\frac{z_{12}}{\Delta_Z} \\ -\frac{z_{21}}{\Delta_Z} & \frac{z_{11}}{\Delta_Z} \end{matrix}$	$\begin{matrix} \frac{1}{h_{11}} & -\frac{h_{12}}{h_{11}} \\ \frac{h_{21}}{h_{11}} & \frac{\Delta_H}{h_{11}} \end{matrix}$	$\begin{matrix} \frac{\Delta_G}{g_{22}} & \frac{g_{12}}{g_{22}} \\ -\frac{g_{21}}{g_{22}} & \frac{1}{g_{22}} \end{matrix}$	$\begin{matrix} y_{11} & y_{12} \\ y_{21} & y_{22} \end{matrix}$	$\begin{matrix} \frac{D}{B} & -\frac{\Delta_A}{B} \\ -\frac{1}{B} & \frac{A}{B} \end{matrix}$	$\begin{matrix} \frac{A'}{B'} & -\frac{1}{B'} \\ -\frac{\Delta_{A'}}{B'} & \frac{D'}{B'} \end{matrix}$
$ABCD$	$\begin{matrix} \frac{z_{11}}{z_{21}} & \frac{\Delta_Z}{z_{21}} \\ \frac{1}{z_{21}} & \frac{z_{22}}{z_{21}} \end{matrix}$	$\begin{matrix} -\frac{\Delta_H}{h_{21}} & -\frac{h_{11}}{h_{21}} \\ -\frac{h_{22}}{h_{21}} & -\frac{1}{h_{21}} \end{matrix}$	$\begin{matrix} \frac{1}{g_{21}} & \frac{g_{22}}{g_{21}} \\ \frac{g_{11}}{g_{21}} & \frac{\Delta_G}{g_{21}} \end{matrix}$	$\begin{matrix} -\frac{y_{22}}{y_{21}} & -\frac{1}{y_{21}} \\ -\frac{\Delta_Y}{y_{21}} & -\frac{y_{11}}{y_{21}} \end{matrix}$	$\begin{matrix} A & B \\ C & D \end{matrix}$	$\begin{matrix} \frac{D'}{\Delta_{A'}} & \frac{B'}{\Delta_{A'}} \\ \frac{C'}{\Delta_{A'}} & \frac{A'}{\Delta_{A'}} \end{matrix}$
$A'B'C'D'$	$\begin{matrix} \frac{z_{22}}{z_{12}} & -\frac{\Delta_Z}{z_{12}} \\ -\frac{1}{z_{12}} & \frac{z_{11}}{z_{12}} \end{matrix}$	$\begin{matrix} \frac{1}{h_{12}} & -\frac{h_{11}}{h_{12}} \\ -\frac{h_{22}}{h_{12}} & \frac{\Delta_H}{h_{12}} \end{matrix}$	$\begin{matrix} -\frac{\Delta_G}{g_{12}} & \frac{g_{22}}{g_{12}} \\ \frac{g_{11}}{g_{12}} & -\frac{1}{g_{12}} \end{matrix}$	$\begin{matrix} -\frac{y_{11}}{y_{12}} & \frac{1}{y_{12}} \\ \frac{\Delta_Y}{y_{12}} & -\frac{y_{22}}{y_{12}} \end{matrix}$	$\begin{matrix} \frac{D}{\Delta_A} & -\frac{B}{\Delta_A} \\ -\frac{C}{\Delta_A} & \frac{A}{\Delta_A} \end{matrix}$	$\begin{matrix} A' & B' \\ C' & D' \end{matrix}$
Reciprocity condition holds if	$z_{12} = z_{21}$	$h_{12} = -h_{21}$	$g_{12} = -g_{21}$	$y_{12} = y_{21}$	$\Delta_A = 1$	$\Delta_{A'} = 1$

11.3. BASIC ASPECTS of FEEDBACK THEORY

Perhaps one the most important contributions to engineering sciences – in particular, electrical engineering – has been the creation and formulation of the feedback theory. The feedback or closed-loop system, as compared to the open-loop system, has many advantageous properties. To give a simple example of these open- and closed-loop situations, consider a case in which an inexperienced person uses a "set and go machine", say, a washing machine. The person with experience knows for each specific garment a correct temperature and cycle settings, because of the prior experience. While, the person with no experience is taking chances by a learning and doing process, and the quality of the final product for the entire process is unpredictable. In all these cases no monitoring scheme has been devised to observe the development of the final product which in this case is the cleanliness and safety of the garment. Both persons use the same machine which operates in an open-loop mode, i.e., an operation in which no mechanism is used to monitor the development of the final product.

In contrast with this open-loop operation, if we have a situation in which the outcome of the process is monitored in the course of its development, then that is called a closed-loop operation. In a way, the main question is: How do we account for the operator's experience? We may consider the scope of experience to play the role of the monitoring scheme in the preceding case. In that sense the result of the case for an inexperienced operator may not be satisfactory let alone perfect, while for the experienced person is satisfactory. These are just typical examples of the open-loop operations, for which the result can either be satisfactory or otherwise and nothing can be done in between. Here, the person with experience is always ahead of the game, because of his/her prior information.

The shortcoming of many open-loop situations to a large extent can be rectified by using a feedback mechanism, or a closed-loop system which is a mechanism that monitors the development of the ultimate product until the operation is completely finished. This closed-loop or feedback operation, of course, is more involved and costly than the open-loop one. For example, developing a circuit to check the cleanliness of a garment in a washing machine is almost impossible. Nevertheless in many practical problems we are able to study the output as it is being developed and use this information to correct the outcome so that it remains within some pre-specified and acceptable conditions. Based on this need and the importance of feedback theory in electronics, we review some of the most elementary aspects of this theory next.

We acknowledge that the following materials are generally taught in the first control course whose prerequisite is this first circuit course. Thus, we need to proceed cautiously, and realize that our approach is to familiarize the readers with all collateral issues rather than distracting them. We are simply planting the seeds.

The schematic diagram of a closed-loop or feedback system is shown in Fig. 11.3-1. Here V_d is a disturbance entering the system and is caused by the measurement devices and/or other plant uncertainties that may exist in a given system. By plant, we mean that part of a dynamic system for which feedback controllers (or circuits) are designed. The combination of plant and feedback controllers and all interconnecting circuits is then called a closed-loop dynamic system. In Fig. 11.3-1, the amplifier with gain $G_{ol}(s)$ is the plant. The term plant uncertainties refers to all unknown parameters of the plant including those parameters that change (because of aging, environmental changes, *etc.*), with or without our knowledge. Furthermore, any time we insert a measuring device, say a voltmeter, into the system (or circuit) we inherently generate noise that will affect the ultimate response of our system. In the case of a voltmeter, *per se*, when we insert this unit into the circuit, that unit becomes a part of the overall network and therefore all the currents and/or voltages of the network are affected because of this meter's circuitries. Although these differences may not be noticeable, they exist. Thus in Fig. 11.3-1, where we show an output by V_o, we mean we have inserted a voltmeter to read this output and as a result the output is *disturbed.* It is generally a common practice to model all such disturbances caused by measuring devices, plant uncertainties, *etc.,* with one external input; for example V_d in Fig. 11.3-1. It is also a common practice to assume that this voltage is *added* to the system, although in reality it may be otherwise.

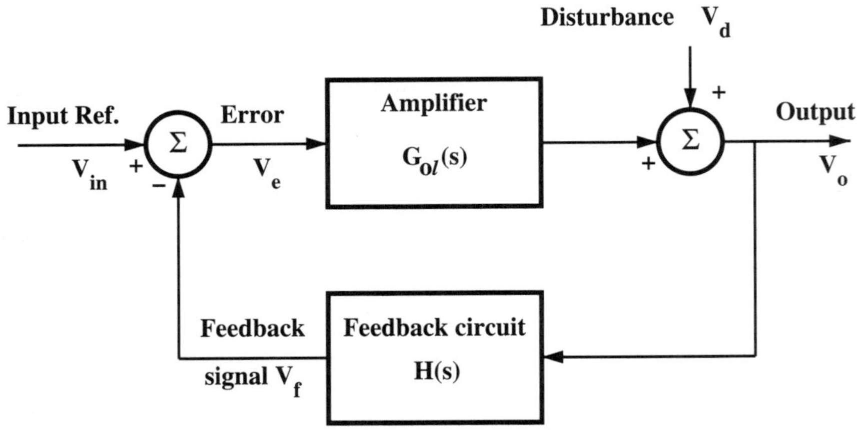

Fig. 11.3-1. A closed-loop (feedback) system.

In Fig. 11.3-1 the first circle on the left is called the comparator unit, which simply compares the reference input signal V_{in} with the feedback signal V_f. Here, $V_e(s) = V_{in}(s) - V_f(s)$. This configuration is called a negative feedback. For each block and based on our discussion about Fig. 9.3-3 and (9.3-37), we have [output(s)] = [transfer function(s)] × [input(s)]. In other words, each signal that enters into an *amplifier* circuit produces a signal that is proportional to the *gain* (or transfer function) of the amplifier expressed in "s", as well as the *value* of the entering signal. The convention has always been to *pre-multiply* the signal value by the corresponding gain. Thus, a simple signal analysis in the next comparator results in

$$V_o(s) = V_d(s) + G_{ol}(s)V_e(s)$$

$$= V_d(s) + G_{ol}(s)[V_{in}(s) - V_f(s)]. \tag{11.3-1}$$

Here, $V_d(s)$ is the measurement and/or plant observation disturbance. Similarly, $V_f(s) \equiv H(s)V_o(s)$, and if we substitute this $V_f(s)$ into the expression for $V_o(s)$ and rearrange the terms we have

$$V_o(s) = \frac{G_{ol}(s)}{1 + G_{ol}(s)H(s)} V_{in}(s) + \frac{1}{1 + G_{ol}(s)H(s)} V_d(s). \tag{11.3-2}$$

For now, we assume that there is no disturbance in our system; that is, $V_d(s) \equiv 0$. This choice results in

$$V_o(s) = \frac{G_{ol}(s)}{1 + G_{ol}(s)H(s)} V_{in}(s) \triangleq G_{cl}(s)V_{in}(s), \tag{11.3-3}$$

where $G_{cl}(s)$ can be considered as a gain factor and correspondingly we call that expression the "closed-loop" gain or transfer function. $G_{ol}(s)$ and $H(s)$ can also be constant as well.

From (11.3-3) we conclude that if $G_{ol} \to \infty$, then $G_{cl} \to 1/H(s)$ which, in general, is finite. This result when we reduce the large open-loop gain with feedback to a manageable level, immediately tells us one of the basic advantages of the feedback configuration over that of an open-loop. That means the feedback tends to stabilize the system at a cost of reducing its output amplitude. The term $G_{ol}(s)H(s)$ is called the *loop gain* and the term $[1 + G_{ol}(s)H(s)]$ is called the *return difference.*

If, instead of negative feedback we had positive feedback in Fig. 1.3-1, which is mostly the case in electronic circuits, then the comparator becomes an adder, i.e., $V_e(s) = V_{in}(s) + V_f(s)$. Thus, upon similar analysis as above, we have

$$V_o(s) = \frac{G_{ol}(s)}{1 - G_{ol}(s)H(s)} V_{in}(s) + \frac{1}{1 - G_{ol}(s)H(s)} V_d(s). \tag{11.3-4}$$

Intuitively, and because of the minus sign in the denominator of (11.3-4), we guess that if the denominator goes to zero, then we will have instability. In other words, if $1 - G_{ol}(s)H(s) \to 0$, then $V_o(s)$ may blow up. Now, the question is, under what conditions will $1 - G_{ol}(s)H(s) \to 0$? Because this is a complex quantity, the answer is when both the real and the imaginary parts of this denominator go to zero. That means equivalently, when we have the following conditions for magnitude and phase.

$$|G_{ol}(j\omega)H(j\omega)| \equiv 1 \quad \text{and} \quad \angle G_{ol}(j\omega)H(j\omega) = 360°k, \quad k=1,2,\cdots. \qquad (11.3\text{-}5)$$

Therefore, as another property of a feedback design, we observe that if we are using a positive feedback, there is a chance of having instability and/or oscillation unless extreme care has been exercised to design the circuit for oscillation purposes. On the other hand, negative feedback tends to stabilize the circuit and is preferable.

Regarding conditions for oscillation (also known as Barkhausen criterion), recall that for any network the node equations in matrix form are I = YV. Since an oscillator circuit works with only active devices (dependent sources), I ≡ 0 (i.e., no independent current source) and further the corresponding Y matrix is not symmetric. For the preceding equation to have a non-trivial solution for V, we must have

$$\det Y \equiv 0. \qquad (11.3\text{-}6)$$

Because Y is a complex matrix, its determinant is also a complex number. Thus, we have two equations to meet, resulting in two parameters for the oscillation condition in the corresponding circuit. (Problem 11.8.10).

Going back to the interconnection networks and realistically, given an electronic circuit to analyze, it is not always easy to recognize or identify its open-loop amplifier, or its feedback circuit part. Nevertheless, with experience this identification can be made, and when one can recognize these units and separate them, the above analysis is easily used to discuss stability and/or any other pertinent question regarding the overall behavior of the main circuit. Recall that the above analysis is valid, under the condition when the ports isolation properties hold. Otherwise, the partitioning of the main circuit into the preceding sub-networks is not exact and may even be incorrect.

11.4. GENERAL DESCRIPTION of FEEDBACK AMPLIFIERS

In this section a general configuration of two-port electrical network is studied. Subsequently, we study this general representation in conjunction with a feedback amplifier network to obtain an *approximated* expression for the overall transfer function of this type amplifiers.

Consider the two-port network in Fig. 11.4-1 with a generalized input source (for example, a current θ_1 in parallel with source impedance Γ_s, or a voltage ϕ_1 in series with source impedance Γ_s), and a generalized output load of Γ_l with its corresponding θ_2 (current) or ϕ_2 (voltage). The γ_{ij}'s are the appropriate set of two-port network parameters. If the γ_{ij}'s are the *Z*-parameters, then both Γ_s and Γ_l are impedances, and if γ_{ij}'s are the *Y*-parameters, then Γ_s and Γ_l are admittances, *etc.*.

For the circuit in Fig. 11.4-1, we can easily write

$$\begin{bmatrix} \phi_1 \\ \phi_2 \end{bmatrix} = \begin{pmatrix} \Gamma_s + \gamma_{11} & \gamma_{12} \\ \gamma_{21} & \gamma_{22} + \Gamma_l \end{pmatrix} \begin{bmatrix} \theta_1 \\ \theta_2 \end{bmatrix}. \tag{11.4-1}$$

This means the input source parameter appears in the (1,1) term and the output source parameter appears in the (2,2) term. Because of the compatibility issue, stated above, we put Γ_s and Γ_l on the left- and right-hand side of the corresponding entries in (11.4-1) to remind us of a proper incorporation of these parameters.

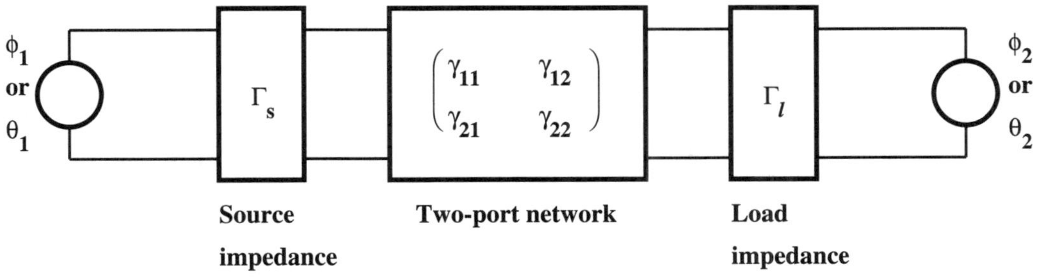

Fig. 11.4-1.

Suppose, we want to compute $A_{11} \triangleq (\theta_1/\phi_1)$ (given $\phi_2 = 0$). Then, from $\phi_2 = 0$ and the second equation in (11.4-1), we have $\gamma_{21}\theta_1 + (\gamma_{22} + \Gamma_l)\theta_2 = 0$. Solving for θ_2 and substituting the result in the first equation in (11.4-1) gives

$$A_{11} = (\gamma_{22} + \Gamma_l)/[(\Gamma_s + \gamma_{11})(\gamma_{22} + \Gamma_l) - \gamma_{12}\gamma_{21}]. \tag{11.4-2}$$

Likewise, we can compute $A_{21} \triangleq (\theta_2/\phi_1)$ (given $\phi_2 = 0$), resulting in

$$A_{21} = -\gamma_{21}/[(\Gamma_s + \gamma_{11})(\gamma_{22} + \Gamma_l) - \gamma_{12}\gamma_{21}]. \tag{11.4-3}$$

In general, we can show that the input and output transfer functions are as follows.

$$\begin{cases} T_{in} \equiv A_{11} = \left.\dfrac{\phi_1}{\theta_1}\right|_{\phi_2=0} = \Gamma_s + \gamma_{11} - \dfrac{\gamma_{12}\gamma_{21}}{\gamma_{22} + \Gamma_l}, \\ T_o \equiv A_{22} = \left.\dfrac{\phi_2}{\theta_2}\right|_{\phi_1=0} = \gamma_{22} + \Gamma_l - \dfrac{\gamma_{12}\gamma_{21}}{\Gamma_s + \gamma_{11}}. \end{cases} \tag{11.4-4}$$

In the terminology of feedback theory, we can have a case such as Fig. 11.4-2a, which is the schematic circuit version of a general positive feedback configuration as shown in Fig. 11.4-2b. In fact, in most feedback amplifier circuits, we have a positive feedback configuration. Also, in the terminology of two-port networks, the interconnecting networks just tell us how these two sub-networks for the feedforward amplifier [a(s) for Fig. 11.4-2a] and the feedback amplifier [f(s) for Fig. 11.4-2a] are connected to each other.

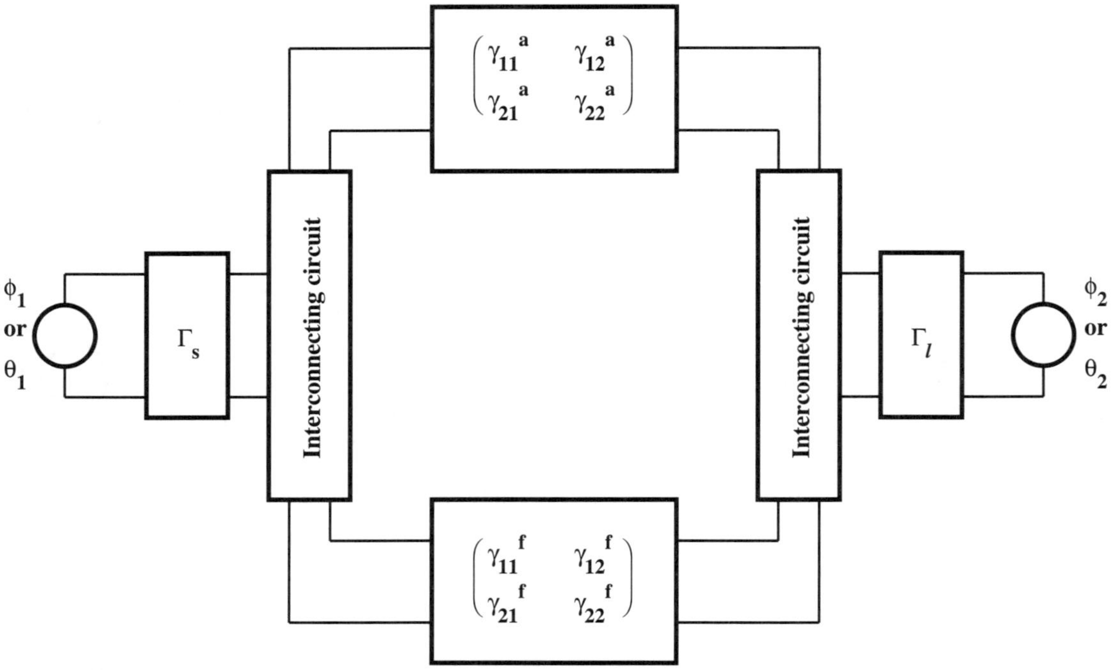

Fig. 11.4-2a.

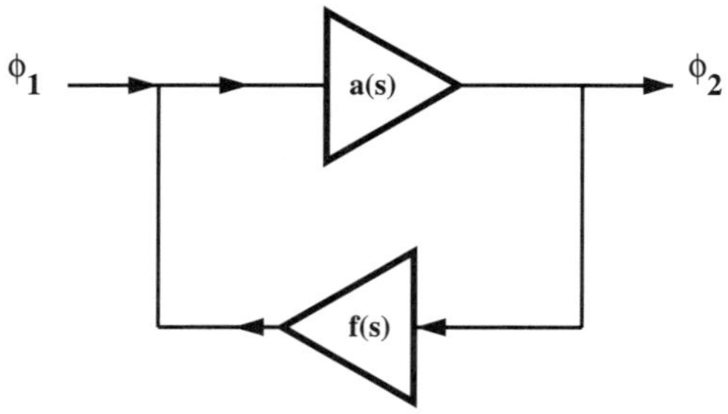

Fig. 11.4-2b.

Based on the interconnecting information in Fig. 11.4-2, and if the γ_{ij}'s are chosen correctly, we can have the following overall parameters for this network:

$$\begin{pmatrix}\gamma_{11}^{'T} & \gamma_{12}^{'T}\\ \gamma_{21}^{'T} & \gamma_{22}^{'T}\end{pmatrix} \triangleq \begin{pmatrix}\Gamma_s+\gamma_{11}^{a}+\gamma_{11}^{f} & \gamma_{12}^{a}+\gamma_{12}^{f}\\ \gamma_{21}^{a}+\gamma_{21}^{f} & \gamma_{22}^{a}+\gamma_{22}^{f}+\Gamma_l\end{pmatrix}$$

$$\triangleq \begin{pmatrix}\Gamma_s+\gamma_{11}^{T} & \gamma_{21}^{T}\\ \gamma_{21}^{T} & \gamma_{22}^{T}+\Gamma_l\end{pmatrix}. \tag{11.4-5}$$

Note that the above algebra is correct if and only if we can separate various sub-networks such that the port currents isolation properties remain valid.

Under the above conditions we may present some practical considerations to simplify the overall parameters for this two-port network to yield a few practical equations as follows.

(a) In the ideal case no reverse transmission of the feedforward amplifier is expected, i.e., $\gamma_{12}^{a} \approx 0$ or $\gamma_{12}^{a} \ll \gamma_{12}^{f}$.

(b) Similarly, we expect $\gamma_{21}^{a} \gg \gamma_{21}^{f}$.

We also note that

$$\frac{\phi_2}{\phi_1} = \frac{\Gamma_l \, \theta_2}{\phi_1} = \frac{-\Gamma_l \, \gamma_{21}^{T}}{(\Gamma_s + \gamma_{11}^{T})(\gamma_{22}^{T} + \Gamma_l) - \gamma_{12}^{T}\gamma_{21}^{T}} . \tag{11.4-6}$$

From Fig. 11.4-2b (and with positive feedback), we have

$$\frac{\phi_2}{\phi_1} = \frac{a(s)}{1 - a(s)f(s)} . \tag{11.4-7}$$

From the above two expressions we get

$$\frac{\phi_2}{\phi_1} = \frac{-\Gamma_l \, \gamma_{21}^{T}/(\Gamma_s + \gamma_{11}^{T})\,(\gamma_{22}^{T} + \Gamma_l)}{1 - \gamma_{12}^{T}\gamma_{21}^{T}/(\Gamma_s + \gamma_{11}^{T})\,(\gamma_{22}^{T} + \Gamma_l)} \triangleq \frac{a(s)}{1 - a(s)f(s)} , \tag{11.4-8}$$

which results in

$$\begin{cases} a(s) \triangleq \dfrac{-\Gamma_l \, \gamma_{21}^{T}}{(\Gamma_s + \gamma_{11}^{T})\,(\gamma_{22}^{T} + \Gamma_l)} , \\ f(s) \triangleq \dfrac{-\gamma_{12}^{T}}{\Gamma_l} . \end{cases} \tag{11.4-9}$$

Incorporating the approximations stated in (a) and (b) above yields

$$\begin{cases} a(s) \triangleq \dfrac{-\Gamma_l \, \gamma_{21}^{a}}{(\Gamma_s + \gamma_{11}^{T})\,(\gamma_{22}^{T} + \Gamma_l)} , \\ f(s) \triangleq \dfrac{-\gamma_{12}^{f}}{\Gamma_l} . \end{cases} \tag{11.4-10}$$

The implication of this algebra is that we must divide a rather complicated feedback problem into two separate sub-networks of feedforward and feedback circuits. Then we must calculate the open-loop transfer function for each block provided that the loading effects of the feedback network in both input and output ports of the feedforward network are appropriately incorporated (recall that $\gamma_{11}^{T} = \gamma_{11}^{a} + \gamma_{11}^{f}$ and

$\gamma_{22}^{T} = \gamma_{22}^{a} + \gamma_{22}^{f}$). Also, the feedback network is not loaded by a(s). Calculating both a(s) and f(s), then the transfer function between input and output becomes $\phi_2/\phi_1 = a(s)/[1 - a(s)f(s)]$. In reality though these conditions are seldom met completely, what we are calculating is a good approximation to the true transfer function. Nevertheless, the above method is easier to use than the true exhaustive and complete circuit analysis of the overall feedback network. A few subtle points regarding the above techniques must yet be explored, but we do not elaborate further.

11.5. COMMON ALGEBRA to ANALYZE CIRCUITS with THREE-TERMINAL DEVICES (TTD)

11.5.1. Introduction

Three-terminal devices (TTD's) in general are nonlinear elements in the sense that their corresponding port relations are nonlinear functions of their arguments. A typical representation of this device is shown in Fig. 11.5-1.

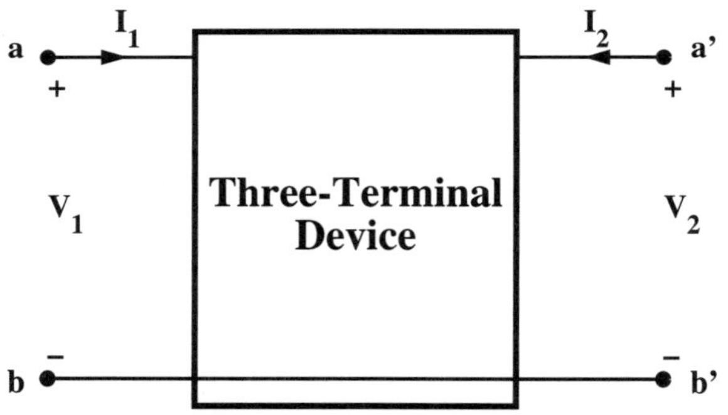

Fig. 11.5-1.

Here we have:

$$v_1 = v_1(i_1, i_2) \text{ or } v_1 = v_1'(i_1, v_2), \tag{11.5-1}$$

$$i_1 = i_1(v_1, i_2) \text{ or } i_1 = i_1'(v_1, v_2), \tag{11.5-2}$$

$$v_2 = v_2(i_2, i_1) \text{ or } v_2 = v_2'(i_2, v_1), \tag{11.5-3}$$

$$i_2 = i_2(v_2, i_1) \text{ or } i_1 = i_2'(v_2, v_1). \tag{11.5-4}$$

All the above functionals are nonlinear with respect to their arguments. To analyze these devices, irrespective of their physical structure, we must first choose one of the above models. Upon choosing this model, we have two different approaches called *analytical* and *graphical*, to complete the analysis.

To explain the above two methods we choose the following input and output relationships which appear in the form of two nonlinear equations.

$$v_1(t) = v_1'[i_1(t), v_2(t)], \tag{11.5-5}$$

$$i_2(t) = i_2[v_2(t), i_1(t)]. \tag{11.5-6}$$

The techniques we present in the following, with slight modifications, are applicable to a large class of circuit problems, including nonlinear circuits.

As is evident in the above model both $v_2(t)$ and $i_1(t)$ play the role of parameters for the input voltage and the output current equations, respectively. How do we analyze circuits involving these elements? Consider a general network N consisting of one TTD and three general sub-networks N_1, N_2 and N_3 as depicted in Fig. 11.5-2.

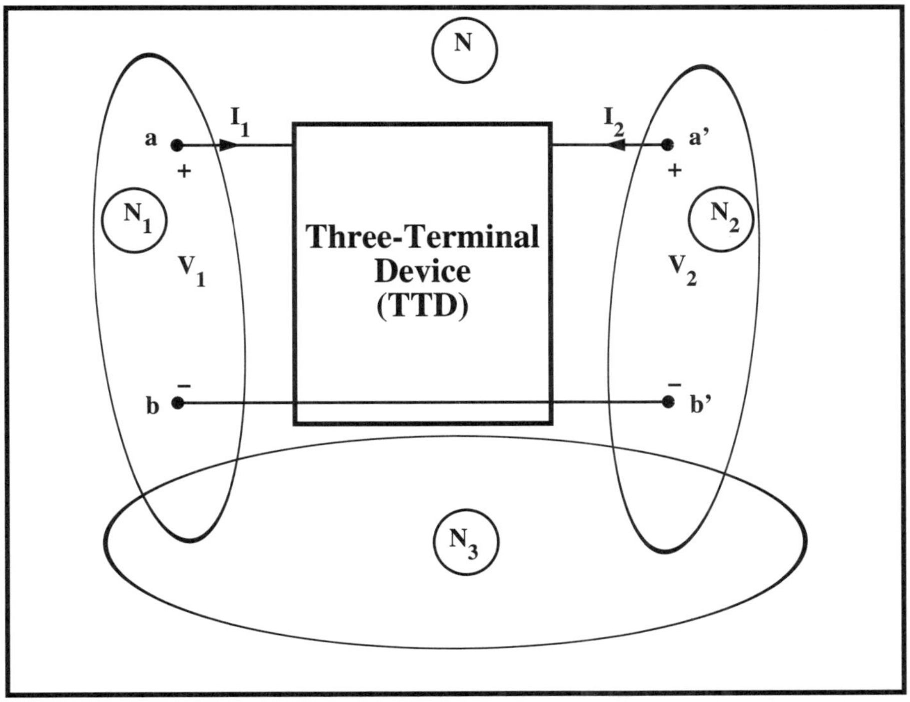

Fig. 11.5-2. The interconnection of a network with TTD.

Clearly, our goal is to analyze N, however, our orientation is to account for the effects of the TTD in various sub-networks in N. If we know the general nonlinear relationships of, say, $v_1(t)$ and $i_2(t)$ for the underlying device, and if we know the corresponding relationships between (i_1, v_1) and (i_2, v_2) from the network N_1 and N_2, respectively, then we can analyze N with computer help. But, in reality, we do not have definite analytical expressions for these equations. What we have at best are some nonlinear curves which are drawn in various coordinate systems depending on the nature of the experiments and the circuit parameters involved. In the following we present two general approaches to handle the analysis of a typical nonlinear network such as N as shown in Fig. 11.5-2.

11.5.2. Analytical Approach

First we recall an active device is the one that requires external dedicated power supply to bring it into an active operational mode. This has not been the case for passive elements such as RLC's. Since we have already assumed that the active TTD is a nonlinear element, and we do not have the exact analytical expressions neither for $v_1(t)$ nor for $i_2(t)$. Therefor we consider a perturbational analysis to tackle this problem. Our main *tool* for this analysis is, of course, the Taylor series expansion approach. We expand each of these input and output nonlinear equations around an *operating* point to derive a *local* or *small-signal* model for the TTD.

The procedure is as follows (we omit the real variable "t" for convenience). Recall that v_1 (or i_2) is a multi-variable function which can be expanded as

$$v_1 = \bar{v}_1 + \left.\frac{\partial v_1}{\partial i_1}\right|_* \Delta i_1 + \left.\frac{\partial v_1}{\partial v_2}\right|_* \Delta v_2 + \cdots, \tag{11.5-7}$$

$$i_2 = \bar{i}_2 + \left.\frac{\partial i_2}{\partial v_2}\right|_* \Delta v_2 + \left.\frac{\partial i_2}{\partial i_1}\right|_* \Delta i_1 + \cdots. \tag{11.5-8}$$

Here the over bar stands for the *nominal* or the operating value, i.e., the value of a function at the point for which we are trying to find its Taylor series expansion, and $|_*$ means evaluated at this operating point. The $\cdots$ means higher-order differentiation terms for v_1 and i_2, multiplied by higher-order terms of Δi_1, Δv_2 and $\Delta i_1 \Delta v_2$, as well as constants. If we assume that these Δi_1 and Δv_2 are *sufficiently small,* i.e., we can neglect Δi_1^2, Δv_2^2 and crossing terms $\Delta i_1 \Delta v_2$, then we are left with the above three terms in each expression, resulting in the incremental variations model for the given TTD as follows.

$$v_1 - \bar{v}_1 \triangleq \Delta v_1 = \left.\frac{\partial v_1}{\partial i_1}\right|_* \Delta i_1 + \left.\frac{\partial v_1}{\partial v_2}\right|_* \Delta v_2, \tag{11.5-9}$$

$$i_2 - \bar{i}_2 \triangleq \Delta i_2 = \left.\frac{\partial i_2}{\partial v_2}\right|_* \Delta v_2 + \left.\frac{\partial i_2}{\partial i_1}\right|_* \Delta i_1. \tag{11.5-10}$$

The implication of Δ in the above equation is just a *warning* sign to remind us that the right-hand side of these equations are only valid when $\bar{v}_1$ (or $\bar{i}_2$) is *sufficiently* close to v_1 (or i_2). As long as this requirement holds, then the Taylor series expansion is valid, otherwise this expansion loses its usefulness. Some textbooks have chosen to show, instead of Δ's, time-varying quantities. We may use time-varying notation as well in the future. We must of course, be alert that these quantities are all evaluated around the nominal points and are valid only locally. Also notice that we could have started with a host of other input and output relationships to come up with several equally valid local representations for the TTD. The procedure is exactly the same as in the above.

Having determined a local model, we must then evaluate the currents and the voltages of the network N for each of these incremental local models of the TTD perhaps with some computing help.

The next question that may be raised is, what are the implications of these various terms in (11.5-9 & 10)? Starting with the input equation and considering the physical dimensions of various terms, we propose to call $r_{in} \triangleq \left.\frac{\partial v_1}{\partial i_1}\right|_*$ (input resistance) and $\mu \triangleq \left.\frac{\partial v_1}{\partial v_2}\right|_*$ (reverse amplification). Similarly, $\frac{1}{r_o} \triangleq \left.\frac{\partial i_2}{\partial v_2}\right|_*$ (output conductance) and $\beta = \left.\frac{\partial i_2}{\partial i_1}\right|_*$ (current gain). Putting together these terms results in an equivalent small-signal model for the TTD as depicted by the circuit in Fig. 11.5-3, which is valid only for small intervals of currents and voltages, i.e., Δi's and Δv's.

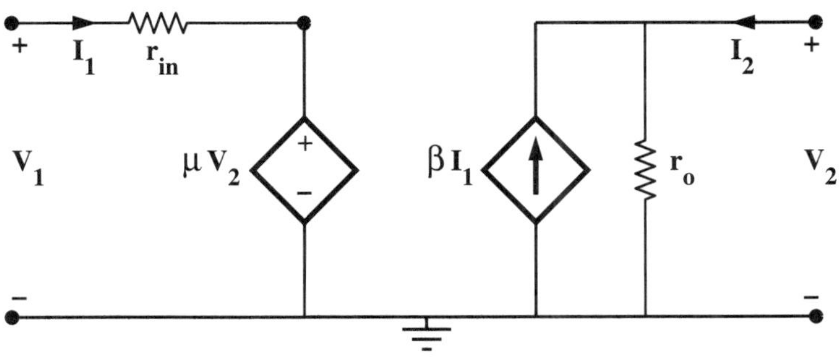

Fig. 11.5-3.

Substituting this equivalent model for the TTD in the network N of Fig. 11.5-2 (provided that all the biasing voltage sources associated with this TTD are shorted – the current sources are opened), we can solve for various currents and voltages of the elements in that network. The result is, of course, local; i.e., it depends on the operating point for which this model is valid. If we want to have a result for a range wider than what we get from this model, then we must seek a new operating point which is the old operating point plus or minus the maximum of Δi_2 or Δv_1, and repeat generating another small-signal model. These are standard techniques given for the analysis of nonlinear networks. We believe reviewing this information at this stage will be very useful in order to prepare students for their future courses. This is the time to reflect and think how we can have a bank of local models for a complicated nonlinear circuit and use that to solve such a problem for as wide an interval as we need. All we need is a computer. Thus, it seems that if we apply ourselves we really can solve any circuit problem no matter how complicated that circuit may become.

11.5.3. Graphical Approach

Another common technique to analyze circuits involving nonlinear TTD's is the graphical approach. The procedure is rather simple and we describe its principal features in the following. Consider network N in Fig. 11.5-2. Suppose, instead of analytical data regarding the input voltage $v_1(t)$ we had the graph as that in Fig. 11.5-4. Similarly, if instead of the analytical output current $i_2(t)$, we had the graph as shown in Fig. 11.5-5. These are the TTD characteristic curves. We can now use the following techniques to find the operating current and voltage values for network N.

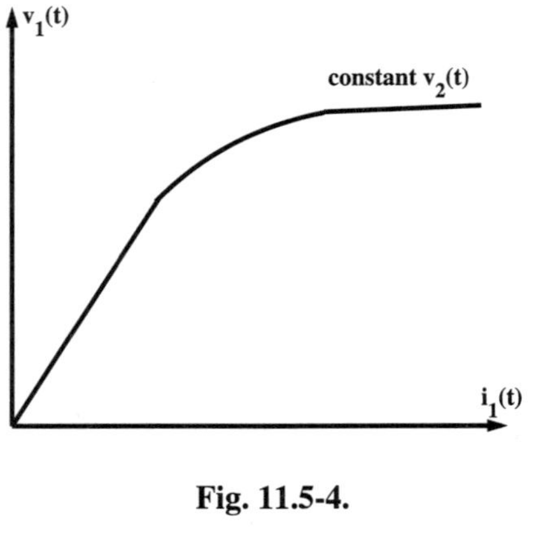

Fig. 11.5-4.

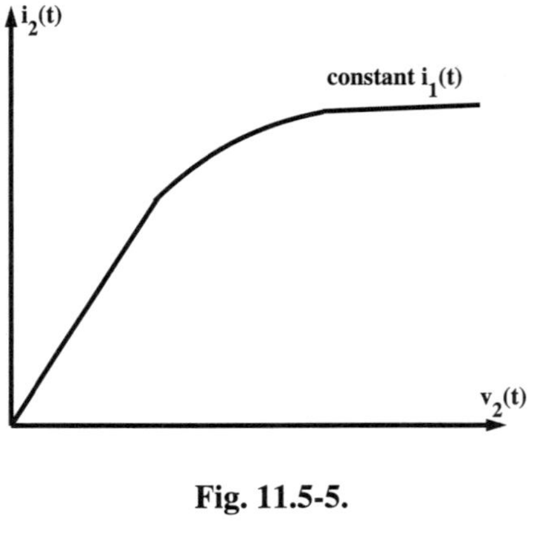

Fig. 11.5-5.

Notice:

Here we are not concerned about any specific sign and/or polarities in these graphs. Our purpose is to demonstrate only the graphical procedure. (Problem 11.8.11).

The next step is to develop a relation between (v_1, i_1) of network N_1, say $v_1(t) = R_1[i_1(t)]$, and (i_2, v_2) of network N_2, say $i_2(t) = G_2[v_2(t)]$. These nonlinear functions are called *load lines.* Obviously the input as well as the output operating points must satisfy both of these equations as well as the characteristic curves concurrently. Thus the problem becomes very manageable when we graph the function R_1 on the coordinates of Fig. 11.5-4, and graph the function G_2 on the coordinates of Fig. 11.5-5, as shown in Figs. 11.5-6 & 7. Intersections of these two curves determine the input and output operating points for network N.

Although we have assumed that both R_1 and G_2 are nonlinear functions, they are generally linear functions. We do not elaborate further on these issues, nevertheless these graphical techniques will be quite useful for this type of problems, which form the essential of electronic circuits to be studied in an electrical engineering program.

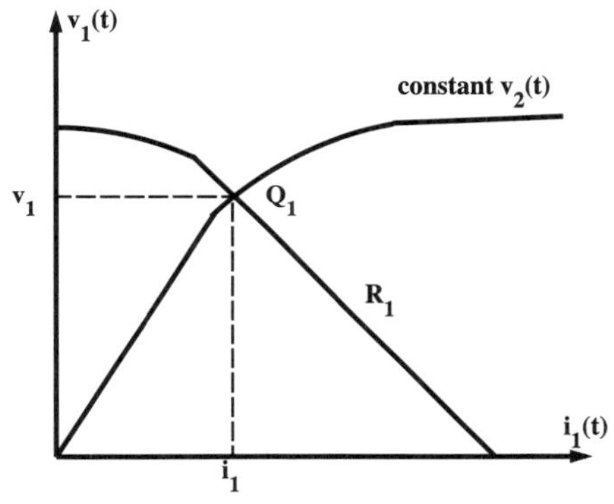

Fig. 11.5-6.

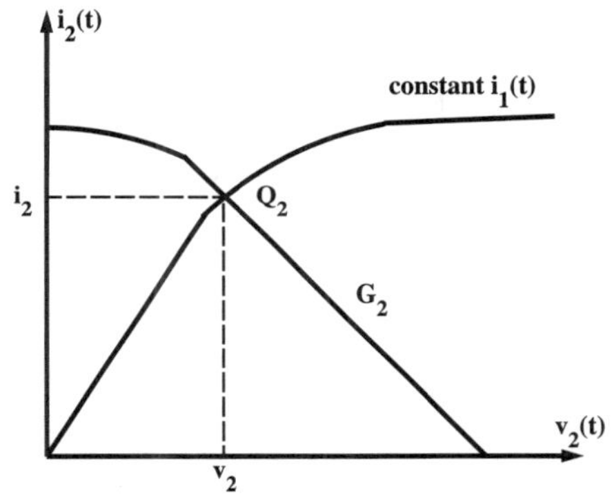

Fig. 11.5-7.

11.6. MULTI-PORT NETWORKS

We close this chapter with the following observation. If we have many ports, say, n-ports, then the analysis in Section 11.2 can be generalized to develop similar set of parameters for an n-port network. For instance in terms of *Z*–parameter, we have the following general case.

$$\begin{Bmatrix} V_1 \\ V_2 \\ \cdot \\ \cdot \\ \cdot \\ V_n \end{Bmatrix} = \begin{pmatrix} z_{11} & z_{12} & \cdots & z_{1n} \\ z_{21} & z_{22} & \cdots & z_{2n} \\ \cdot & \cdot & \cdot & \cdot \\ \cdot & \cdot & \cdot & \cdot \\ \cdot & \cdot & \cdot & \cdot \\ z_{n1} & z_{n2} & \cdots & z_{nn} \end{pmatrix} \times \begin{Bmatrix} I_1 \\ I_2 \\ \cdot \\ \cdot \\ \cdot \\ I_n \end{Bmatrix}. \tag{11.6-1}$$

This equation can be partitioned in a number of ways. Suppose, V_3 to V_n are zero, then we have

$$\left\{ \begin{matrix} \begin{pmatrix} V_1 \\ V_2 \end{pmatrix} \\ \begin{pmatrix} 0 \\ \cdot \\ \cdot \\ \cdot \\ 0 \end{pmatrix} \end{matrix} \right\} = \begin{pmatrix} Z_{11} & Z_{12} \\ Z_{21} & Z_{22} \end{pmatrix} \times \left\{ \begin{matrix} \begin{pmatrix} I_1 \\ I_2 \end{pmatrix} \\ \begin{pmatrix} I_3 \\ \cdot \\ \cdot \\ \cdot \\ I_n \end{pmatrix} \end{matrix} \right\}, \tag{11.6-2}$$

where Z_{ij}'s are partition matrices of appropriate dimension. Or equivalently we have the following two equations instead of (11.6-2).

$$\begin{Bmatrix} V_1 \\ V_2 \end{Bmatrix} = Z_{11} \times \begin{Bmatrix} I_1 \\ I_2 \end{Bmatrix} + Z_{12} \times \begin{Bmatrix} I_3 \\ \cdot \\ \cdot \\ \cdot \\ I_n \end{Bmatrix}, \tag{11.6-3}$$

$$0 = Z_{21} \times \begin{Bmatrix} I_1 \\ I_2 \end{Bmatrix} + Z_{22} \times \begin{Bmatrix} I_3 \\ \cdot \\ \cdot \\ \cdot \\ I_n \end{Bmatrix}. \tag{11.6-4}$$

Clearly, Z_{22} is a square matrix, and if its inverse exists, then we have

$$\begin{Bmatrix} I_3 \\ \cdot \\ \cdot \\ \cdot \\ I_n \end{Bmatrix} = -Z_{22}^{-1} Z_{21} \times \begin{Bmatrix} I_1 \\ I_2 \end{Bmatrix}. \tag{11.6-5}$$

Substituting (11.6-5) in (11.6-3) yields

$$\begin{Bmatrix} V_1 \\ V_2 \end{Bmatrix} = Z_{11} \times \begin{Bmatrix} I_1 \\ I_2 \end{Bmatrix} - Z_{12} Z_{22}^{-1} Z_{21} \times \begin{Bmatrix} I_1 \\ I_2 \end{Bmatrix}$$

$$= \left[Z_{11} - Z_{12} Z_{22}^{-1} Z_{21} \right] \times \begin{Bmatrix} I_1 \\ I_2 \end{Bmatrix}. \tag{11.6-6}$$

The above is just one flavor of this type analysis where the quantity in bracket is similar to the Z–parameter for the corresponding network. This is a very rich area of learning and we invite the reader to study further these issues.

11.7. OPERATIONAL AMPLIFIERS

Operational amplifiers are complex, two-port, active devices, where active means they need dedicated power supply to be functional. From circuit theory point of view, to dissect strictly the internal circuit of a typical operational amplifier requires a basic knowledge of properties of transistors in conjunction with two-port active networks. Such analysis as needed before venturing in the realm of circuits with operational amplifiers is not the subject of this textbook. Also, in today's complex electronics design, while circuits with just a few transistors and discrete elements have limited applications, the advent of integrated circuits design has enabled us to construct circuits which are many times more complicated than those built using discrete elements and transistors. Furthermore, these circuits take less space and are more reliable with more diverse functionality than ever before. The technology that makes such *miniaturization* possible is called integrated circuits (microelectronic) technology. These integrated circuits (or chips) are abbreviated as IC's. The only IC that is of interest is called an operational amplifier or Op-Amp for short.

Although Op-Amp's as stated above are active two-port devices in which their internal circuits consist of many transistors, diodes (diodes are a class of one-port nonlinear circuit elements), resistors and capacitors, these devices are considered as a new class of network elements. Trends in the contemporary engineering programs are to review aspects of the input-output (terminal) characteristics of these devices in the first circuit course. Indeed, for many engineering students this may be their only chance to learn about these widely used devices. Clearly, some functional behaviors of these devices when they are working in their linear mode as defined subsequently can be studied in a first circuit course. In fact, to elucidate examples for at least one class of two-port networks, the essential properties of an operational amplifier can be reviewed in the first circuit course. Because, this family of two-port active devices has an important role in the overall study of basic electronic circuits design. Also, as an active two-port device, circuits with operational amplifiers have all the relevant properties of such two-port networks. In particular, this device represents a good example of applications for the concept of feedback amplifiers, as described in Section 11.4. However, it would have been out of character for the rigorous development of the materials in this textbook, had operational amplifiers were presented in earlier chapters as is the case elsewhere. Because, we strongly believe that in order to be accurate and rigorous, even the linear modes of these devices must be reviewed in the context of two-port networks and after such topics as frequency response and feedback amplifier are established.

Thus, in the following, upon some additional remarks on these devices, we study their fundamental properties from their input-output (terminal) characteristics points of view. We review their input-output (terminal) characteristics as linear devices, when such a linear operation is possible. On that note and when we learn their linear characteristics, these active devices are not any different from other elements which we have studied their terminal behaviors in this textbook.

Having said the above general remarks regarding the input-output (terminal) characteristic, as well as the internal circuitry of an operational amplifier, we explore some of the general characteristics of this device in the context of an active two-port network. We bear in mind that each manufacturer or vendor of these devices has specific instructions about the general properties of the terminals or pins for their products. Therefore it is not essential for us to be too specific about the exact physical structure of the underlying device as far as the pin numbering or similar manufacturing matters are concerned. Since these details may differ from one brand to another. But rather we must be concerned with the generic properties of a typical Op-Amp, and to gain that knowledge is the purpose of our study. To do so, we present our case by first introducing a purely mathematical or an ideal model for these devices. Followed by demonstrating its real behaviors and/or its deviations from this perfect model, in order to develop an accurate practical model for this device. Our presentation closely follows [2], where many practical circuits using Op-Amp's with manufacturers' specification data sheets are displayed.

Each Op-Amp has at least eight terminals or external pins that are numbered, and depending on the vendors, each pin may have a specific meaning. Also, there are chips which contain more than one Op-Amp in a package. However, we are not concerned with the manufacturing aspects of the design of a chip, but rather with one typical or generic device. On that note, each of these chips has five important pins as shown in Fig. 11.7-1a. The dedicated power supply which is needed for the operation of an Op-Amp usually must provide voltages in the range of ±20 volts and less. Usually these supply voltage sources are understood and are not shown in the figures. For an input voltage that results in an output voltage in this range, the Op-Amp acts as a linear device. If the input voltage yields an output voltage that theoretically exceeds the $|V_{supply}|$, or for that matter if the input voltage theoretically exceeds $|V_{supply}|/G^{v}_{ol}$

(G^v_{ol} is the Op-Amp's open-loop gain), then the device *saturates,* and nonlinearity will occur, Fig. 11.7-1b.

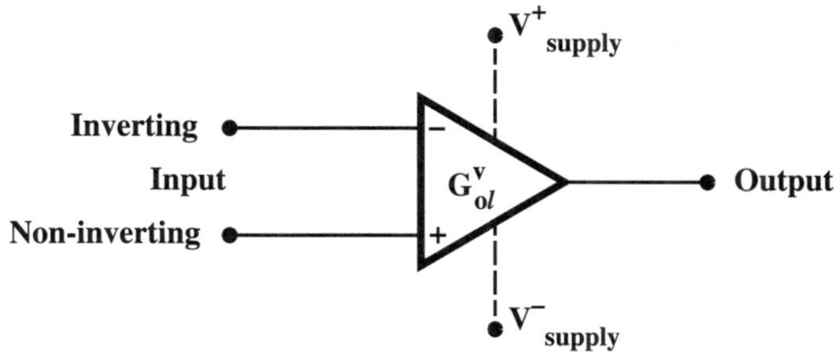

Fig. 11.7-1a. Schematic diagram of an Op-Amp.

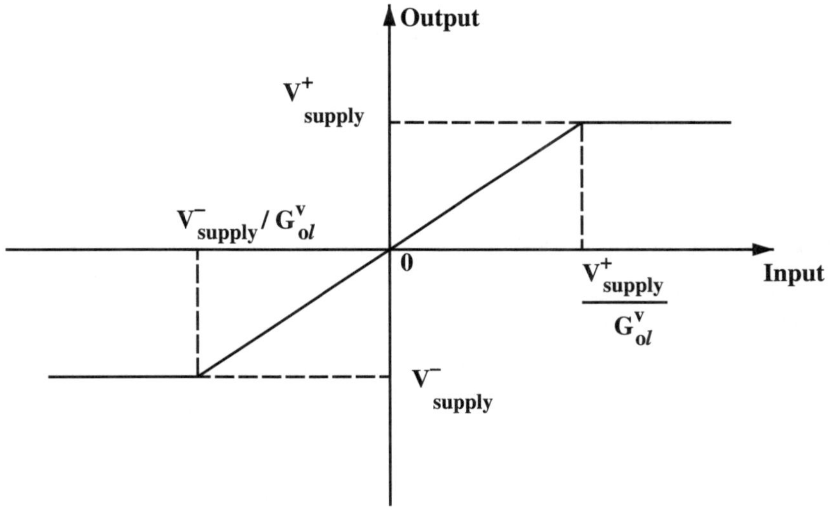

Fig. 11.7-1b. Input-output characteristic of an Op-Amp.

In the following we first define a mathematical or an ideal model for this device. Then we study similarly an actual or physical Op-Amp with all its limitations and/or non-idealness problems. *Both studies are at the terminal level,* i.e., without dissecting the internal circuitry of the device. Because of an extremely high gain factor, these devices are not generally used in an open-loop fashion. Subsequently, we study their applications in circuits with feedback configurations.

11.7.1. Mathematical Model for an Ideal Open-loop Op-Amp

A mathematical model for an ideal Op-Amp is depicted in Fig. 11.7-2, with a list of properties and parameters that follows. This mathematical model should be considered as the limiting configuration for an ideal element that does not exist as such physically, nevertheless the model serves as a frame of reference to compare the properties of a practical Op-Amp with an ideal one.

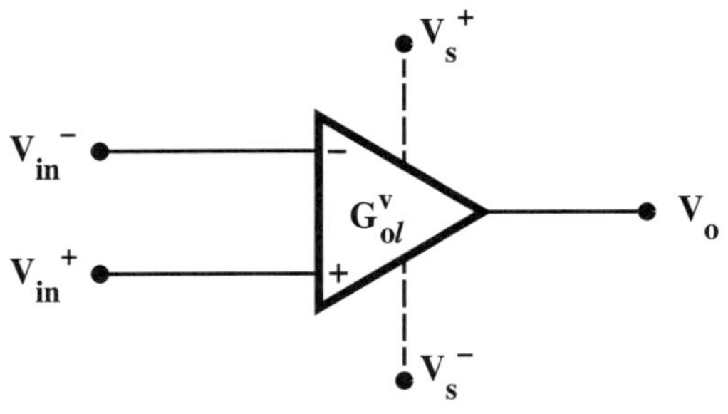

Fig. 11.7-2. An ideal and open-loop Op-Amp.

A. Infinite Open-Loop DC Gain ($G_{ol}^v = \infty$)

This gain is the ratio of the maximum output voltage swing with load to the change in the input voltage required to derive the output from zero to this value. For a practical Op-Amp this gain is finite, though a large number. A typical value for this gain is in the order of 200,000. This finite and in most cases frequency-dependent, open-loop gain is the source of many errors (i.e., deviations from its idealness) in Op-Amp. Also, this gain is responsible for the range of the maximum output voltage that the underlying device produces. For linear operation of an Op-Amp, the input should not exceed $|V_{supply}|/G_{ol}^v$. In most applications we consider this device as a unit for which a small differential input of $(V_{in}^- - V_{in}^+)$ yields an output voltage which is

$$V_o = -G_{ol}^v(V_{in}^- - V_{in}^+). \tag{11.7-1}$$

This relationship is the fundamental definition for the operation and functionality of an operational amplifier. In the case of an ideal device this gain approaches infinity as the input difference goes to zero, i.e.,

$$G_{ol}^v \rightarrow \infty \text{ as } (V_{in}^- - V_{in}^+) \rightarrow 0. \tag{11.7-2}$$

The implication of this definition is that for an ideal device with unusually high gain factor a very small input differential signal is amplified to a level which is measurable and that property is the beauty of this device. Because any instrumentation or signal processing starts with a very low-level signal that must be amplified for a meaningful processing of any kind and this device does exactly that. Based on this assumption for the existence of such a large gain factor, the following set of parameters seems reasonable and justifiable. Thus, when the device is not functioning as defined, we realize that there is a discrepancy in this gain and such discrepancy is resulted from the fact that this gain though very large is finite, and often is less than what we have assumed.

B. Infinite Input Resistance ($R_{in} = \infty$)

An input resistance is the ratio of input voltage over input current. In general this quantity is finite, though is very large – say in the order of mega-ohms. Equivalently, this statement is the same as saying no current can enter an ideal Op-Amp.

C. Zero Output Resistance ($R_o = 0$)

This definition is similar to the input resistance except in terms of the ratio of output voltage and current. In most practical situations this quantity is in the order of a few ohms.

D. Zero Input Offset Voltage ($V_{io} = 0$)

This value means there is no offset voltage or simply as $(V_{in}^{-} - V_{in}^{+}) \rightarrow 0$, then $V_o \rightarrow 0$.

E. Infinite Bandwidth ($BW = \infty$)

Infinite bandwidth implies that all signals will be processed equally well, without regard to their frequency contents, which is not the case in the real situations. In other words, real Op-Amp's are essentially DC devices, operating best when the frequency of the input signal is zero.

Putting together all the above definitions, we have the equivalent circuit for the model of an ideal Op-Amp as shown in Fig. 11.7-3. Although the assumptions that we have made often simplify our subsequent analysis drastically, in practice there is no such device as an ideal Op-Amp. The main source of discrepancy in the application of the Op-Amp as an ideal device is the fact that G_{ol}^{v} is large but not infinite. Also, due to this large open-loop gain, it is desirable to use this device in conjunction with a feedback configuration. Thus, we briefly review the concept of circuits with operational amplifiers in feedback configurations next.

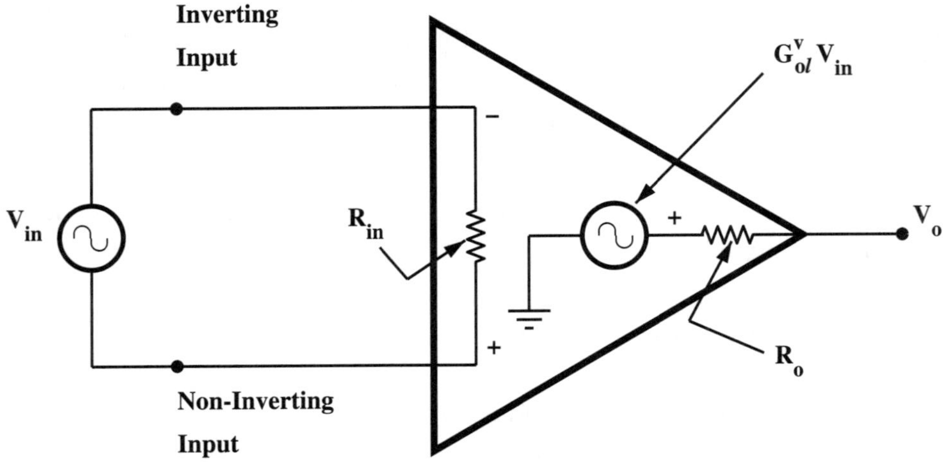

Fig. 11.7-3. Equivalent circuit for an ideal Op-Amp.

11.7.2. Mathematical Models for Feedback Configurations of an Ideal Op-Amp

We now study two commonly used circuits with Op-Amp's in feedback configurations. Using their mathematical models, in each category we choose the associated Op-Amp to be an ideal device in the sense that as $(V^-_{in} - V^+_{in}) \to 0$, $G^v_{ol} \to \infty$. Also, it is assumed that no current enters this device.

Starting with an amplifier circuit shown in Fig. 11.7-4, a few words on the existence of feedback in this circuit are in order. In Fig. 11.7-4, a portion of the output voltage is fed back to the node SP (summing point) through $Z_f(s)$. Consider Z_1, Z_f as a voltage divider. We continue our circuit analysis using definitions and properties of an ideal Op-Amp. Here, we have $V^+_{in} = 0$. From the idealness of the Op-Amp, we conclude that $V^-_{in} = 0$. This value defines what is known as a *virtual ground*. In other words, at node SP, $V^-_{in} = V^+_{in} = 0$. Also, we have $I_{in} = I_f$, since no current enters the Op-Amp. Node analysis based on these assumptions yields the following fundamental relationship for an ideal Op-Amp in a configuration known as an *inverting* feedback:

$$\frac{V_o(s)}{V_{in}(s)} = -\frac{Z_f(s)}{Z_1(s)} \triangleq G_{cl}(s). \qquad (11.7\text{-}3)$$

Here, the term inverting refers to the minus sign in this equation. Also, $G_{cl}(s)$ refers to the closed-loop gain in the context of a block-diagram representation for this circuit constructed similar to that in Fig. 11.3-1, by calculating the corresponding $G_{ol}(s)$ and $H(s)$. To do so, the materials in § 11.7.4, where we perform the necessary steps for such a construction, but without explicitly building a block-diagram representation, should be consulted. (Problem 11.8.17). The beauty of feedback is that we now can control within an acceptable range the value of $G_{cl}(s)$ despite uncertainty in $G_{ol}(s)$. (Problem 11.8.20).

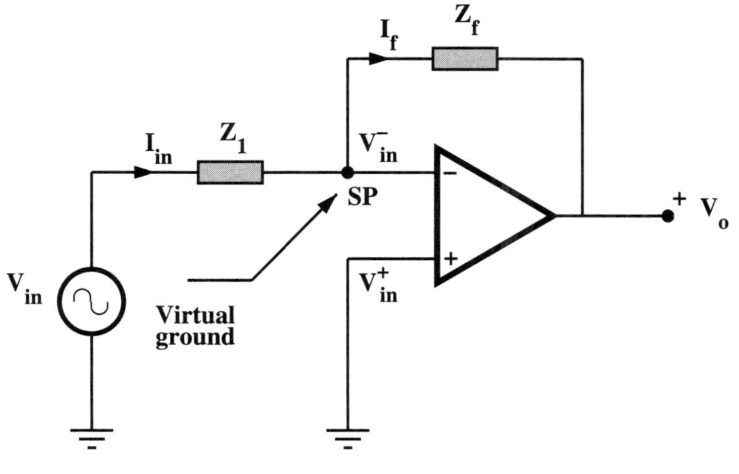

Fig. 11.7-4. An inverting feedback amplifier.

Now, consider the *non-inverting* circuit shown in Fig. 11.7-5. From the idealness properties of the Op-Amp, we conclude $V_{in} = V_{in}^{+} = V_{in}^{-}$, and $I_{in} = I_f$, since no current enters the Op-Amp. Node analysis for this circuit also yields a manageable gain.

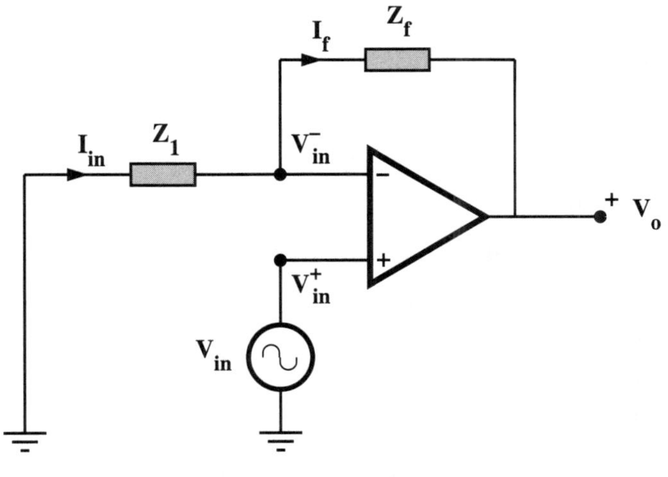

Fig. 11.7-5. A non-inverting (positive) feedback amplifier.

$$\frac{V_o(s)}{V_{in}(s)} = \frac{Z_1(s) + Z_f(s)}{Z_1(s)}. \tag{11.7-4}$$

In (11.7-4), if we let $Z_f(s) \rightarrow 0$ (or theoretically let $Z_1(s) \rightarrow \infty$ although this is not recommended), then we have $V_o \equiv V_{in}$. This circuit is called a *unity-gain feedback amplifier,* or a *voltage-follower*, which has many applications in design, such as serving as a buffer. One such circuit is shown in Fig. 11.7-6a, below.

11.7.3. Parameters of a Practical Open-Loop Op-Amp

The source of all discrepancies for non-idealness of an Op-Amp is the fact that G_{ol}^{v} is not infinite, nor is a constant. This gain or transfer function is indeed a frequency-dependent quantity. For instance, it is commonly known that such a gain or transfer function can be described at least by a low-pass filter such as

$$G_{ol}^{v}(s) = \frac{G_0}{1+\tau_0 s}. \tag{11.7-5}$$

This class of transfer functions or filters are studied rigorously in Chapter Ten. However, for the sake of continuity we recall that $s = j\omega$ is the Laplace variable, with $j = \sqrt{-1}$. In (11.7-5), τ_0 is the device's time constant, which is, of course, very small, nevertheless is not zero. If τ_0 were zero, then we could have had an ideal device. Correspondingly, to develop a relationship between the properties of an ideal amplifier and those of a practical one, we must use the above expression for the gain of the Op-Amp, instead of the constant G_{ol}^{v}, which we have used customarily in the preceding § 11.7.1. Based on this new expression, we can re-evaluate the Op-Amp parameters that we have studied in § 11.7.1 as follows.

A. The Open-Loop Gain $G_{ol}^{v}(s)$

As mentioned above, we now have a first-order, low-pass, transfer function to describe the open-loop gain of an Op-Amp. The *true* open-loop gain or transfer function of an Op-Amp, however, is more complicated than that presented approximately in (11.7-5). Further elaboration on exact expression is beyond the scope of our analysis. In any event, our main objective is to use the fundamental definition for the operation or functionality of an Op-Amp in a given circuit, in order to complete the analysis and/or design of the underlying circuit. The only change that occurs in this fundamental relationship is in the expression for open-loop gain term that must now be replaced by that for our practical Op-Amp, i.e., a transfer function similar to (11.7-5). This substitution is the only change we must make whenever we are using a practical Op-Amp. We may interpret this situation by considering the gain of our practical amplifier to be the transfer function of a feedback amplifier consisting of an ideal Op-Amp in a feedback configuration, such that the overall feedback gain becomes what we call $G_{ol}^{v}(s)$ of our practical amplifier.

B. The Input Resistance R_{in}

Here, this parameter is generally a frequency-dependent quantity, whose amplitude is in the order of mega-ohms.

C. The Output Resistance R_o

This parameter is also a frequency-dependent quantity, whose magnitude is in the order of tens of ohms. Because of this low output impedance, if by some mistake the output be shorted to ground, then an extremely large current will pass through this impedance that most likely will destroy the entire circuit. Usually manufacturers provide additional circuitry inside the device to limit the occurrence of this short-circuit current. If this protection has not been made for us, then we must use additional external circuitry to protect the device, but generally this addition will degrade the overall performance of the amplifier. This output impedance and the short-circuit current limiter must be consistent for proper functioning of the device. To show the importance of the proper choice of this output resistance, consider the *unity-gain feedback amplifier* circuit shown in Fig. 11.7-6a.

Fig. 11.7-6a.

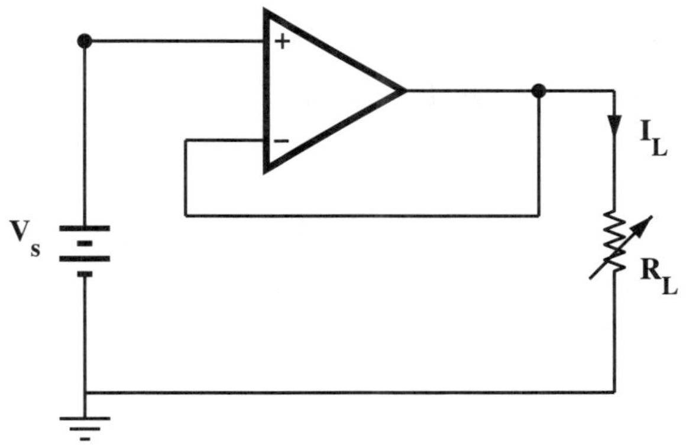

Fig. 11.7-6b.

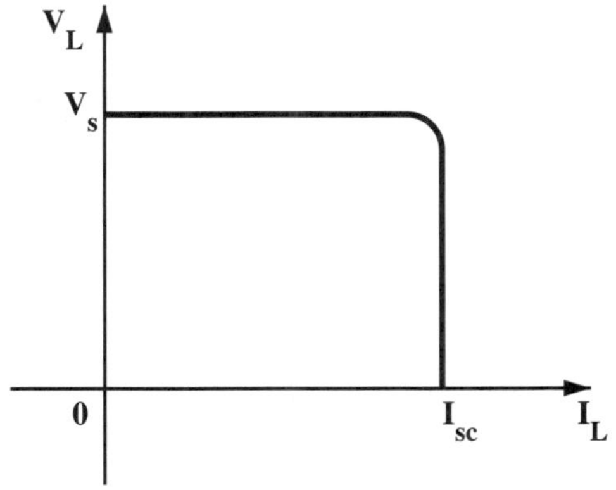

In Fig. 11.7-6a, we can show that as R_L decreases the current I_L increases, while the output voltage remains almost constant (cf., Fig. 11.7-6b). Gradually as R_L drops to a value corresponding to the short-circuit current I_{sc}, then further decrease will not change this short-circuit current. The current that is statically being limited actually saves the device. If instead of the constant DC voltage source we use $v_s(t) = V_s \sin \omega t$

and choose a small loading resistor, then it is quite possible that the alternating V_L becomes that shown in Fig. 11.7-6c, which is clipped on both the positive and negative portions of the cycle.

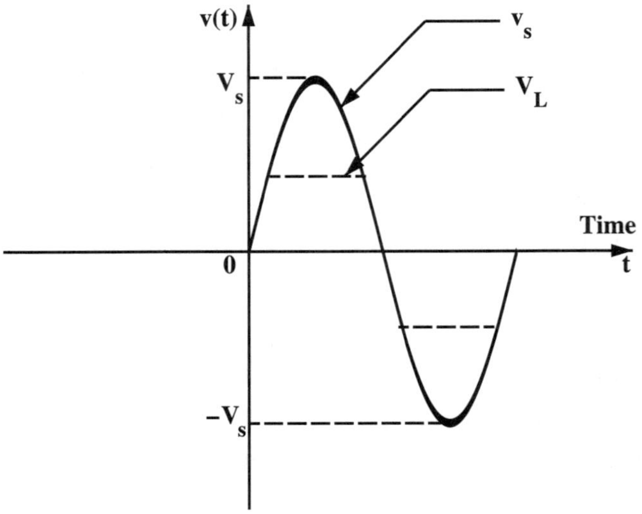

Fig. 11.7-6c.

D. The Input Offset Voltage V_{io}

In a practical amplifier, various offset voltages are not zero. The input offset V_{io}, *per se,* is the voltage that must be applied between the input terminals through two equal resistors to obtain zero output voltage. This voltage may have different polarities. The output offset voltage V_{oo}, is the voltage which we have for an Op-Amp with $V_{in}^- = V_{in}^+$.

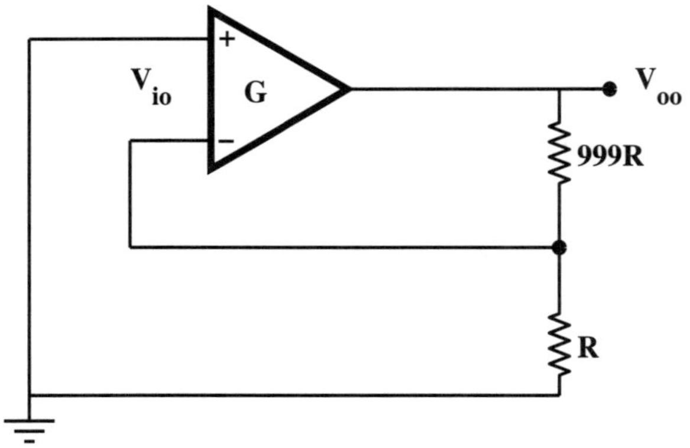

Fig. 11.7-7.

The circuit in Fig. 11.7-7 gives a simple procedure to measure this offset voltage. The resistor R is chosen such that $|V_{io}| \gg |I^- R|$, where I^- is the current that enters the device because of non-idealness. Thus $V_{oo} = G\,(V_{io} - 10^{-3}\,V_{oo})$, where the voltage across R is fed back to the input. This equation can also be written as

$$V_{oo} = GV_{io}/(1 + 10^{-3}G) \approx 10^3\,V_{io}. \qquad (11.7\text{-}6)$$

Using this expression and the fact that it is easy to read V_{oo} with a voltmeter, therefore V_{io} can be determined readily.

E. The Bandwidth (BW)

We have noticed that for the most practical purposes we must use the Op-Amp in a feedback configuration due to its very high open-loop gain. In general, even for an open-loop Op-Amp the BW is not infinite. To measure this parameter we must follow the general approach of drawing the amplitude for the transfer function versus the operational frequency from which the BW becomes easily derived (cf., Fig. 10.7-2, for a typical such graph). For instance, with the choice of a first-order low-pass filter model for an Op-Amp, say (11.7-5), we can draw its Bode diagram with the corresponding corner frequency to find the overall BW of the device.

For a negative closed-loop case, we have $G^{v}_{cl}(s) = G^{v}_{ol}(s)/[1 + G^{v}_{ol}(s)H(s)]$. If $|G^{v}_{ol}(j\omega)H(j\omega)| >> 1$, then $G^{v}_{cl}(j\omega) \approx 1/H(j\omega)$, and for $|G^{v}_{ol}(j\omega)H(\omega)| \ll 1$, we have $G^{v}_{cl}(j\omega) \approx G^{v}_{ol}(j\omega)$. At the corner frequency of ω_c, these two quantities become the same. Substituting for $|G^{v}_{ol}(j\omega_c)|$, and for $\omega_c > \omega_0$, results in

$$\frac{1}{|H(j\omega_c)|} = \frac{G_0\omega_0}{\omega_c} \quad \text{or} \quad G_0\omega_0 = \omega_c/|H(j\omega_c)|. \tag{11.7-7}$$

The above expression shows that the Gain-Bandwidth product (GBW) for our amplifier is constant, and we call this product as $\omega_u . 1$. Here, ω_u stands for the unity-gain angular frequency.

There are several other parameters associated with a practical Op-Amp, which are described next. In the case of an ideal Op-Amp, these parameters take on their limiting values which are also described in each case below.

F. The Input-Bias Current I_B

We must realize that in a practical Op-Amp some bias current must enter the device. The current that is required at the input of a single-input amplifier or the average of the two input currents for a *differential*-input amplifier, when the output voltage is at zero, is called the input-bias current. In the complete balance case these two input currents are zero. In general however,

$$I_B = (I_B^- + I_B^+)/2. \tag{11.7-8}$$

G. The Input-Offset Current I_{io}

The difference in currents entering into the two input terminals of a differential amplifier when the output voltage is at zero, is called the input-offset current. But in general the dependence of both bias and offset current on the output voltage is minimal. Thus,

$$I_{io} = I_B^- - I_B^+. \qquad (11.7\text{-}9)$$

We notice that I_B in (11.7-8) is of one polarity, while I_{io} in (11.7-9) is of either polarities.

We may use the above two relations to model I_B^- and I_B^+ as follows.

$$I_B^- = I_B + I_{io}/2 \quad \text{and} \quad I_B^+ = I_B - I_{io}/2\,. \qquad (11.7\text{-}10)$$

To measure these quantities we look into the circuits in Figs. 11.7-8a & b. In Fig. 11.7-8a, we have $V_{oo} = I_B^- R$, and in Fig. 11.7-8b, we have $V_{oo} = -I_B^+ R$. With a right choice of R we can measure V_{oo}, and thus determine the I_B^- and I_B^+. Then from (11.7-8) we have I_B, and from (11.7-9) we compute I_{io}.

Fig. 11.7-8a.

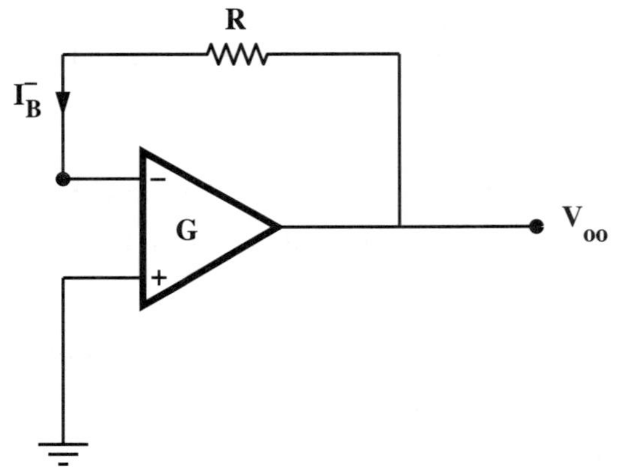

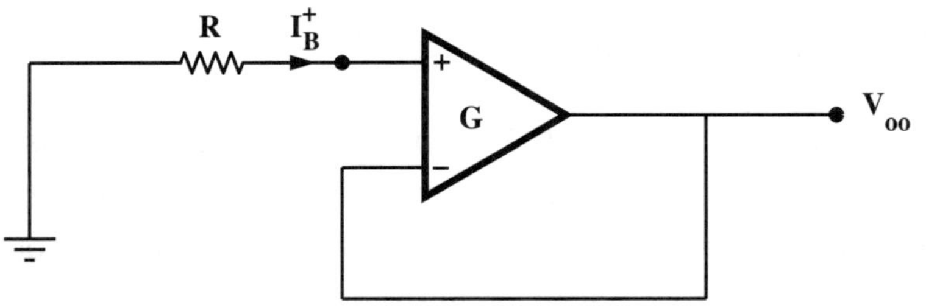

Fig. 11.7-8b.

H. The Common-Mode Rejection Ratio CMRR

It is a well-known fact that not only do amplifiers respond to the differential or input signal, but experience has shown that they also respond to another signal that is called the common-mode signal. In other words, if we want to amplify the difference between two sets of signals, say one set is 10^{mV} and 12^{mV}, and the other set is 50^{mV} and 52^{mv}, which the difference between these two signals in both sets is 2^{mV}, but one set is averaged at 11^{mV}, while the other set is averaged at 51^{mV}, then the result for amplification of the difference signal will be different in these two cases. In other words, the level of the amplitude for the signal or its common mode has also something to do about the final outcome for amplification. To be clear about this concept let us define the following parameters:

$$V_{dm}(\text{differential} - \text{mode}) = V_{in}^{-} - V_{in}^{+}, \tag{11.7-11a}$$

$$V_{cm}(\text{common} - \text{mode}) = (V_{in}^{-} + V_{in}^{+})/2. \tag{11.7-11b}$$

Equivalently, we can state that:

$$V_{in}^{-} = V_{cm} + V_{dm}/2, \tag{11.7-12a}$$

$$V_{in}^{+} = V_{cm} - V_{dm}/2. \tag{11.7-12b}$$

Since V_o is a function of both V_{in}^{-} and V_{in}^{+} and these terms per (11.7-12) are functions of both V_{dm} and V_{cm}, we can state that

$$V_o = G_{dm} V_{dm} + G_{cm} V_{cm}. \tag{11.7-13}$$

In other words, the output signal has two corresponding components, while each has its own gain factor, but together they form the output signal. Since V_{dm} is the signal that we have more control over, or perhaps interest in, than V_{cm}, it is natural to expect that we want to have G_{dm} substantially larger than G_{cm}. In other words, in order that the effect of V_{cm} be minimized, the ratio of G_{dm}/G_{cm}, must be maximized. This ratio for all intents and purposes is extremely large in any design, so much so that is expressed in dB, and is called the CMRR for the given amplifier.

I. The Slew Rate (SR) and Transient Response

This parameter is another limiting factor in the application of an Op-Amp. Slew rate refers to the maximum rate of change for the output voltage under the large signal conditions. This quantity is usually measured and specified for the closed-loop condition. This phenomenon is caused by current limiting and saturation occurring in some internal circuitry; in particular, the capacitive elements of the Op-Amp due to high frequency input signals. Some typical outputs of an amplifier considering slew rate limiting is shown in Fig. 11.7-9.

Fig. 11.7-9a.

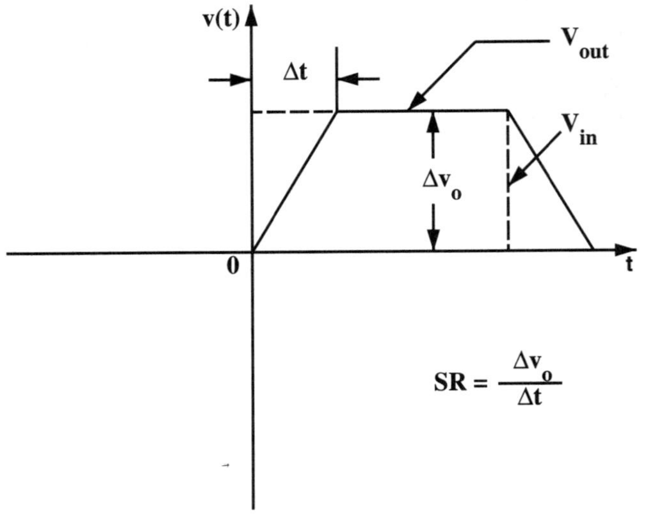

Fig. 11.7-9b.

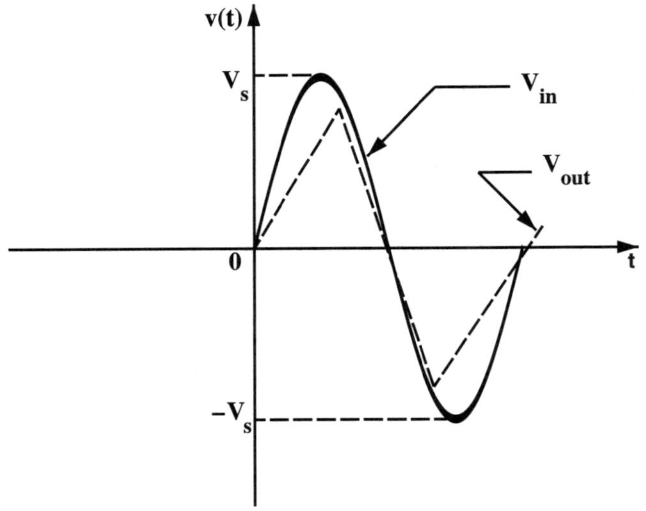

We may now define the slew rate as $SR = \Delta V_o/\Delta t$. For a sinusoidal input signal $v(t) = V \sin \omega t$ the $SR = |dv(t)/dt|_{Max} = |V \omega \cos \omega t|_{Max} = V\omega$, or $f_{Max} \leq (SR)/2\pi V_{peak}$ to avoid distortion. This is called the full-power response. The physical dimension of $\Delta V_o/\Delta t$ reinforces the idea that the slew rate is caused by changes in voltage of some capacitance. The capacitance that is responsible for limiting the SR is generally a compensation capacitance. At higher frequencies or higher rates of signal changes, the current available to charge and discharge this capacitance becomes exhausted and the slew rate limiting will occur. Thus, we must be keenly aware of the limitation that f_{Max} or SR imposes on our device.

In regard to the transient response of the Op-Amp, we should recall that a more realistic transfer function for an Op-Amp than a constant gain is the one with at least one pole, for instance (11.7-5). This transfer function must then be substituted for the open-loop gain of Op-Amp, in order to find the overall network transfer function. From the inverse (time-domain) response corresponding to this transfer function we may find the relevant transient parameters of the network. Further analysis of these issues is outside the scope of this textbook.

11.7.4. Feedback Configurations with a Practical Op-Amp

To develop a comparable result for the circuit in Fig. 11.7-4, using a practical Op-Amp, we use the following equations:

$$V_o(s) = -G^v_{ol}(s)\,[V^-_{in}(s) - V^+_{in}(s)] = -G^v_{ol}(s)\,V^-_{in}(s), \tag{11.7-14}$$

$$[V_{in}(s) - V^-_{in}(s)]/Z_1(s) = [V^-_{in}(s) - V_o(s)]/Z_f(s)\,. \tag{11.7-15}$$

Assuming $H(s) = Z_1(s)/[Z_1(s) + Z_f(s)]$ (the feedback factor), and upon eliminating V^-_{in}, we have

$$\frac{V_o(s)}{V_{in}(s)} = \left(-\frac{Z_f(s)}{Z_1(s)}\right)\left(\frac{1}{1+\dfrac{1}{H(s)\,G^v_{ol}(s)}}\right). \tag{11.7-16}$$

Here, the term corresponding to $Z_f(s)/Z_1(s)$ refers to the ideal case, and the next term on the right-hand side of (11.7-16) corresponds to the contribution of the non-idealness of the device, because as $G^v_{ol} \to \infty$, (11.7-16) is left with only the ideal gain.

If for all s, $|G^v_{ol}(s)|$ is sufficiently large, then we can approximate (11.7-16) to get

$$\frac{V_o(s)}{V_{in}(s)} = \left(-\frac{Z_f(s)}{Z_1(s)}\right)\left(1-\frac{1}{H(s)\,G^v_{ol}(s)}\right), \tag{11.7-17}$$

where the term $(1 - 1/HG^v_{ol})$ is called the gain error. The readers are invited to develop the parallel equation for the circuit in Fig. 11.7-5. (Problem 11.8.15).

11.7.5. General Comments on Second-Order Active Networks

Consider the circuit of Fig. 11.7-10, which is a typical active filter we have in practice for wave shaping applications. In order to analyze this circuit as a filter, we need to compute its overall transfer function as follows.

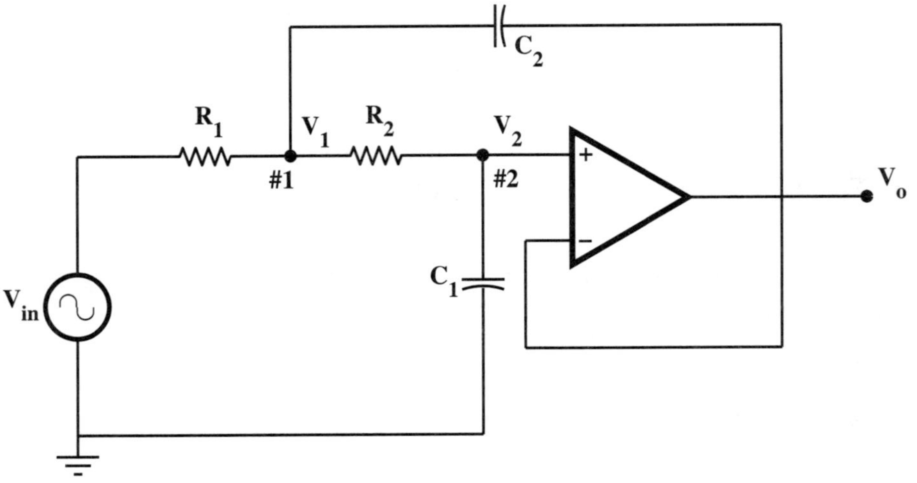

Fig. 11.7-10. A second-order, low-pass, active filter.

To analyze this type of circuit, customarily we use node equations. To perform a node analysis, we must first identify all the existing nodes. In our circuit, if we know the node voltages, say, at node #1, node #2, and the output node, then we can analyze this circuit according to the Definition 2.1-1 completely. In general, we must write as many node equations as we have unknown node voltages, say three here, to analyze a given network. In this type of active network, however, one equation must always be written about each Op-Amp, relating its input and output node voltages. Therefore one less node equation is needed to analyze this network. To solve for $V_o(s)$, as our primary objective, we have the following set of equations to contend with, where G_1 and G_2 are the reciprocal of R_1 and R_2, respectively.

For node #1

$$G_1\, V_{in}(s)\ (\triangleq \text{total current coming to node \#1}) =$$

$$+(G_1 + G_2 + sC_2)\ (\triangleq \text{total admittance of node \#1}) \times V_1(s)$$

$$-(G_2)\ (\triangleq \text{total admittance between nodes \#1 and \#2}) \times V_2(s)$$

$$-(sC_2)\ (\triangleq \text{total admittance between nodes \#1 and (output)}) \times V_o(s)\ . \qquad (11.7\text{-}18)$$

For node #2

$$0 = +(G_2 + sC_1)\,(\triangleq \text{total admittance of node \#2}) \times V_{\#2}(s)$$

$$-(G_2)\,(\triangleq \text{total admittance between nodes \#2 and \#1}) \times V_1(s)$$

$$-(\text{zero})\,(\triangleq \text{total admittance between nodes \#2 and (output)}) \times V_o(s)\ . \tag{11.7-19}$$

For the single Op-Amp,

$$V_2 = V_o. \tag{11.7-20}$$

This is true because of the idealness property of the Op-Amp and the fact that $V_2 = V_{in}^+ = V_{in}^-$, which together with $V_{in}^- = V_o$ (direct connection) yields (11.7-20).

Eliminating $V_1(s)$ and $V_2(s)$ from the above three equations yields the following transfer function which is a second-order, low-pass filter.

$$\frac{V_o(s)}{V_{in}(s)} = \frac{1}{R_1R_2C_1C_2s^2 + (R_1 + R_2)C_1s + 1}\ . \tag{11.7-21}$$

Comparing this transfer function with a standard second-order, low-pass filter

$$\frac{V_o(s)}{V_{in}(s)} = \frac{\omega_n^2}{s^2 + 2\zeta\omega_n s + \omega_n^2}\ , \tag{11.7-22}$$

yields

$$\omega_n = 1/\sqrt{R_1R_2C_1C_2}\ , \tag{11.7-23a}$$

$$\zeta = \frac{R_1 + R_2}{2\sqrt{R_1R_2}}\sqrt{\frac{C_1}{C_2}}\ . \tag{11.7-23b}$$

If we let $R_1 = R_2 = R$, then we have

$$\omega_n = 1/R\sqrt{C_1C_2}\ , \text{ and } \zeta = \sqrt{C_1/C_2}\ . \tag{11.7-24}$$

Similarly, if we consider the circuit in Fig. 11.7-11.

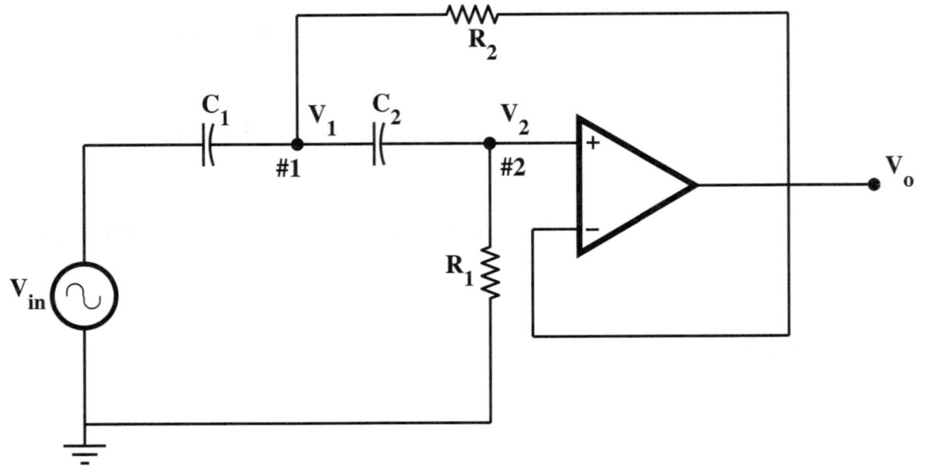

Fig. 11.7-11. A second-order, high-pass, active filter.

Then we can show that

$$\frac{V_o(s)}{V_{in}(s)} = \frac{R_1R_2C_1C_2s^2}{R_1R_2C_1C_2s^2 + R_2(C_1+C_2)s+1}. \tag{11.7-25}$$

This transfer function is in a standard second-order, high-pass form of

$$\frac{V_o(s)}{V_{in}(s)} = \frac{(s/\omega_n)^2}{(s/\omega_n)^2 + 2\zeta(s/\omega_n)+1}. \tag{11.7-26}$$

If we let $C_1 = C_2 = C$, then we get

$$\omega_n = 1/C\sqrt{R_1R_2} \text{ and } \zeta = \sqrt{R_2/R_1}. \tag{11.7-27}$$

We invite the readers to show (11.7-25). (Problem 11.8.23).

11.7.6 Basic Applications of Op-Amp

The Op-Amp allows someone with a limited knowledge of electronic circuits to perform subtle manipulations with electrical signals. For instance, consider the inverting feedback amplifier circuit depicted in Fig. 11.7-12. We have seen that for this circuit and for sufficiently large gain we have:

$$\frac{V_o(s)}{V_{in}(s)} = -\frac{Z_f(s)}{Z_1(s)}. \tag{11.7-28}$$

We now propose a few choices for $Z_f(s)$ and $Z_1(s)$ and evaluate the outcomes.

Fig. 11.7-12.

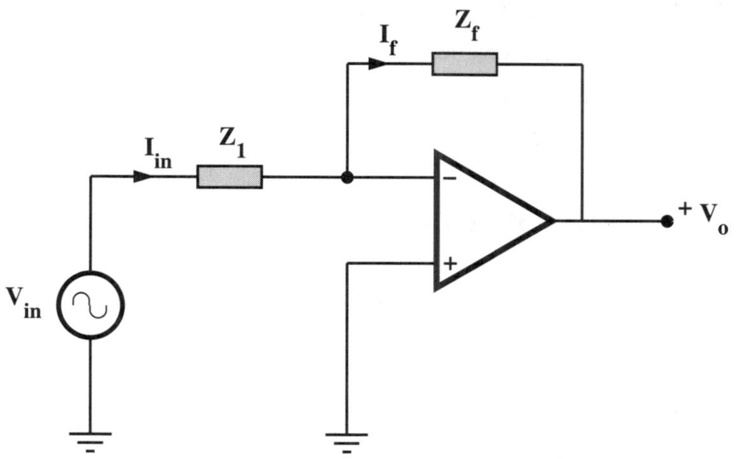

A. An Integrator

Suppose $Z_1(s) = R$ and $Z_f(s) = 1/sC$, then we have

$$V_o(s) = -\frac{1}{RC}\left[\frac{V_{in}(s)}{s}\right]. \tag{11.7-29}$$

We recall that "1/s" is an integrator operator. Thus, this relation shows that the simple inverting feedback amplifier of Fig. 11.7-12 with a proper arrangement can be made to act as an integrator of the input signal.

B. A Differentiator

Suppose $Z_1(s) = 1/sC$ and $Z_f(s) = R$, then we have

$$V_o(s) = -RC[sV_{in}(s)]. \tag{11.7-30}$$

We recall that "s" is the differentiator operator. Thus this relation shows that the inverting feedback amplifier of Fig. 11.7-12, with the above arrangement can now act as a differentiator of the input signal.

Building a differentiator is usually undesirable in practice, due to its noise amplification. Maybe we should point out how this amplification takes place. Associated with each signal of frequency, say f, there may be a noise signal which is of much higher frequency than f, say 1000f. This high frequency signal will become extremely large in amplitude when it is being differentiated, while the integration of the same signal will result in a signal with very small amplitude. For example, let $x(t) = \sin 1000t$, (amplitude is one), then $dx/dt = 1000 \cos 1000t$, while $\int x\,dt = -\cos 1000t/1000$. Clearly, one can see how small the amplitude of the integrated signal is compared to the original signal, and how large the amplitude of the differentiated signal is also compared to the original signal. This is why integration is considered to be a stable operation, but differentiation is always subject to a possible difficulty and/or instability.

C. A Constant Multiplier

In the circuit of Fig. 11.7-12, if we let $Z_1(s)=R_1$ and $Z_f(s)=R_f$, then we have

$$V_o(s)=(-R_f/R_1)V_{in}(s)\ . \tag{11.7-31}$$

This relation shows that the output signal is a constant multiple of the input signal with 180° phase shift.

D. An Adder or Summer

If an Op-Amp is used in its linear operation range, then the superposition rule can be applied. Using this concept we can add several signals with their corresponding gain factor as depicted in the circuit of Fig. 11.7-13. The result is as follows.

$$V_o(s)=\sum_{j=1}^{n}-\frac{R_f}{R_j}V_{in}^{j}(s). \tag{11.7-32}$$

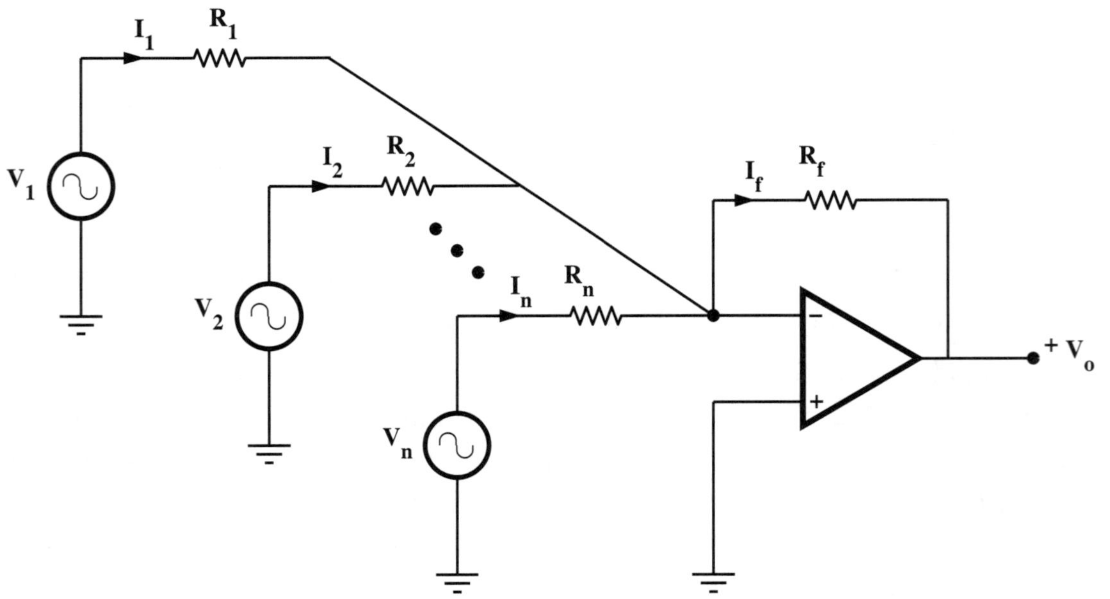

Fig. 11.7-13.

E. Analog Computations

One of the most important applications of the Op-Amp is in the area of analog computations. Although not so widely used as before, there are still many cases where we must use an analog computer to simulate a trajectory. We may do so quite easily thanks to the above four operations that can be performed with a basic inverting feedback amplifier using Op-Amp's and simple R's and C's. Thus, in the following we propose a simple technique to solve an nth-order linear differential equation with constant coefficients, as a point of view on how these simple circuits can be used to perform a rather interesting mathematical operation.

Suppose we want to solve the following linear differential equation with constant coefficients in terms of x(t).

$$\frac{d^n x(t)}{dt^n} + a_1 \frac{d^{n-1} x(t)}{dt^{n-1}} + \ldots + a_j \frac{d^{n-j} x(t)}{dt^{n-j}} + \cdots + a_n x(t) = f(t). \tag{11.7-33}$$

We also assume for the sake of simplicity that $\frac{d^{n-j} x(0)}{dt^{n-j}} = 0$, for j = 1 to n. Rearranging the terms yields

$$\frac{d^n x(t)}{dt^n} = -\sum_{j=1}^{n} a_j \frac{d^{n-j} x(t)}{dt^{n-j}} + f(t). \tag{11.7-34}$$

In this equation we have expressed the highest-order derivative of the response (or unknown) signal x(t) in terms of its lower-order derivatives and the forcing (or input) function f(t). Using integrators, constant multipliers, and adder circuits that we have studied in this section, we can simulate (11.7-34) to find x(t) as follows.

Suppose we have the $\frac{d^n x(t)}{dt^n}$, i.e., the highest-order derivative of the unknown signal in our differential equation. Then by successive integration of this signal, we can generate x(t). This concept is the sole idea behind an analog computer programming. Having at first artificially generating x(t), $\frac{dx(t)}{dt}$, ..., $\frac{d^{n-1} x(t)}{dt^{n-1}}$, we can complete the simulation process by inserting the appropriate adder and/or multiplier circuits to finish the integration process incorporating all pertinent initial conditions as well.

As an example, consider the following third-order differential equation:

$$\frac{d^3 x}{dt^3} + a_1 \frac{d^2 x}{dt^2} + a_2 \frac{dx}{dt} + a_3 x = f(t), \tag{11.7-35}$$

where the initial conditions are $x(0) = 0$, $\frac{dx(0)}{dt} = 0$, and $\frac{d^2 x(0)}{dt^2} = 0$.

Using the above procedure, we can express the third-order derivative of x(t) in terms of the remaining terms in this differential equation as follows.

$$\frac{d^3x}{dt^3} = -a_1\frac{d^2x}{dt^2} - a_2\frac{dx}{dt} - a_3x + f(t)\ . \qquad (11.7\text{-}36)$$

If we had $\frac{d^3x}{dt^3}$, then we could have successively integrated this signal to generate x(t). The procedure is shown in Fig. 11.7-14, where the approximation signs are used to emphasize that the actual signal is different in value or is scaled, though is proportional to the particular displayed value.

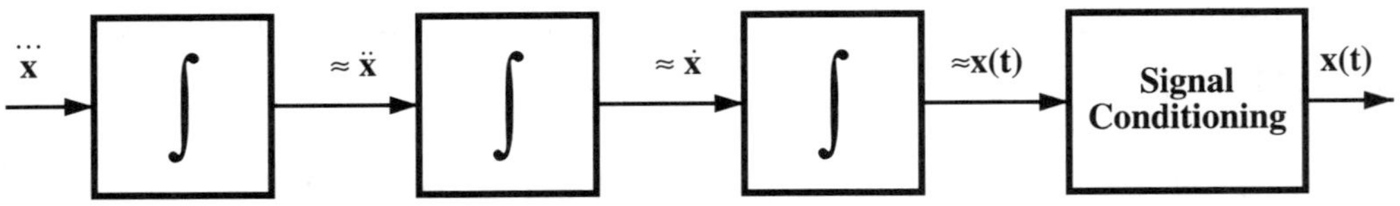

Fig. 11.7-14.

As illustrated in Fig. 11.7-14, the output of each integrator, in general, is *proportional* to the integral of the input signal. Therefore, we must do the necessary signal conditionings in order to make this simulation meaningful. This simulation is carried out next, as shown in Fig. 11.7-15, where we have kept the approximation signs as explained subsequently.

We must point out that the block diagram in Fig. 11.7-15, is not intended to serve as a simulator diagram in a real analog computer setting. It is only for the purpose of illustrating the concept. In real situations the output of each integrator has a minus sign plus a specific gain that must be accounted for, and that is why we have kept the approximation signs.

This sort of integration of a differential equation can be done by an analog computer. Each analog computer has several operational amplifiers that can be used with the help of capacitors and resistors already housed in the computer to build integrators, constant multipliers, summers or adders. Therefore, programming on an analog computer is simply an easy way to simulate a differential equation through the use of these operational blocks. Provisions have been made in these computers to have certain signal multipliers in order to simulate some nonlinear differential equations. Also we can incorporate the initial conditions of the differential equation in the integrating process by inserting these signals into the appropriate integrators.

We must mention that in using analog computers the independent variable is always time and the dependent variable is voltage. Thus if we have a differential equation in terms of other physical quantities we must rescale our vertical and horizontal readings from any output devices such as an oscilloscope, or X-Y plotter, *etc.*, into the corresponding physical quantities of our interest. Also we must point out that in certain cases the amplitude of the signal may become very large which may saturate the Op-Amp. There is a technique called amplitude scaling which rectifies this problem by scaling down the x(t) to below the saturation limit. Finally, if the speed of response is not convenient, and it may cause us difficulty, we have the option of frequency scaling to modify the differential equation to avoid this difficulty.

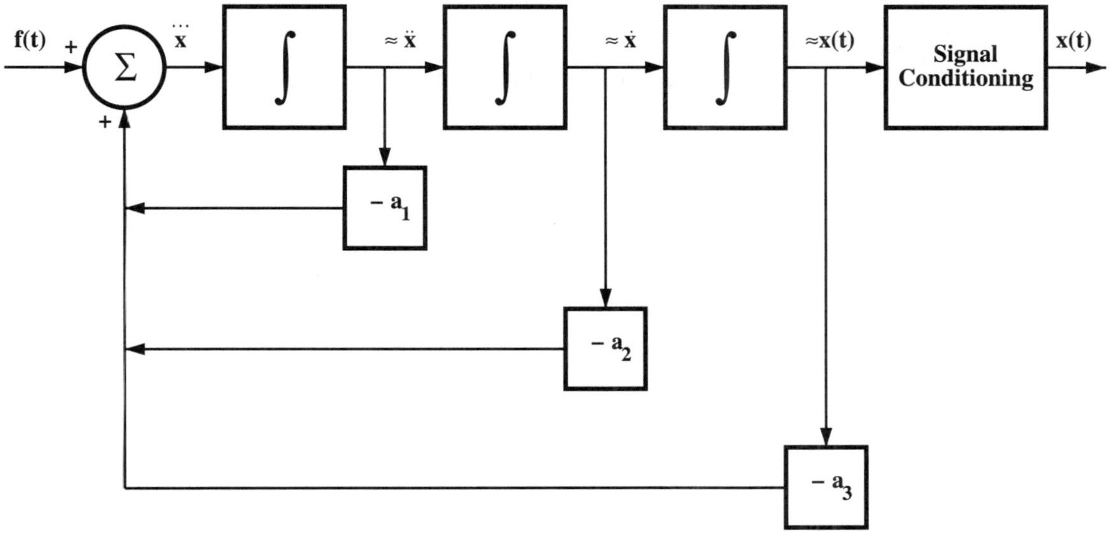

Fig. 11.7-15.

F. A General Synthesis Procedure

To show another important application of the Op-Amp, we present the following general synthesis procedure, where by synthesis we mean constructing a circuit whose transfer function matches a desired specification. Although the following approach is a straightforward method that serves to elucidate a general synthesis concept, there are many other ways from system theoretical points of view, which result in just the same final product.

Consider the following rational transfer function,

$$TF \triangleq \frac{N(s)}{D(s)} = K\frac{s^m + b_1 s^{m-1} + \ldots + b_j s^{m-j} + \ldots + b_m}{s^n + a_1 s^{n-1} + \ldots + a_j s^{n-j} + \ldots + a_n}, \qquad (11.7\text{-}37)$$

where K is a constant and $m \leq n$ and all coefficients are real. The N(s) and the D(s) are two arbitrary signals. Our objective is to find a circuit whose transfer function is (11.7-37). To do so, we introduce an auxiliary signal X(s) and multiply both numerator and denominator of the right-hand side of (11.7-37) with X(s), then we may equate the numerator and denominator in each side to get

$$N(s) = K \sum_{i=0}^{m} b_i s^{m-i} X(s), \quad b_0 \equiv 1, \tag{11.7-38}$$

$$D(s) = \sum_{j=0}^{n} a_j \, s^{n-j} X(s), \quad a_0 \equiv 1. \tag{11.7-39}$$

Each of the above two equations is a linear differential equation in its own right. We have studied their simulation in the previous example (cf., Fig. 11.7-15). Because X(s) is an auxiliary signal and plays no role in our final synthesis, we may assume all of its initial conditions are zero. Since in general we assume $m \le n$, we therefore make the following rearrangement for an input signal, D(s).

$$s^n X(s) = D(s) - \sum_{j=1}^{n} a_j s^{n-j} X(s). \tag{11.7-40}$$

Having made the above simplifications, we can present the block diagram of Fig. 11.7-16, as a synthesis approach to simulate (11.7-37). We notice that N(s) has terms that are already generated in (11.7-40).

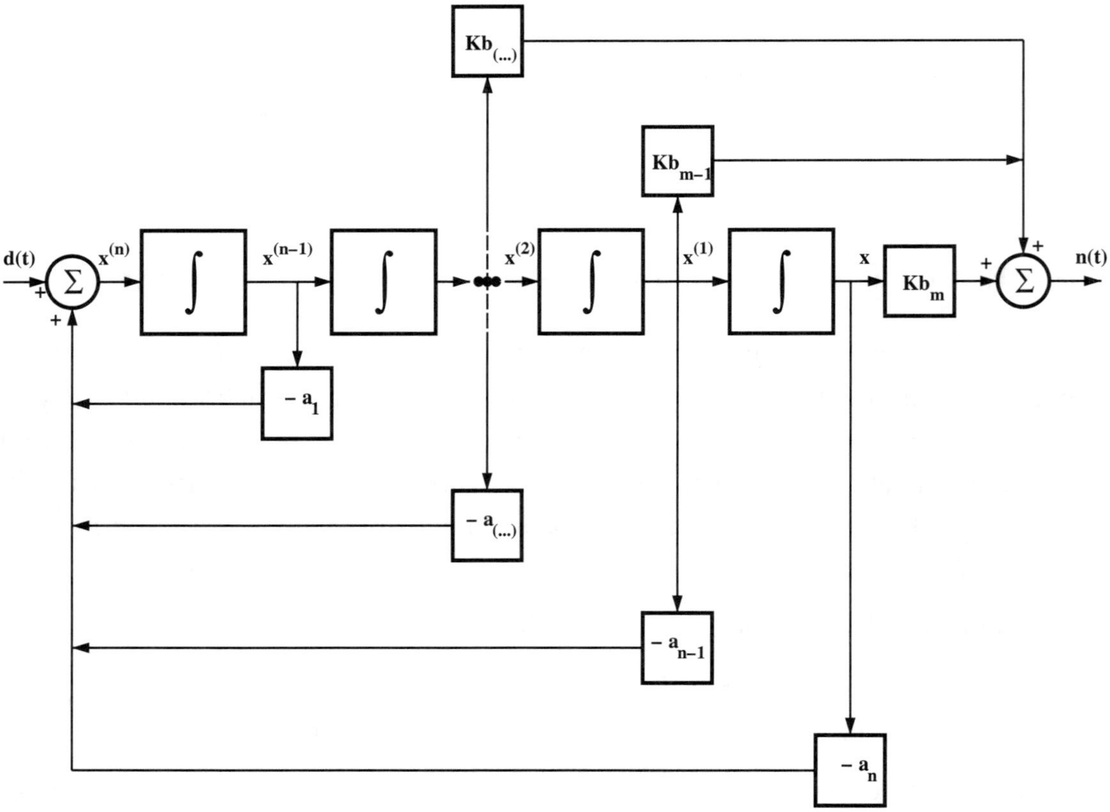

Fig. 11.7-16. A general simulation diagram of a transfer function.

11.8. PROBLEM SET

Two-Port Network Parameters

11.8.1. Find *a proper* set of two-port network parameters for each circuit in Fig. 11.8-1.

Fig. 11.8-1.

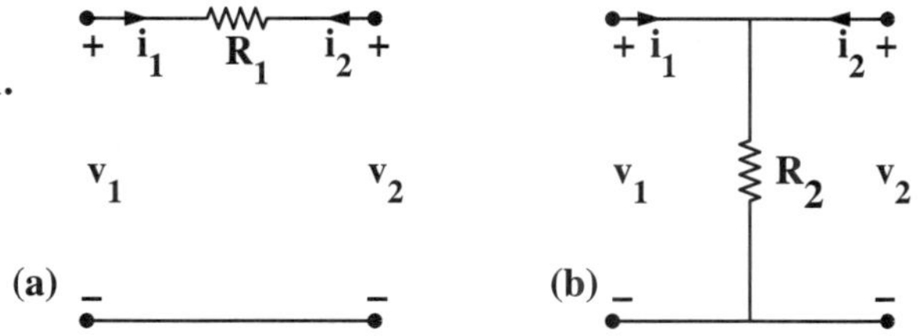

11.8.2. Find a set of *ABCD*-parameters for the ideal transformer in Fig. 8.3-1.

11.8.3. Derive the *G*-, *Y*- and the ABCD-parameters for the circuit in Fig. 11.2-3.

11.8.4. What is the *H*-parameters, for the circuit in Fig. 11.8-2.

Fig. 11.8-2.

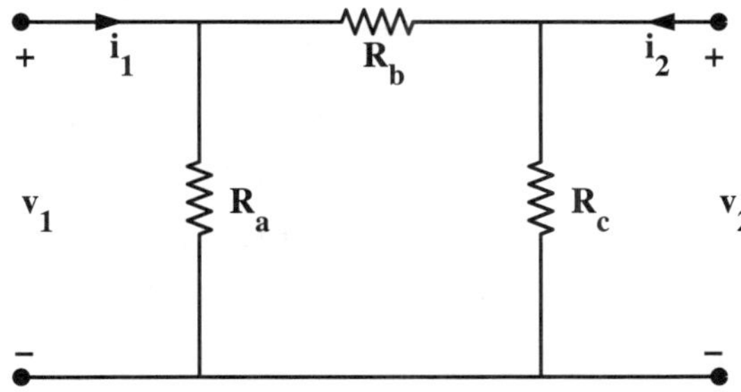

11.8.5. Consider the circuit in Fig. 11.8-3, which is called a *ladder network*. Using the input impedance $Z_{in}(s)$ as shown, prove that $Z_{in}(s)$ is expressed by the following continued fraction expansion.

$$Z_{in}(s) = Z_1(s) + \cfrac{1}{Y_2(s) + \cfrac{1}{Z_3(s) + \cfrac{1}{Y_4(s) + \cfrac{1}{Z_5(s) + \cfrac{1}{Y_6(s)}}}}}. \tag{11.8-1}$$

Fig. 11.8-3.

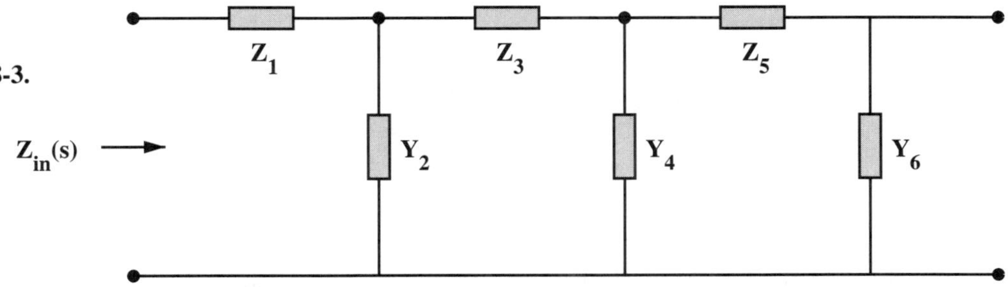

11.8.6. Verify at least three entries in Table 11.2-2 which describes conversions among various two-port network parameters. For instance, suppose one has *Z*-parameters, then from this set the corresponding *H*-, *G*- and *ABCD*-parameters can be generated for the same two-port network if no divisor is zero.

11.8.7. Consider the feedback configuration in circuit of Fig. 11.8-4.

Fig. 11.8-4.

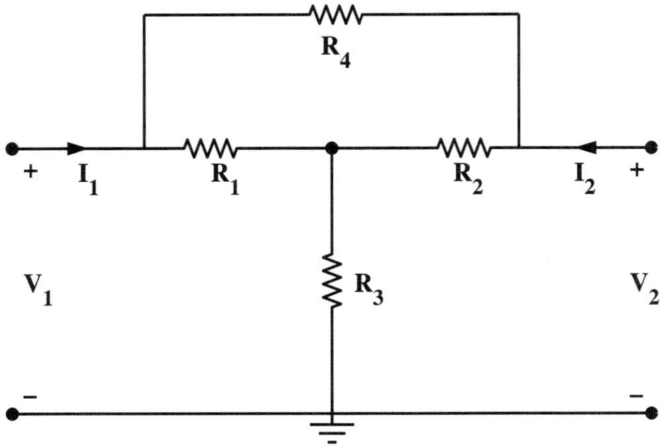

Find the overall two-port parameter set in a traditional sense, then in the terminologies introduced in this chapter. Because there is no buffer or isolation transformers, we have the loading effect in this setting. Thus, when the nature of feedback is identified and the overall parameter for each block is developed, we cannot use that information to find the feedforward and feedback portion of the system according to the result in this chapter. Nevertheless, we may approximate such an interconnection and develop an approximated model. Discuss how closely this approximated model is relative to the exact parameters which are calculated from the standard procedure.

11.8.8. Consider the circuit in Fig. 5.8-21 of Problem 5.8.16, repeated below for convenience.

Fig. 11.8-5. The Fig. 5.8-21 revisited.

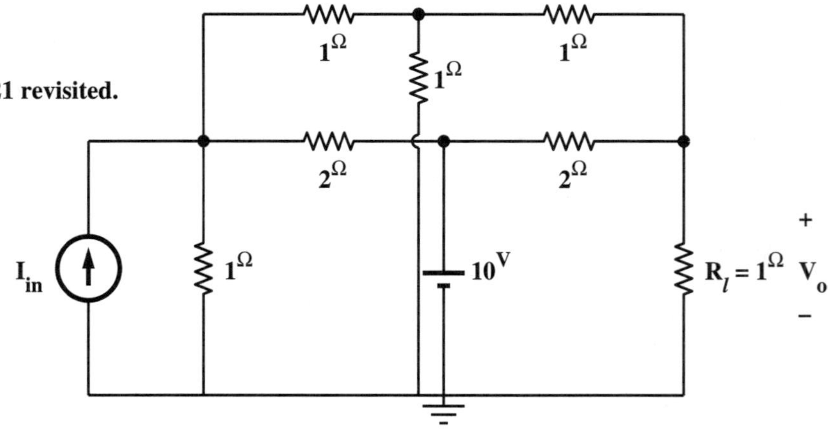

In Problem 5.8.16, you are asked to compute V_o/I_{in}, using standard procedures up to

Chapter Five. In the context of materials in this chapter you may reformulate this circuit as a feedback circuit whose interconnection is yet to be identified. Recalculate this problem with the approximated methods presented in this chapter, and compare your results with the exact values you have calculated in Problem 5.8.16. We suggest you compare these two approaches step by step to see how much calculations were needed in each approach and what are the final results for each approach.

Dependent Sources and TTD

11.8.9. Find $v_o(t)$ for the circuit in Fig. 11.8-6, where we have a dependent current source as shown.

Fig. 11.8-6.

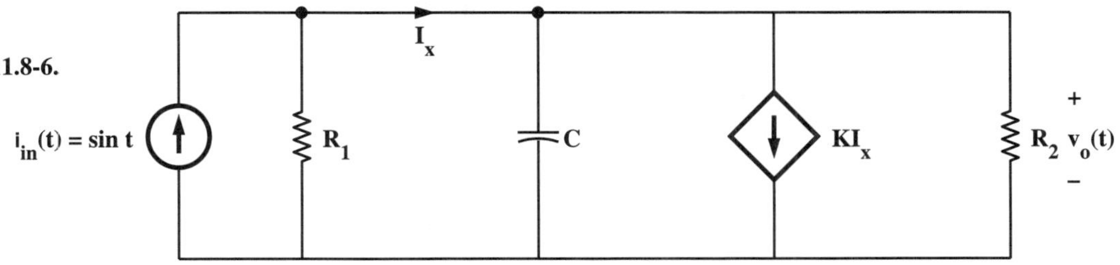

Remark 11.8-1: In this textbook we generally use one dot to display the voltage for each connection of two or more elements, except in Figs. 3.2-2 & 4.2-16 where we use connected rectangles, and in Fig. 11.7-5 where we use another dot at V_{in}^{+} for the sake of demonstrations. Many textbooks however use drawing as depicted in Fig. 11.8-6 to show a current like I_x. This current is supposedly circulating in the node. To remove any ambiguity or impediment for the analysis, we suggest to imbed a resistor, say R_x in the line associated with I_x, as a catalyst, in order to parameterize the circuit. When the analysis is completed and a parameterized version of the desired quantity is determined, let $R_x \to 0$ and compute its limit. This parameterization makes the problem systematic and amenable with the analysis in this textbook.

11.8.10. A phase-shift oscillator is shown in Fig. 11.8-7. Find conditions necessary for the oscillation and its oscillation frequency.

Fig. 11.8-7.

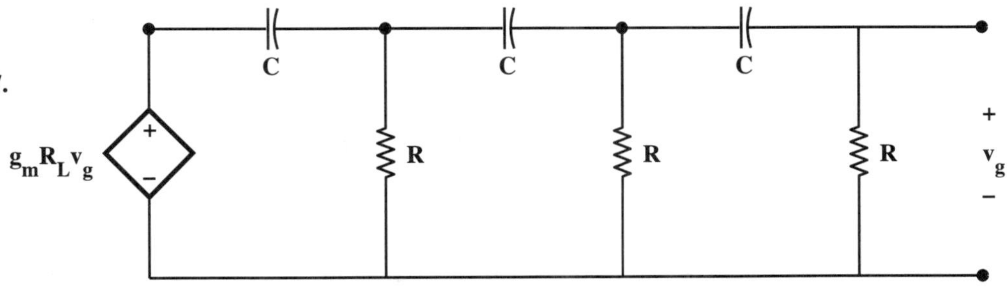

11.8.11. Consider two linearized characteristic curves as shown in Figs. 11.8-8 & 9 corresponding to typical Figs. 11.5-4 & 5 for a TTD. This device is used in the circuit of Fig. 11.8-10. Analyze this circuit assuming an input-output operating points as suggested by P and Q on the linearized curves. Choose a range of values for both R_1 and R_2 for a proper tuning of this circuit, in order that its operational conditions will remain in the neighborhood of P and Q.

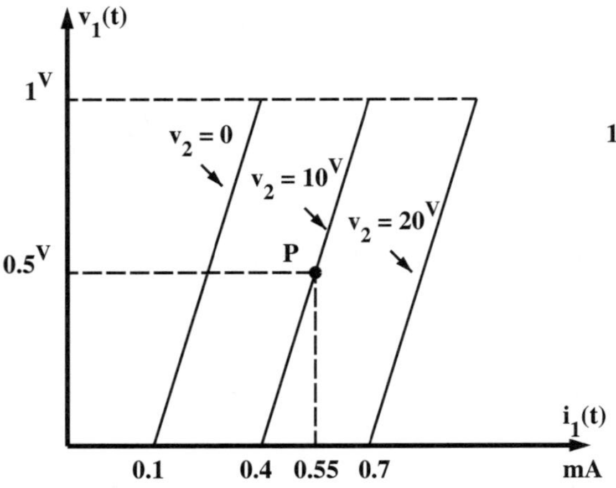

Fig. 11.8-8.

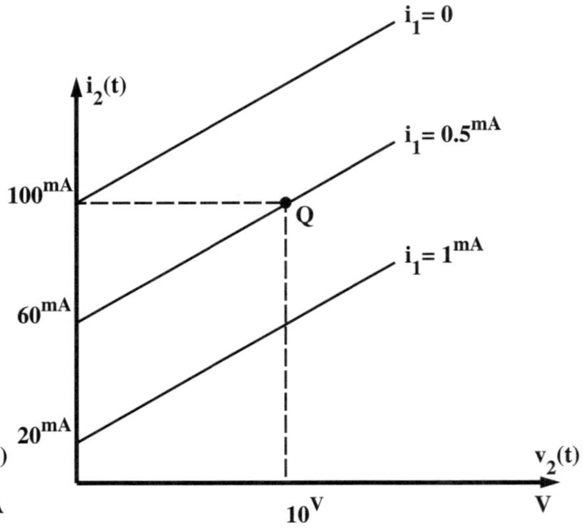

Fig. 11.8-9.

Fig. 11.8-10.

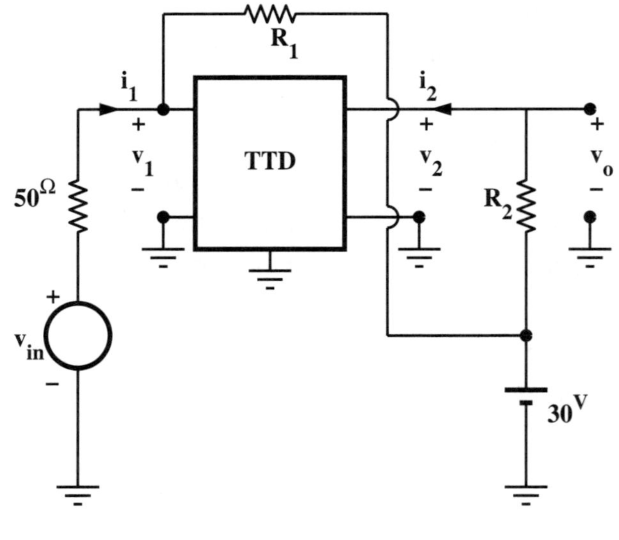

Basic Ideal Op-Amp Circuits

11.8.12. For an ideal Op-Amp, and with an assumption that the circuit in Fig. 11.8-11 operates in its linear mode, find V_o if $V_a = 2^V$ and $V_b = 1^V$. The supply voltage range is $\pm 15^V$.

Fig. 11.8-11.

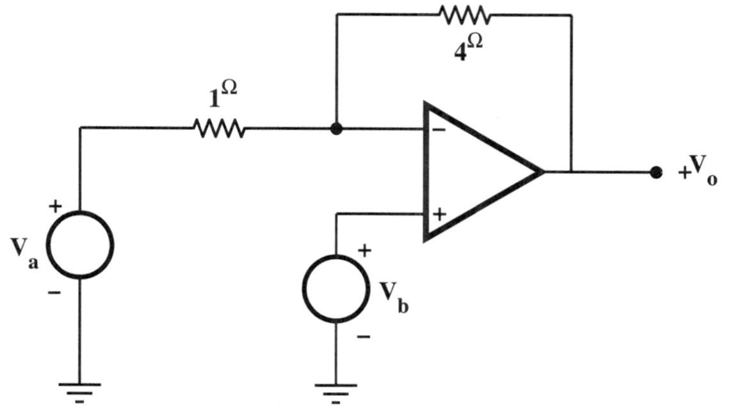

11.8.13. For an ideal Op-Amp, and with a supply voltage in the range of $\pm 10^{V}$, find V_o in the circuit of Fig. 11.8-12, if V_{in} changes within a range of -2^{V} to 4^{V}. Does the system remain linear in this entire range of input voltage, or will it saturate?

Fig. 11.8-12.

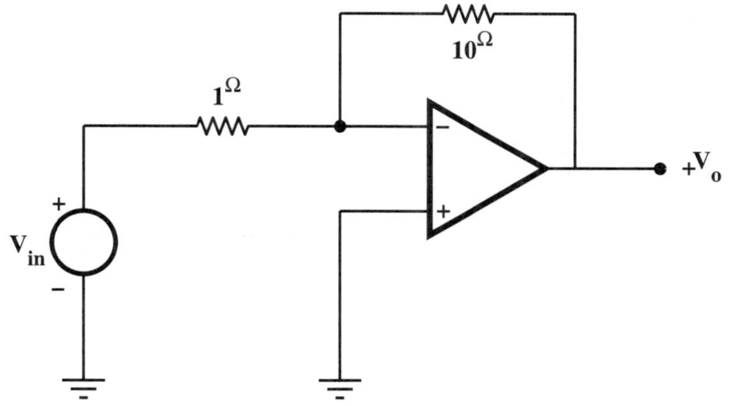

11.8.14. For an ideal Op-Amp, and with a supply voltage in the range of $\pm 20^{V}$, find R_f in the circuit of Fig. 11.8-13, such that V_o remains within the linear range of operation. Let $V_{in} = 10^{mV}$.

Fig. 11.8-13.

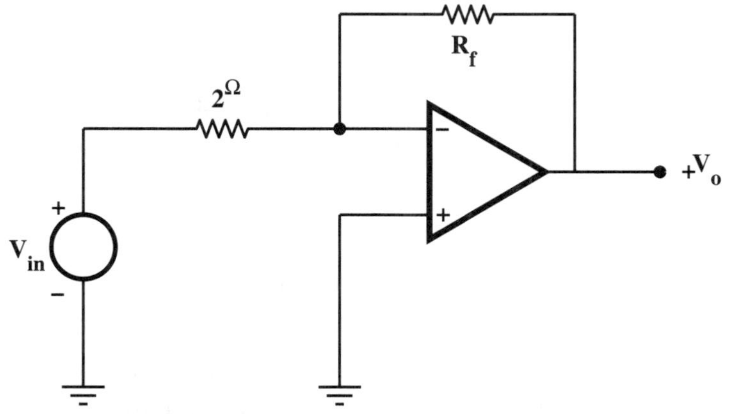

11.8.15. For an ideal Op-Amp, and with a supply voltage in the range of $\pm 20^{V}$, find V_o for the circuit in Fig. 11.8-14, when $V_1 = V_2 = 10^{mV}$ and $V_3 = 25^{mV}$. Find the maximum range for V_3 for which the circuit remains in the linear mode.

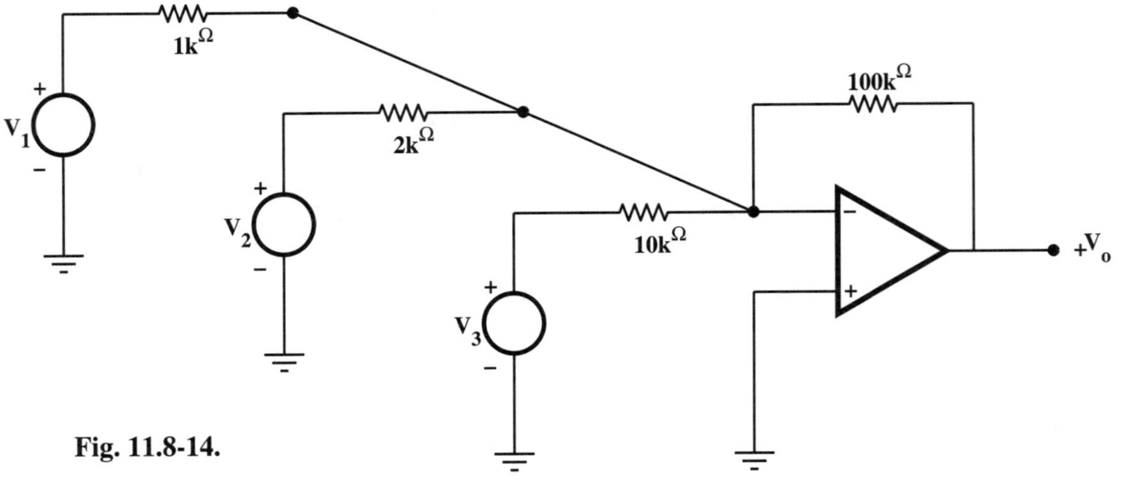

Fig. 11.8-14.

11.8.16. For an ideal Op-Amp, and with a supply voltage in the range of $\pm 15^V$, find the range of R_x in the circuit of Fig. 11.8-15, such that V_o remains within the linear range of operation.

Fig. 11.8-15.

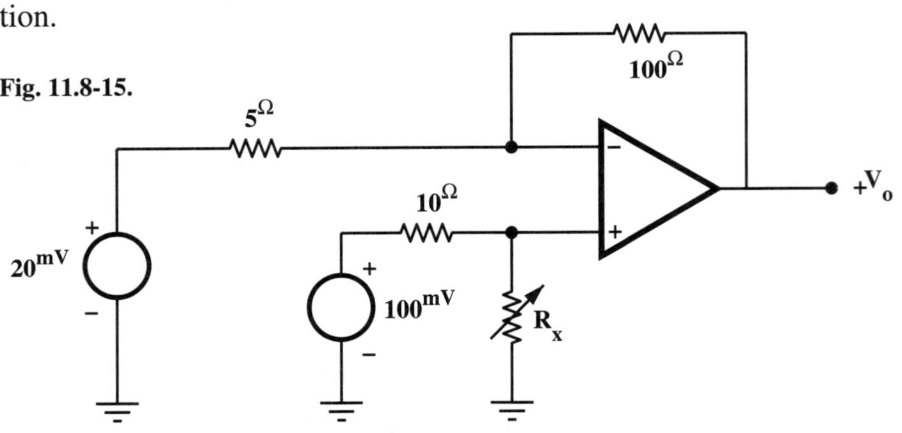

Basic Practical Op-Amp Circuits

11.8.17. Develop an expression similar to (11.7-16) for the non-inverting feedback amplifier in Fig. 11.7-5, but with a practical Op-Amp, i.e., follow similar steps as those leading to (11.7-16) but for the non-inverting amplifier.

11.8.18. Construct a block-diagram representation similar to that in Fig. 11.8-16 corresponding to the closed-loop non-inverting feedback amplifier in Problem 11.8.17, by finding all of its relevant components. [*Comment:* In this type construction one may assume an entire block to be as a unity-gain item, or certain negative sign is carried inside the box if it can be justified. Similar comments also apply to Problem 11.8.19.]

11.8.19. Construct a block-diagram representation similar to that in Problem 11.8.18 (Fig. 11.8-16) corresponding to the closed-loop inverting feedback amplifier represented by (11.7-16). Find all of its relevant components.

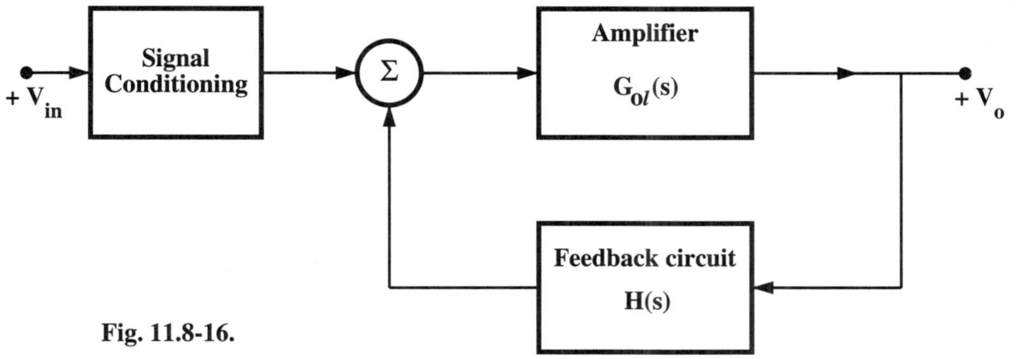

Fig. 11.8-16.

11.8.20. Consider the circuit in Fig. 11.8-17. Find V_o for the following cases: (a) $G^1_{ol} = \infty$, (b) $G^1_{ol} = 10^5$, (c) $G^1_{ol} = 1000$, (d) $G^1_{ol} = 10^5/(1 + 10^{-3}s)$, while $G^2_{ol} = \infty$. Repeat the analysis for $G^2_{ol} = 1000$. Discuss the effects or consequences of the non-idealness of Op-Amp's on the overall performance of the circuit.

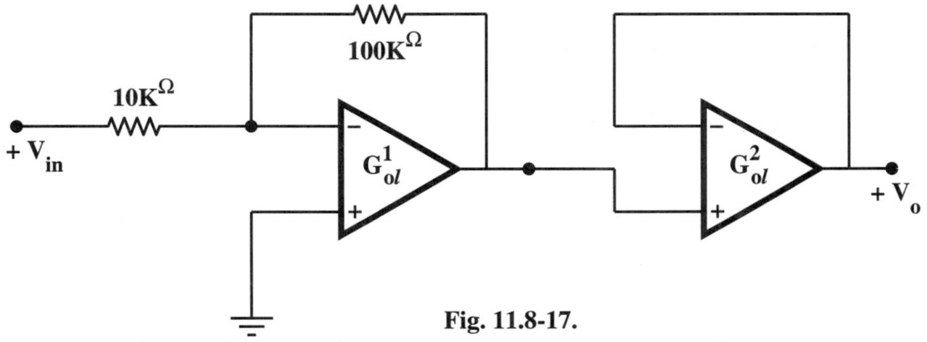

Fig. 11.8-17.

11.8.21. For a difference amplifier with common- and differential-mode input voltages and an ideal Op-Amp as shown in the circuit of Fig. 11.8-18, find V_o in terms of G_{cm} and G_{dm}. Then compute an expression for CMRR in terms of various circuit's parameters.

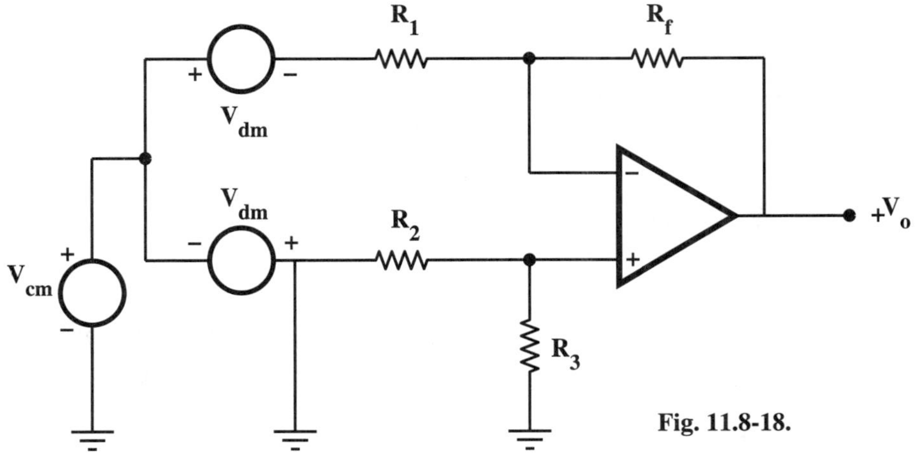

Fig. 11.8-18.

11.8.22. In the circuit of Fig. 11.8-19, what is the range of R_x in order to have at least 40dB CMRR? Compute CMRR when $R_x = 1k^{\Omega}$

Fig. 11.8-19.

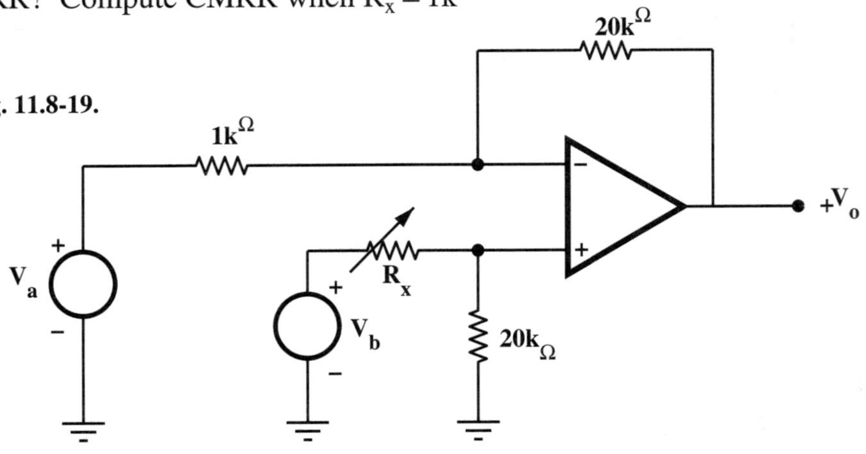

Active Networks with Ideal Op-Amps

11.8.23. Prove expression (11.7-25) for circuit in Fig. 11.7-11.

11.8.24. For the circuit in Fig. 11.8-20, find V_o/V_{in}.

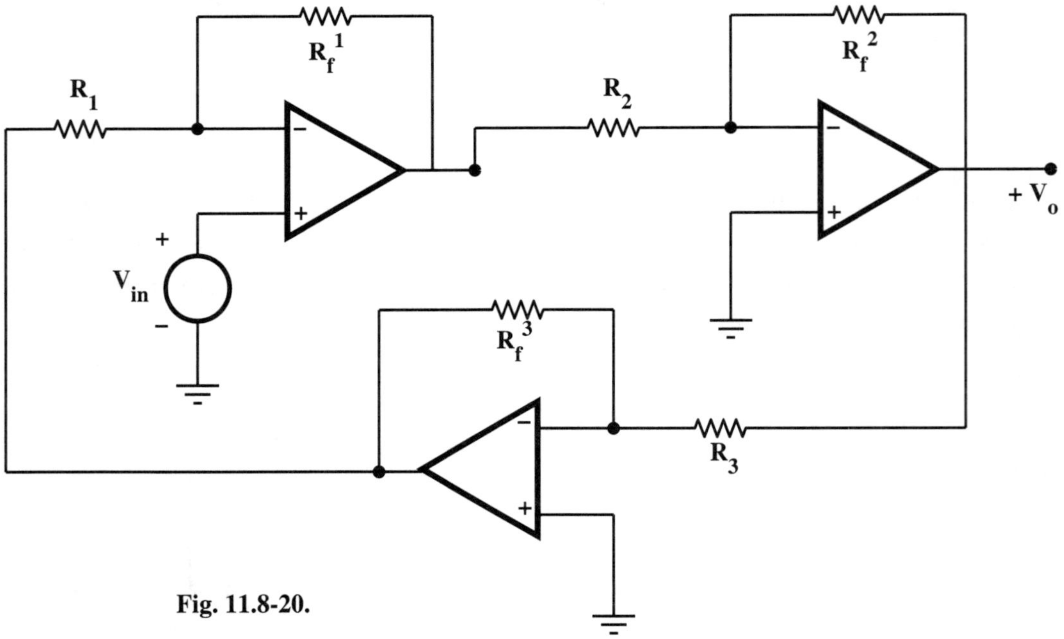

Fig. 11.8-20.

11.8.25. Consider the circuit in Fig. 11.8-21. Find V_o as a function of V_1, V_2 and the remaining circuit parameters. Simplify your results for the case where $R_1 = R_2$ and $R_3 = R_f$.

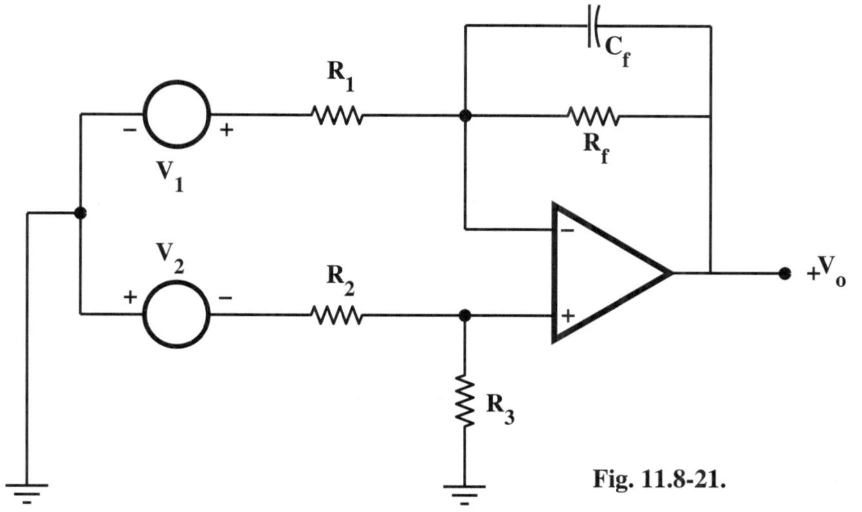

Fig. 11.8-21.

11.8.26. Consider the Twin-T filter of Problem 10.9.5 shown in Fig. 10.9-3 with all of its relevant parameters. Insert that filter in the following two configurations, Figs. 11.8-22 & 23. Find the overall transfer function for each of the circuit in these figures, and draw their corresponding magnitude Bode diagrams. Discuss the overall behavior of each circuit in a comparative way. For instance, when do these filters act as band-pass or band-reject and their other potential applications.

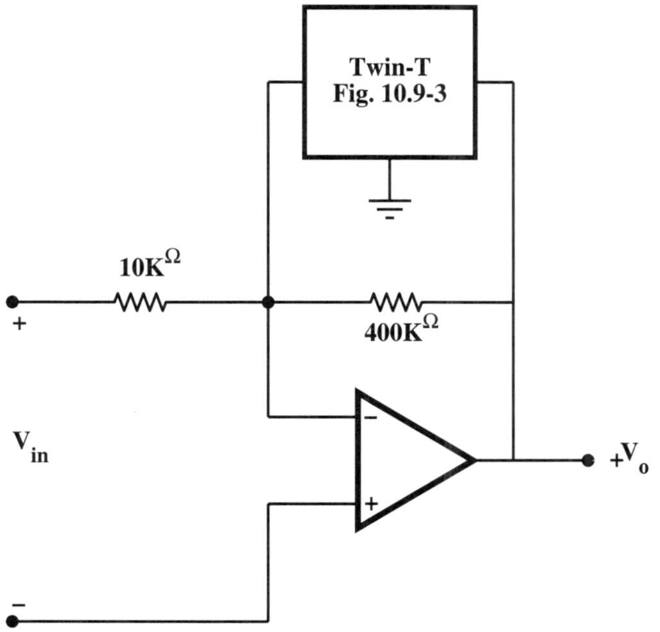

Fig. 11.8-22.

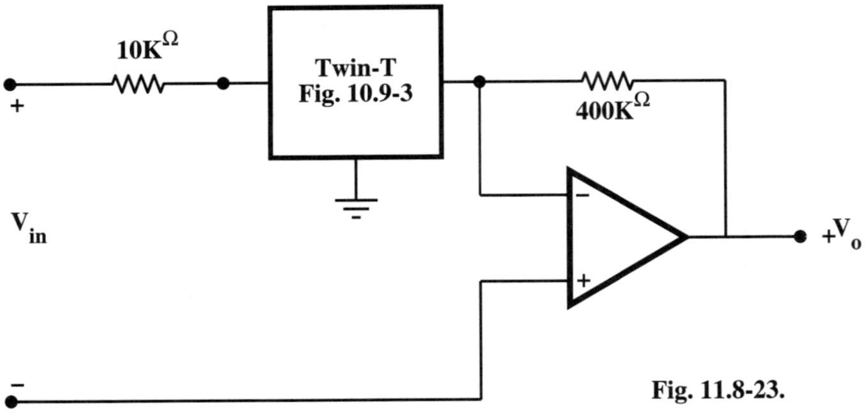

Fig. 11.8-23.

11.8.27. Find the transfer function of $V_2(s)/V_1(s)$ for circuits in Figs. 11.8-24 & 25 with ideal Op-Amps.

Fig. 11.8-24.

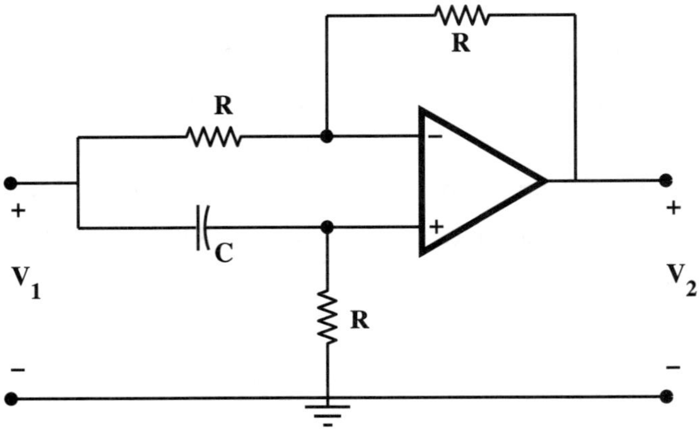

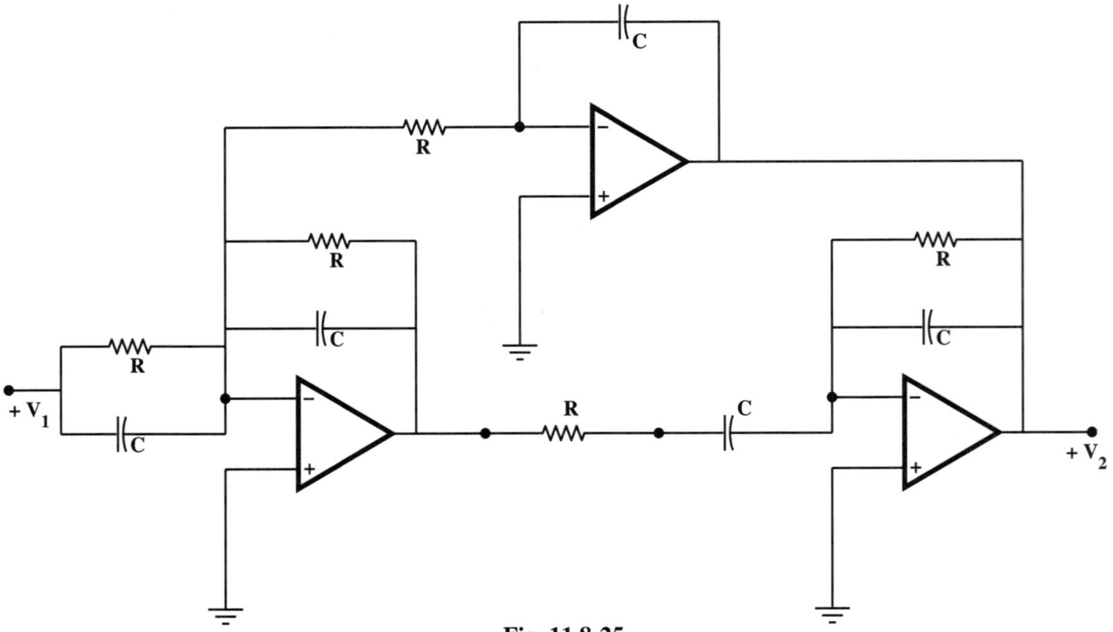

Fig. 11.8-25.

11.8.28. For the circuit in Fig. 11.8-26, and with an ideal Op-Amp, find $v_o(t)$. Redraw this circuit by applying $v_{in}^{+}(t) = \sin t$, then find $v_o(t)$.

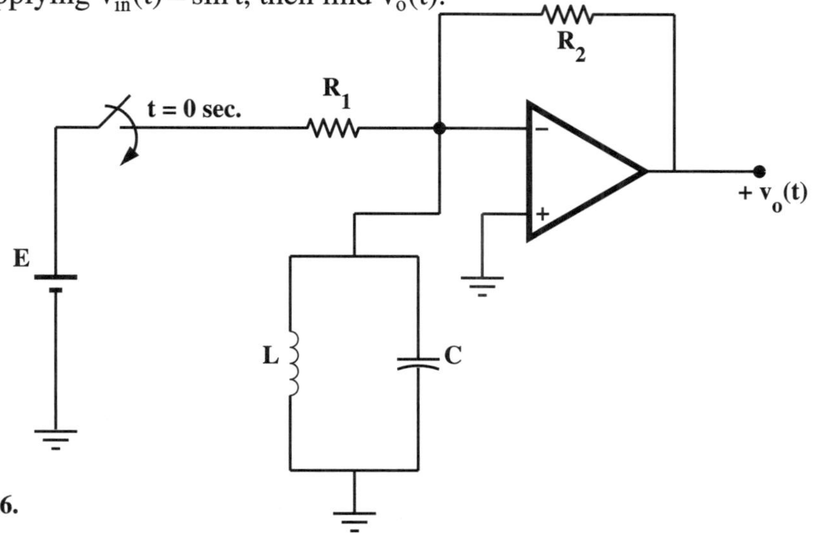

Fig. 11.8-26.

11.8.29. In the circuit of Fig. 11.8-27 known as an *instrumentation amplifier* with a tuning resistor R, find K, such that $V_o = K(V_2 - V_1)$.

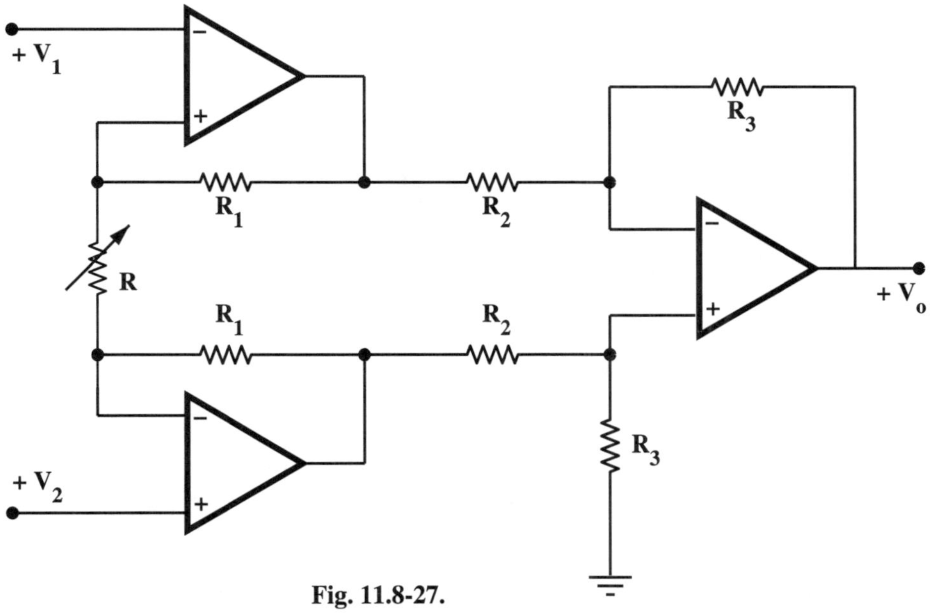

Fig. 11.8-27.

Comment 11.8-1: One of the most interesting circuits of the early stage of active networks with Op-Amp's is called a *gyrator.* A gyrator is an ideal two-port network (Fig. 11.8-28) with the following characteristics equations.

$$\begin{cases} i_1 = Gv_2 \\ i_2 = -Gv_1 \end{cases} \Rightarrow \begin{Bmatrix} i_1 \\ i_2 \end{Bmatrix} = \begin{pmatrix} 0 & G \\ -G & 0 \end{pmatrix} \times \begin{Bmatrix} v_1 \\ v_2 \end{Bmatrix} \Rightarrow \tag{11.8-2}$$

$$\frac{v_1}{i_1} = \frac{1}{G^2} = \frac{-i_2}{v_2}. \tag{11.8-3}$$

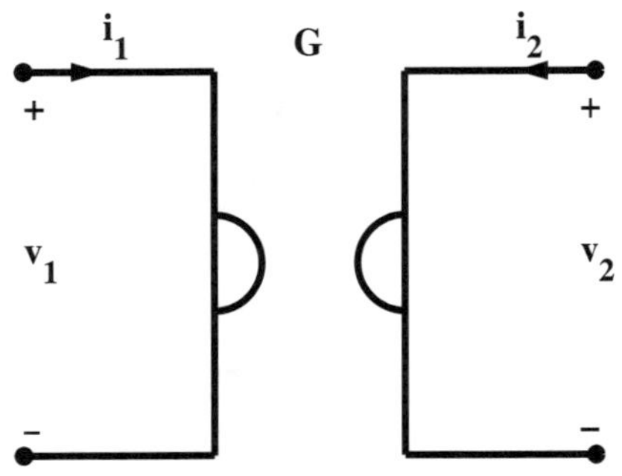

Fig. 11.8-28.

We can show that the power delivered to an ideal gyrator is zero. The universal appeal of this device stems from the above relationship which shows a gyrator is a perfect converter. In the sense that, if it is terminated by a capacitor ($\frac{i_2}{v_2} = sC$), then the input acts as an inductor since $\frac{v_1}{i_1} = -sC$. The idea that a bulky inductor can be replaced with a tiny capacitor and a device yet to be built, then created tremendous interest in active networks.

11.8.30. Show that the circuit in Fig. 11.8-29 acts as a gyrator and find $\frac{V_2}{V_1}$.

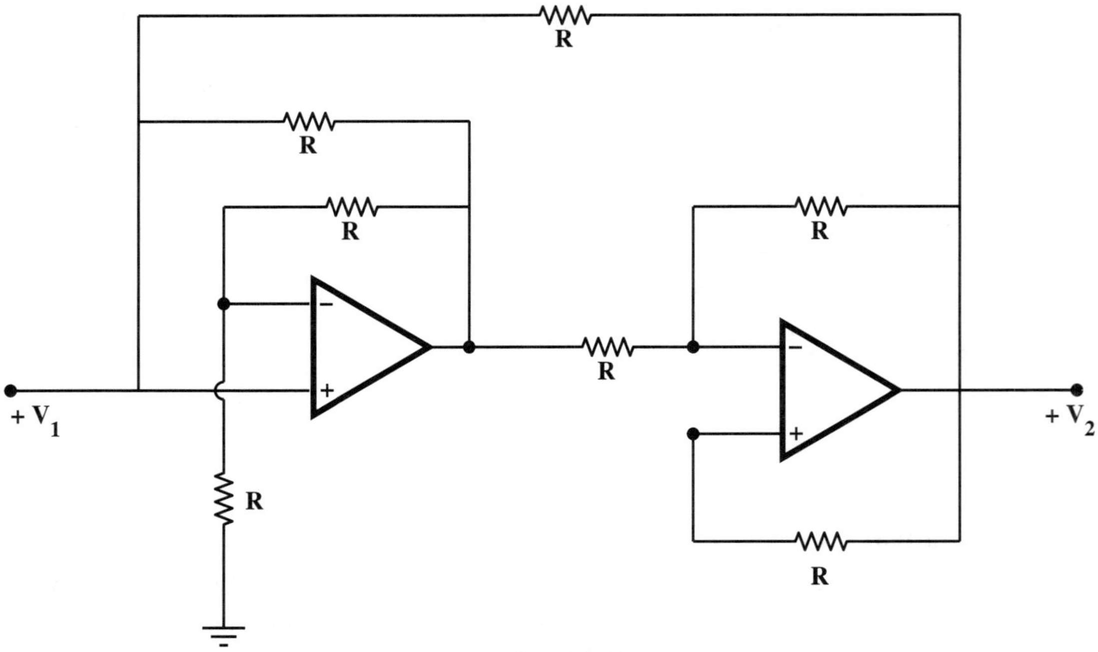

Fig. 11.8-29.

11.8.31. Is a gyrator a reciprocal circuit? Is an ideal transformer a reciprocal circuit?

Simulation with Ideal Op-Amps

11.8.32. Set up an ideal Op-Amp circuit to solve the following differential equation with zero initial conditions. (Example 7.2-1).

$$\frac{d^3x(t)}{dt^3} + 6\frac{d^2x(t)}{dt^2} + 11\frac{dx(t)}{dt} + 6x(t) = 1. \tag{11.8-4}$$

11.8.33. Set up an ideal Op-Amp circuit to solve the following three coupled differential equations with zero initial conditions. Recall that $D \equiv d/dt$.

$$-i_1 + i_2 + (D+1)i_3 = 0, \tag{11.8-5}$$

$$(2D+1)i_1 + (D+1)i_3 = D\sin\omega t, \tag{11.8-6}$$

$$(D^2 + D + 1)i_2 - (D^2+1)i_3 = D\cos\omega t. \tag{11.8-7}$$

11.8.34. Draw a simulation diagram for the transfer function of the circuit in Problem 11.8.27 (Fig. 11.8-25). Here, we expect a different circuit from what we have started in that figure. Interpret your result.

11.8.35. Draw a simulation diagram for the following transfer function.

$$T(s) \triangleq \frac{s^3 + 3s^2 + s + 1}{s^4 + 2s^3 + 3s^2 + s + 1}. \tag{11.8-8}$$

11.8.36. Draw a simulation diagram for the following transfer function.

$$T(s) \triangleq \frac{s^4 + 3s^3 + 2s^2 + s + 1}{s^4 + 6s^3 + 11s^2 + 5s + 1}. \tag{11.8-9}$$

General Two-Port Circuits with Ideal Op-Amps

11.8.37. For the circuit in Fig. 11.8-30 with ideal Op-Amps, find an overall two-port network parameters, where $r_{in} = 5K^{\Omega}$, $\mu = 10^{-2}$, $\beta = 100$, $r_o = 1K^{\Omega}$, $R_1 = 1^{\Omega}$, $R_1^f = 10^{\Omega}$, $R_2 = 2^{\Omega}$, $R_2^f = 50^{\Omega}$, $R_3 = 10^{\Omega}$, $R_{in}^1 = 100^{\Omega}$ and $R_{in}^2 = 250^{\Omega}$. *Hint:* Look deeply inside the circuit and work outwardly.

11.8.38. For the circuit in Fig. 11.8-30 with ideal Op-Amps and as described in Problem 11.8.37, find $v_o(t)$ when $v_{in}(t) = 1(t)$.

11.8.39. In the circuit of Problem 11.8.38, replace R_1^f by C_1 and R_2^f by C_2, and find $V_o(s)$.

11.8.40. Find an equivalent simulation diagram to represent $V_o(s)/V_{in}(s)$ for the modified circuit in Problem 11.8.39.

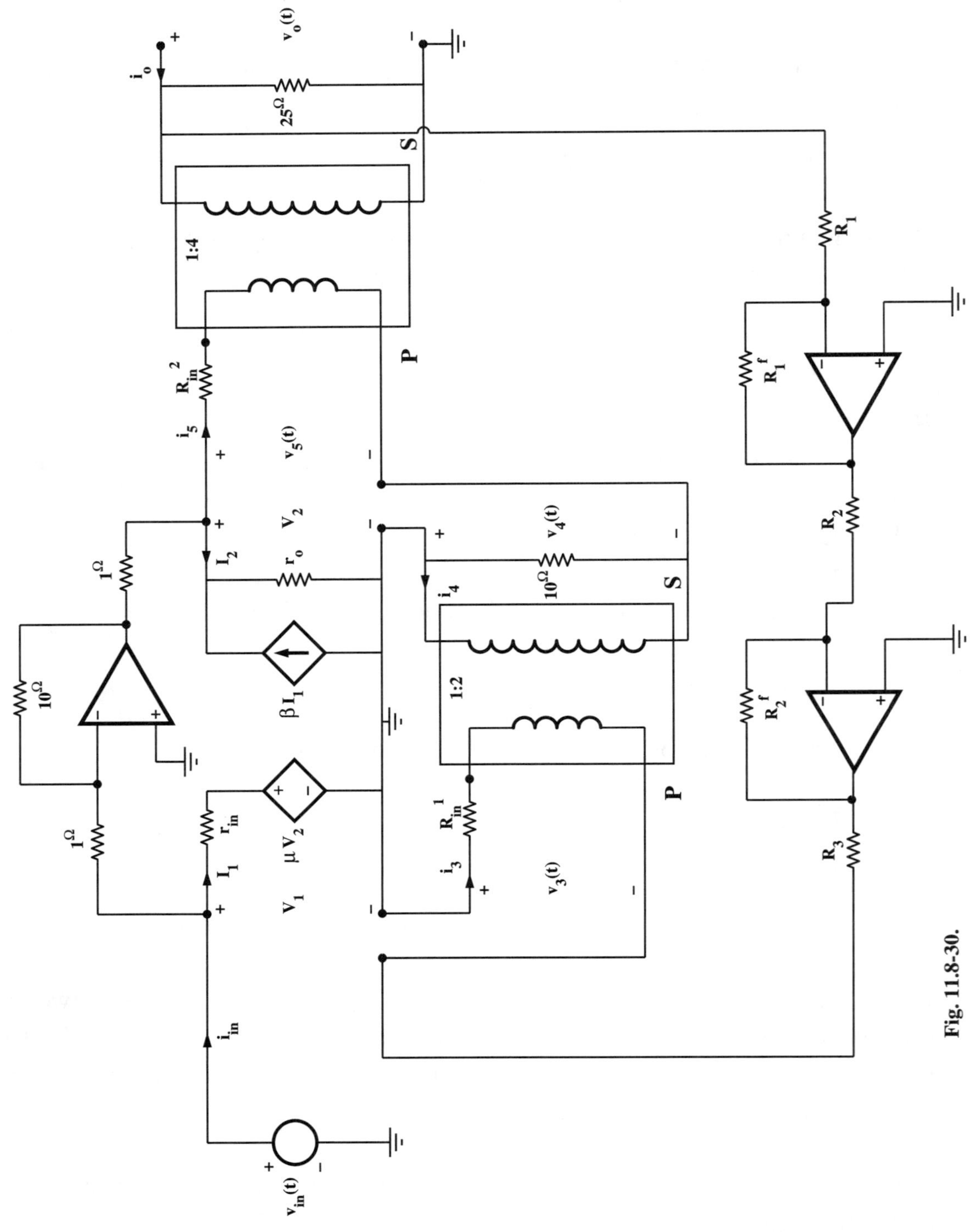

Fig. 11.8-30.

In conclusion, the active network theory plays an extremely important role in the overall circuit design, and in general it depends heavily on applications and analysis of circuits with Op-Amp's. Clearly, there is a wide range of other applications for Op-Amp's in electronic circuits design. A few examples that we have described in the preceding subsections serve only as a good starting point for someone interested to learn more about this device and its potential applications.

12 POLYPHASE CIRCUITS

12.1. INTRODUCTION

For many students this course is the only circuit course they take and this is their last chance to review the very basics of electrical power and become familiar with its essential features. The electricity produced by an electric power company is in the form of balanced sinusoidal *polyphase* sources due to the economical advantages of producing and transmitting electricity in this fashion. For instance, the inevitable industrial applications of electrically driven rotating machines and motors require the use of polyphase power sources to generate a constant torque, in order that this torque will cause the least vibration and thus yield the maximum mechanical durability for these machines. In the following sections we study some basic properties of several commonly used polyphase systems.

12.2. POLYPHASE SOURCES

In the following we review a single-phase source as the building block for polyphase sources, followed by three major types of polyphase sources.

A. Single-Phase Sources: Perhaps it sounds strange that we now mention a single-phase source. Nevertheless we realize that such a source forms the first block of a polyphase source. Furthermore, most of our daily applications involves using this type of power sources. This source is referred to a perfect sinusoidal waveform of 60 Hz frequency having an amplitude of about 110 volts of alternating current (AC). While this source may be produced using a regular single-phase generator, in reality it is generated as part of a multi-phase source. A single-phase source can be represented by a source similar to that in Fig. 12.2-1, with perfect sinusoidal waveform shown in Fig. 12.2-5, say by $v_{an}(t)$. As is evident from Fig. 12.2-1, it is convenient to use a phasor diagram to represent this type of source.

B. Single-Phase, Three-Wire Sources (SPTWS): In many applications the current and/or voltage of a regular single-phase source is not enough. For instance, many home appliances use SPTWS for their normal operations. In these and similar cases we may use SPTWS, which is the combination of two identical regular single-phase sources in series. A typical diagram for this source is shown in Fig. 12.2-2. If the source is connected to two identical loads, then the diagram becomes that shown in Fig. 12.2-3.

Fig. 12.2-1.

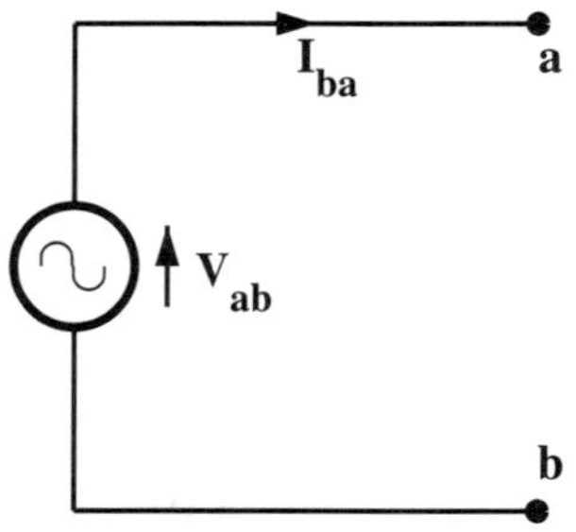

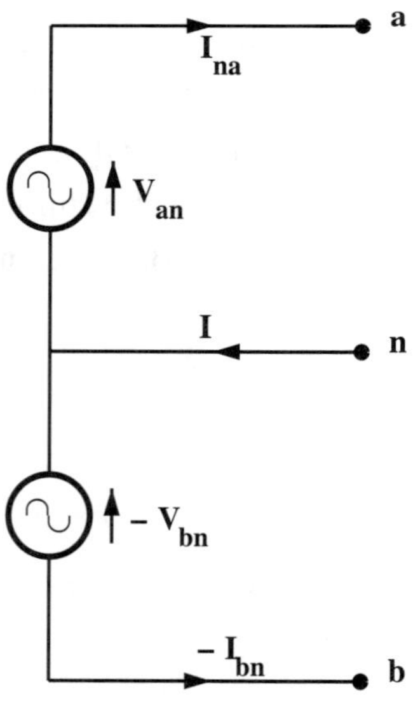

Fig. 12.2-2.

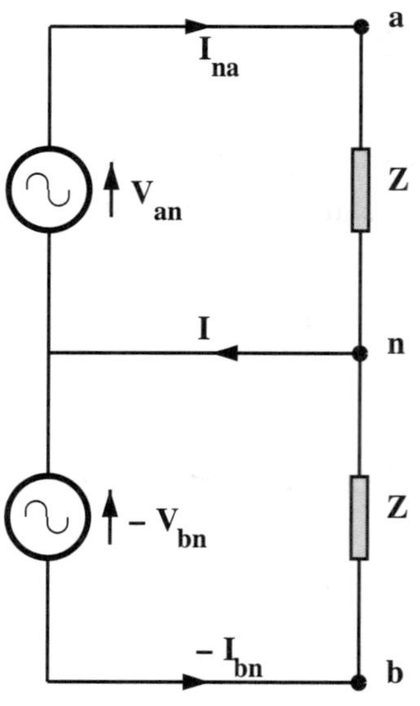

Fig. 12.2-3.

In Fig. 12.2-3, and by definition $V_{an} = -V_{bn}$. Thus $V_{an} + V_{bn} = 0$. The current through each load is $I_{na} = \dfrac{V_{an}}{Z}$ and $-I_{bn} = \dfrac{V_{bn}}{Z}$, or $I_{bn} = -\dfrac{V_{bn}}{Z} = \dfrac{V_{an}}{Z} = I_{na}$. In this condition the current through the center line is zero, because $I = I_{na} + (-I_{bn}) = I_{na} - I_{na} = 0$. In a practical situation, of course, we have neither two identical sources nor two identical loads. Thus, some unbalanced current always exists in the center line. One other source of such unbalancedness is the impedances of top and bottom wirings which are not always equal. In general, we expect a non-zero current in the center line.

C. Two-Phase, Three-Wire Sources (TPTWS): This is the more general case of a SPTWS. Here we have a source diagram similar to that in Fig. 12.2-2, but with $V_{an} = \pm jV_{bn}$. The main application of this source is in the two-phase servomotors which are designed such that the magnetic field of one coil becomes perpendicular to that of the other coil, because of j in the $V_{an} = \pm jV_{bn}$. Here, the voltage between top and bottom lines is:

$$V_{ab} = V_{an} + V_{nb} = V_{an} - V_{bn} = V_{an}\,(1 \pm j)$$

$$= \sqrt{2}\,V_{an}\,\angle(\pm 45°). \tag{12.2-1}$$

The corresponding current can be computed for any load under study.

D. Three-Phase Balanced Sources (TPBS): Three-phase balanced sources refer to systems that in general have three high-current carrying wires called *lines* (also known as hot wires) and one *neutral* wire. These sources are generated such that the voltage between each two consecutive lines are the same in amplitude and with 120° phase difference from one to the next, respectively. We generally select one line as reference point for phase measurements. This choice makes the analysis straightforward. A typical model for this system is shown in Fig. 12.2-4 (the lower case letter "n" is the neutral node for source, while we save "N" for the neutral node of the load), with three consecutive sinusoidal waveforms shown in Fig. 11.2-5.

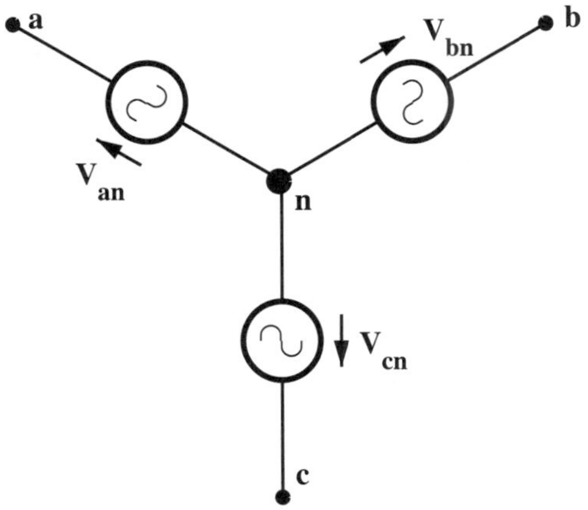

Fig. 12.2-4.

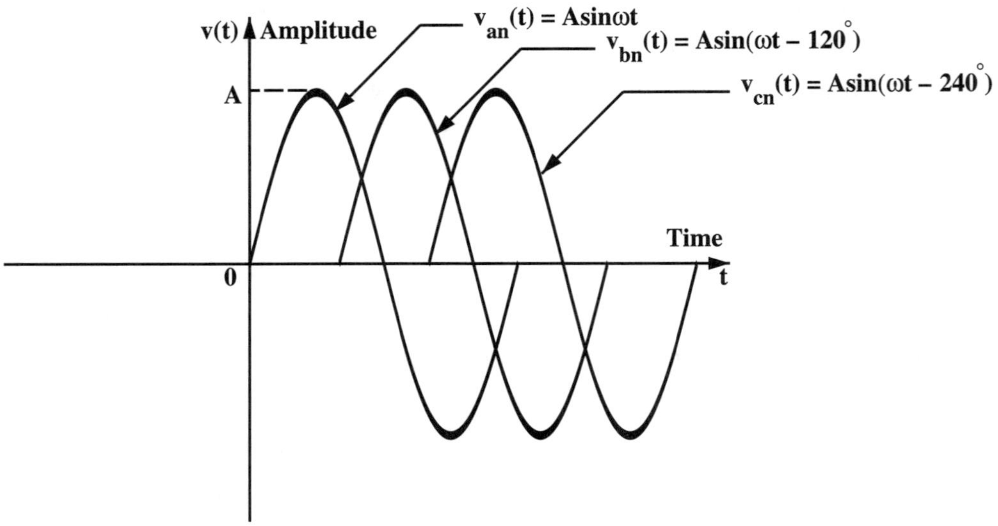

Fig. 12.2-5. Three-phase balanced source.

The connection in Fig. 12.2-4 is called a Wye (Y) (or star) connection. Here, we arbitrarily choose $V_{an} = |V_{an}|\angle 0° \triangleq A\angle 0°$, where A is the amplitude as shown in Fig. 12.2-5. Then we may have either:

$$V_{bn} = A\angle(-120°), \text{ and } V_{cn} = A\angle(-240°), \tag{12.2-2}$$

or

$$V_{bn} = A\angle 120°, \text{ and } V_{cn} = A\angle 240°. \tag{12.2-3}$$

We recall that the positive direction of rotation is counterclockwise. Each of these two cases of (12.2-2) or (12.2-3) is equally useful and is usually predefined in a given problem. We choose the first case of (12.2-2) to draw the phasor diagram of voltages between the three phases and that between the lines as depicted in Fig. 12.2-6. Here, we note that the amplitude of voltage between the lines is $\sqrt{3}$ A. We also note that for a sinusoidal waveform the amplitude and magnitude are the same.

In the context of Table 1.1-1, when we apply the above sources to a proper RLC load, we extend our circuit complexity on Level 2 horizontally first, then to that of what we have called Level 2.1. This symbolic vertical extension is inevitable. In other words, the issues which we face in polyphase circuits are more complex than just those on Level 2, and that is all we mean symbolically by Level 2.1.

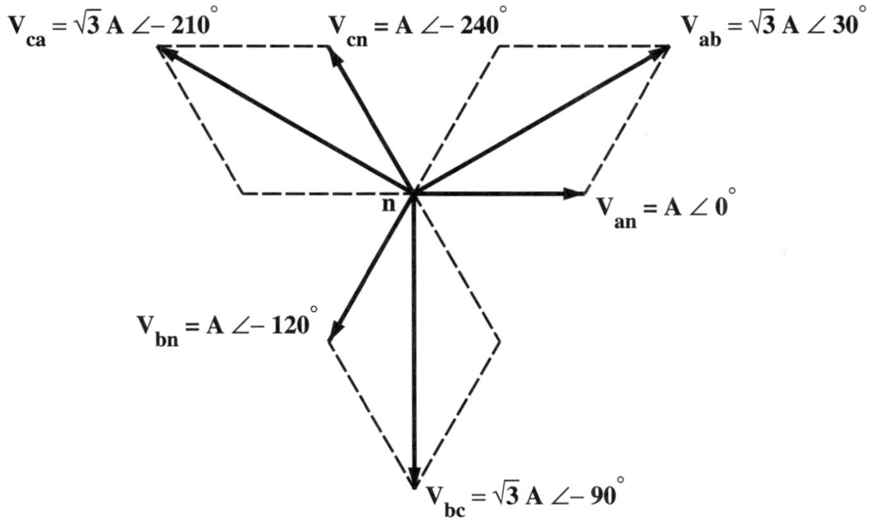

Fig. 12.2-6. Phasor diagrams for a three-phase balanced source.

12.3. APPLICATIONS of TPBS

There are two different ways to connect a load to a TPBS. These are known as the Y- and Δ-connection respectively, as shown in Figs. 12.3-1 & 2.

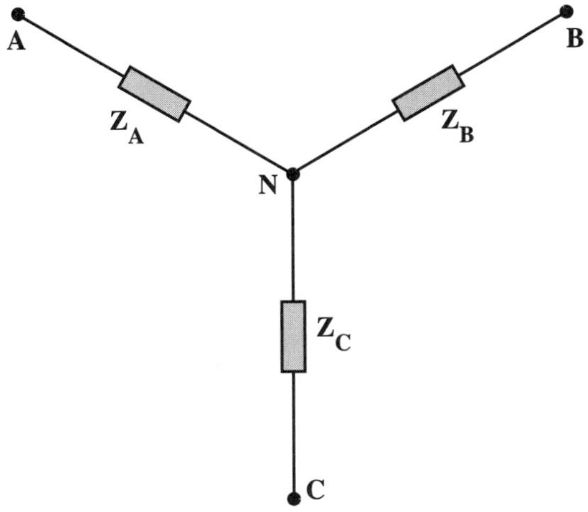

Fig. 12.3-1. A Y/star connection.

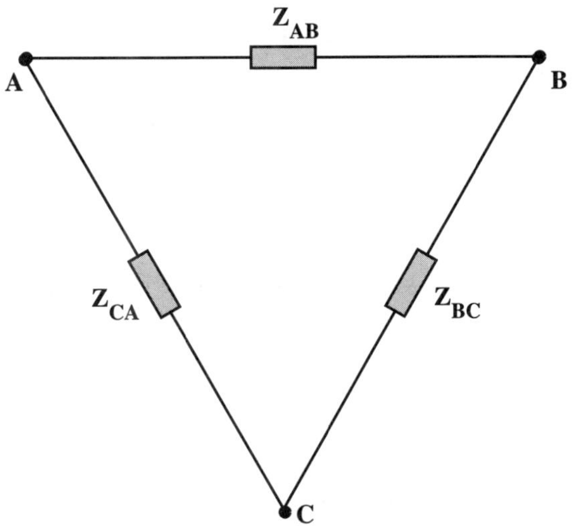

Fig. 12.3-2. A Δ connection.

These sets of loads may be balanced depending on whether the three loads in each phase are equal or not. To see some applications of the TPBS we connect these sources to any one of the above sets of loads in the following manner. First we connect the source in Fig. 12.2-4 to the load in Fig. 12.3-1. This arrangement is called a Y-Y connection, Fig. 12.3-3. In this figure the load is drawn slightly different from that in Fig. 12.3-1, to avoid overlapping the wires and therefore to simplify the presentation for our analysis.

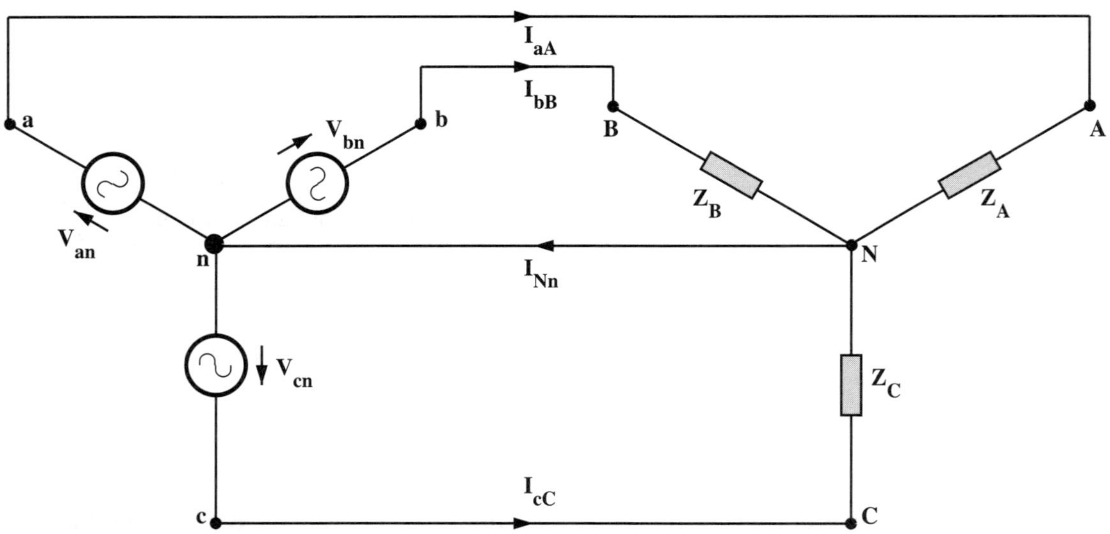

Fig. 12.3-3.

Independent of what the sources and the loads may be (i.e., balanced or not), we have the following set of equations for line currents, provided that the line connecting n to N has zero (small) impedance.

$$I_{aA} = V_{an}/Z_A, \quad I_{bB} = V_{bn}/Z_B, \quad \text{and } I_{cC} = V_{cn}/Z_C. \tag{12.3-1}$$

Let $V_{an} = A\angle 0°$, $V_{bn} = A\angle(-120°)$, and $V_{cn} = A\angle(-240°)$, then we have

$$I_{aA} = (A/|Z_A|)\angle(-\phi_A), \tag{12.3-2}$$

$$I_{bB} = (A/|Z_B|)\angle(-120° - \phi_B), \tag{12.3-3}$$

$$I_{cC} = (A/|Z_C|)\angle(-240° - \phi_C). \tag{12.3-4}$$

For a set of balanced loads with $Z_A = Z_B = Z_C = |Z| \angle \phi_Z$, we have

$$I_{aA} = (A / |Z|) \angle(-\phi_Z) \triangleq I_l \angle(-\phi_Z), \tag{12.3-5}$$

where I_l is the load current. Similarly,

$$I_{bB} = I_l \angle(-120° - \phi_Z), \tag{12.3-6}$$

$$I_{cC} = I_l \angle(-240° - \phi_Z). \tag{12.3-7}$$

Clearly, we always have

$$I_{Nn} = I_{aA} + I_{bB} + I_{cC}. \tag{12.3-8}$$

For a set of balanced loads (and balanced sources) this current becomes (Problem 12.8.1).

$$I_{Nn} = I_l [\angle(-\phi_Z) + \angle(-120° - \phi_Z) + \angle(-240° - \phi_Z)] \equiv 0. \tag{12.3-9}$$

Only for this set of balanced loads and balanced sources (and, of course, a set of balanced connecting wirings), this neutral current I_{Nn} is equal to zero. Only in this special case we may ignore the neutral wire all together. But in reality, we always have some I_{Nn} even for a seemingly balanced operation. We do not, however, need to use a very low resistance wire due to its cost, since the power loss in this wire is small enough to justify a less expensive wire. Although the above analysis is straightforward, it gives the main idea as to what must be done in similar situations. There are a host of computerized techniques to apply to the most realistic problems in which the balanced operation is not the case. Even for a large number of balanced loads the analysis soon becomes complicated requiring the utilization of computer-aided algorithm to finish the task.

To analyze a load similar to that shown in Fig. 12.3-2, we must study the ways that a source is connected to this load. Here the easiest way to analyze this problem is to assume both source and load are connected in the form of Δ-Δ configuration. If the source is given as a Y-connection TPBS, without the neutral wire, then we have the circuit in Fig. 12.3-4. But the circuit in Fig. 12.3-4, and for the purpose of analysis, can be transformed into its equivalent form, which is amenable for analysis with a Δ-Δ type configuration as depicted in Fig. 12.3-5.

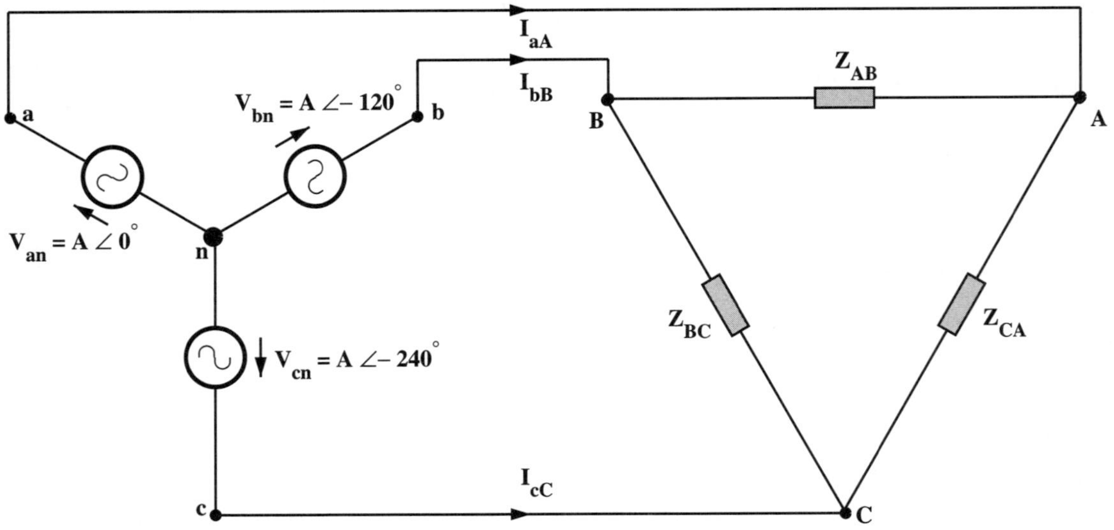

Fig. 12.3-4.

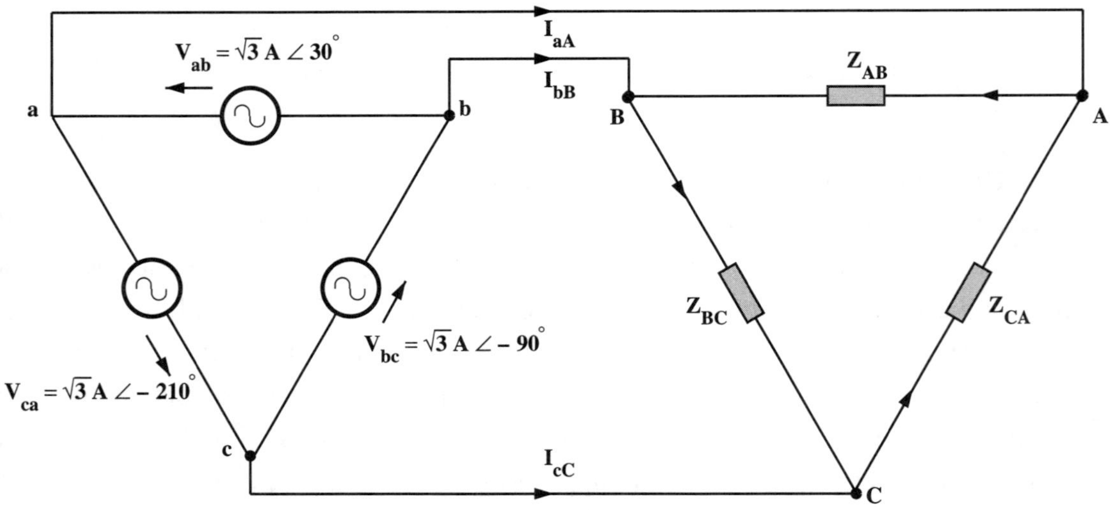

Fig. 12.3-5.

In Fig. 12.3-5, we can show that

$$I_{AB} = V_{ab}/Z_{AB} = (\sqrt{3}A\angle(30°))/Z_{AB}, \tag{12.3-10}$$

$$I_{BC} = V_{bc}/Z_{BC} = (\sqrt{3}\,A\angle(-90°))/Z_{BC}, \tag{12.3-11}$$

$$I_{CA} = V_{ca}/Z_{CA} = (\sqrt{3}\,A\angle(-210°))/Z_{CA}. \quad (12.3\text{-}12)$$

If $Z_{AB} = Z_{BC} = Z_{CA} = |Z|\angle\phi_z$, i.e., a set of balanced loads, and we call $\sqrt{3}A|Z| \triangleq I_p$, then each phase current becomes

$$I_{AB} = I_p\angle(30° - \phi_z), \quad (12.3\text{-}13)$$

$$I_{BC} = I_p\angle(-90° - \phi_z), \quad (12.3\text{-}14)$$

$$I_{CA} = I_p\angle(-210° - \phi_z). \quad (12.3\text{-}15)$$

Clearly in this special case we have

$$|I_{AB}| = |I_{BC}| = |I_{CA}| = I_p. \quad (12.3\text{-}16)$$

To compute the current through a source line we have, for example, $I_{aA} = I_{AB} - I_{CA}$, and similarly for I_{bB} and I_{cC}, we can show that (Problem 12.8.2)

$$I_{aA} = I_p[\angle(30° - \phi_z) - \angle(-210° - \phi_z)] = \sqrt{3}\,I_p\angle(-\phi_z), \quad (12.3\text{-}17)$$

$$I_{bB} = \sqrt{3}I_p\angle(-120° - \phi_z), \quad (12.3\text{-}18)$$

$$I_{cC} = \sqrt{3}I_p\angle(-240° - \phi_z). \quad (12.3\text{-}19)$$

12.4. LOAD TRANSFORMATIONS

We can find equivalent impedances between any two corresponding ports in circuits shown in Figs. 12.3-1 & 2. Thus, we have the following set of equations:

$$Z_A + Z_B = Z_{AB} \,||\, (Z_{BC} + Z_{CA}), \quad (12.4\text{-}1)$$

$$Z_B + Z_C = Z_{BC} \,||\, (Z_{CA} + Z_{AB}), \quad (12.4\text{-}2)$$

$$Z_C + Z_A = Z_{CA} \,||\, (Z_{AB} + Z_{BC}). \quad (12.4\text{-}3)$$

Solving for Z_A, Z_B, Z_C, results in (Problem 12.8.3)

$$Z_A = Z_{AB} Z_{CA} / (Z_{AB} + Z_{BC} + Z_{CA}), \tag{12.4-4}$$

$$Z_B = Z_{BC} Z_{AB} / (Z_{AB} + Z_{BC} + Z_{CA}), \tag{12.4-5}$$

$$Z_C = Z_{CA} Z_{BC} / (Z_{AB} + Z_{BC} + Z_{CA}). \tag{12.4-6}$$

Similarly, we can show that (Problem 12.8.4)

$$Z_{AB} = Z_A + Z_B + Z_A Z_B / Z_C, \tag{12.4-7}$$

$$Z_{BC} = Z_B + Z_C + Z_B Z_C / Z_A, \tag{12.4-8}$$

$$Z_{CA} = Z_C + Z_A + Z_C Z_A / Z_B. \tag{12.4-9}$$

12.5. POWER MEASUREMENTS

To measure the power dissipated in a polyphase system we must measure the power dissipated in each element of each phase in this system. This measurement requires complete analysis of the system, i.e., complete determination of the voltage across and the current through every element in the system. Having found all of these quantities, the power is then measured in terms of an average value of the instantaneous power for each element k, namely an average of $p_k(t) = v_k(t)i_k(t)$. Since most of these instantaneous powers are periodic, the average value can be calculated for any given cycle according to the following definition.

Definition 12.5-1: The *average value* of a signal s(t) over a period T is defined as $S_{Avg} = \frac{1}{T}\int_0^T s(\tau)d\tau$.

The physical implication of this average for a well-defined continuous function s(t) is that we can replace the area under the integral by a rectangular whose dimensions are T and S_{Ave}. For a perfect sinusoidal waveform $s(t) = A\sin(\omega t)$ this average is always zero, and therefore this computation becomes problematic and is not a good choice for further study. To remedy this problem the next definition describes a new measure to quantify a signal.

Definition 12.5-2: The *root-mean square (RMS)* of a signal s(t) over a period T is defined as $S_{RMS} = \sqrt{\frac{1}{T}\int_0^T s^2(\tau)d\tau}$.

The physical interpretation of this value is as follows. Given a wire that is immersed in a tank of water and for a period of T seconds, the temperature variation in this tank caused by applying a time-varying signal s(t) across this wire is the same as that resulted by applying a DC source whose corresponding value is equal to the S_{RMS} of the alternating signal during the same period of time. For instance, for the above sinusoidal s(t) we have,

$$S_{RMS} = \sqrt{\frac{1}{T}\int_0^T A^2 \sin^2 \omega\tau d\tau} = \frac{A}{\sqrt{2}}. \tag{12.5-1}$$

When s(t) is periodic, as in (12.5-1), there is another terminology called *effective value* for this waveform shown by S_{eff} which is the same as S_{RMS}.

Now, consider the circuit in Fig. 12.3-3 and suppose $Z_A = Z_B = Z_C = |Z|\angle(\phi_z)$. We make this assumption only to have the balance operation. To compute the total instantaneous power which is delivered from the source we assume

$$V_{an} = V_m \cos\omega t, \quad \text{and} \quad i_{aA}(t) = I_m \cos(\omega t - \phi_z). \tag{12.5-2}$$

If we replace each of these V_m and I_m by their RMS values we have

$$p_a(t) = 2V_{RMS}I_{RMS}\cos\omega t\cos(\omega t - \phi_z)$$

$$= V_{RMS}I_{RMS}\cos\phi_z + V_{RMS}I_{RMS}\cos(2\omega t - \phi_z). \tag{12.5-3}$$

Similarly,

$$p_b(t) = V_{RMS}I_{RMS}\cos\phi_z + V_{RMS}I_{RMS}\cos(2\omega t - 120° - \phi_z), \tag{12.5-4}$$

$$p_c(t) = V_{RMS}I_{RMS}\cos\phi_z + V_{RMS}I_{RMS}\cos(2\omega t - 240° - \phi_z). \tag{12.5-5}$$

Then adding (12.5-3) to (12.5-5) yields the total instantaneous power delivered from the source as follows.

$$p(t) = 3\,V_{RMS}I_{RMS}\cos\phi_z, \tag{12.5-6}$$

because the remaining terms which are similar to (12.3-9) vanish. Thus

$$P_{Ave} = \frac{1}{T}\int_0^T p(t)dt = 3V_{RMS}\,I_{RMS}\cos\phi_z. \tag{12.5-7}$$

This is a constant quantity which is three times the dissipated power in each phase.

We may equivalently show this dissipated power in terms of the line voltage and/or current quantity (single load) as follows. To avoid any confusion regarding this calculation, but at the expense of being repetitive, we present the following.

Suppose we show the current and voltage in a line/load as follows.

$$v(t) = V_{max} \sin(\omega t + \theta_V), \tag{12.5-8a}$$

$$i(t) = I_{max} \sin(\omega t + \theta_I), \tag{12.5-8b}$$

where the subscript max stands for maximum and θ_V and θ_I are phase difference with respect to the common reference. We call $\phi_z = \theta_V - \theta_I$. Then the instantaneous power in Watts for this line is

$$p(t) = v(i)\, i(t), \tag{12.5-9}$$

while the energy in Joules is

$$w(t) = \int_{t_1}^{t_2} v(t)\, i(t)dt. \tag{12.5-10}$$

Using (12.5-8), (12.5-9) becomes

$$p(t) = V_{max} I_{max} \sin(\omega t + \theta_V)\sin(\omega t + \theta_I). \tag{12.5-11}$$

Using

$$\cos(A + B) = \cos A \cos B - \sin A \sin B, \tag{12.5-12a}$$

$$\cos(A - B) = \cos A \cos B + \sin A \sin B, \tag{12.5-12b}$$

$$\sin A \sin B = ½[\cos(A - B) - \cos(A + B)]. \tag{12.5-12c}$$

in (12.5-11) results in

$$p(t) = \frac{1}{2} V_{max} I_{max}[\cos(\theta_V - \theta_I) - \cos(2\omega t + \theta_V + \theta_I)]. \tag{12.5-13}$$

For a periodic waveform, and using Definition 12.5-1 we have:

$$[p(t)]_{Ave} = \frac{V_{max}}{\sqrt{2}} \frac{I_{max}}{\sqrt{2}} \cos(\theta_V - \theta_I) \triangleq V_{RMS} I_{RMS} \cos \phi_z. \tag{12.5-14}$$

As expected (12.5-14) is only one third of (12.5-7), because is calculated for only one phase out of three phases.

Finally, if we complete the above analysis for every element in the polyphase system, then we have a general relation like (6.5-1) for each element, repeated below for convenience. Recall that "*" means complex conjugate.

$$P_{element} = V_{element}\, I^*_{element} \triangleq R_{element} + jQ_{element}. \tag{12.5-15}$$

The system's total dissipated power can be calculated from its elements as follows.

$$R_{Ave}^{Total} \triangleq \sum_{elements} R_{Ave}, \qquad \text{W (Watts).} \qquad (12.5\text{-}16)$$

$$Q^{Total} \triangleq \sum_{elements} Q, \qquad \text{VAR (volt – ampere reactive).} \qquad (12.5\text{-}17)$$

$$|P|^{Total} \triangleq \sum_{elements} \sqrt{R^2 + Q^2}, \qquad \text{VA (volt ampere).} \qquad (12.5\text{-}18)$$

We show in the next section that instead of making the measurements for all three phases, we need to make one measurement less than the total number of phases. We also show how to use wattmeters to make these measurements.

12.6. POWER MEASUREMENTS USING WATTMETERS

One obvious way of measuring power in a three-phase system is to connect a wattmeter to each load in each phase and measure the consumed power as we often do for a single-load as shown in Fig. 12.6-1.

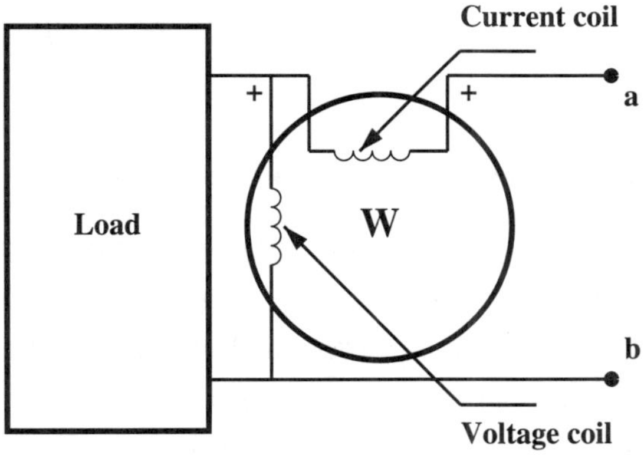

Fig. 12.6-1. A single-load wattmeter.

As depicted in Fig. 12.6-1, each wattmeter has two coils – one for the line current and the other one for the line voltage. In some units the voltage coil are adjacent to the load as in Fig. 12.6-1, while in some units the voltage coil are adjacent to the reference (source) as in Fig. 12.6-2. Both arrangements should read the same, although one may argue that these two configurations give different readings. Either arrangement is acceptable for our discussion to follow.

Now, going back to the main premises of measuring power in a three-phase system, we note that in practice we have no direct access to the internal circuitry of the load. In other words, we know how the lines are connected to the load externally. Therefore real measurements will not be easy to make as the theoretical procedures suggest. However, we start by connecting wattmeters to a three-phase load externally as shown in Fig. 12.6-2. Followed by a discussion on how we can simplify the configuration in this figure using one less wattmeter.

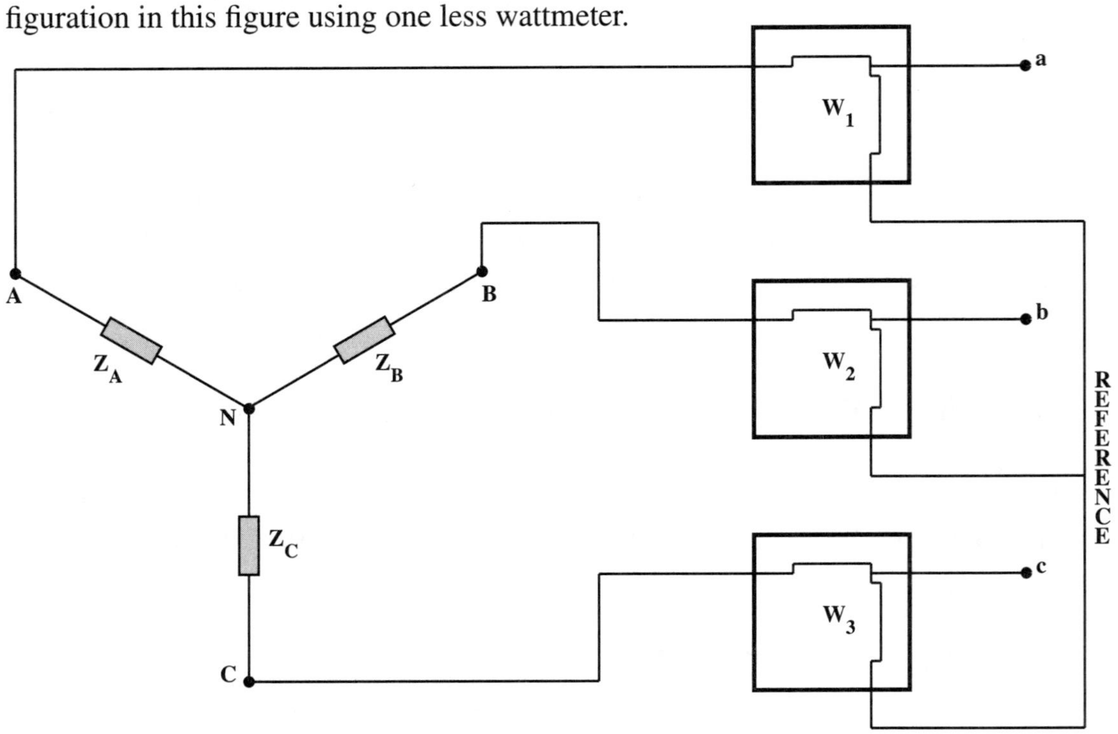

Fig. 12.6-2. Power measurements using three wattmeters.

In Fig. 12.6-2, we have assumed that the load is connected in the Y-form. Then the sum of the three readings becomes the sum of the total power consumed in the three-phase load. A quick look at wiring in Fig. 12.6-2 shows that we can have the reference voltage point anywhere that we please. To prove this fact we notice that if the reference voltage is v_x relative to the neutral point n, then

$$p_a(t) = v_{ax}(t)\, i_a(t), \tag{12.6-1}$$

$$p_b(t) = v_{bx}(t)\, i_b(t), \tag{12.6-2}$$

$$p_c(t) = v_{cx}(t)\, i_c(t). \tag{12.6-3}$$

While the average over one period becomes

$$P_{a,Ave} = \frac{1}{T}\int_0^T v_{ax}(t)\, i_a(t)dt, \tag{12.6-4}$$

$$P_{b,Ave} = \frac{1}{T}\int_0^T v_{bx}(t)\, i_b(t)dt, \tag{12.6-5}$$

$$P_{c,Ave} = \frac{1}{T}\int_0^T v_{cx}(t)\, i_c(t)dt. \qquad (12.6\text{-}6)$$

From $v_{xy} = v_{xz} + v_{zy}$, we may choose

$$v_{ax} = v_{an} + v_{nx},\ v_{bx} = v_{bn} + v_{nx}, \text{ and } v_{cx} = v_{cn} + v_{nx}. \qquad (12.6\text{-}7)$$

Thus

$$P_{a,Ave} + P_{b,Ave} + P_{c,Ave} = \frac{1}{T}\int_0^T (v_{an}\, i_a + v_{bn}\, i_b + v_{cn}\, i_c)dt$$
$$+ \frac{1}{T}\int_0^T v_{nx}\,(i_a + i_b + i_c)dt. \qquad (12.6\text{-}8)$$

Here, $v_{nx} = -v_{xn}$, with v_{xn} as the reference voltage with respect to n. Because, $i_a + i_b + i_c = 0$, where the load is a super node, (12.6-8) is independent of the reference voltage value v_{nx}. Thus, we can choose this reference voltage at any value as we wish. That is, we may choose as a reference, one of the line voltages. If so, such an arrangement yields the voltage reading of one of the wattmeters to be zero which means we can eliminate that wattmeter as shown in Fig. 12.6-3. This figure shows how two wattmeters are used to measure the consumed power in a typical Y-connected load.

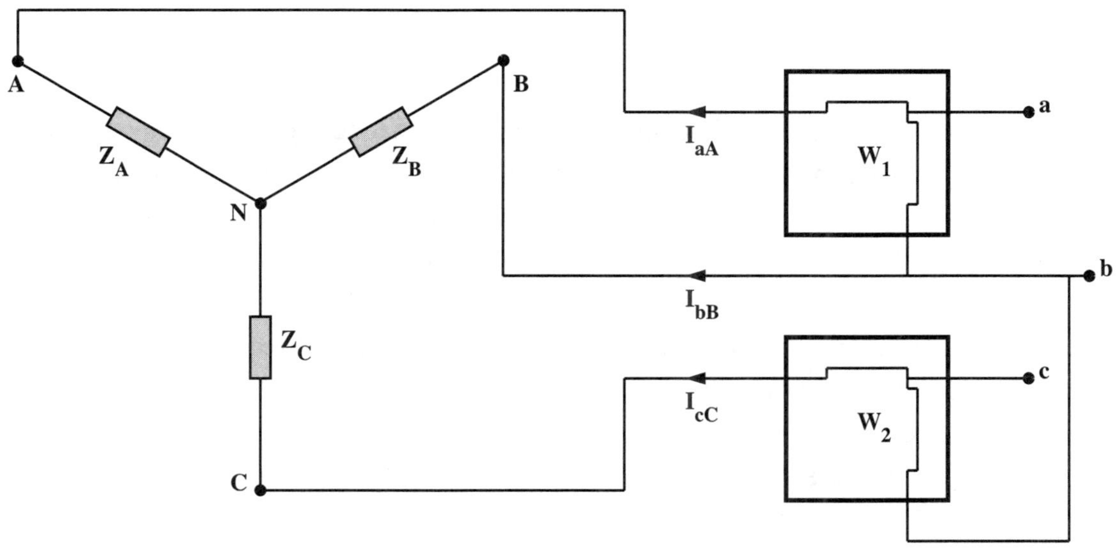

Fig. 12.6-3. Power measurements using two wattmeters.

12.7. UNBALANCED LOAD

One of the important issues we must address is the fact that we often face unbalanced load to analyze. For instance, consider the load shown in Fig. 12.7-1. We assume that the source is symmetric and balanced, which makes sense and is a reasonable assumption. Also, the line to line voltage is 460^{V}. In this configuration, suppose we want to compute the total power delivered to the load in terms of its real and imaginary parts, i.e., active and reactive power which requires us to find the line and phase current and phase voltage first. We also assume that in this setting the inductive components of the load are physically separated from each other. In the sense that there would be no mutual inductance or magnetic coupling among them.

To analyze this problem is straightforward if we apply our knowledge gained in this chapter and try to reason based on that rationally and systematically. First of all, when we have a line to line voltage of say 460^{V}, that means the amplitude of the line voltage is $A = 460^{V}/\sqrt{3} = 265.5^{V}$. Thus, we may now utilize a balanced source configuration similar to Fig. 12.2-4, where "n" is the neutral point inside the source which we have no access to, and choose the following phase arrangement since no other restrictions are imposed on us.

$$V_{an} = 265.5^{V} \angle 0°, \tag{12.7-1}$$

$$V_{bn} = 265.5^{V} \angle(-120°), \tag{12.7-2}$$

$$V_{cn} = 265.5^{V} \angle(-240°). \tag{12.7-3}$$

The above equivalent Y-connected source is attached to the unbalanced load in Fig. 12.7-1 in a configuration similar to that in Fig. 12.3-3, but the main difference between these two figures is that in practice we do not have a wire connecting "n", the source neutral node, to "N", the load neutral point. This situation in fact makes our analysis in the case of unbalanced load simpler than before as is described subsequently. The reason that the analysis in the case of both source and load being symmetric is straightforward is because in such a case we only have to compute the voltage and current for one phase. For the other two phases we can conclude the results by appropriate phase shifts. That is why we seldom use a connecting wire between "n" and "N", since it has no effect on the analysis. In reality, however, we cannot have this wire because we have no access to the "n", which is inside the power company's generator. In the case of unbalanced load not having this wire as stated above makes our analysis even simpler than otherwise.

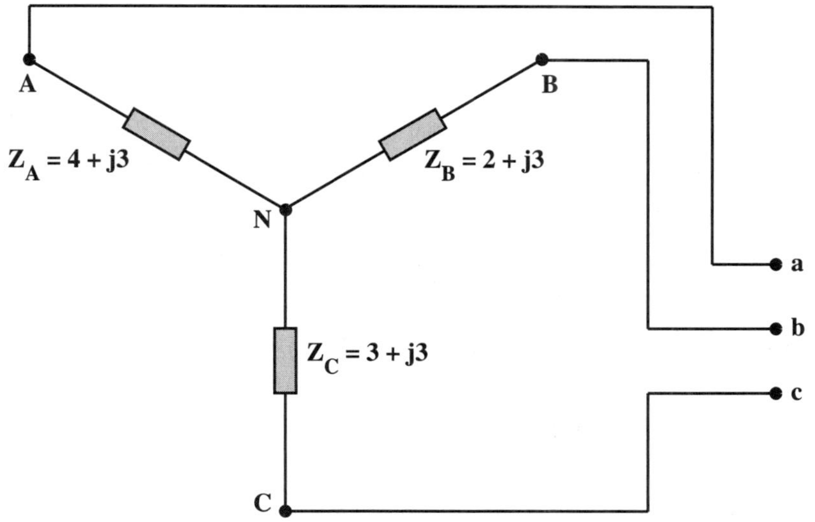

Fig. 12.7-1. An unbalanced Y-connected load.

Before proceeding further, let us pause and review the physics of an unbalanced load. If we study Fig. 12.7-1, we note that in this academic example we have three equal inductors but three different resistors as the components of the unbalanced load. No matter what is the shape (physics) of each component, we can always let two out of the three components be described in terms of the first one in parallel with some new component, Fig. 12.7-2.

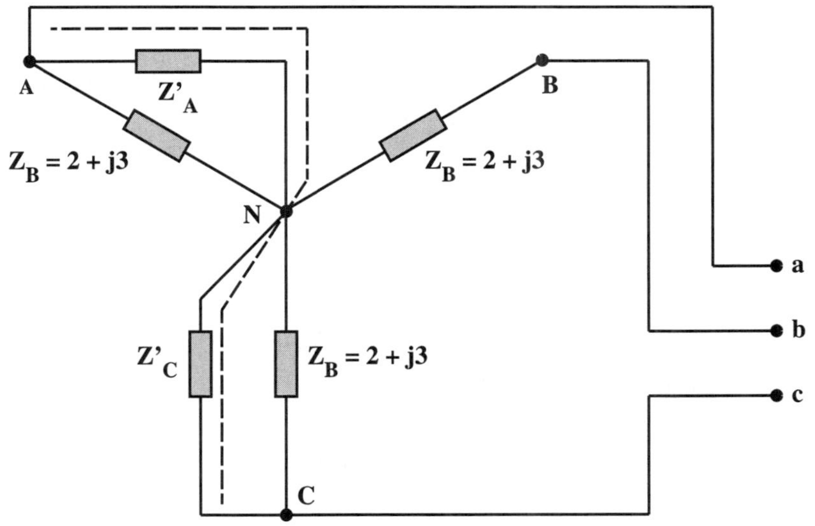

Fig. 12.7-2. An unbalanced Y-connected load.

For instance, referring to Fig. 12.7-2, we choose Z_B as the reference, and say that there are two other components called Z'_A and Z'_C such that:

$$Z_A = Z_B \,||\, Z'_A = Z_B Z'_A / (Z_B + Z'_A), \tag{12.7-4}$$

$$Z_C = Z_B \,||\, Z'_C = Z_B Z'_C / (Z_B + Z'_C). \tag{12.7-5}$$

If we want numerical values for Z'_A or Z'_C, then from (12.7-4) and (12.7-5) we can solve for these unknowns respectively. But we are not interested in their numerical values since we can always do this partitioning no matter what are the values of three components. We only want to show what is happening when we have an unbalanced load. However, when we look at Fig. 12.7-2, we realize that contrary to a balanced load with three equal components in which the corresponding three currents that are coming to node "N" will be absorbed by that node. Thus, the voltage at node "N" becomes the same as that in the reference node "n" which its voltage is zero by definition. In (12.7-2), because of the unbalanced situation some current is passing through the fictitious dotted line which reflects pictorially the contribution of the unbalancedness of the load. After all we expect that these three balanced components shown by Z_B will act exactly the same as what we have studied before. To understand the fact that some "unbalanced" current is *passing through* "N", and as a result of that phenomenon we have some voltage associated with node "N" relative to the global reference point "n", is the key to solve an unbalanced load problem. Now that we understand $V_N \neq V_n$, where V_n is the source reference point, we can use the first law in this textbook, namely the KCL to solve our problem. Thus, we have the following equation for V_N, which reads as the total currents entering node N is equal to zero.

$$\frac{V_a - V_N}{Z_A} + \frac{V_b - V_N}{Z_B} + \frac{V_c - V_N}{Z_C} = 0. \tag{12.7-6}$$

Upon solving the above equation for V_N, the rest of the analysis is straightforward and we leave the details as an exercise. (Problem 12.8.13).

12.8. PROBLEM SET

Fundamentals / Signals

12.8.1. Prove that indeed (12.3-9) is valid.

12.8.2. Prove the validity of (12.3-17) to (12.3-19).

12.8.3. Prove the load transformation formulae (12.4-4) to (12.4-6).

12.8.4. Prove the load transformation formulae (12.4-7) to (12.4-9).

12.8.5. Find the average and the RMS values for the periodic waveforms in Figs.

12.8-1 to 4.

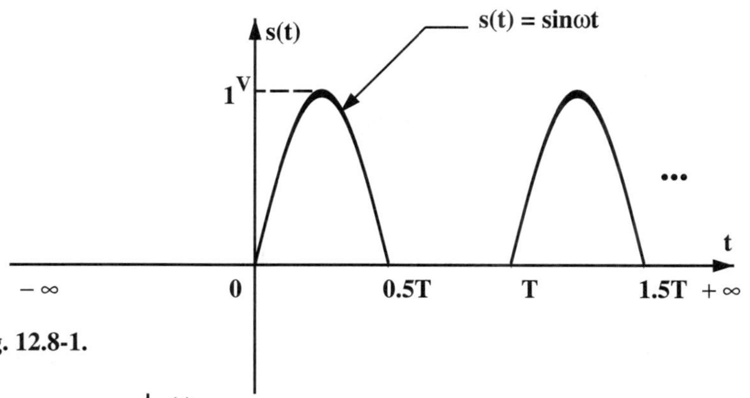

Fig. 12.8-1.

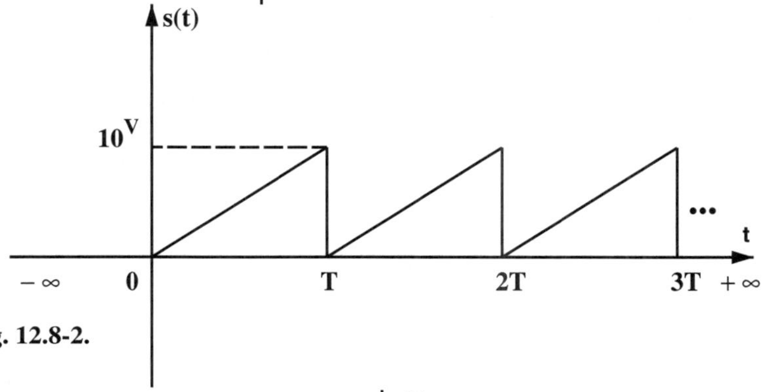

Fig. 12.8-2.

Fig. 12.8-3.

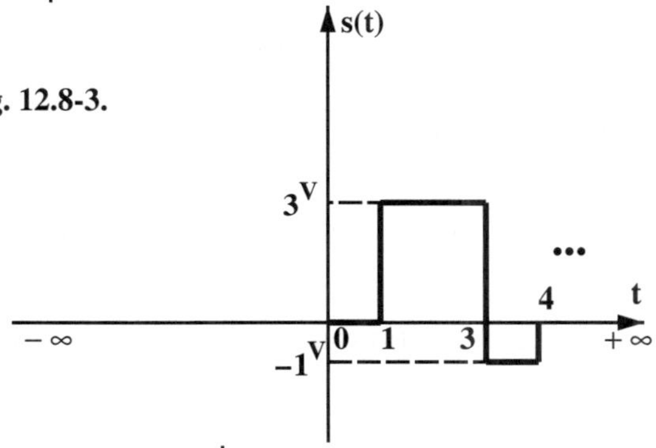

Fig. 12.8-4.

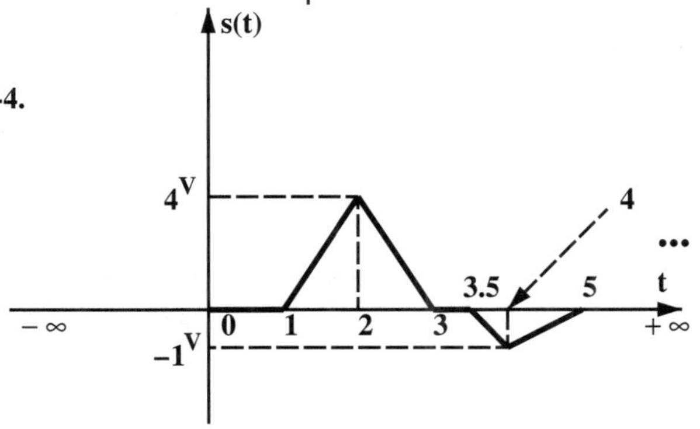

Power / Single Phase

12.8.6. What is the average power delivered to the elements in the circuit of Fig. 12.8-5.

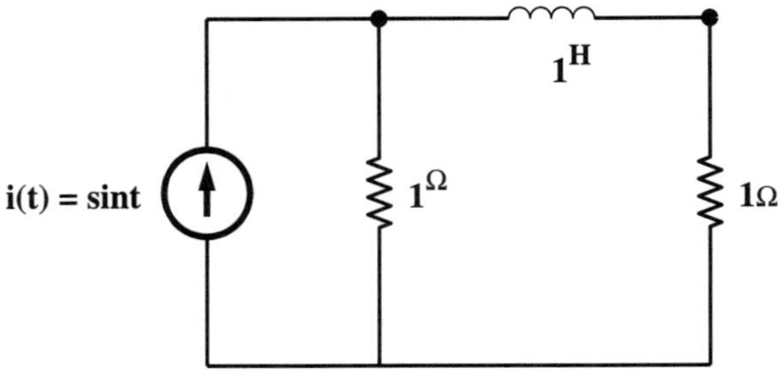

Fig. 12.8-5.

12.8.7. For each element in the single-phase circuit of Fig. 12.8-6, find power received in the three categories of real, reactive and apparent. Compare the total power received by elements in each category to that delivered by the current source $i(t)=10\sqrt{2}\sin 10t$. Repeat the analysis for the transfer energy from the source to the elements.

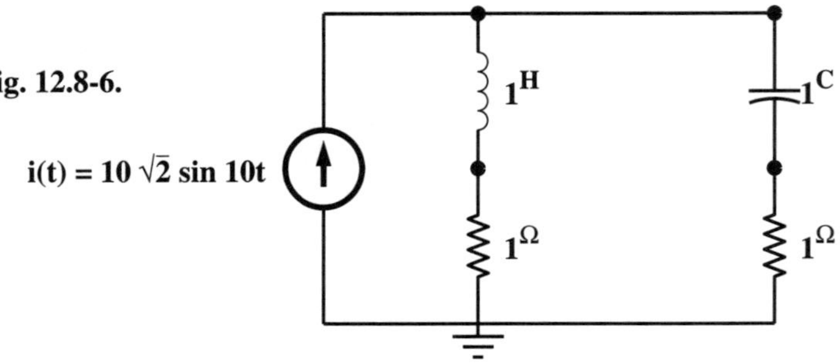

Fig. 12.8-6.

12.8.8. Compute the power factors for circuits in Figs 10.6-1 & 10.7-1.

Single-Phase, Three Wire

12.8.9. For circuit in Fig. 12.8-7, find the total dissipated power at 60 Hz frequency.

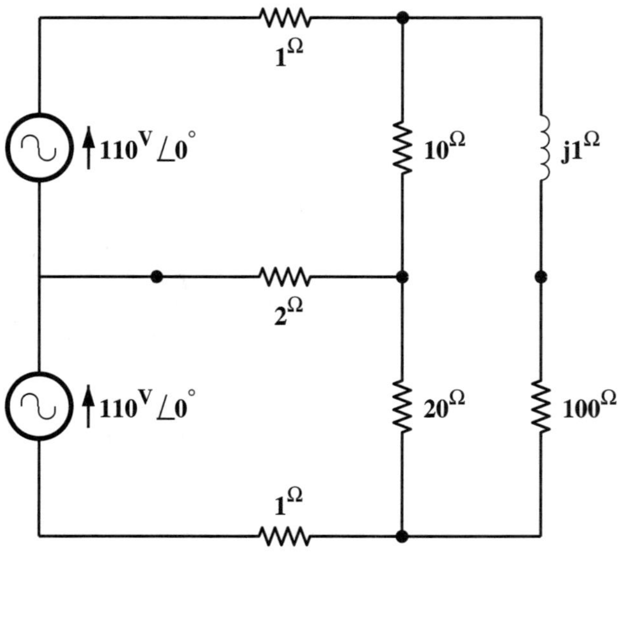

Fig. 12.8-7.

Three-Phase Balanced Sources

12.8.10. For the circuit in Fig. 12.8-8, if $V_{ab} = 100\sqrt{3}\angle(30°)$, $V_{bc} = 100\sqrt{3}\angle(-90°)$, and $V_{ca} = 100\sqrt{3}\angle(-210°)$, then find the currents through each load and the total dissipated power in this system. What is the power factor and the total reactive power in this system?

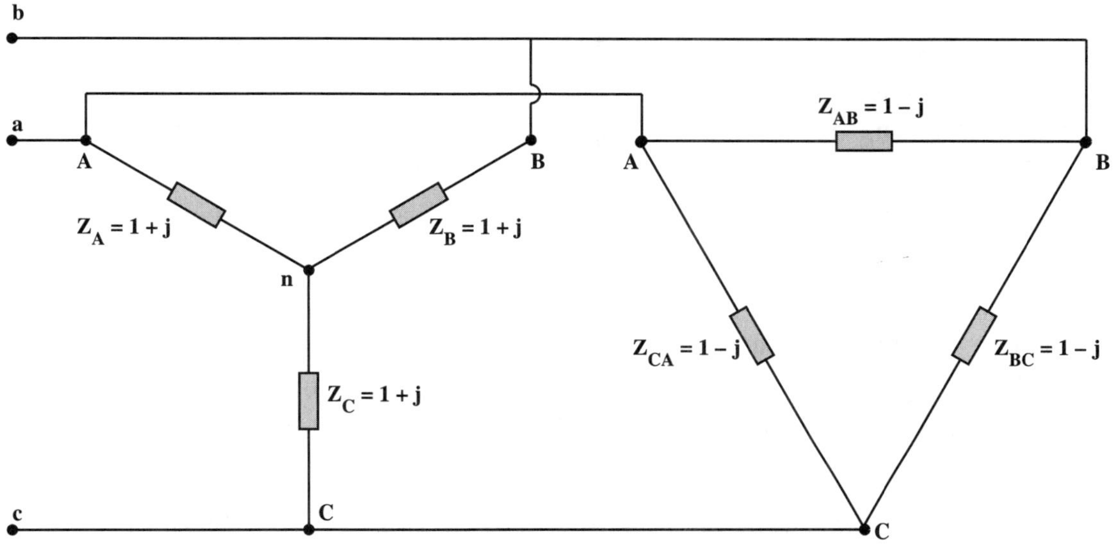

Fig. 12.8-8.

12.8.11. For the circuit in Fig. 12.8-8, draw a phasor diagram similar to that in Fig. 12.2-6 for the overall currents and voltages of each load in each phase.

12.8.12. In the circuit of Fig. 12.8-8, keep Z_{AB}, Z_{BC} and Z_{CA} as they are. Choose Z_A, Z_B and Z_C such that the overall load becomes purely resistive. Recall that the operating frequency is 60 Hz.

Unbalanced Load

12.8.13. Complete the analysis for the unbalanced Y-connected load in the circuit of Fig. 12.7-1 by calculating its total power delivered to the load in terms of its active and reactive parts.

12.8.14. Repeat Problem 12.8.13 by considering the impedances for each line as follows. $Z_{Aa} = (0.05 + j0.1)^{\Omega}$, $Z_{Bb} = (0.06 - j0.1)^{\Omega}$, and $Z_{Cc} = (0.01 + j0.2)^{\Omega}$. Here, we need to calculate all currents and voltages for every element in the line.

12.8.15. Repeat Problem 12.8.10 for $Z_{AB} = (1 - j)^{\Omega}$, $Z_{BC} = (2 - j)^{\Omega}$, and $Z_{CA} = (3 - j)^{\Omega}$.

12.8.16. Repeat Problem 1.7.5 by checking to see all relevant topics in Table 1.1-1 are covered in the course. If anything is yet outstanding, then seek ways to explore that topic.

Farewell Message: It is that time again when I must say farewell and close this project. I hope the students will find this book useful. Those who have taken courses with me will probably hear my voice in this book. I certainly hope that the book is read and analyzed cover to cover and all topics are thoroughly reviewed, because this is the minimum amount of knowledge expected from an electrical engineering student. Regarding questions about topics which are covered in the course (Problem 12.8.16 and the like), these are asked because I strongly believe that to ask students only numerical problems without challenging them as to what has been covered in the course is a disservice to them. Students must know they are expected to remember the covered topics not only in this course, but also in other courses to follow. I hope this new attitude will transcend to all courses. I am absolutely certain that when this level of competence is achieved within a disciplined structure, strong confidence will be built and inspired in students which will carry them through their studies. Yes, there does not exist a circuit that you cannot analyze.

Well, they are calling my flight (returning to Chicago from Amsterdam, Jan. 30, 2005) and a project (the actual typing of the manuscript) that had started at Los Angles International Airport, passing through St. Louis, Detroit, Minneapolis, Vancouver, Seattle, San Francisco, New York and had crossed 16 times the Atlantic ocean, passing through London, Amsterdam, Frankfort, Teheran, Shiraz, Mashhad, as well as many other places in between is about to finish this semester. I sincerely hope this course has been a pleasant experience, and if students remember my name, then I will be very happy. If however they remember major parts of the book, then certainly that will be my great bonus. Farewell my friends, it has been an honor and a great pleasure to work on this project. We will meet again.

REFERENCES

[1] A. Budak, *Circuit Theory Fundamentals and Applications.* Englewood Cleffs, NJ: Prentice Hall, 1978.

[2] M. Eslami, *Analog and Digital Circuits Theory and Experimentation.* Melbourne, FL: R.E. Krieger Publishing Company (in two volumes), 1987, 3rd edition 1993.

The pertinent references for the above source are as follows.

- A.J. Diefenderfer, *Principles of Electronic Instrumentation.* Philadelphia: W.B. Sauders Company, Second Edition, 1979.
- L.P. Huelsman and P.E. Allen, *Introduction to the Theory and Design of Active Filters.* New York: McGraw-Hill, 1980.
- W.G. Jung, *IC Op-Amp Cookbook.* Indianapolis, IN: Howard W. Sams & Co., Inc., 1974.
- J.K. Roberge, *Operational Amplifiers: Theory and Practice.* New York: Wiley, 1975.
- D.L. Schilling and C. Belove, *Electronic Circuits, discrete and integrated.* New York: McGraw-Hill, 1979.
- G.E. Tobey, J.G. Graeme and L.P. Huelsman, Eds., *Operational Amplifiers Design and Applications.* New York: McGraw-Hill, 1971.
- M.E. Van Valkenburg, *Introduction to Modern Network Synthesis.* New York: Wiley, 1967.
- M.E. Van Valkenburg, *Analog Filter Design.* New York: Holt, Rinehart and Winston, 1982.
- J.V. Wait, L.P. Huelsman and G.A. Korn, *Introduction to Operational Amplifier Theory and Applications.* New York: McGraw-Hill, 1975.

[3] W.H. Hayt, Jr., and J.E. Kemmerly, *Engineering Circuit Analysis.* New York: McGraw-Hill, 2nd edition 1971.

[4] P. Penfield, Jr., R. Stence, and S. Duinker, *Tellegen's Theorem and Electrical Networks.* Cambridge, MA: MIT Press, 1970.

[5] M.E. Van Valkenburg, *Network Analysis.* Englewood Cleffs, NJ: Prentice Hall, 3rd edition 1974.

Other textbooks used to teach the first circuit course.

[6] J.W. Nilsson and S.A. Riedel, *Electric Circuits.* Englewood Cliffs, NJ: Prentice-Hall, 2000.

[7] M.E. Van Valkenburg and B.J. Kinariwala, *Linear Circuits.* Englewood Cliffs, NJ: Prentice-Hall, 1982.

Two classical textbooks for basic and intermediate circuit courses.

[8] C.A. Desoer and E.S. Kuh, *Basic Circuit Theory.* New York: McGraw-Hill, 1969

[9] L.O. Chua, C.A. Desoer and E.S. Kuh, *Linear and Nonlinear Circuits.* New York: McGraw-Hill, 1987.

Circuit Simulation with SPICE.

Some instructors prefer to teach concurrently simulation in the first circuit course. We do not share this view. However for those students who want to explore circuit simulation as early as possible, any one of the following references, in particular [12], is a good starting point.

[10] W. Banzhaf, *Computer-Aided Circuit Analysis Using SPICE.* Englewood Cliffs, NJ: Prentice-Hall, 1989.

[11] T.W. Thorpe, *Computerized Circuit Analysis with SPICE: A Complete Guide to SPICE, with applications.* New York: Wiley, 1992.

[12] P.W. Tuinenga, *SPICE: A Guide to Circuit Simulation and Analysis Using PSpice™.* Englewood Cliffs, NJ: Prentice-Hall, 1995. (PSpice™ is a registered trademark of MicroSim Corporation.)

[13] A. Vladimirescu, *The SPICE Book.* New York: Wiley, 1994.

[14] B.M. Wilamowski and R.C. Jaeger, *Computerized Circuit Analysis Using SPICE Programs.* New York: McGraw-Hill, 1997.

Index